S0-BLX-567

MATHEMATICAL MODELING IN EXPERIMENTAL NUTRITION

ADVANCES IN EXPERIMENTAL MEDICINE AND BIOLOGY

Recent Volumes in this Series

A Continuation Order Plan is available for this series. A continuation order will bring delivery of each new volume immediately upon publication. Volumes are billed only upon actual shipment. For further information please contact the publisher.

MATHEMATICAL MODELING IN EXPERIMENTAL NUTRITION

Edited by

Andrew J. Clifford

University of California at Davis
Davis, California

and

Hans-Georg Müller

University of California at Davis
Davis, California

PLENUM PRESS • NEW YORK AND LONDON

Library of Congress Cataloging-in-Publication Data

Mathematical modeling in experimental nutrition / edited by Andrew J.
Clifford and Hans-Georg Müller.
p. cm. -- (Advances in experimental medicine and biology ; v.
445)
"Proceedings of a conference held August 17-20, in Davis, Calif."-
-CIP t.p. verso.
Includes bibliographical references and index.
ISBN 0-306-46020-3
1. Nutrition--Mathematical models--Congresses. 2. Nutrition-
-Statistical methods--Congresses. I. Clifford, Andrew J.
II. Müller, Hans-Georg. III. Series.
QP143.5.M37M38 1998
612.3'9'015118--dc21 98-8315
CIP

Proceedings of a conference on Mathematical Modeling in Experimental Nutrition,
held August 17–20, 1997, in Davis, California

ISBN 0-306-46020-3

A Division of Plenum Publishing Corporation
233 Spring Street, New York, N.Y. 10013

http://www.plenum.com

10 9 8 7 6 5 4 3 2 1

Printed in the United States of America

PREFACE

Nutrients have been recognized as essential for maximum growth, successful reproduction, and infection prevention since the 1940s; since that time, the lion's share of nutrient research has focused on defining their role in these processes. Around 1990, however, a major shift began in the way that researchers viewed some nutrients—particularly the vitamins.

This shift was motivated by the discovery that modest declines in vitamin nutritional status are associated with an increased risk of ill-health and disease (such as neural tube defects, heart disease, and cancer), especially in those populations or individuals who are genetically predisposed. In an effort to expand upon this new understanding of nutrient action, nutritionists are increasingly turning their focus to the mathematical modeling of nutrient kinetic data.

The availability of suitably-tagged (isotope) nutrients (such as ß-carotene, vitamin A, folate, among others), sensitive analytical methods to trace them in humans (mass spectrometry and accelerator mass spectrometry), and powerful software (capable of solving and manipulating differential equations efficiently and accurately), has allowed researchers to construct mathematical models aimed at characterizing the dynamic and kinetic behavior of key nutrients *in vivo* in humans at an unparalleled level of detail. Organizing and interpreting this detail will lead to a better and more fundamental understanding of this behavior in health and disease, the factors that affect it, and the health consequences of its alteration. Because all facets of nutrition—environmental, organismic, cellular, and molecular—are becoming accessible to chemical, physical, mathematical, and statistical approaches, modeling is an increasingly important partner of experimental work.

This book was developed from the papers presented at the sixth in a series of conferences centered on advancing nutrition research through modern statistical/ mathematical modeling and state–of–the–art physico/chemico analytic techniques. Only in recent years have advanced statistical methods such as generalized linear models, error-in-variables models, and nonparametric methods been applied to problems in the field of nutrition. Application of accelerator mass spectrometry to nutrition research is a major advance in tracer analysis. This volume represents intellectual exchange between scientists from the disparate fields of biological and physical science. Some of the insights gained by this interdisciplinary effort are reflected in this volume.

This conference series began as two general conferences (Canolty and Cain, 1985; Hoover-Plow and Chandra, 1988). A third conference (Abumrad, 1991) focused on amino acid and carbohydrate metabolism, the fourth (Siva Subramanian and Wastney, 1995) centered on mineral metabolism, while the fifth (Coburn and Townsend, 1996) concentrated on vitamins, proteins, data acquisition, and the computational aspects of modeling.

The sixth conference was held at the University of California, Davis on August 17 through 20, 1997. The first part of the conference focused on the background and theoretical aspects of statistical and physiological compartmental modeling. The second discussed the application of statistical and compartmental modeling methods to the design and interpretation of studies of vitamin, mineral, amino acid, and protein metabolism. Vitamin A, ß-carotene, folic acid, calcium, zinc, copper, and magnesium were chosen as focus nutrients because they are currently under intense investigation by means of tracer methodologies and compartmental modeling. Other presentations centered on the dynamics of energy, growth, lactation, aging, and survival and mortality. The final part of the conference included a discussion of state-of-the-art physical and chemical analysis methods for use in measuring tagged nutrients, and the introduction of a new application of accelerator mass spectrometry to human nutrition research. All articles included in this volume have been refereed.

The editors wish to honor the memory of Loren A. Zech, MD, a pioneer in the area of compartmental modeling and the initiator of this series of conferences.

ACKNOWLEDGEMENTS

The conference was made possible by the generous financial support of NIH R13 DK 53081, NSF DMS 9510511, the USDA-Western Human Nutrition Research Center, the Center for Accelerator Mass Spectrometry at the Lawrence Livermore National Laboratory, F. Hoffman-La Roche & Co. Ltd, Ralston Purina, and the conference registrants. At the University of California, Davis, we would also like to recognize the financial support provided by the Vice Chancellor of Research and the Deans of the Division of Biological Sciences, the College of Agricultural and Environmental Sciences, and the Division of Mathematical and Physical Sciences of the College of Letters and Sciences.

We would also like to recognize the numerous referees for their valuable service in reviewing the submitted manuscripts.

The conference was made successful by its speakers, the UC Davis Conference & Events Center staff, the Conference Organizing Committee, and Dr. Peg Hardaway, who organized and conducted the technical editing of the conference proceedings.

REFERENCES

Abumrad N; Ed. Mathematical models in experimental nutrition. *JPEN*, 1991, 15:44s-98s.

Canolty N; Cain TP; Eds. *Mathematical Models in Experimental Nutrition.* University of Georgia: Athens, GA. 1985. pp. 1-144.

Coburn SP; Townsend D; Eds. Mathematical modeling in experimental nutrition. *Adv Food Nutr Res*, 1996, 40:1-362.

Hoover-Plow J; Chandra RK; Eds. Mathematical modeling in experimental nutrition. *Prog Food Nutr Sci*, 1988, 12:211-338.

Siva Subramanian KN; Wastney ME; Eds. *Kinetic Models of Trace Element and Mineral Metabolism During Development.* CRC Press: New York. 1995. pp 1-389.

CONTENTS

Part I. Theoretical Considerations in Compartmental Modeling

Part II. Statistical Modeling in Nutrition

Part III. Applications of Modeling

Part IV. Chemical / Physical Analytical Methods

MATHEMATICAL MODELING IN EXPERIMENTAL NUTRITION

Part I

THEORETICAL CONSIDERATIONS IN COMPARTMENTAL MODELING

BALANCING NEEDS, EFFICIENCY, AND FUNCTIONALITY IN THE PROVISION OF MODELING SOFTWARE: A PERSPECTIVE OF THE NIH WINSAAM PROJECT.

Peter Greif,[1] Meryl Wastney,[2] Oscar Linares,[3] Ray Boston[4]

[1]Laboratory of Experimental and Computational Biology
National Institutes of Health, Washington DC
[2]Department of Neonatology
Georgetown University Medical Center, Washington, DC
[3]Department of Geriatrics
University of Michigan Medical School, Ann Arbor, MI
[4]Clinical Studies Department, New Bolton Center
University of Pennsylvania

ABSTRACT

The development of new software or the refinement of existing software for new operating environments each calls for judicious balancing. On the one hand, we strive for simplicity, predictability, and operational protection as it is well recognized that software with these attributes will attract an audience of satisfied users. But, on the other hand, these attributes do not conjure a sense of power, efficiency, or flexibility, and these other properties are also appreciated by users, albeit a somewhat different group of users. The goal is to achieve a blend which isolates critical functionality, flexible control, and user support while meeting the needs of the broadest collection of serious users. In this chapter, we discuss the issues impacting the migration of SAAM to the Windows environment, the NIH WinSAAM Project, and we outline the steps taken to ensure its feasibility. In addition, we describe a new paradigm for software development and use which ensures the durability of the software for modeling.

INTRODUCTION

Mathematical models have been used extensively as tools for nutritional research. Indeed, it is unlikely that any serious examination of a nutritional subsystem could be advanced without recourse to current related models. The reason for this is clear: the systems explored involve multiple tissues and organs, multiple metabolites and nutrients,

Mathematical Modeling in Experimental Nutrition
Edited by Clifford and Müller, Plenum Press, New York, 1998

and multiple processes and mechanisms, and it is not possible to explore the complex effects involved here without systematically formulating the nature of each from a mathematical perspective.

Nutritional models have been developed to describe whole-animal level functions (Baldwin, 1995), organ levels functions (Dijkstra, 1993), tissue level functions (Gill, 1984), and single substrate level functions (Faichney, 1985). The have also been used for processes embracing digestion (France, 1982), absorption (Pettigrew, 1992), metabolism (Hanigan, 1994), and production (France, 1984). Indeed, models in the nutritional area have even been developed at the cellular level to explain production consequences of metabolism there (Neal, 1984).

Thanks to the work of Thornley (1991), France (1984), and Baldwin (1987) on the one hand, and Berman (1983), Jacquez (1996), and Cobelli (1983) on the other, fairly clear paths now exist for the development of models. The first group has refined the so-called 'transactional' perspective of metabolic systems in which only the exposed or limiting aspects of metabolic pathways or digestive processes are incorporated into models. From this integrated perspective, they adopt the position that if an adequate representation of the fundamental processes (transactions) of subsystems are evident, then integration across subsystems should lead to realistic system-wide response predictions. To help with their realistic (effective) portrayal of subsystems (transactions), they have developed a number of ingenious mathematical representations of metabolic steps including substrate limited processes, allosteric binding (also referred to as threshold processes), various forms of inhibition, and hormonal sensitization. A number of large systems involving 17 (Dijkstra, 1992) to approximately 60 (Baldwin, 1995) state variables have been modeled in this way.

Large-scale modeling, particularly transactionally-based, large-scale modeling, needs access to the results of investigations which define the properties of the rate-limiting processes in order to incorporate them into the representation of the sub-systems... The results of what might be referred to as micro-scale modeling. It is this area of modeling which has benefited from the work of the second group of investigators. They have systematically refined approaches which permit the accurate characterization of sub-system processes. These processes include model decoupling, model reduction, identifiability, minimal modeling, and constrained modeling (see for example, Boston, 1984; Jacquez, 1996; Cobelli, 1983). Used appropriately, these techniques can greatly enhance the quality of a system investigation, especially where the nature of this data masks simple direct data fitting and interpretation.

To enable these strategies to yield their full potential, the tools, techniques, and methodological efforts refined by both groups have had to be matched with powerful computer software. Initially, this was provided by SAAM (1978) software, where an ingenious approach to model-fitting and data analysis enabled direct fitting of complex systems to data. This software was originally developed in the sixties for the analysis of tracer data, but rapidly evolved to the tool of choice for micro-scale modeling groups. Around the same time, SAAM was evolving, a large-scale system simulator, CSMP (Fugazi, 74), was also taking shape which provided comprehensive computer support for investigators involved with the development of vary large models. In the seventies, CSMP was replaced by ACSL (1992), and this package has clearly emerged as the tool of choice for modelers building large systems.

Since the eighties, we have witnessed, on the one hand transformations, and enhancements to the older modeling software, and on the other hand, the emergence of a series of new 'look and feel' software with the attendant buttons, windows, icons, and menu-based services we now currently expect of computer software. In this latter group we include Stella, Scientist, and SAAM II as outstanding examples of how modeling functions can be incorporated into the 'quasi-mode-less' application setting of typical Windows programs. The first group of Windows software, the ported software, has had to address many more and complex issues in regard to its target setting than those worked into this setting from the ground up. These included such critical issues as whether to re-deploy function access by capitalizing on the built-in tools of the Windows environment, or to displace command lines and command-based processing flow in favor of buttons and menu icons in order to deliver essentially the same operations.

It seems at first flush unconscionable that we could ask these types of questions, for there is no doubt that those without prior (and perhaps extensive) exposure to mode-less software would be lost without the attendant widgets to guide them through their agendas. But by the same token, skilled and efficient software users (a stage to which we almost all eventually aspire) who are comfortable with the CLI (command line interface) paradigm, feel cheated and impeded by this sometimes cumbersome and always less flexible form of operational control. This chapter will explore these issues and reconcile our approaches and implementations as we recount efforts to port Consam to the Windows setting ...The NIH WinSAAM project.

METHODS

Because of the difference between this chapter and the others in this series, we will describe our approach in terms of relating the Methods, Results, and Conclusion sections of our work. In the Methods section and under appropriate sub-headings, we will present the WinSAAM Design Objectives and Design Basis, as well as the software architecture. In the Results section, we will present some implementation details and outline our major new directions for modeling software per se. Finally, in our Conclusion, we will describe the state of our current and planned rates of progress and how interested groups might participate in the WinSAAM project.

Design Objectives

Whereas higher aspirations such as 'goals' might be conceived as the target of a software refinement effort, Press (1972) has cautioned against this. Indeed, he stresses that the outcome, or production, needs to be evaluated against a concrete set of criteria and hence endorses 'objectives' for this purpose. We have isolated twelve objectives of our approach that ought to be evident in and through the operational style of our final product.

Incorporate the total functionality of SAAM and Consam into WinSAAM.

We believe that if SAAM and Consam are to become obsolete through WinSAAM, then the SAAM user-community needs access to a software tool with at least the combined capabilities of the former versions. Thus, WinSAAM will functions as a superset of the functionality of the earlier versions. In some respects WinSAAM will incorporate additional functionality, however, at this stage extended functionality will primarily relate to access to Windows tools to manage processing more flexibly in the Windows setting.

Preserve the operational flow of Consam.

Because of the wide interest Consam has attracted to SAAM-style modeling, and because of the need to embrace new WinSAAM users in a familiar and productive fashion, WinSAAM will support a 'terminal' window to provide instant access to the processing capabilities and operational flow of Consam. Indeed, in our estimation a former Consam user will be immediately able to run WinSAAM in a productive way.

Integrate the 'look and feel' of Windows products.

If the target operating environment of WinSAAM is Windows, than clearly it would be most inappropriate not to incorporate a Windows-type 'look and feel' into the system. Here, for example, we intend that file management, text editing support, clipboard access, and allied attendant Windows widgets and tools will be routinely available and efficiently deployable while running the system.

Built on a third-party component basis.

A clear advantage in developing a product in the Windows environment is that one immediately gets access to the vast array of 'software add-ins.' These are special purpose tools (editors, spreadsheets, graphics tools, plotting tools, etc.) which, upon purchase, can be smoothly incorporated into a product on a component-by-component basis. With the savings offered to us by adopting this strategy, we decided that as much as possible of the new interactive tools for WinSAAM would be comprised of third-party components.

Prototype incrementally.

Because of the limited man-power of the WinSAAM project team, we proposed an incremental prototyping paradigm for development. As each new component of the software is foreshadowed for inclusion, an iterative implementation and refinement approach was adopted (see also 'design basis' below). In this setting, once a component functions properly and integrates smoothly with the other components, it is considered a part of the system. Regular integration testing of the product, at any stage, can be undertaken to confirm its reliability and processing flow.

Design for an open style of operation.

Since the first release of Consam in 1978, users have enjoyed considerable freedom in controlling their modeling sessions. Indeed, what we witnessed was an attempt to create a fully-protected, command-based, mode-less environment. We have set ourselves this objective again with WinSAAM. That is, as opposed to tightly controlling and synchronizing user actions, we will attempt to cater for a more 'free wheeling' (mode-less), but protected operational flow.

Design for extensibility.

In keeping with the development deployment of third-party software and, at the same time, recognizing the valuable contributions users and other software developers may offer, we propose that WinSAAM demonstrate and capitalize on two key areas of extensibility: modeling libraries and extended functionality. We propose a means whereby focus groups in various areas of modeling (e.g. lipid metabolism, or glucose/insulin kinetics) can be registered to enable them to provide model 'add-ins.' Other WinSAAM users will then be able to subsequently access and use these models. In addition, we intend to facilitate the functional extension of WinSAAM via 'DLLs' (dynamic linking libraries) in much the same fashion as is done by Scientist. As new processing algorithms are developed, we can explore their incorporation directly into WinSAAM without actually modifying the underlying WinSAAM code.

Accommodate communication with other software.

One of the great problems associated with developing software is defining the extent of its functionality. For example, software providing data-fitting capabilities could potentially be extended to allow, at one end, data preparation support, and, at the other end, statistical tools for analysis and review of the model fit. However with the powerful data management and statistical processing software available, simply providing access to these from the data-fitting software may be enough to address most needs. We intend to implement the most flexible and most common array of data transportation tools into WinSAAM to enable user-preferred access to pre-processing and post-processing software.

Accommodation of an efficient help facility.

As was common to its time, SAAM was poorly documented and difficult to use. Consam possessed some of these traits as well, largely because of its close relationship to SAAM. With the pervasive availability of powerful 'help generators,' we intend that WinSAAM will be comprehensively and efficiently documented for user convenience.

Ensure that WinSAAM is consistent, reliable, and efficient.

The fact that, on the one hand, WinSAAM is built around the well-worn operational paradigm of SAAM and Consam and, on the other hand, its kernel (SAAM) has been subject to many years of rigorous testing and evaluation, gives us a competitive edge in assuring the overall robustness of the new WinSAAM software. Nevertheless, we intend to deploy all the same quality control checks with WinSAAM that we have submitted both SAAM and Consam to in the past. We will also ensure that WinSAAM's flow and mode-less operational style are efficient in delivering user needs in a consistent style across the operational interface.

Distribute application activities amongst the user community.

As we have mentioned, application of the new software will be driven through the incorporation of 'domain' libraries. Users skilled in particular domains will be able to be registered as library domain contributors; the incorporation of their library domain into the WinSAAM application library will provide a vehicle for the robust extention of the routine application of the software. This will expedite evolution of the software application base and, at the same time, ensure that topical applications are routinely available to WinSAAM users.

Preservation of the SAAM kernel; architecture, and organization.

To ensure that all of SAAM's functionality is available in WinSAAM, and to maintain momentum with WinSAAM delivery, the SAAM kernel as incorporated in Consam will, in its current form, provide the processing heart of WinSAAM. Whereas restructuring SAAM's architecture may eventually expedite maintenance and enhancement efforts for WinSAAM, the net yields to the user community from our diverting efforts to this purpose were not justified at this stage. In fact, as seen above, maintenance and enhancement efforts, which give the highest returns for effort, have been identified as the critical modules we will be adding and linking into the SAAM processing environment.

Design Basis

As each step in the development of WinSAAM software is negotiated, it is carefully evaluated against the following criteria:

- Does it fulfill or comply with our design objectives?
- Is there a preferred implementation approach which may facilitate its use or access?

Based on the consequences of exploring these issues, a prototypical implementation of the step (e.g., the incorporation of a service or refinement of a display) is created. After some routine modeling analysis to exercise the new facility, aspects are reviewed not only against the above criteria but also in regard to the ease with which they fit into the overall WinSAAM setting.

It is hoped that by using β-testing and further iterative refinement, faster and smoother closure on the first release version will be achieved.

WinSAAM Architecture

In Figure 1, we present an architectural perspective of WinSAAM. We see that there are essentially four classes of software components: locally-developed components which largely collect and coordinate related WinSAAM functions; third-party 'add-ins' or component software, each providing a complete aspect of interface processing needs; library modules, including a base library and an integrated set of domain-specific libraries; and a set of DLLs, including the SAAM/Consam system, and a set of functional enhancements.

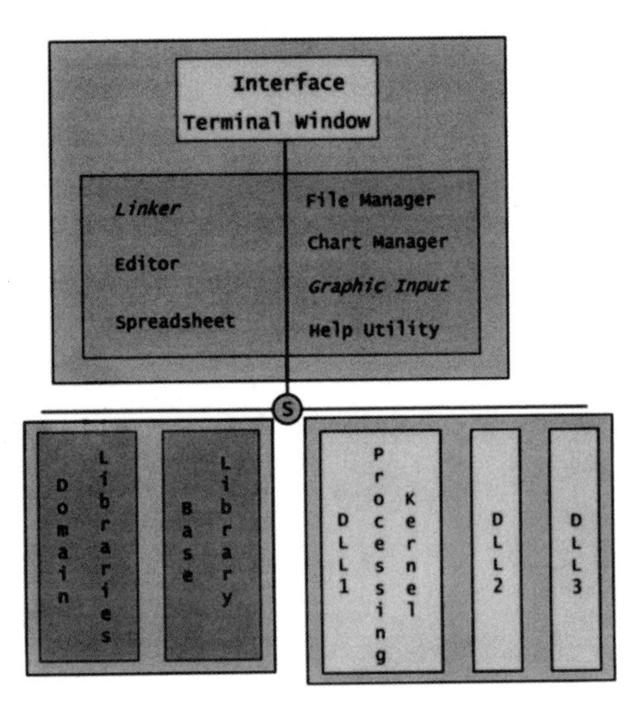

Figure 1. The architectural organization of WinSAAM. There are four types of functional modules incorporated into three activity domains: (1) in the interface, the terminal window (with a supervisory role as indicated), and the linker, the editor, the spreadsheet manager, the file manager, the chart manager, the graphical input manager, and the help utility; (2) in the library area, the base library, and the domain libraries; and (3) in the processing area, the kernel, and an array of client dlls. Those areas currently underway are italicized.

To clarify the functional capabilities and processing integration of WinSAAM, we present the purpose and capabilities of its interface components in Table 1.

Table 1. WinSAAM interface objects and their functions.

Interface Object	Functions
Terminal Window	Terminal input Line-oriented text editor CLI-style processing control
Editor	Graphic text management Tex input editing (graphic) Text output editing (graphic)
Spreadsheet	Tabular-oriented input Tabular-oriented processing Tabular-oriented output Tabular-oriented formatting Tabular-oriented communication
File Manager	Coordinate user files Facilitate file access Check integrity of user files
Chart Manager	Plotting results Configuring plots Chart-based communication
Linker	Assemble input components Edit project specification Manage projects Facilitate project associations
Graphics Module	Graphical model input Graphical-lexical model translation

RESULTS

Aspects of the implementation of WinSAAM can be gleaned from Figures 2 to 6, which present screen dumps of various states of WinSAAM processing.

Figure 2 shows the so-called 'terminal' or introductory window, with the familiar lexical display of a pre-assembled WinSAAM input. Also shown is the simple menu scheme which provides access to the integrated array of support tools embracing the file manager, the spreadsheet analyzer, and the editor.

Figure 3 shows the integrated graphical text editor provided to enable users to rapidly edit lexical (text) information including WinSAAM input files, components (or segments) of input, and familiar full processing logs.

In Figure 4, we show the datasheet of our notebook-oriented spreadsheet output. Sheets are compiled into pages labeled 'data,' 'parameters,' 'covariance,' and 'partials' to provide the user access to all processing results available from system solving and iterative parameter adjustment. Results sheets can be easily reformatted, transformed, saved as EXCEL compatible inputs, or captured for post-processing analysis in other statistical and related software.

In Figure 5, we present a plot of some WinSAAM processing results captured in PowerPoint from WinSAAM's charting subsystem, via the clipboard. With the capacity to scale, zoom, and rotate plots, as well as label and algorithmically modify axes, this subsystem fits the new demands of the WinSAAM environment extremely well. Indeed, this utility has been targeted to provide all the basic support needed for users to take WinSAAM results from a research investigation level to publication quality output.

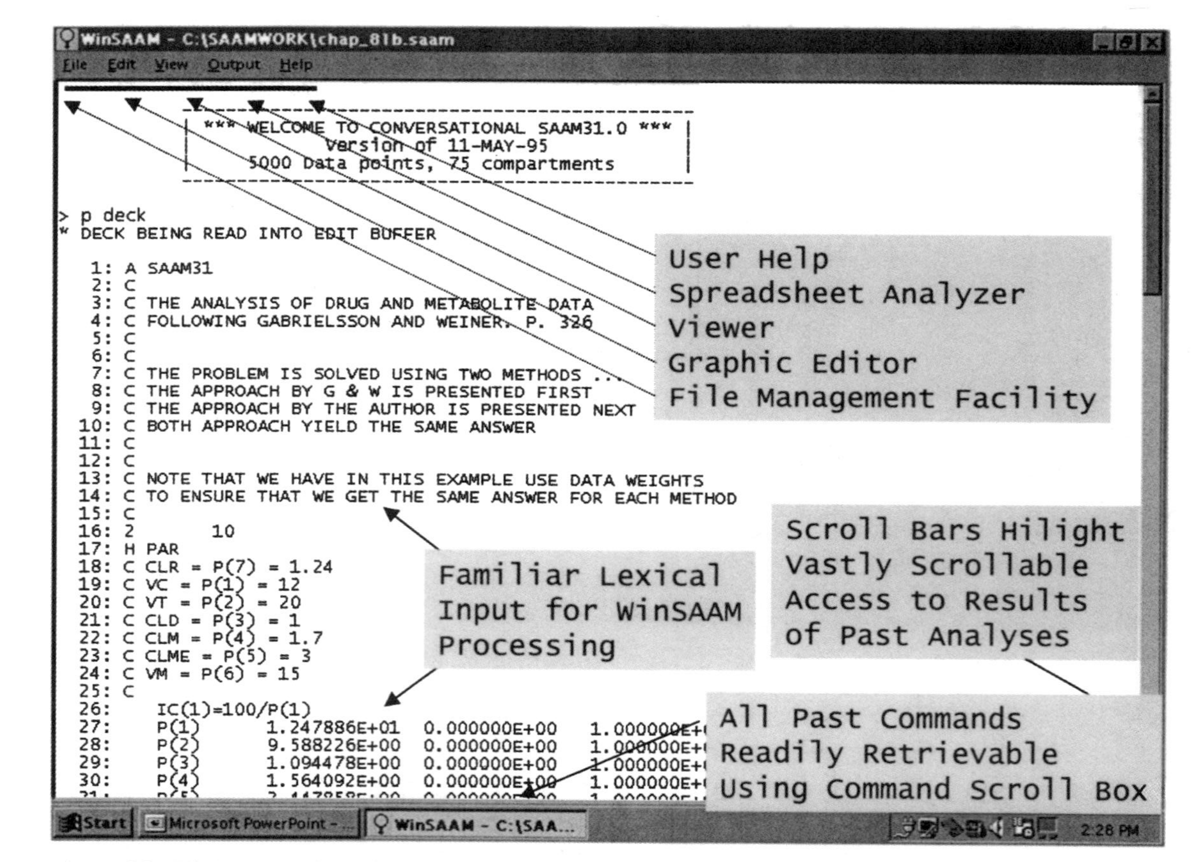

Figure 2. A screen dump of the WinSAAM terminal window showing aspects of a modeling session captured here and the operational flow supported from this point.

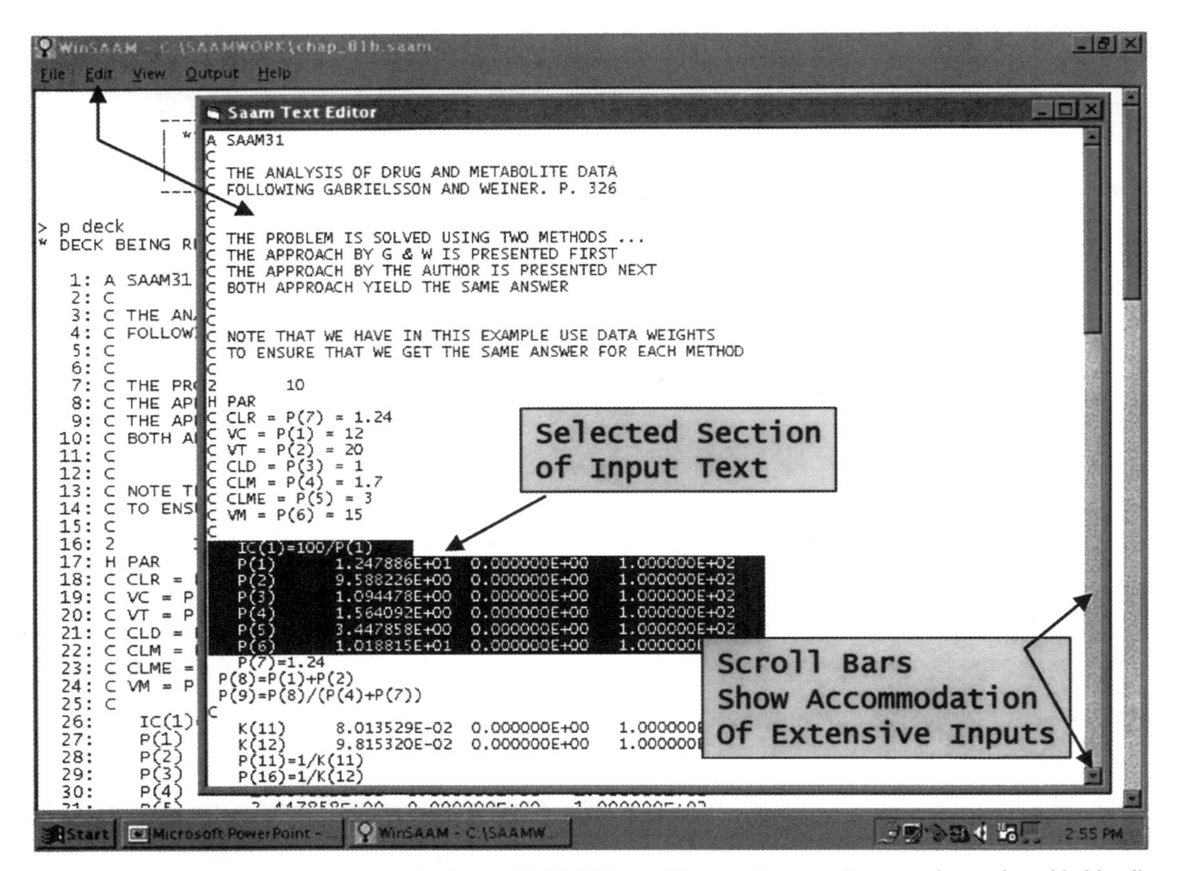

Figure 3. A screen dump showing the steps leading to WinSAAM's graphic text editor as well as some interaction with this editor.

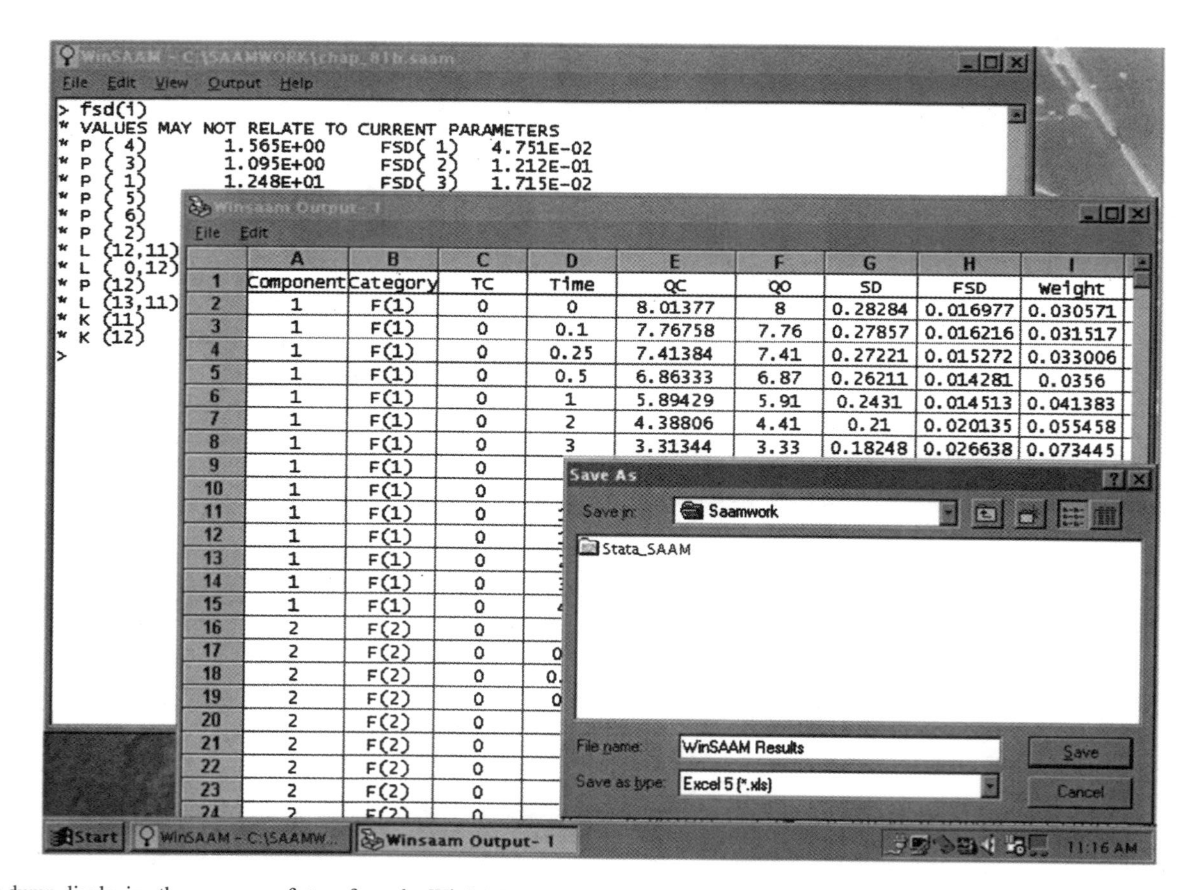

Figure 4. A screen dump displaying the sequence of steps from the WinSAAM terminal window to the preparation of the WinSAAM results for output as an EXCEL 5 spreadsheet. The intermediate spreadsheet provides for the highly flexible re-organization of the WinSAAM processing results as needed prior to transfer of them to an EXCEL file, and presumably to EXCEL for further analysis.

Figures 5A-5C. This series of three figures illustrate the flexibility of the WinSAAM graphics system. In each case, the WinSAAM plot was captured in the clipboard as a Windows metafile and transferred directly to PowerPoint for improvement, color enhancement, or more comprehensive annotation.

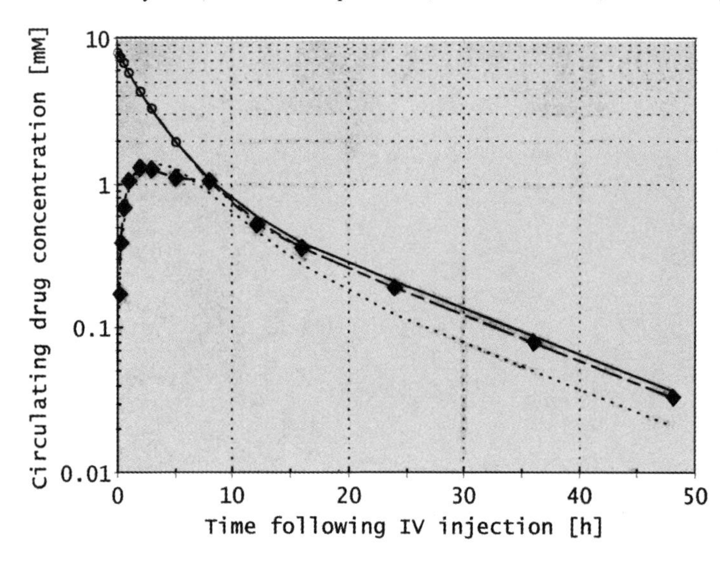

Figure 5A: A standard two-response plot in which we have also joined the data points (long and short dashes) to help clarify the consistency between the observed and fitted points. Joining the observation set in a data series is very easily accomplished with the WinSAAM charting system.

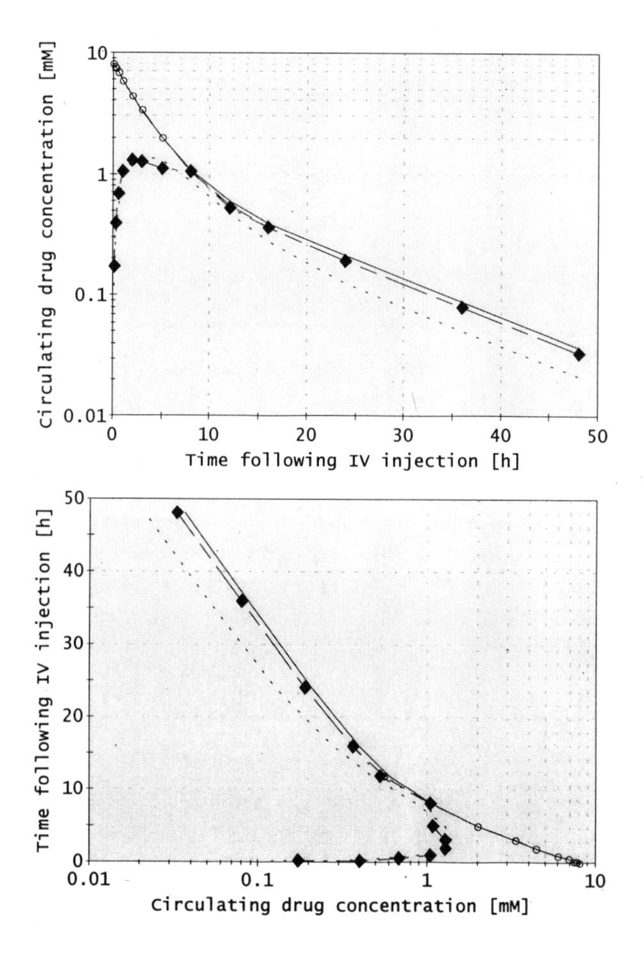

Figure 5B: The above plotted first as per Fig. 5A (upper plot), and secondly plotted on its side (lower plot). Changing the x-y assignment of plots often helps clarify the nature of a relationship between two variables. It requires no more than a one button activation to accomplish the role reversal in WinSAAM's charting system.

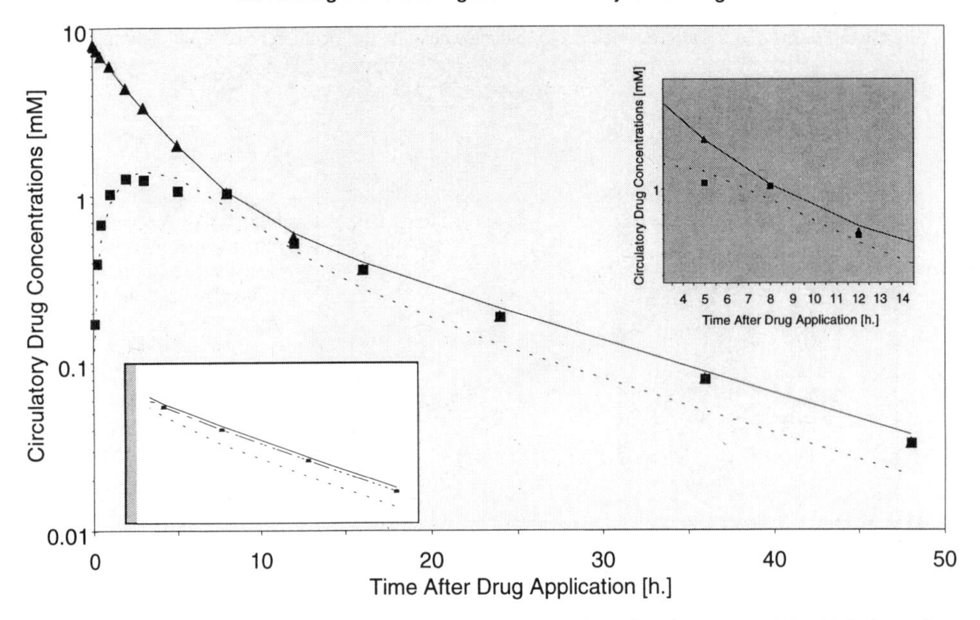

Figure 5C: The same plot as in Fig. 5A, but in addition we have here incorporated detailed views of (a) the values near the peak of the large plot (upper inset), and (b) values near the tail of the large plot (lower inset). It requires no more that a simple drag to accomplish either zooming (upper inset) or re-scaling (lower inset), and restoring the original plot (foreground plot) involves only a single key press. Features such as this can be of enormous assistance to investigators.

Finally, Figure 6 illustrates the ease with which WinSAAM results are made available for further statistical or related examinations using a typical general purpose statistical package. Specifically, in the top left corner of the figure, we see the WinSAAM data spreadsheet from which the results were exported. Next, as a spreadsheet overlaying the WinSAAM spreadsheet, we see STATA's data entry spreadsheet with the WinSAAM results automatically transferred into it. In the foreground, we see the typical STATA terminal interface with a **de**scription (STATA command) of the WinSAAM data, followed by a strategic **su**mmary (STATA command) of the same data. Finally, in the bottom left corner of the figure we see a STATA plot of the observed and calculated values exported from WinSAAM to STATA. This entire process involved no more than four button activations and the invocation of a couple of simple STATA commands. A particular feature of the process is the automatic naming of WinSAAM variables as the data is compiled into the STATA spreadsheet.

NEW DIRECTIONS

The movement of the SAAM/Consam software to Windows has enabled us to explore and refine a number of new directions. While these may pertain most particularly to SAAM and Consam, to some extent they may apply to modeling methodology per se, specifically with regard to how software may better meet the needs of investigators.

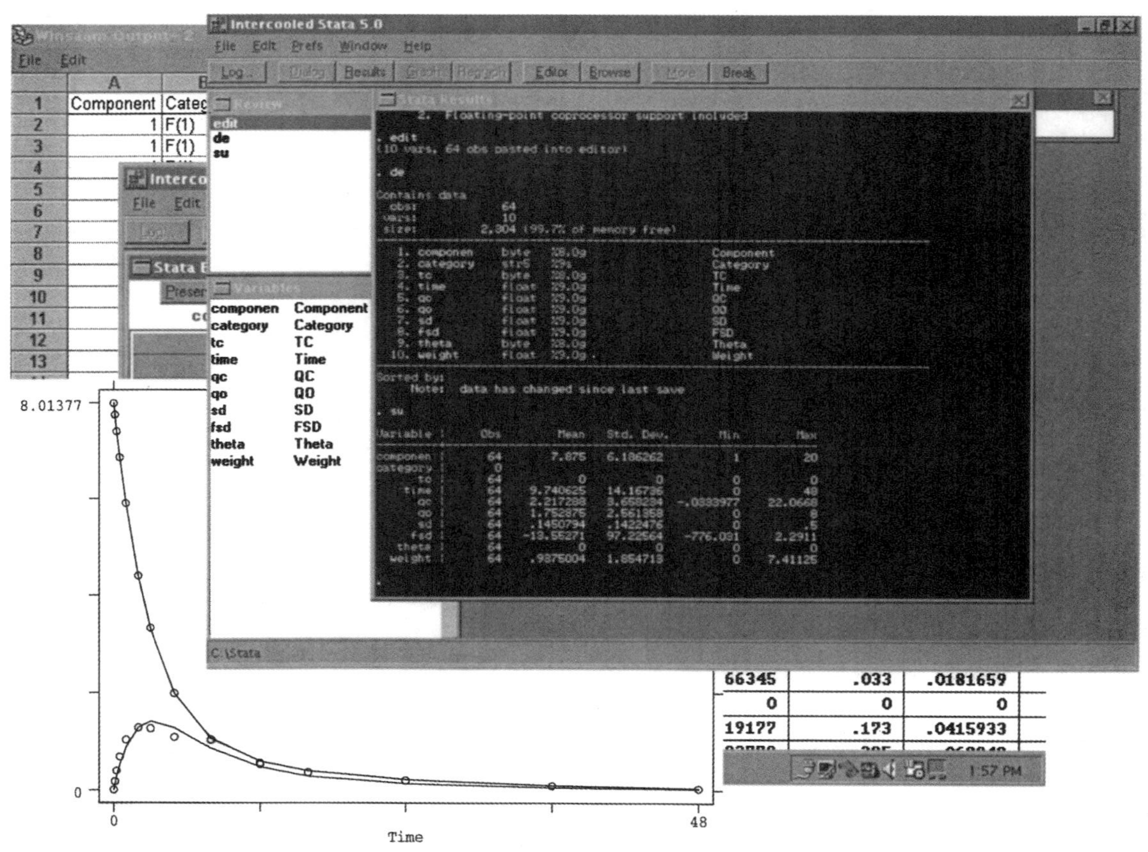

Figure 6. The steps associated with exporting WinSAAM results to an advanced statistical package (STATA) for further analysis. See text for details.

Input Units, the Modeling Project, and the Linker

During the redesign of Consam's file management facility for WinSAAM, it became clear to us that the opportunity had now arisen in which we could radically re-think the entire method by which objects can be assembled and linked in preparation for processing. Specifically, whereas inputs to Consam were historically made on a 'job' basis in which a single input data set was combined with a single model for a single analysis, we could now entertain the idea of alternate data sets and models from different bases linked for analysis. Indeed, we stumbled on the idea of *Input Units*, *Modeling Projects*, and the *Linker*.

An *input unit* represents the atomic or minimal entity from which information can be complied into a modeling investigation. This could be the data associated with a particular stage of an experiment, or it could be a model acquired from the internet considered worthy of testing against a local set of data (which would also be considered an input unit). An input unit could also be as large as the total set of information needed for a single WinSAAM run, i.e. a self-contained unit, or a saved modeling project.

A *modeling project* constitutes all the files essential for a modeling session. Thus, more fully, the project comprises essential input units, as well as essential file-based information generated in what we may refer to as the 'instance of a modeling session.' An instance of a modeling session represents all the external resources needed (file names and locations, file assignments and associations, for example) for WinSAAM interaction to take place in association with a particular modeling project.

Finally, to enable a better integration of input units and management of modeling projects, we have created the notion of a *linker*. In addition to displaying linkable items in their array of classes, in preparation for project creation, the linker will facilitate both their class-wise association (i.e. naming or identifying global functions within the model units, portraying segments within the data units or within time segments of the data units), and their physical collection for project processing.

Derived Variables and Finesse Fitting

Conceivably the most efficient approach for processing data from several identical studies is with the use of Zeger's (1994) derived variable analysis, or 'finesse' fitting. This method derives its strengths from its two-step approach to analysis viz:

Step One: Each study is examined to isolate a *single* model structure which consistently characterizes all measured responses. Each study's data units are fitted appropriately to yield

$$\mu_1 = \mu + r_1$$
$$r_i = a_i\, r \qquad (E(a_1) = 1) \tag{1}$$

where μ_1 is the estimate of μ for the *i*th individual; μ, the population mean value; r_i, the estimate of r from the *i*th individual; and r, the population covariance.

Step Two: μ and r are estimated using either the method of maximum likelihood or its restricted version (to avoid bias).

The strengths of this two-step approach are:

- It commits the modeler to preserving the consistency between the target population model and each specific study.
- It is extremely flexible and virtually limitlessly extensible.
- It easily accommodates a cumulative development of the covariance base.

- It can readily be modified to incorporate Bayesian methods concerning future or prospective applications of the model.
- It leads to immediate isolation of aberrant individuals.

A limitation of this approach is the restrictive representation of individual study variances. Users need to be aware that response bias (heterogeneity) and response heteroscedasticity (over-variability) can each lead to inefficient estimators and hence impaired inference. However, the first step imposed by this method forces the user to reflect carefully on the data sets to be aggregated.

It should be clear to the reader that there are strong connections between our last discussion item involving the linker and the modeling project and this particular topic. The use of the above methodology in SAAM, under the title of multiple studies analysis (Lyne, 1993), required that users link and associate studies manually using arcane and fixed formatted data structures. By contrast, by using the linker in WinSAAM, almost all of the lexical and other manual issues will be automatically coordinated, as the procedure is invoked interactively. Indeed, linked studies for population modeling will emerge as a discrete and highly coordinated function specific to the linker.

Data Exchange with Other Software

Publication preparation, presentation preparation, data preparation and screening, advanced statistical analysis, and advanced graphical management are all activities investigators often need support with but clearly should fall only partly within the scope of kinetic modeling software. We have elected, in designing and implementing WinSAAM, to distribute these function in ways that allow flexibility for users to explore their own array of productivity tools in conjunction with data analysis. Specifically, through opening data exchange channels between WinSAAM and word processors via the clipboard, between WinSAAM and spreadsheet software via the spreadsheet manager, between WinSAAM and graphics processors via either metafile or bitmap encapsulated graphic renditions, and between WinSAAM and advance statistical software via the spreadsheet manager, we have orchestrated absolute flexibility within WinSAAM for leveraging the pertinent processing strengths of related software.

Of course, another way to look at the data exchange issue is to bring a more realistic and manageable closure to the software development and maintenance effort. Rather than duplicating features in other software, if we simply ensure that the opportunity exists to exchange data at an appropriate state with other software in order to facilitate specific processing needs, then we have a clearer and more manageable delineation of the functional boundaries of our evolving system, WinSAAM. We have exercised considerable care in identifying target exchange software and data exchange modes, and at the heart of our consideration have been the basic concerns … "Is the target software from a setting familiar to the user?" … "Is the mode of data exchange direct and flexible?" and, to a lesser degree … "Does the target software have some similarity to and compatibility with the WinSAAM system?"

Third-Party Component Modules

We recognized early in the WinSAAM development project that our ideas could never be realized if all the work were left solely to us. A project of this magnitude calls for multi-faceted skills and contributions well beyond what we were capable of. Accordingly, we designed the system to admit contributions from both the software community and the user community, and also to avoid stalling the system's evolution as these contributions evolved and were themselves incorporated into it.

Specifically we have identified three areas where third-party component modules can be dynamically incorporated into the system:

Functional enhancements to the interface.

Great progress and power enhancement for the user interface has been made possible through our incorporation of third-party software modules. For example, the graphics, spreadsheet, and help utilities have all been refined from commercial tools privately supplied. Other tools which have the potential to enhance the interface power and flexibility will be explored.

Processing enhancement to the kernel.

At the heart of WinSAAM is the SAAM/CONSAM system, encapsulated as a dynamic linking library (DLL). To accomplish this integration, we were required to carefully define the architectural interface between the kernel and the user interface component of the software to ensure that all results from a kernel activation were accessible to the user via the user interface (GUI). Going through this exercise has now provided us with the background, information, and operational details needed to enable us to build on the dynamic linking library concept. As new integrators, optimizers, and allied processing algorithms become available, they will be able to be incorporated into the WinSAAM environment and use repertoire via linking libraries. The two great advantages of pursuing this type of development are, firstly, we can again capitalize on expertise beyond that of the WinSAAM team, and, secondly, we can enhance WinSAAM's capabilities without risking corrupting the kernel code.

Domain Libraries.

It is our belief that the capacity of modeling software such as WinSAAM to support model development constitutes just a small percentage of its return to the scientific community. Considerably greater benefit comes from it being able to deliver the power from the models generated to laboratories, clinics, and educational settings. Indeed, the contributions from investigators such as Cobelli and coworkers have been largely through their unique skill in evolving models, most particularly in the glucose-insulin area (in their case) which address key clinical and allied issues.

In addition to supporting a 'base' model library architecture, WinSAAM will support a comprehensive set of 'domain' libraries which will each include salient models from important application areas. Experts such as Cobelli and his group could be registered as authorized suppliers of a specific domain, in their case glucose-insulin models. Users with access to that domain could then explore the applications of the models encapsulated within it.

Again we have tried to assemble a strategy for extending the application and functionality of WinSAAM by distributing the effort amongst the greater community of biomedical investigators.

Open Architecture

In moving from the CLI (command line interface) style of operation of CONSAM and Mlab (1996) towards the GUI (graphical user interface) style of operation of Stella II (1992) and SAAM II (1997), we are immediately confronted by what are known as 'mode' issues. *Mode-less* software is software that asserts only limited control over the processing flow available for the user. In essence, except for blocking outlandish steps (such as editing a binary file), the mode-less software user is pretty much free to chart her own course. Mode-less software is often described as 'flat,' as the windowing hierarchy is usually fairly shallow.

In contrast to this, *moded* software is somewhat more rigidly controlling in that Windows dispensing services only become available following a particular activation sequence. Another sequence needs to be followed to move hierarchically up through the windows to other tasks and services.

Each type of architecture serves its own niche: moded software tends to be more helpful for use by novices for example, while the mode-less, free-wheeling approach is greatly appreciated by advanced investigators who need to capitalize on their own skills in order to expedite their modeling efforts.

The design of WinSAAM has tended to favor the mode-less approach by providing flexible access to many of its services through a fairly flat architecture. Nevertheless, where necessary and where appropriate or helpful, we have also embraced a moded operation. We feel that a fine balance has to be struck between these diametric provisions. The greater development effort required by a mode-less operation is usually balanced by its commensurate returns to the user, as a consequence of the freedom it allows to explore the ensuing open architecture.

CONCLUSION

We have discussed many of the issues motivating the development of the NIH WinSAAM project. Specifically, we have stressed the care with which we have balanced our development efforts against the expected returns to the user community for that effort. We have identified a series of efficient steps to ensure not only the widest possible participation in this project, but also the highest utility end-product we can possibly evolve.

From the development perspective, and for the user community, *evolve* is the key word. As a very small development team, the only practical way for us to effectively advance the project is by deploying the evolutionary or incremental prototyping paradigm. At the completion of each stage in the evolution of WinSAAM, the next step is negotiated subject to (1) it is subject to incremental incorporation, (2) it demonstrably advances the utility of the software, and (3) its initial implementation is subject to revision against extensive testing and evaluation. Adopting this approach has greatly accelerated the delivery of a 'working' product.

As this chapter was completed, work is underway on WinSAAM's file management and help facilities and, indeed, a first version of the file management facility will only support the integrated WinSAAM input file. Based on the above criteria, we have established a development sequence plan which will see all major components of WinSAAM in place this year.

Because of the current state of the utility of the WinSAAM software and the greater ease with which it fits into the Windows operating environment than either SAAM or Consam, we now invite Consam users to fully participate in the WinSAAM project as registered β-site testers of our new system.

CORRESPONDING AUTHOR

Please address all correspondence to:
Dr. Ray C. Boston
University of Pennsylvania
New Bolton Center
346 West Street Road
Kennettt Square, PA 19348

REFERENCES

ACSL: Advanced Continuous Simulation Language, Version 10.2. Mitchell and Gauthier Asscociates: Concord, MA. 1992.

Baldwin RL; France J; Beever DE; Gill M; Thornley JHM. Metabolism of the Lactating Cow III: Properties of Mechanistic Models Suitable for Evaluation of Energetic Relationships and Factors Involved in the Partition of Nutrients. *J Dairy Res,* 1987, 54:133-145.

Baldwin RL. *Modeling Ruminant Digestion and Metabolism.* Chapman Hall:London. 1995.

Berman M; Weiss MF. *SAAM 27 Manual.* DHEW Publ. (NIH) 78-180. Laboratory of Theoretical Biology, NCI, NIH: Bethesda, MD. 1963-1978.

Berman M; Grundy SM; Howard BV. *Lipoprotein Kinetics and Modeling.* Academic Press: New York. 1982.

Boston RC; Greif PC; Berman M. Conversational SAAM-an interactive program for the kinetic analysis of biological systems. *Comput Prog Biomed,* 1981, 13:111-119.

Boston RC; Weber KM. Modeling with SAAM and its advancement in association with the study of mineral metabolism. *Math Biosci,* 1984, 72(2):181-198.

Carson ER; Cobelli C; Finkelstein L. *The Mathematical Modeling of Metabolic and Endocrine Systems.* John Wiley: New York. 1983.

Diggle PJ; Liang KV; Zeger SL. *Analysis of Longitudinal Data.* Oxford University Press: Oxford. 1994.

Dijkstra J; Neal HDStC; Beever DE; France J. Simulation of nutrient digestion, absorption and outflow in the rumen: Model description. *J Nutr,* 1992, 122:2239-2256.

Dijkstra J. *Mathematical Modeling and Integration of Rumen Fermentation Processes.* Univ. Wageningen Press: Netherlands. 1993.

Faichney GJ; Boston RC. Movement of water within the body of sheep fed at maintenance under thermoneutral conditions. *Aust J Biol Sci,* 1985, 38:85-94.

France J; Thornley JHM; Beever DE. A mathematical model of the rumen. *J Agric Sci Camb,* 1982, 99:343-353.

France J; Thornley JHM. *Mathematical Models in Agriculture.* Butterworths: London. 1984.

Fugazi K. *Continuous System Modeling for the B6700.* Computer Center: University of California, Davis, CA. 1974.

Gill M; Thornley JHM; Black JL; Oldham JD; Beever DE. Simulation of the metabolism of absorbed energy-yielding nutrients in young sheep. *B J Nutr,* 1984, 52:621-649.

Hanigan MD; Baldwin RL. A mechanistic model of mammary gland metabolism in the lactating cow. *Agric Systems,* 1994, 45:369-393.

Jacquez JA. *Compartmental Analysis in Biology and Medicine.* Biomedware: Ann Arbor. 1996.

Lyne A; Boston R; Pettigrew K; Zech L. EMSA – A SAAM service for the estimation of population parameters based on model fits to identically replicated experiments. *Comp Mthds Programs Biomed,* 1992, 38:117-151.

Mlab: A Mathematical Laboratory. Civilized Software, Inc.: Bethesda, MD. Revision Date: Oct. 8, 1996.

Neal HDStC; Thornley JHM. The lactation curve in cattle: A mathematical model of the mammary gland. *J Agric Sci Camb,* 1983, 101:389-400.

Pettigrew FE; Gill M; France J; Close WH. A mathematical integration of energy and amino acid metabolism of lactating sows. *J Ani Sci,* 1992, 70:3742-3761.

Pressman RS. *Software Engineering: A Practioner's Approach.* McGraw Hill: New York. 1982.

SAAM II: A Program for Kinetic Modeling. Resource Facility for Kinetic Analysis: University of Washington, Seattle, WA. 1997.

Scientist: For Experimental Data Fitting, Version 2.0. MicroMath Scientific Software: Salt Lake City, Utah. 1995.

STATA: Release 5.0, Statistics, Data Management, Graphics. StataCorp: College Station, TX. 1997.

Stella II: An Introduction to Systems Thinking. High Performance Systems, Inc.: Hanover, NH. 1992.

Thornley JHM; Johnson IR. *Plant and Crop Modeling: A Mathematical Approach to Plant and Crop Physiology.* Oxford University Press: Oxford. 1990.

COMPARTMENTAL MODELING OF HUMAN LACTATION

Janet A. Novotny[1] and Benjamin Caballero[2]

[1]U.S. Department of Agriculture
Agricultural Research Service
Beltsville Human Nutrition Research Center
Diet and Human Performance Laboratory
Beltsville, MD 20705
[2]Center for Human Nutrition
School of Hygiene and Public Health
Johns Hopkins University
615 North Wolfe Street
Baltimore MD 21205

INTRODUCTION

Mathematical modeling is a powerful and exciting tool for the quantitative analysis of complex systems. In nutrition and other biological sciences, the systems studied are extremely complex with many processes difficult to observe directly. The rigors of mathematics provide a framework to indirectly study processes which cannot be observed directly. In this way, mathematical modeling offers scientists a window with which to view aspects of systems which would otherwise be inaccessible.

For a new modeler, it can be difficult to know where to begin to learn the principles of modeling and its capabilities. Often, an unfamiliar discipline is most easily approached initially by means of a very clear example. To introduce modeling to the beginning modeler, we present here a description of the modeling process through an example of human lactation. Because this model of human lactation is simple and does not require in-depth knowledge of a particular research area, it serves as a useful example to illustrate the modeling process.

EXPERIMENTAL BACKGROUND

Lactation is a very important source of nutrients to infants, and the accurate measurement of lactation through modeling has important nutritional applications. For example, if a nutrient is supplemented to a mother, and the appearance of the nutrient in the

Mathematical Modeling in Experimental Nutrition
Edited by Clifford and Müller, Plenum Press, New York, 1998

breast milk is measured, then an accurate prediction of infant breast milk intake would reveal the exposure of the infant to that nutrient.

The traditional means for determining breast milk intake is test-weighing (Jensen and Neville, 1985). In this case, the infant is weighed before and after every feeding. This is cumbersome and inconvenient, and for field studies, it is difficult to assure compliance.

An alternative is isotope dilution. Isotope dilution for the determination of breast milk intake was first introduced by Coward and colleagues in 1979 (Coward et al., 1979). Their original method involved dosing an infant with deuterium oxide and collecting the infant's urine for two weeks following the dose. The method included the assumption that the infant exclusively consumed breast milk. Because this is usually not the case, this method overestimated breast milk intake. They subsequently reported a method in which the deuterium oxide was administered to the mother (Coward et al., 1982). Work by Butte and colleagues later validated the method against the test-weighing technique (Butte et al., 1988).

The data presented here were generated as part of a study to evaluate the efficacy of using isotope dilution and modeling to determine breast milk intake in rural field studies. The data used in the model presented here were collected from a mother-infant pair after the mother received a single 30 gram dose of deuterium oxide. Samples of breast milk were collected on days 1, 2, 12, and 14 following the dose, and samples of infant urine were collected on days 1, 2, 3, 4, 6, 8, 10, 12, and 14 following the dose. For this model, it was assumed that total body water is one well-mixed pool, and thus it was assumed that samples of urine and breast milk can be used to determine the enrichment of the total body water pools of the infant and mother, respectively. Deuterium enrichment of the samples was determined by isotope ratio mass spectrometry. Figure 1 shows the deuterium enrichment data from a sample mother-infant pair. Deuterium enrichment is expressed as percent above baseline in the following manner:

$$\text{Enrichment} = (^2\text{H Mass / Dilution Volume}) \times 100 \qquad (1)$$

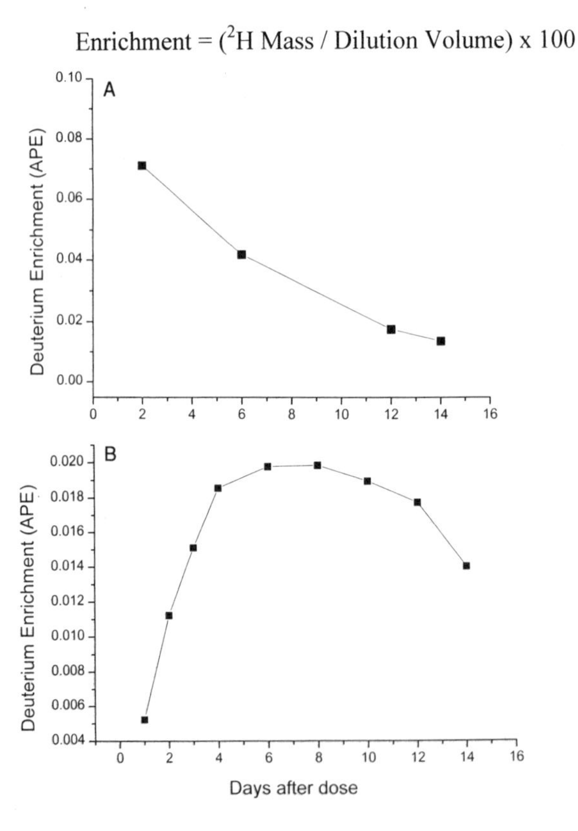

Figures 1A and 1B. Sample deuterium enrichment data for one mother-infant pair after the mother received a 30 g dose of deuterium oxide. (1A) Mother deuterium oxide enrichment as percent above baseline. (1B) Infant deuterium oxide enrichment as percent above baseline.

THE MODELING PROCESS

The steps of modeling are delineated below. Different modelers may have slightly different versions of these steps or may alter the modeling approach to fit a particular modeling task, but the basic steps remain nearly the same from case to case. These steps serve as a general guideline rather than a recipe, because each modeling endeavor is unique, and creativity and originality are required in every case.

The steps used to model this system were as follows:

1. Definition of an objective.
2. Collection of system information.
3. Development of a system diagram.
4. Derivation of model equations.
5. Estimation of parameter values and initial conditions.
6. Incorporation of the model structure and values into a computer program or code.
7. Fitting the model to the data.
   A. Comparison of model prediction to experimental data.
   B. Modification of model structure or parameter values to improve fit.

Step 1. Definition of an Objective

A clearly defined objective is important in every scientific endeavor. Upon initiating a modeling effort, a clearly defined objective helps to determine the model structure as well as the experimental sampling and measurement scheme. A poorly defined objective may lead to an overly simple or complex model as well as excess or insufficient sampling. The objective of the lactation model was to determine the breast milk and complementary water intake of infants from the decay kinetics of deuterium oxide in mothers and infants.

Step 2. Collection Of System Information

The modeler next gathers information about the system. This includes a broad range of information, including methods for measuring the analyte of interest, analyte pool sizes, favorable tracer dose sizes for the specified objective, estimated flow rates and half-times, and so forth.

Our initial calculations showed that a dose of 30 grams of deuterium oxide to a mother would produce isotopomer enrichments in the mother and infant that were within the range of sensitivity of an isotope ratio mass spectrometer. In addition, ingestion of 30 grams of deuterium oxide would not perturb flow of body water as to compromise our results. We further collected information regarding the expected intake of breast milk and complimentary water intake by the infant, which was expected to be several hundred milliliters each per day, and water intake by the mothers, which was expected to be a few liters per day.

Step 3. Development of a System Diagram

Next, the modeler develops a diagram of the system of interest including all the relevant components and the connections between them. The schematic in Figure 2 represents an initial model structure for lactation. Each circle represents a body pool, and the arrows represent flows of water into or out of a pool. In this two-pool model, water flows into the mother from external sources, and leaves the mother via urine, perspiration, and insensible water loss (shown collectively as one arrow) or leaves the mother via lactation to enter the infant. Water also enters the infant from external sources and leaves the infant via urine, perspiration, and insensible water loss.

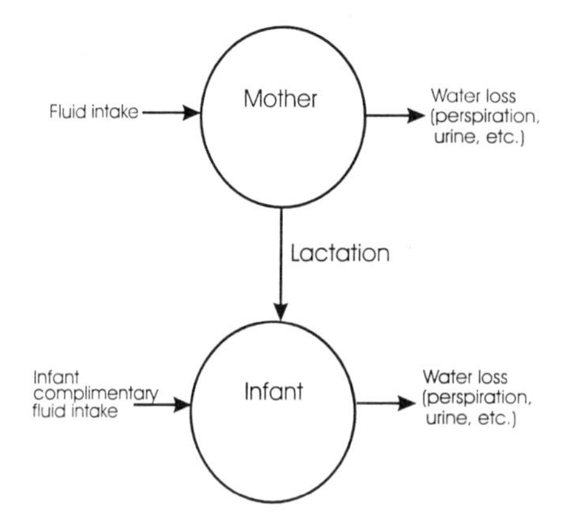

Figure 2. Schematic which serves as an initial compartmental model of human lactation.

Step 4. Derivation of Model Equations

One derives the model equations from the system diagram, beginning with the summation of the inputs and outputs for each compartment.

$$d(D_2O_M)/dt = -D_2O\ Loss_M - D_2O\ Lactation \quad (2)$$

$$d(D_2O_I)/dt = D_2O\ Lactation - D_2O\ Loss_I \quad (3)$$

where D_2O_M is the mass of D_2O in the mother, D_2O $Loss_M$ is the rate of loss of D_2O from the mother via urine, perspiration, and insensible water loss, D_2O Lactation is the rate of transfer of deuterium from the mother to the infant via lactation, D_2O_I is the mass of D_2O in the infant, and D_2O $Loss_I$ is the rate of loss of D_2O from the infant via urine, perspiration, and insensible water loss. Because the deuterium was given to the mother as a bolus, that bolus is incorporated into the model as an initial condition of the Mother-D_2O compartment.

The equation for each flow is written under the assumption that this system behaves according to the law of mass action. In other words, the system can be described by first order linear differential equations. If we call the mother's deuterium pool "Compartment 1" and the infant's deuterium pool "Compartment 2," then the equations can be written as follows:

$$dF(1)/dt = -L(0,1) \times F(1) - L(2,1) \times F(1) \quad (4)$$

$$dF(2)/dt = L(2,1) \times F(1) - L(0,2) \times F(2) \quad (5)$$

where F(1) is the mass of tracer in the Mother-D_2O compartment (Compartment 1), L(0,1) is the rate coefficient (also called fractional transfer rate) describing flow of D_2O out of the system from the mother, L(2,1) is the rate coefficient describing flow of D_2O from the mother to the infant, F(2) is the mass of tracer in the Infant-D_2O compartment (Compartment 2), and L(0,2) is the rate coefficient describing flow of D_2O out of the system from the infant. The notation used for these equations is that of the computer program SAAM31 (Berman and Weiss, 1978; Berman et al., 1983). SAAM31, which stands for Simulation, Analysis, and Modeling, is a computer program designed for the analysis of data in terms of compartmental models.

In SAAM31 notation, the letter L refers to a fractional transfer rate. L(A, B) refers to the fractional transfer of material from Compartment B (the donor compartment) to Compartment A (the recipient compartment) per unit time. In other words, L(A, B) is the fraction of material in compartment B which is transferred to compartment A per unit time. Using SAAM31 notation, the letter F refers to the quantity of tracer in a given compartment. Therefore, in this example F(1) is the quantity of D_2O in Compartment 1, the Mother-D_2O compartment. Figure 3A shows the tracer model with the notation for pool sizes (quantities in compartments) and flows.

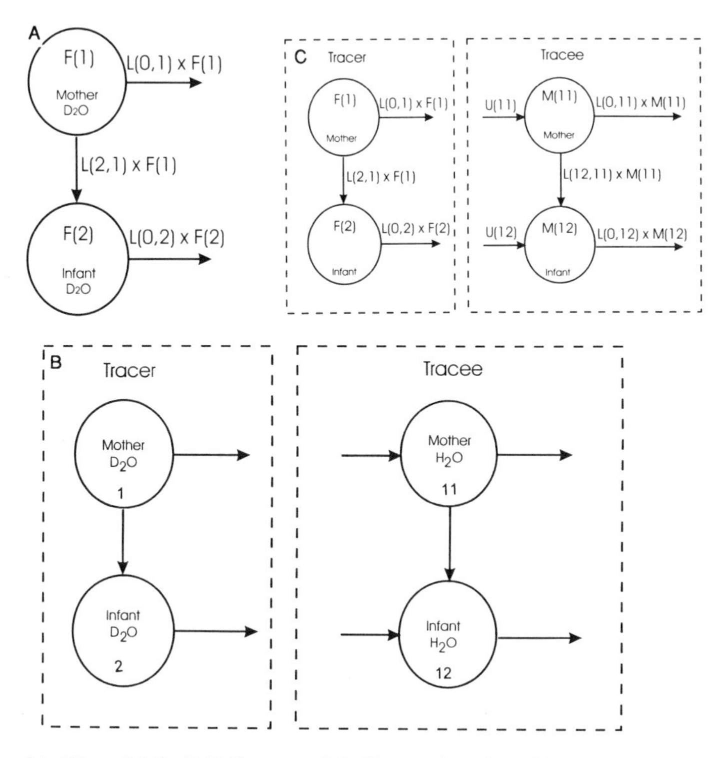

Figures 3A, 3B, and 3C. (3A) Tracer model of human lactation with notation for pool sizes (quantity in compartments) and flow equations. (3B) Parallel models of human lactation. The left portion of the diagram represents deuterium oxide flow through the system, and the right portion of the diagram represents water flow through the system. (3C) Parallel models of human lactation including parameters associated with compartments and fractional transfer rates.

Note that the pair of equations above represents deuterium oxide flow rather than water flow. To account for the tracer (D_2O) and the tracee (H_2O) simultaneously, the modeler can create *parallel models*. Figure 3B shows parallel models for the lactation system. For the deuterium portion of this system, there are no flows into either compartment from external sources, because the bolus dose is incorporated into the model as an initial condition. Because it is assumed that the tracer and tracee behaviors are kinetically identical, the rate coefficients describing the flows for the tracer and the tracee are equal. As shown in Figure 3B, the Mother-H_2O compartment and the Infant-H_2O compartment are labeled Compartment 11 and 12, respectively. The equations for the tracee (H_2O) model can be written as follows:

$$dM(11)/dt = U(11) - L(0,11) \times M(11) - L(12,11) \times M(11) \quad (6)$$

$$dM(12)/dt = U(12) + L(12,11) \times M(11) - L(0,12) \times M(12) \quad (7)$$

where M(11) is the mass of tracee in the Mother-H_2O compartment (Compartment 11), M(12) is the mass of tracee in the Infant-H_2O compartment (Compartment 12), L(0,11) is the rate coefficient describing flow of H_2O out of the system from the mother, L(0,12) is the rate coefficient describing flow of H_2O out of the system from the infant, and L(12,11) is the rate coefficient describing flow of H_2O from the mother to the infant. In the tracee equations, SAAM31 uses the letter M to designate steady-state masses. U(11) and U(12) refer to the steady-state flow of water into Compartments 11 and 12, respectively. Figure 3C shows the parallel compartmental models with notation for pool sizes (quantities in compartments) and the equations for transfer of material.

Because the tracer and tracee are assumed to be kinetically indistinguishable, the rate coefficients describing the tracer movement are equal to those for the tracee movement. Thus, L(0,11) = L(0,1); L(0,12) = L(0,2); and L(12,11) = L(2,1)

Step 5. Estimation of Parameter Values and Initial Conditions

Before beginning the fitting of the model prediction to the experimental data, the modeler must provide initial conditions for the masses in each compartment. For this example, it will be assumed that the background deuterium enrichment is negligible. Therefore, the initial mass of the D_2O in the Mother-D_2O compartment (Compartment 1) is 30 grams (the mass of the D_2O dose), and the initial mass of the D_2O in the Infant-D_2O compartment is 0 grams.

The initial mass of the Mother-H_2O compartment (the mother's total body water) can be calculated from the decay of D_2O enrichment as a function of time. The mother's total body water can be calculated from the dilution volume at time 0 had the dose mixed instantaneously. Figure 4 shows the D_2O enrichment of the mother's water pool as a function of time plotted on a semilog plot. Linear regression produces a y-intercept which represents the D_2O enrichment if the dose had mixed instantaneously. The specific equation used to calculate total body water from the calculated time 0 enrichment will depend on how the enrichment is measured and expressed. In this case, the enrichment was measured in units of δ and converted to percent above baseline using a standard curve to units such that the following equation was true:

$$(\text{Dose} / \text{Dilution Volume}) \times 100 = \text{Enrichment} \quad (8)$$

Thus, in this case, Dilution Volume (total body water) can be calculated as

$$\text{Dilution Volume (kg)} = (\text{Dose (kg)} / \text{Enrichment}) \times 100 \quad (9)$$

From the dose and the enrichment, we can calculate the mother's total body water, which is 31.6 kg in this example. Note that the calculation of TBW has been simplified slightly for this example. For a discussion of calculation of TBW by isotope enrichment, see Schoeller et al. (1982). For further discussion of enrichment terminology, see Wolfe (1992).

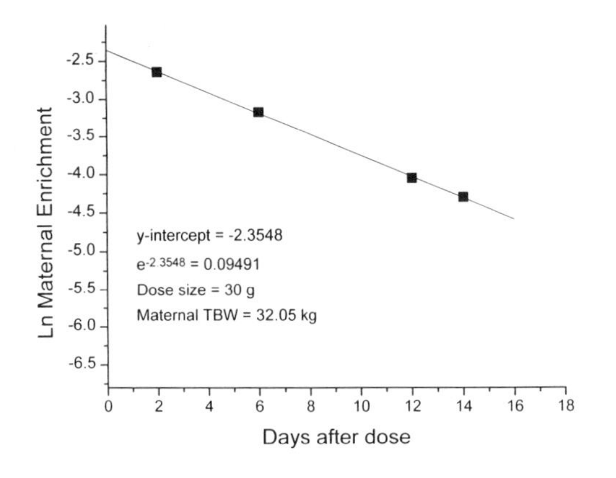

Figure 4. Mother's deuterium oxide enrichment as a function of time. The y-intercept represents the deuterium oxide enrichment had the dose mixed with the mother's body water pool instantaneously.

The infant's total body water can be calculated from an equation derived by Butte and colleagues (1992) based on oxygen-18 and deuterium dilution spaces in infant subjects: Infant Total Body Water (kg) = 1.132 x $\text{Weight}^{0.650}$. For this example, the infant's total body water is 5 kg.

The modeler must also provide initial estimates of rate coefficients. This can be accomplished by reviewing previously collected knowledge of flow rates and half-times of the system under investigation or a similar biological system. In this case, one may begin by estimating the flow rates for the tracee.

An adult would be expected to consume two or three liters of water daily. An infant would be expected to consume several hundred milliliters of water from both breast milk and complimentary sources each day. Using this knowledge, initial flow estimates are shown in Figure 5. The mother is assumed to consume 2.3 kg of water daily, lose 2.0 kg of water daily via urine, perspiration, and insensible water loss, and transfer 0.3 kg of water to the infant daily via lactation. The infant is assumed to consume 0.5 kg of water from complimentary sources, and to lose 0.8 kg of water daily via urine, perspiration, and insensible water loss. Note that because this system is assumed to be in steady-state, total flow into a given compartment must be equal to total flow out of that compartment.

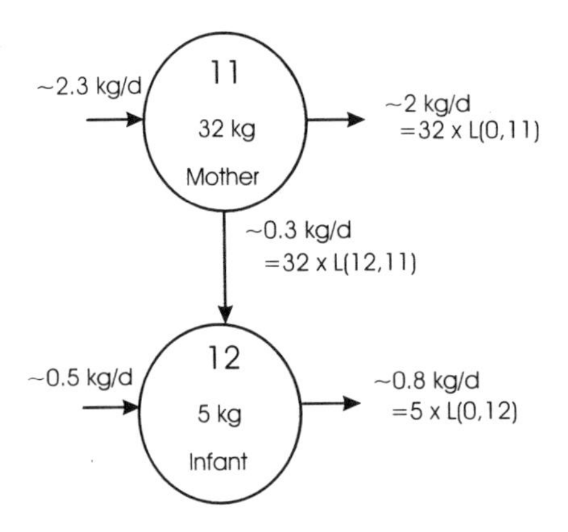

Figure 5. Compartmental model of water flow with model equations and estimated flow rates included.

From the estimated flows and the initial values of total body water, initial values for rate coefficients can be estimated, as shown in Figure 5. The rate coefficients are equal to fractional transfer rates, and are equal to the fraction of material transported from the donor compartment per unit time.

$$L(0,1) = L(0,11) = (\ 2.0\ \text{kg/day}\ )\ /\ (\ 31.6\ \text{kg}\ ) = 0.063\ \text{day}^{-1} \tag{10}$$

$$L(0,2) = L(0,12) = (\ 0.8\ \text{kg/day}\ )\ /\ (\ 5\ \text{kg}\ ) = 0.16\ \text{day}^{-1} \tag{11}$$

$$L(2,1) = L(12,11) = (\ 0.3\ \text{kg/day}\ )\ /\ (\ 31.6\ \text{kg}) = 0.0095\ \text{day}^{-1} \tag{12}$$

Step 6. Incorporation of the Model Structure into a Computer Program

Before simulations can be performed, the model structure and parameters must be incorporated into a computer program so that calculations may be performed rapidly and accurately. There are many types of mathematical software and computer resources available to assist modelers in the analysis of data. Some tools are specifically designed for compartmental modeling, like ACSL, SAAM31, and SAAM II, while others are helpful for more general mathematical analysis, like Mathematica and Maple. The model presented here was incorporated into the SAAM31 software package, and modeling was performed using the interactive CONSAM interface (Berman and Weiss, 1978; Berman et al., 1983). A sample input file for this model with data for a second mother-infant pair is shown in the Appendix.

Step 7. Fitting the Model to the Data

In fitting the model to the data, the modeler begins by comparing the model's predictions to the observed values. In this case, the prediction of mother and infant D_2O enrichment is compared to observed D_2O enrichment in breast milk and in infant urine. Differences between the model prediction and the experimental observation are reduced iteratively by alternately modifying model parameters or structure and comparing the model prediction to observation. The nature of the differences between prediction and observation are useful in directing the modifications.

Figure 6A shows the comparison between model prediction and the infant's experimentally measured deuterium oxide mass for the initial model and parameter estimates. Figure 6B shows the comparison between model prediction and experimentally measured deuterium oxide mass for the mother as generated with the initial model and parameter estimates. Examination of Figure 6B shows that the model predicts a rate of decay of D_2O content which is slower than that observed experimentally. This decay rate can be increased by increasing urinary water loss or lactational water loss. Because urinary loss of water is of greater quantity than lactation loss of water, it is best to adjust that parameter first.

Increasing the rate coefficient for maternal urinary water loss by approximately 60%, such that $L(0,1)=0.10\ \text{day}^{-1}$, improves the fit of the model prediction to the mother's D_2O level, as shown in Figure 6C, but the decay of the D_2O level is still too slow. An additional 25% increase in the rate coefficient for maternal urinary water loss such that $L(0,1)=0.125\ \text{day}^{-1}$ produces a fairly good fit of the model prediction for the mother's D_2O level to the experimental data, as shown in Figure 6D.

Once the model predicts the mother's D_2O level well, the infant's D_2O data is compared to the model prediction. Figure 6E shows that with the current set of parameters, the model predicts insufficient transfer of water from the mother to the infant. Increasing the rate coefficient for lactation (L(2,1)) approximately 50% from 0.0095 day^{-1} to 0.0145 day^{-1} substantially improves the fit of the model to the data, as shown in Figure 6F.

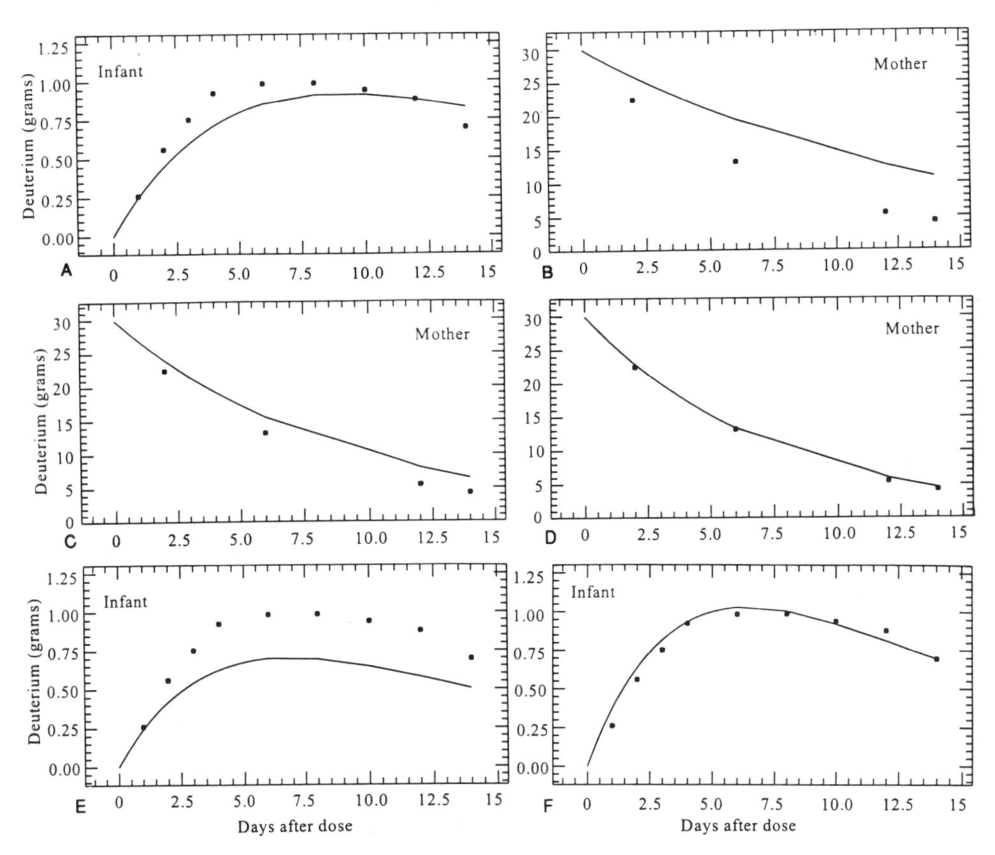

Figures 6A–6F. Progressive fitting of model prediction to experimental data. The line represents the model prediction, and the points represent experimental observation. Note that D_2O level is now expressed in grams. (6A) Model prediction and experimental data for infant's deuterium oxide enrichment when simulations were performed using initial estimates of rate coefficients [L(0,1)=0.063 d^{1}, L(0,2)=0.16 d^{-1}, L(2,1)=0.0095 d^{-1}]. (6B) Model prediction and experimental data for mother's deuterium oxide enrichment when simulations were performed using initial estimates of rate coefficients [L(0,1)=0.063 d^{-1}, L(0,2)=0.16 d^{-1}, L(2,1)=0.0095 d^{-1}]. (6C) Model prediction and experimental data for mother deuterium oxide enrichment when simulations were performed after mother-urine rate coefficient was increased approximately 60% [L(0,1)=0.1 d^{-1}, L(0,2)=0.16 d^{-1}, L(2,1)=0.0095 d^{-1}]. (6D) Model prediction and experimental data for mother deuterium oxide enrichment when simulations were performed after mother-urine rate coefficient was increased by an additional 25% [L(0,1)=0.125 d^{-1}, L(0,2)=0.16 d^{-1}, L(2,1)=0.0095 d^{-1}]. (6E) Model prediction and experimental data for infant deuterium oxide enrichment when simulations were performed after mother-urine rate coefficient was increased by an additional 25% [L(0,1)=0.125 d^{-1}, L(0,2)=0.16 d^{-1}, L(2,1)=0.0095 d^{-1}]. (Companion graph to 6D). (6F) Model prediction and experimental data for infant deuterium oxide enrichment when simulations were performed after lactation rate coefficient was increased by approximately 50% [L(0,1)=0.125 d^{-1}, L(0,2)=0.16 d^{-1}, L(2,1)=0.0145 d^{-1}].

Close examination of Figure 6F demonstrates that the model curve must be shifted to the right. This can be achieved by the addition of a delay between the mother and the infant. The addition of a delay is physiologically sensible in that such a delay would represent the time necessary for the mixing of the dose and for the onset of the next infant feeding. This exemplifies the need in modeling not only for modification of parameters in fitting of model prediction to data, but also in the modification of model structure.

Figure 7 shows the model structure with the delay element inserted between the mother and infant water pools. Continued fitting of the new model structure by modification of rate coefficients and the delay time result in model predictions which are in good agreement with both mother and infant D_2O data. Finally, least squares iteration is used to minimize the distances between the observed data points and the model prediction. The final fit of model prediction to the experimental data is shown in Figures 8A and 8B.

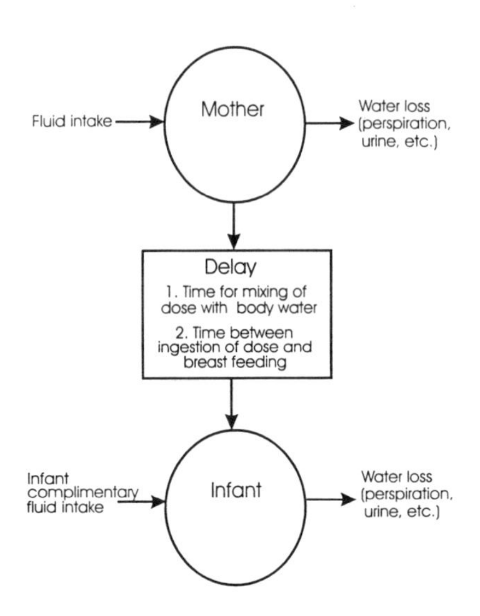

Figure 7. Modified compartmental model of human lactation which includes a delay between transfer of water from the mother to the infant.

CALCULATION OF BREAST MILK INTAKE

The model results predict that the infant examined with this model consumed 0.45 kg of water daily from breast milk and 0.84 kg of water daily from other sources. The mother consumed 4.42 kg of water daily. Because breast milk is 87% water (Butte et al., 1988), this infant consumed 0.52 kg breast milk daily.

SUMMARY AND CONCLUSION

A difficult task for a novice modeler is beginning a modeling project. The methods described here delineate the modeling process step by step. Aristotle is quoted as having said "One learns to play the flute by playing the flute," and the same principle holds somewhat for modeling. An excellent means for understanding modeling is to work with example models, altering rate coefficients and model structure, to learn first-hand the mechanics of the process. An excellent resource for viewing and downloading published models is the Library of Mathematical Models of Biological Systems, and can be accessed via the World Wide Web at the address http://gopher.dml.georgetown.edu/model/model.html. This modeling library provides access to working models of biological systems. Experimenting with available examples of models, such as those presented in this volume and those available through the Modeling Library, provides an excellent opportunity for new modelers to get started in compartmental modeling.

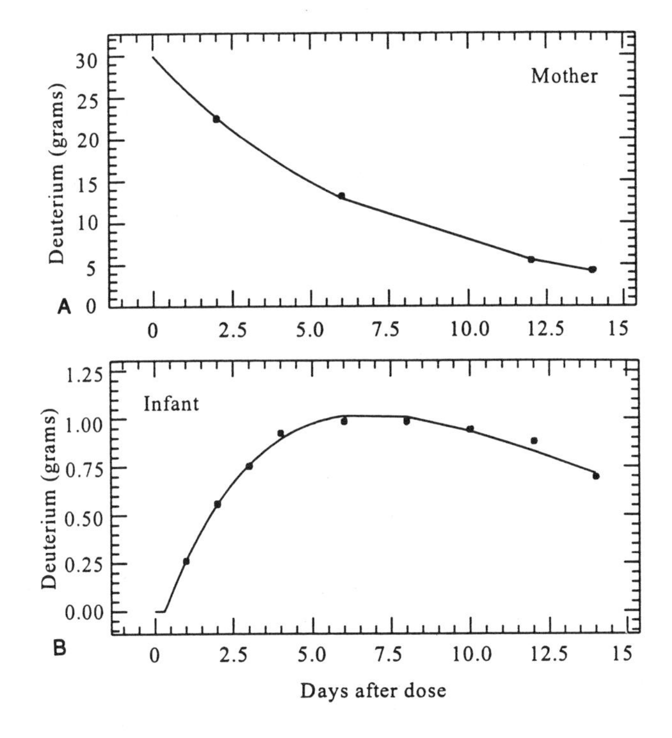

Figures 8A and 8B. Final comparison of mother (8A) and infant (8B) deuterium oxide enrichment. The line represents the model prediction, and the points represent experimental observation. The model produces results which are in good agreement with experimental observation.

CORRESPONDING AUTHOR

Please address all correspondence to:
Janet Novotny
USDA BHNRC DHPL
Bldg 308, BARC-East
Beltsville, MD 20705

REFERENCES

Berman M; Weiss MF. *SAAM Manual.* DHEW Publ. No. (NIH) 78-180. US Government. Printing Office, Washington, DC. 1978.

Berman M; Beltz WF;Greif PC; Chabay R; Boston RC. *CONSAM User's Guide.* PHS Publ. No. 1983-421. US Government. Printing Office, Washington, DC. 1983.

Butte NF; Wong WW; Patterson BW; Garza C; Klein PD. Human-milk intake measured by administration of deuterium oxide to the mother: A comparison with the test-weighing technique. *Am J Clin Nutr,* 1988, 47:815-821.

Butte NF; Wong WW; Garza C. Prediction equations for total body water during early infancy. *Acta Paediatr,* 1988, 81:264-265.

Coward WA; Sawyer MB; Whitehead RG; Prentice AM. New method for measuring milk intakes in breast-fed babies. *Lancet,* 1979, 2:13-14.

Coward WA; Cole TJ; Sawyer MB; Prentice AM. Breast-milk intake measurement in mixed-fed infants by administration of deuterium oxide to their mothers. *Hum. Nutr. Clin. Nutr,* 1982, 36C:141-148.

Jensen RG; Neville MC. Human lactation, in*: Human Lactation: Milk Components and Methodologies.* Jensen RG; Neville MC; Eds. Plenum Press, New York. 1985.

Wolfe RR. *Radioactive and Stable Isotope Tracers in Biomedicine.* Wiley-Liss, Inc.: New York. 1992.

APPENDIX

```
A SAAM31 INFANT MILK INTAKE
C  EXAMPLE SAAM INPUT FILE
c2        20
c3
c4  2       1
C*****************************************************************************
C
C DEFINITION OF PARAMETERS
C  IC(1)  D2O dose in kg                  L(3,1)  Rate coeff. from Mom to Delay
C  P(1)   Mother's total body water       L(2,3)  Rate coeff. from Delay to Infant
C  P(2)   Infant's total body water       L(0,1)  Rate coeff. out of Mom
C  DT(3)  Delay time                      L(0,2)  Rate coeff. out of Infant
C  DN(3)  Describes delay sharpness
C
C-----------------------------------------------------------------------------
H PAR
   IC(1)     .030
   P(1)     31.55
   P(2)     4.99
   L(3,1)    1.427216E-02  0.000000E+00   1.000000E+01
   L(2,3)   1.
   DT(3)     3.199061E-01  9.999998E-04   1.500000E+01
   DN(3)    2.
   L(0,1)    1.257309E-01  0.000000E+00   5.000000E+01
   L(0,2)    1.680049E-01  0.000000E+00   5.000000E+01
C
C
C****************************************************************************
C  EXPERIMENTAL DATA SECTION
C
C----------------------------------------------------------------------------
H DAT
C
C  Note below, columns for enrichment are converted to kg.  Enrichment is expressed
C     as percent above baseline.  Multiplication of the enrichment times body water
C      divided by 100 yields grams of D2O.
C
C  Mother enrichment data
C         Days          Enrichment times body water
101F(1)                   *.3155        FSD=.2
            2             .071058
            6             .041736
            12            .017359
            14            .013484
C Extra points calculated for a smoother curve
101F(1)                                 WT=0.
            0.
2           .1                          60
C
C  Infant enrichment data
C        Days            Enrichment times body water
102F(2)                   *.0499        FSD=.2
            1             .005222
            2             .011205
            3             .015082
            4             .018508
            6             .019709
            8             .019766
            10            .018855
            12            .017628
            14            .01394
C  Extra points calculated for a smoother curve
102F(2)                                 WT=0.
            0.
2           .1                          60
C
C
C***************************************************************************
C  DESCRIPTION OF TRACEE IN STEADY STATE
C
C---------------------------------------------------------------------------
H STE
C U(11)  Steady state water intake by mother
C U(12)  Steady state water intake by infant
C M(11)  SS mass of mother's water (set equal to P(1), mother's body water)
C M(12)  SS mass of infant's water (set equal to P(2), infant's body water)
C
C
   U(11)      4.417098E+00 .0             100.
   U(12)      3.880580E-01 .00            100.
   M(12)=P(2)
   M(11)=P(1)
C
C
C**************************************************************************
C
C STEADY STATE PARAMETER SECTION
C
C--------------------------------------------------------------------------
H PAR
```

```
C  All steady state parameters are set equal to corresponding tracer parameters
   IC(11)=P(1)
   IC(12)=P(2)
   L(0,11)=L(0,1)
   L(0,12)=L(0,2)
   L(13,11)=L(3,1)
   L(12,13)=L(2,3)
   DT(13)=DT(3)
   DN(13)=DN(3)
C************************************************************************
H DAT
C Instructions for SAAM31 to remember results for R(13,11) {water flow
C    from the mother to the infant}, R(0,11) {water flow out of the
C    mother}, R(0,12) {water flow out of the infant}, U(11) {Mother's
C    water intake}, U(12) {Infant's complementary water intake}.
105                                        WT=0.
   R(13,11)
   R(0,11)
   R(0,12)
   U(11)
   U(12)
Y
```

MODELING PROTEIN TURNOVER: A MODULE FOR TEACHING MODELING

Heidi A. Johnson

Animal Science Department
University of California at Davis
Davis, CA 95616

INTRODUCTION

The tendency for science is to reduce large questions or systems into smaller postulates which can be examined through experimentation. But in order to understand how the system works, the experimenter must be able to take bits of information from experiments and piece them back together to define the system. A model can provide the framework to describe the system, to determine what questions to ask, which information is needed, and if the new knowledge improves the understanding or prediction of how the system functions. A simple model of protein turnover by Waterlow et al. (1978) is used in this chapter to illustrate the modeling process.

Protein fractional synthesis rate (FSR) is estimated by measuring the incorporation of radiolabeled amino acid into protein relative to the proportion of radiolabeled amino acid in the precursor pool per unit time. The precursor for protein is aminoacyl tRNA. However, because the specific radioactivity of aminoacyl tRNA is difficult to measure, the extra-cellular or intracellular specific radioactivity is assumed to approximate the aminoacyl tRNA specific radioactivity. There are two common methods for estimating FSR: flooding dose and continuous infusion. The flooding dose technique involves giving a large dose of unlabeled amino acid (AA) with the radiolabeled AA.

This chapter will determine if the assumptions of the continuous infusion method apply to the flooding dose method. First, current knowledge of protein turnover in a mouse is summarized by using previous data to build the model. The model is then used to analyze the behavior of the system due to the flooding dose and continuous infusion protocols. Two independent data sets are fit to the model to test model predictions of specific radioactivities and the results are evaluated to determine if the original assumptions were valid.

Mathematical Modeling in Experimental Nutrition
Edited by Clifford and Müller, Plenum Press, New York, 1998

MODEL BACKGROUND

In their book, *Protein Turnover in Mammalian Tissues and in the Whole Body,* Waterlow et al. (1978) described a method for measuring protein FSR using a continuous infusion of a radiolabeled amino acid (leucine). As radiolabeled leucine is infused, plasma specific radioactivity is monitored until it reaches a plateau. At plateau, the specific radioactivity of leucine in the plasma has equilibrated with the specific radioactivity of leucine in the cell and therefore is assumed to be the specific radioactivity of the precursor pool for protein synthesis. However, because there is a lag on the specific radioactivity of the plasma equilibrating with the specific radioactivity of the precursor pool, the plasma-precursor specific radioactivity must be corrected for the rate of rise of the plasma specific radioactivity to plateau. Therefore FSR is based on three measurements: the rate of rise of the specific radioactivity of the plasma; the plateau specific radioactivity of the plasma; and the specific radioactivity of the protein pool when the specific radioactivity of the plasma reaches plateau.

Several assumptions about the state of the system needed to be made in order to use the continuous infusion method. First, because the estimation of protein degradation is difficult, only protein synthesis rate was measured. Protein degradation could then be assumed to equal protein synthesis. This assumption required the protein pool to be in steady-state therefore the animal or tissue could not be growing. Secondly, changes in specific radioactivity measurements of the plasma and protein pools had to be due to equilibration of radiolabeled amino acid and protein synthesis *only*. An increase or decrease in specific radioactivity of the plasma could not be due to a decrease or increase in the free amino acid pool size. Therefore, amino acid intake had to equal amino acid oxidation.

MODEL DESCRIPTION

To clarify these assumptions, Waterlow et al. (1978) created a whole-body model of protein turnover with two compartments: a free amino acid pool which contained all of the leucine in the extracellular, plasma, intracellular, and tRNA pools; and a protein pool which contained all of the leucine bound in protein. Amino acids (e.g., leucine) from intake went directly into the free amino acid pool or entered from protein degradation. Amino acids exited the free pool by going to protein synthesis (e.g., the amino acids in protein compartment) or were oxidized and exited the system. Figure 1 shows a diagram of the model.

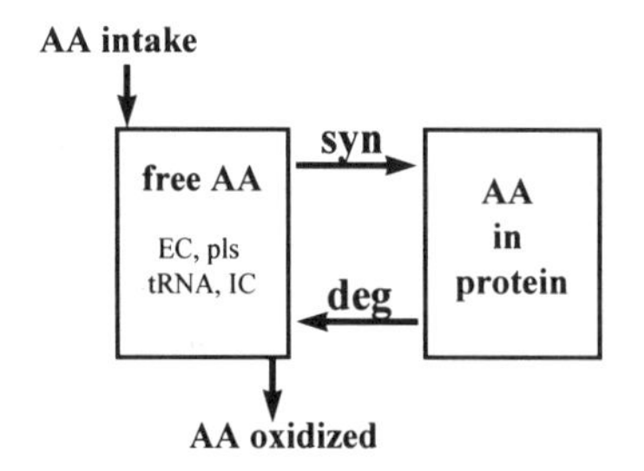

Figure 1. Diagram of protein turnover in the whole body. AA is amino acid, syn is protein synthesis, deg is protein degradation, EC is extracellular, pls is plasma, and IC is intracellular.

THE MODELING PROCESS

The two-pool model of protein turnover will be used as an example of the modeling process. Figure 2 outlines the steps followed in model development. The equations and code will be described in the format used for the ACSL (Advanced Continuous Simulation Language; MGA 1995) program. A complete copy of the program file (WG.CSL), command file (WG.CMD), sample output (ACSL Builder sample output), independent data set (WG.DAT), and instructions for using ACSL Builder and ACSL Math are included in Appendices A, B, C, D, and E, respectively, of this chapter. The program file (WG.CSL) contains definitions and descriptions (preceded by !) of common ACSL program categories and settings.

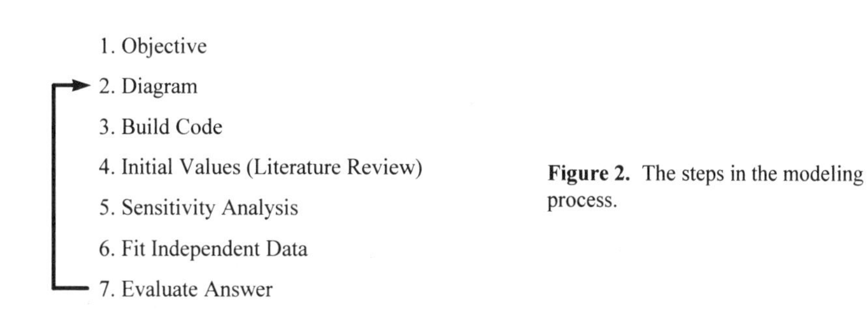

Figure 2. The steps in the modeling process.

Step 1. Objective

There are two objectives to be examined. The first is to determine if the protein synthesis rate equals the protein degradation rate; the second is to determine if the intake of leucine equals the oxidation of leucine in the flooding dose technique. The model defined by Waterlow et al. (1978) was originally developed for the continuous infusion technique of estimating FSR. However, the flooding dose technique is a simpler technique to measure FSR and is based on assumptions similar to those of the continuous infusion method.

The animal (mouse) is assumed to be non-growing over the experimental time period. Because there is no net protein accretion, protein synthesis should be equal to protein degradation and amino acid intake should equal amino acid oxidation. The plasma specific radioactivity is also assumed to approximate the precursor specific radioactivity because a large bolus of amino acid is given at the beginning of the experiment which acts to 'swell' or maintain the specific radioactivity of the precursor pool. The estimation of FSR, therefore, is based on the ratio of the specific radioactivity of the protein pool and the average specific radioactivity of the plasma pool over the experimental time period.

A clearly defined objective provides guidance for model creation and development. All decisions such as adding pools, defining flux equations, determining data to collect, etc., are based on the purpose of the model. For instance, because the objectives of this model are to determine if protein synthesis equals protein degradation and if intake equals oxidation, it would add unnecessary detail to the model to include multiple protein pools turning over at different rates. At this point, the number of protein pools does not matter so long as protein synthesis is set equal to degradation. However, if the objective of this model was to determine the effect of multiple protein pools turning over at different rates on estimates of FSR over different experimental time lengths, then multiple protein pools would be justified.

Step 2. Model Diagram

The model diagram is a helpful tool to aid in writing the differential equations which describe the model, to define the notation and units used, and to troubleshoot the model by filling in the diagram with input and output. Because this model represents changes in specific radioactivities in the free amino acid and protein pools, two diagrams are needed. The first (Figure 3) represents the whole-body dynamics of unlabeled leucine in the free leucine pool (Aa) and leucine bound to protein pool (Pb). The intake of leucine is constant and is represented by FdAa (feed to free Aa pool), oxidation is represented by AaCd (free Aa pool to carbon dioxide), leucine synthesized into protein by AaPb (free Aa pool to Protein pool), and leucine from protein degradation by PbAa (leucine from degraded protein to free leucine pool). The units for pool sizes (Aa, Pb) are μmoles of leucine and the fluxes (FdAa, AaCd, AaPb, PbAa) are μmoles of leucine per minute.

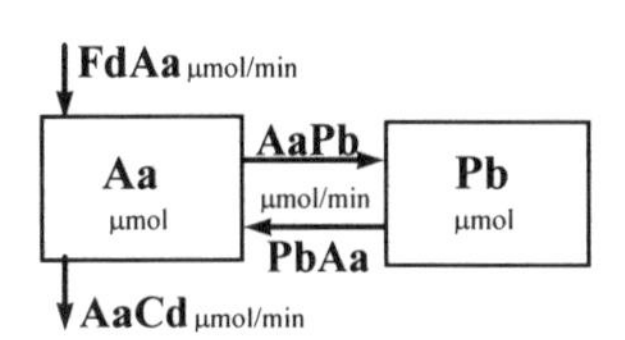

Figure 3. Diagram of unlabeled leucine dynamics. Aa is free leucine pool (μmoles), Pb is leucine bounds to protein pool (μmoles), FdAa is leucine intake (μmoles/min), AaCd is leucine oxidation (μmoles/min), AaPb is μmoles leucine incorporated into protein/min, and PbAa is μmoles leucine from protein degradation to the free amino acid pool/min.

The second portion of the model representing radiolabeled leucine dynamics is the same structure as the unlabeled portion of the model. However, this diagram (Figure 4) has no intake of leucine (FdAa) because the experiment or model run begins when all of the tracer has reached the Aa pool. The same fluxes control the flow of ^{14}C leucine between pools, but the specific radioactivity of the pool of origination must be used to express the concentration of label flowing from pool to pool. Therefore, the units of the second portion of the model are μCi of leucine in the free leucine pool (Aa14) and μCi of leucine bound to protein (Pb14). The units of radiolabeled leucine fluxes are μmoles leucine/min * μCi leucine/μmol leucine or μCi of leucine per minute.

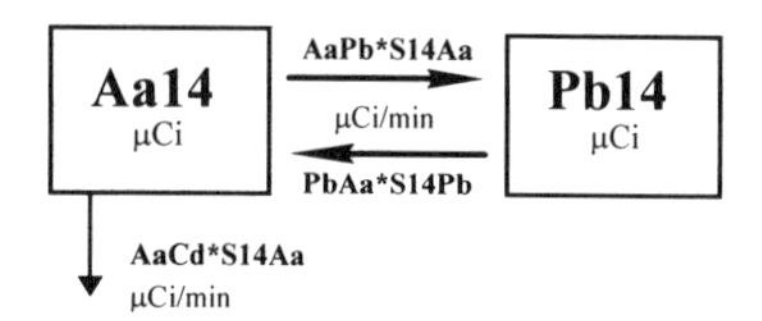

Figure 4. Diagram of radiolabeled leucine dynamics. Aa14 and Pb14 are the μCi of ^{14}C leucine in the Aa and Pb pools, respectively. S14Aa and S14Pb represent the specific radioactivity of the Aa and Pb pools.

Step 3. Build Code

Now that the notation and units of pools and fluxes have been defined and the structure of both the unlabeled and radiolabeled portions of the model has been determined, the model equations can be written. These differential equations describe how the amount of leucine or ^{14}C leucine changes in the Aa and Pb (or Aa14 and Pb14) pools from minute to minute or how much leucine is flowing in and out of the pools each minute. Therefore the net amount of leucine in and out of the pool which the equations are describing is the sum of the fluxes. For the unlabeled portion of the model (from Figure 3) the differential equations for the Aa pool (dAa) and protein pool (dPb) are,

$$dAa = FdAa + PbAa - AaCd - AaPb \tag{1}$$

$$dPb = AaPb - PbAa \tag{2}$$

The units of dAa and dPb are μmoles of leucine per minute. The differential equations can then be integrated (by the INTEG function in ACSL) based on an initial pool size (iAa and iPb for the free leucine and leucine bound to protein pools, respectively) to yield the μmoles of leucine in each of the pools.

$$Aa = INTEG(dAa,iAa) \tag{3}$$

$$Pb = INTEG(dPb,iPb) \tag{4}$$

For the isotope portion of the model, the same process is followed. However, the fluxes must be 'corrected' for the concentration of radiolabeled leucine relative to unlabeled leucine in the pool from which the flux originates. The μCi/min of radiolabeled leucine in the Aa pool flowing in or out per minute is represented (Eq. 5), and the μCi/minute flowing in or out of the leucine bound to protein pool is represented (Eq. 6). These equations are then integrated (using INTEG) based on initial values for the pool sizes of dose14 and iPb14 for the Aa14 and Pb14 pools respectively.

$$dAa14 = (PbAa*S14Pb) - S14Aa*(AaCd + AaPb) \tag{5}$$

$$dPb14 = (AaPb*S14Aa) - (PbAa*S14Pb) \tag{6}$$

$$Aa14 = INTEG(dAa14, dose14) \tag{7}$$

$$Pb14 = INTEG(dPb14, iPb14) \tag{8}$$

The purpose of isotope experiments is to use a tracer to identify the dynamics of the tracee. Therefore the two portions of the model are tied together through the ratio of the tracer to the tracee (specific radioactivity).

$$S14Aa = Aa14/Aa \tag{9}$$

$$S14Pb = Pb14/Pb \tag{10}$$

The specific radioactivities of the Aa and Pb pools (μCi/μmol) are our experimental data. We can then use this data to fit to the model and test our objectives: that, in a flooding dose experiment, the protein synthesis rate equals the protein degradation rate and that intake equals oxidation. Equations 1-10 are written in ACSL format and can be seen incorporated into an ACSL program in Appendix A. The only other tasks needed in the program are to set the values (using CONSTANT in ACSL terminology) for the initial pool sizes (iAa, iPb, dose14, and iPb14) and the fluxes (FdAa, AaCd, AaPb, and PbAa).

Step 4. Initial Values - Literature Review

Initial pool sizes were based on data for a 30 g mouse. The amount of protein in the whole body of a mouse was determined by summing organ protein contents from several papers in the literature (Johnson, 1997). The amount of leucine in protein was assumed to be constant at 2% of the protein (Waterlow et al., 1978). This equals 982 μmoles of leucine in a 30 g mouse. The initial free leucine pool size was based on free leucine concentration in the plasma/extracellular pool (Waterlow et al., 1978), in the intracellular pool (Obled et al., 1989), and that associated with tRNA (Vinayak, 1987; Palmiter, 1975). Together with the unlabeled leucine in the flooding dose (30 μmoles), the sum of these indicated an initial free leucine pool size of 38.6 μmoles. Intake of leucine (John and Bell, 1976) is equal to oxidation (0.385 μmoles/min). Protein synthesis is equal to protein degradation (0.216 μmol/min) and is based on a FSR of 32% (Johnson, 1997).

Initial pool sizes in the tracer portion of the model are based on the flooding dose protocol of Bernier and Calvert (1987). The ^{14}C leucine dose was 3.0 μCi, which corresponds to the 30 μmoles of unlabeled leucine in the tracee portion of the model. The initial amount of tracer in the protein pool is zero (1D-10) because there is no endogenous tracer at the beginning of the experiment.

Once the equations, starting values, and fluxes have been input into the model program, the model can be run using ACSL Builder. It will then output specific radioactivity changes as predicted by the model and based on the literature values (Figure 6).

Using the continuous infusion technique, the model predicts that the free amino acid pool specific radioactivity reaches plateau around 40 minutes but the protein specific radioactivity rises continuously. For the flooding dose technique, the specific radioactivity of the free amino acid pool decreases but the specific radioactivity of the protein pool rises to a plateau just after 120 minutes.

Step 5. Sensitivity Analysis

Sensitivity analysis is done by comparing the changes observed in model output variables such as S14Aa and S14Pb with changes in model parameters. If a small increase or decrease in a model parameter or initial value causes a large increase or decrease in a model output, then the model is sensitive to the parameter. If the model is very sensitive to a particular parameter, that parameter may need to be better characterized by further experimentation or described by more than one equation/function. However, if a small change in a model parameter results in small or no change in an output, then the model is insensitive to the parameter. Eliminating the parameter would probably not have any effect on the model's ability to function or predict.

Sensitivity analysis is a very important process that allows us to identify weaknesses in our representation of the system and any assumptions made in building the model, to detect unimportant model elements (which will produce an over-parameterized model), and to retain only those model elements which are important. Sensitivity analysis can also be used to set those model parameters not be found in the literature to either their most sensitive range or to a range which gives physiological output values.

There are several ways to conduct sensitivity analyses. The appropriate method depends on the objectives of the model and what information is needed from the analysis. Two examples are presented in Table 1 and Figure 7.

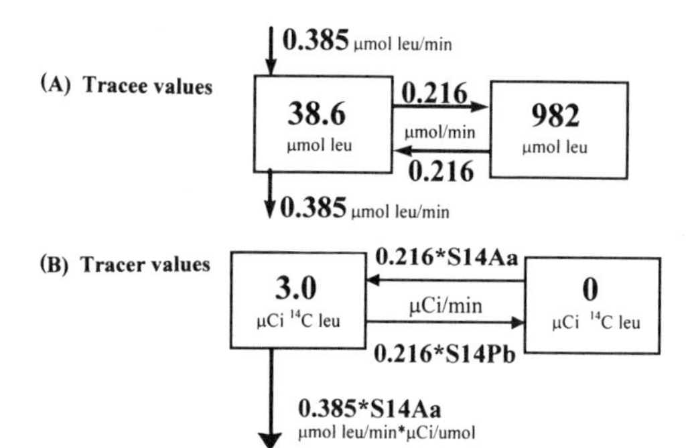

Figure 5. Initial pool sizes and flux values for the tracee (5A) and tracer (5B) portions of the model.

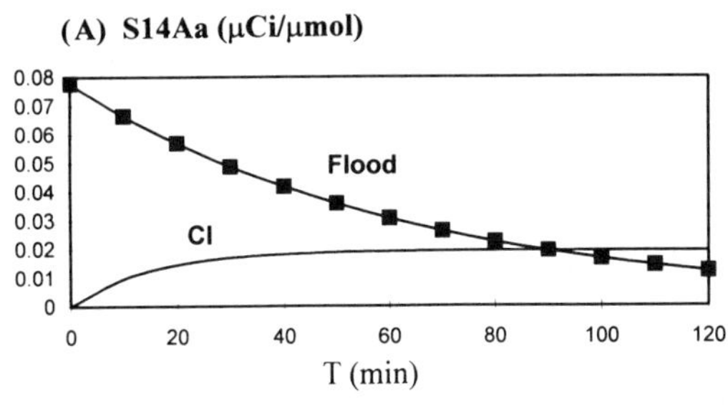

Figure 6. Changes in specific radioactivities of the free leucine pool (6A) and leucine in protein pool (6B) over 120 minutes with the continuous infusion (CI) and flooding dose (Flood) methods.

Table 1. Percent changes in the specific radioactivities of the free leucine and leucine in protein pools due to changes in model parameters.

Model Parameter (± 25% change)	% Change in S14Aa	% Change in S14Pb
iAa	-2.9 / 3.0	-4.2 / 4.6
iPb	-4.1 / 4.3	-1.9 / 2.0
dose14	25 / -25	25 / -25
FdAa	-5.4 / 6.1	-2.8 / 3.1
AaCd	-1.8 / 1.7	-0.53 / 0.53
AaPb	-0.10 / 0.95	24 / -25
PbAa	-3.1 / 3.3	-1.6 / 1.7

The greatest changes in specific radioactivities are observed with changes in the dose of ^{14}C leucine (dose14) and protein synthesis rate (AaPb). Because the initial dose of tracer is based on the method used (flooding dose) and is directly related to the specific radioactivity of the amino acid and protein pools, it would be expected that there is a one-to-one relationship. The sensitivity of the specific radioactivity of the protein pool to the protein synthesis rate is expected because the protein synthesis rate controls how much radiolabeled leucine gets into the protein pool. All other model parameters appear to affect specific radioactivities very little. Either the model may not be a unique representation of protein turnover (a unique predictor of specific radioactivities) or the model may not be sensitive to 25% changes in parameters. The determination of the proper percent to use to change the parameters should be based on how much the specific radioactivities in the amino acid and protein pools change over the experimental time period.

Changing model parameters by a fixed amount will identify those parameters which have the least and greatest influences on the model output. This procedure does not, however, show how sensitivities may change over a range of parameter values. For instance, a model output may be insensitive (1-2% change) to a 25% change in a parameter, but if the parameter is increased by a 100%, the model output may become very sensitive to the model parameter (120% change). Figure 7 shows a different sensitivity analysis where a parameter is varied over a wide range of values to determine a sensitive range for the model output.

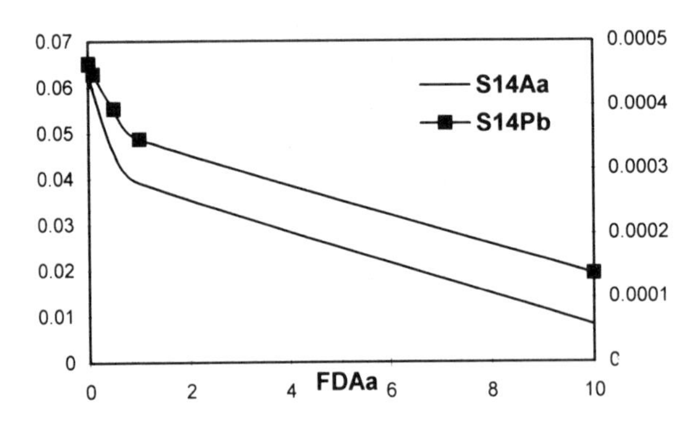

Figure 7. The response of specific radioactivities of free leucine and leucine in protein pools to a wide range of leucine intakes (FdAa in μmol/min). The right axis is S14Aa (μCi/μmol) and left axis is S14Pb (μCi/μmol).

The specific radioactivity values output by the model are most sensitive to changes in leucine intake at levels below 1 μmol/min. Because intake is set to 0.385 μmol/min, the intake parameter is set at that region where the specific radioactivities would be more sensitive to intake.

Step 6. Fit Independent Data

In order to determine if synthesis equals degradation and intake equals oxidation in the flooding dose method, the model predictions need to be tested with an independent data set. Protein synthesis will be forced to equal protein degradation and intake will be forced to equal oxidation. Then intake and protein synthesis will be adjusted to fit the model output to equal the specific radioactivity data from Bernier and Calvert (1987; Appendix D). The Math-Optimize part of ACSL used the generalized reduced gradient algorithm (GRG) to search for intake and protein synthesis values to produce the 'most likely' specific radioactivities in the dataset by the maximum likelihood method. Instructions for using

ACSL Math-Optimize are in Appendix E. Figure 8 shows the model predictions vs. the data specific radioactivities before intake and protein synthesis were fit and Figure 9 shows the predictions of specific radioactivities after fitting.

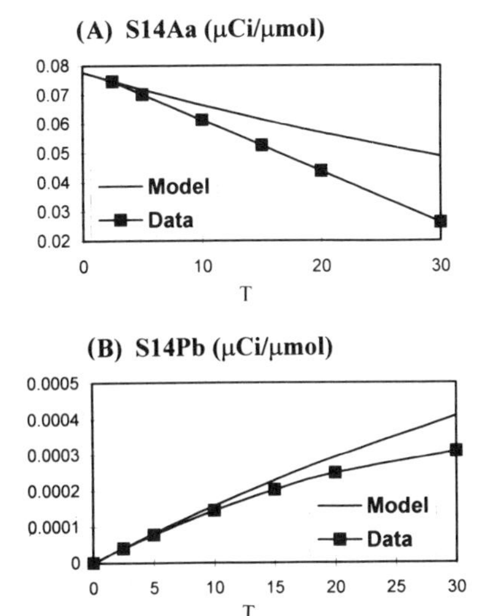

Figure 8. Comparison of specific radioactivities of the free leucine pool (8A) and leucine in the protein pool (8B) from the model to the data before data were fit.

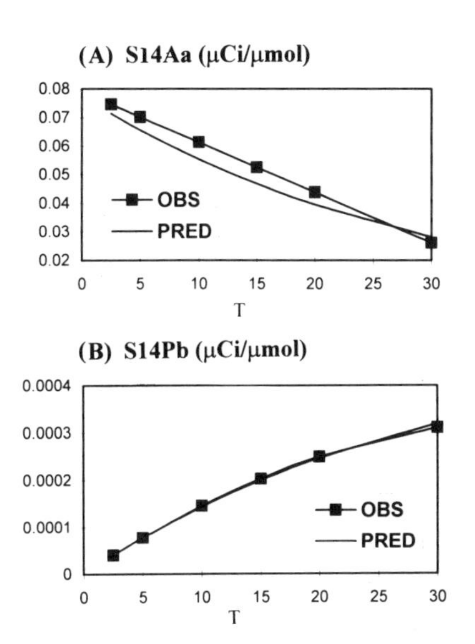

Figure 9. Comparison of specific radioactivities of the free leucine pool (9A) and leucine in protein pool (9B) from the model (PRED-predicted) to the data (OBS-observed) after the data was fit.

As illustrated in Figure 9, ACSL Math was successful in fitting the data to the model. Percent errors on predicted specific radioactivities from adjusting intake and protein synthesis were less than 11% for the S14Aa and less than 3% for S14Pb. Several different combinations of FdAa, AaCd, and AaPb, PbAa were able to generate specific radio-activities which were close to the specific radioactivities of the data. However, only one solution (the combination of FdAa, AaCd, and PbAa, AaPb) was able to fit the data when the restriction of synthesis equaling degradation and intake equaling oxidation was placed on the solution. The standard deviations of the adjusted parameters (intake and protein synthesis) were also very low in comparison to the final parameter values. Figure 10 shows part of the printout of the model solution generated by ACSL Math-Optimize.

DESCRIPTION	PARAMETER ESTIMATES		
	INITIAL	FINAL	STANDARD DEVIATION
LOG LIKELIHOOD FUNCTION	69.692	93.865	
FDAA	0.385	1.0897	0.06238
FSR	0.317	0.31432	0.0037613

Figure 10. Initial and final parameter values and standard deviations from fitting data to the model using ACSL Math-Optimize. FdAa was equal to AaCd, and FSR (the percent leucine which is incorporated into protein/day and which is used to calculate AaPb) was equal to protein degradation rate.

Confidence in solutions from a numerical, iterative method (such as this one) should be based on making sure the solution obtained is the 'best' fit and that the solution is unique. To determine if the solution is the 'best fit,' a wide range of starting values for the parameters to be fit need to be used to see if the algorithm which searches for the solution is able to come up with the same solution for each starting value. The algorithm (GRG) searches over a response surface dependent on the parameters which are being adjusted and the iteration step size. Obtaining different solutions with different starting values (for FdAa and AaPb) indicates that the algorithm maybe stuck on a local minima or maxima and/or the iteration step size may be too small. Using different starting values (over a physiological range) will ensure that the best fit is also the most likely solution.

A unique solution means that only one model structure and set of parameters will produce the specific radioactivities predicted by the model. In this example, the solution was not unique if FdAa, AaCd, AaPb, and PbAa were all allowed to vary to fit the data. However, when the parameters were restricted (i.e., FdAa = AaCd and AaPb = PbAa), the solution appeared to be unique and the best fit. The sensitivity analysis in Figure 6 indicated that the model may not be a unique representation of protein synthesis because the specific radioactivities of the free leucine and leucine in protein pools were insensitive to almost all of the model parameters. In addition, the standard deviation of prediction of the parameters in the ACSL Math-Optimize printout will be very large compared to the final values of the adjusted parameters if the solution is not unique. Fitting more than one data set will also build confidence in the parameter estimates obtained.

Step 7. Evaluate Answer

For the flooding dose method, the model was able to reproduce the specific radioactivities in the data set by setting protein synthesis equal to protein degradation and leucine intake equal to leucine oxidation. The protein synthesis/degradation rates were also very close to the values from the literature review: literature values were 31.7%/day and the predicted values were 31.4%/day. But the predicted rates of intake and oxidation were three times the amount from the literature: 1.09 μmol/min and 0.385 μmol/min, respectively. This would correspond to a 30 g mouse eating 8.6 g of protein/day. It appears that protein synthesis equals protein degradation, but leucine intake cannot equal oxidation.

The model fails at being a good representation of protein turnover for the flooding dose method. Before that conclusion can be accepted, the data, the model assumptions, and the model structure should be examined to determine if there are other factors which may be influencing the results.

The model failed to predict the specific radioactivity of the free leucine pool. Figures 8 and 9 showed that the model under-predicted S14Aa when literature values were used, and over-predicted S14Aa when values from fits of the data were used. The only way the model would be able to bring S14Aa down to fit the data would be to increase the amount of unlabeled leucine coming into the free leucine pool or increase the leucine oxidized. Only two sources of unlabeled leucine were available: increased leucine intake or increased protein degradation. Increased protein degradation was an unlikely source of unlabeled leucine: when degradation was set equal to synthesis, the model was able to predict S14Pb (Figure 9). Thus any increase in degradation would also increase synthesis, which would change the specific radioactivity of the protein pool, and the model would no longer be able to predict S14Pb. Thus the relationship between intake and oxidation needs to be examined.

Go Back To Step 2...

CONCLUSIONS

Model predictions are based on being able to repeat the conditions of the experiment with the model. Several facts are unknown about the data which could affect the solutions predicted by the model. If the mice were fed just before the experiment was conducted, the free leucine pool would be larger than the literature values used in the model. These mice were also given the flooding dose with an interperitoneal (IP) injection instead of an intravenous (IV) injection. Absorption of the flooding dose may not be as rapid with an IP as with an IV injection. In addition, because a large, unphysiological dose of unlabeled leucine is given in the flooding dose method, cells may not be able to absorb all of the dose and much of it may be oxidized immediately. Intake may be constant and oxidation may be very large and increasing over the experimental time period.

The model does not account for a small, rapidly turning over tRNA pool or an intracellular pool which may expand in the presence of a large influx of leucine. Additional pools representing tRNA, and the intracellular and extracellular leucine could be added to the model to determine if intake could be made to equal oxidation. But before more pools are added, it must be determined whether the model failure was due to the structure of the model or to the data or experimental conditions which the model was trying to reproduce. Therefore better control over the data set used (e.g., information on the food intake of the mice just before the experiment, measurement of the oxidation of leucine in response to the flood dose, etc.) would confirm if the model can be used in the flooding dose technique and if the assumptions of intake equaling oxidation and synthesis equaling degradation would apply to the flooding dose technique.

REFERENCES

Bernier JF; Calvert CC. Effect of a major gene for growth on protein synthesis in mice. *J Anim Sci,* 1987, 65:982-995.

John A; Bell JM. Amino acid requirements of the growing mouse. *J Nutr,* 1976, 106:1361-1367.

Johnson HA. A modeling investigation of whole body protein turnover based on leucine kinetics in rodents. Doctoral thesis, University of California, Davis.1997.

Mitchell and Gauthier Assoc. Inc. *ACSL: Advanced Continuous Simulation Language.* MGA Inc.: Concord, MA. 1993.
Obled C; Barre F; Millward DJ; Arnal M. Whole body protein synthesis: Studies with different amino acids in the rat. *Am J Physiol,* 1989, 257:E639-E646.
Palmiter RD. Quantitation of parameters that determine the rate of ovalbumin synthesis. *Cell,* 1975, 4:189-197.
Vinayak M. A comparison of tRNA populations of rat liver and skeletal muscle during aging. *Biochem Inter,* 1987, 15:279-285.
Waterlow JC; Garlick PJ; Millward DJ. *Protein Turnover in Mammalian Tissues and in the Whole Body.* North-Holland: Amsterdam. 1978. p. 119.

Appendix A. ACSL Program:WG.CSL

```
PROGRAM WG
!ACSL program follows
'Whole mouse body protein TO model over 30 minutes'
'uses Bernier data'
'synth=degrade,intake=oxidation (no growth)'
'time in minutes,pools in umoles,radioiso in uCi'
'1 radioisotope;14=14C leucine (flood)'
'Aa=1 amino pool,Pb=1 protein pool'
'Fd=intake,Cd=CO2 or oxidation'
'd=differential equation,S=specific radioactivity,i=initial'

INITIAL
! For setting values (initial conditions) which are only
! evaluated/calculated once before run begins.

'-----Initial Pool sizes & factors'
      CONSTANT iAa=38.6  $'initial leu pool size,umol'
                          '8.6+unlabeled leu in flood(30)'
      CONSTANT iPb=982   $'initial leu in protein pool,umol'
      CONSTANT iCd=1E-10 $'initial leu oxidized,umol'

      CONSTANT iPb14=1E-10  $'initial 14C leu in protein,uCi'
      CONSTANT iCd14=1E-10  $'initial 14C leu oxidized,uCi'
      CONSTANT dose14=3.0    $'10uCi 14C leu/100umol/100g BW'
                             '30 g mouse,JAS(1987)65:982-995,uCi'

'---Initial rates'
      CONSTANT FdAa=0.385 $'leu intake, umol/min'
      CONSTANT AaCd=0.385 $'leu intake=leu oxidized'
      CONSTANT RF=0.317   $'FSR=31.7%/day'
      CONSTANT DF=0.317   $'protein synthesis rate=degradation rate'
                           'DF=RF'

! Program settings
! Algorithm for derivative section is Runge-Kutta fixed step, 4th order
! Nsteps is number of steps across an interval to determine derivative
! Maxterval,Minterval control step size (1/100 of a min)
! Communication interval is the data logging rate (every minute)

      ALGORITHM IALG = 5
      NSTEPS    NSTP = 1
      MAXTERVAL MAXT = 0.01
      MINTERVAL MINT = 0.0001
      CINTERVAL CINT = 1

END ! of initial

DYNAMIC
! Section which moves forward in time at data logging rate
! equations not sorted, can use to convert variable units for output,
! for calculations which do not affect equations in derivative section

DERIVATIVE
! Contains d and integrations
! equations are sorted and calculated at every integration step

'----Unlabeled Leucine in Aa, Pb'
```

```
    dAa=FdAa+PbAa-AaCd-AaPb $'d describing change of umol leu'
                             'in pool/min'
    Aa=INTEG(dAa,iAa) $'integration of d describing leu'
                      'in pool at any one minute,umol'

    AaPb=(RF/1440)*Pb $'convert FSR (%/day) to umols leu in protein'
                      'synthesized per min,umol/min'
    PbAa=(DF/1440)*Pb $'fraction leu in protein degraded/min,umol/min'

    dPb=AaPb-PbAa $'d describing change of umol leu in protein pool/min'
    Pb=INTEG(dPb,iPb)  $'integration of d,umol'
    Cd=INTEG(AaCd,iCd) $'integration of d describing leu oxidized,umol'

'----14C Leucine in Aa, Pb using Flooding dose'
    dAa14=(S14Pb*PbAa)-S14Aa*(AaCd+AaPb)
                             'd describing change in 14C leu pool'
                             'uCi/min'
    Aa14=INTEG(dAa14,dose14)  $'integration of d describing 14C'
                              'leu in pool at any one minute,uCi'
    dPb14=(S14Aa*AaPb)-(S14Pb*PbAa) $'d describing change in 14C'
                                    'leu in protein pool,uCi/min'
    Pb14=INTEG(dPb14,iPb14)  $'integration of d,uCi/min'
    dCd14=S14Aa*AaCd         $'d describing 14C leu oxidized,uCi/min'
    Cd14=INTEG(dCd14,iCd14)  $'integration of d,uCi'

    S14Cd=Cd14/Cd            $'S of 14C leu oxidized,uCi/umol'
    S14Pb=Pb14/Pb            $'S of protein pool,uCi/unmol'
    S14Aa=Aa14/Aa            $'S of leu pool,uCi/umol'

'---AVG SA with 14C for each Aa pool'
           btAa14=((dose14/iAa)+S14Aa)/2

'---FSR'
           FSR=S14Pb/(btAa14*(t+1D-10)/1440)
           WBFSR=AaPb*1440/Pb       $'true FSR calculation'

END ! of derivative
! Model runs for 30 minutes

      CONSTANT  TSTOP = 30
      TERMT( T .GE. TSTOP )

END ! of dynamic
TERMINAL
! Section executed once after time has stopped
! for statistical calculations
END ! of terminal
END ! of program
```

Appendix B. ACSL Command file:WG.CMD

```
!File:WG.cmd
PROCEDURE MTH
SET WEDITG=.FALSE.
SET HVDPRN=.F.
FILE/PRNFILE='NUL'
END

PROCEDURE OUT
SPARE
SET HVDPRN=.T.
OUTPUT t,S14Pb,S14Aa,Aa14,Pb14
OUTPUT iAa,iPb,Aa,Pb
OUTPUT FdAa,AaCd,AaPb,PbAa,Rf,Df
OUTPUT S14Aa,S14Pb,FSR,WbFSR
PREPARE t,S14Pb,S14Aa,Aa14,Pb14
PREPARE iAa,iPb,Aa,Pb
PREPARE FdAa,AaCd,AaPb,PbAa,Rf,Df
PREPARE S14Aa,S14Pb,FSR,WbFSR
set ndbug=1
s pcwprn=120
```

```
START
print /nciprn=1 t,S14Aa,S14Pb,FSR,AaPb,PbAa,AaCd,FdAa,Rf,Df,Aa,Pb
END

PROCEDURE DP
d Aa Pb RF DF.
d FdAa AaCd AaPb PbAa
d S14Aa S14Pb FSR,WBFSR
END
```

Appendix C. ACSL Builder Sample Output

```
Switching CMD unit to 4 to read wng.cmd

 !File:WG.cmd
 PROCEDURE MTH
 SET WEDITG=.FALSE.
 SET HVDPRN=.F.
 FILE/PRNFILE='NUL'
 END

 PROCEDURE OUT
 SPARE
 SET HVDPRN=.T.
 OUTPUT t,S14Pb,S14Aa
 OUTPUT iAa,iPb,Aa,Pb
 OUTPUT Aa14,Pb14
 OUTPUT FdAa,AaCd,AaPb,PbAa,Rf,Df
 OUTPUT S14Aa,S14Pb,FSR,WbFSR
 PREPARE t,S14Pb,S14Aa
 PREPARE iAa,iPb,Aa,Pb
 PREPARE Aa14,Pb14
 PREPARE FdAa,AaCd,AaPb,PbAa,Rf,Df
 PREPARE S14Aa,S14Pb,FSR,WbFSR
 set ndbug=1
 s pcwprn=120
 START
 print /nciprn=1 t,S14Aa,S14Pb,FSR,wbfsr,AaPb,PbAa,AaCd,FdAa,Rf,Df,Aa,Pb
 END

 PROCEDURE DP
 d Aa Pb RF DF
 d FdAa AaCd AaPb PbAa
 d S14Aa S14Pb FSR wbFSR
 END
 End of file found on unit 4
 Reverting to logical unit number 5
 out
 SPARE
 Accumulated cp time 5.360000. Elapsed cp time 0.
 SET HVDPRN=.T.
 OUTPUT t,S14Pb,S14Aa
 OUTPUT iAa,iPb,Aa,Pb
 OUTPUT Aa14,Pb14
 OUTPUT FdAa,AaCd,AaPb,PbAa,Rf,Df
 OUTPUT S14Aa,S14Pb,FSR,WbFSR
 PREPARE t,S14Pb,S14Aa
 PREPARE iAa,iPb,Aa,Pb
 PREPARE Aa14,Pb14
 PREPARE FdAa,AaCd,AaPb,PbAa,Rf,Df
 PREPARE S14Aa,S14Pb,FSR,WbFSR
 set ndbug=1
 s pcwprn=120
 START
 ....Debug dump - System Variables. NDBUG is 1, block number 1
           T 0.                  ZZTICG 0.                  CINT 1.00000000
      ZZIERR    F                ZZNBLK       1           ZZICON        1
      ZZSTFL    F                ZZFRFL      T            ZZICFL      T
      ZZRNFL    F                ZZJEFL     F             ZZNIST        6
      ZZNDST      0              ZZNAVR       0             IALG        5
        NSTP      1                MAXT 0.01000000          MINT 1.0000D-04
```

```
 State Variables                  Derivatives          Initial Conditions
          AA14 3.00000000           Z99995-0.04672355        DOSE14 3.00000000
            AA 38.5999985           Z99998-2.7756D-17           IAA 38.5999985
          CD14 1.0000D-10           Z99993 0.02992228         ICD14 1.0000D-10
            CD 1.0000D-10           Z99996 0.38499999           ICD 1.0000D-10
          PB14 1.0000D-10           Z99994 0.01680127         IPB14 1.0000D-10
            PB 982.000000           Z99997 0.                   IPB 982.000000

 Algebraic Variables

Common Block /ZZCOMU/
      ZZSEED    55555555
Common Block /ZZCOMP/
          AACD 0.38499999             AAPB 0.21617639        BTAA14 0.07772021
        DAA14-0.04672355               DAA-2.7756D-17         DCD14 0.02992228
            DF 0.31700000            DPB14 0.01680127           DPB 0.
          FDAA 0.38499999              FSR 18.8676166          PBAA 0.21617639
            RF 0.31700000            S14AA 0.07772021         S14CD 1.00000000
         S14PB 1.0183D-13            TSTOP 30.0000000         WBFSR 0.31700000

             T 0.                    S14PB 1.0183D-13         S14AA 0.07772021
           IAA 38.5999985              IPB 982.000000            AA 38.5999985
            PB 982.000000             AA14 3.00000000          PB14 1.0000D-10
          FDAA 0.38499999             AACD 0.38499999          AAPB 0.21617639
          PBAA 0.21617639               RF 0.31700000            DF 0.31700000
         S14AA 0.07772021            S14PB 1.0183D-13           FSR 18.8676166
         WBFSR 0.31700000

             T 1.00000000            S14PB 1.6975D-05         S14AA 0.07651918
           IAA 38.5999985              IPB 982.000000            AA 38.5999985
            PB 982.000000             AA14 2.95364024          PB14 0.01666928
          FDAA 0.38499999             AACD 0.38499999          AAPB 0.21617639
          PBAA 0.21617639               RF 0.31700000            DF 0.31700000
         S14AA 0.07651918            S14PB 1.6975D-05           FSR 0.31695858
         WBFSR 0.31700000

             T 2.00000000            S14PB 3.3684D-05         S14AA 0.07533680
           IAA 38.5999985              IPB 982.000000            AA 38.5999985
            PB 982.000000             AA14 2.90800053          PB14 0.03307730
          FDAA 0.38499999             AACD 0.38499999          AAPB 0.21617639
          PBAA 0.21617639               RF 0.31700000            DF 0.31700000
         S14AA 0.07533680            S14PB 3.3684D-05           FSR 0.31690412
         WBFSR 0.31700000

             T 3.00000000            S14PB 5.0130D-05         S14AA 0.07417279
           IAA 38.5999985              IPB 982.000000            AA 38.5999985
            PB 982.000000             AA14 2.86306969          PB14 0.04922813
          FDAA 0.38499999             AACD 0.38499999          AAPB 0.21617639
          PBAA 0.21617639               RF 0.31700000            DF 0.31700000
         S14AA 0.07417279            S14PB 5.0130D-05           FSR 0.31683661
         WBFSR 0.31700000

             T 4.00000000            S14PB 6.6320D-05         S14AA 0.07302686
           IAA 38.5999985              IPB 982.000000            AA 38.5999985
            PB 982.000000             AA14 2.81883670          PB14 0.06512577
          FDAA 0.38499999             AACD 0.38499999          AAPB 0.21617639
          PBAA 0.21617639               RF 0.31700000            DF 0.31700000
         S14AA 0.07302686            S14PB 6.6320D-05           FSR 0.31675608
         WBFSR 0.31700000

             T 5.00000000            S14PB 8.2255D-05         S14AA 0.07189873
           IAA 38.5999985              IPB 982.000000            AA 38.5999985
            PB 982.000000             AA14 2.77529073          PB14 0.08077413
          FDAA 0.38499999             AACD 0.38499999          AAPB 0.21617639
          PBAA 0.21617639               RF 0.31700000            DF 0.31700000
         S14AA 0.07189873            S14PB 8.2255D-05           FSR 0.31666256
         WBFSR 0.31700000

             T 6.00000000            S14PB 9.7940D-05         S14AA 0.07078811
           IAA 38.5999985              IPB 982.000000            AA 38.5999985
            PB 982.000000             AA14 2.73242111          PB14 0.09617710
          FDAA 0.38499999             AACD 0.38499999          AAPB 0.21617639
          PBAA 0.21617639               RF 0.31700000            DF 0.31700000
         S14AA 0.07078811            S14PB 9.7940D-05           FSR 0.31655605
         WBFSR 0.31700000
```

```
      T 7.00000000        S14PB 1.1338D-04        S14AA 0.06969475
    IAA 38.5999985          IPB 982.000000           AA 38.5999985
     PB 982.000000         AA14 2.69021732         PB14 0.11133848
   FDAA 0.38499999         AACD 0.38499999         AAPB 0.21617639
   PBAA 0.21617639           RF 0.31700000           DF 0.31700000
  S14AA 0.06969475        S14PB 1.1338D-04          FSR 0.31643659
  WBFSR 0.31700000

      T 8.00000000        S14PB 1.2858D-04        S14AA 0.06861837
    IAA 38.5999985          IPB 982.000000           AA 38.5999985
     PB 982.000000         AA14 2.64866903         PB14 0.12626204
   FDAA 0.38499999         AACD 0.38499999         AAPB 0.21617639
   PBAA 0.21617639           RF 0.31700000           DF 0.31700000
  S14AA 0.06861837        S14PB 1.2858D-04          FSR 0.31630421
  WBFSR 0.31700000

      T 9.00000000        S14PB 1.4354D-04        S14AA 0.06755871
    IAA 38.5999985          IPB 982.000000           AA 38.5999985
     PB 982.000000         AA14 2.60776605         PB14 0.14095146
   FDAA 0.38499999         AACD 0.38499999         AAPB 0.21617639
   PBAA 0.21617639           RF 0.31700000           DF 0.31700000
  S14AA 0.06755871        S14PB 1.4354D-04          FSR 0.31615894
  WBFSR 0.31700000

      T 10.0000000        S14PB 1.5826D-04        S14AA 0.06651550
    IAA 38.5999985          IPB 982.000000           AA 38.5999985
     PB 982.000000         AA14 2.56749837         PB14 0.15541038
   FDAA 0.38499999         AACD 0.38499999         AAPB 0.21617639
   PBAA 0.21617639           RF 0.31700000           DF 0.31700000
  S14AA 0.06651550        S14PB 1.5826D-04          FSR 0.31600082
  WBFSR 0.31700000

      T 11.0000000        S14PB 1.7275D-04        S14AA 0.06548850
    IAA 38.5999985          IPB 982.000000           AA 38.5999985
     PB 982.000000         AA14 2.52785611         PB14 0.16964238
   FDAA 0.38499999         AACD 0.38499999         AAPB 0.21617639
   PBAA 0.21617639           RF 0.31700000           DF 0.31700000
  S14AA 0.06548850        S14PB 1.7275D-04          FSR 0.31582989
  WBFSR 0.31700000

      T 12.0000000        S14PB 1.8702D-04        S14AA 0.06447745
    IAA 38.5999985          IPB 982.000000           AA 38.5999985
     PB 982.000000         AA14 2.48882957         PB14 0.18365099
   FDAA 0.38499999         AACD 0.38499999         AAPB 0.21617639
   PBAA 0.21617639           RF 0.31700000           DF 0.31700000
  S14AA 0.06447745        S14PB 1.8702D-04          FSR 0.31564620
  WBFSR 0.31700000

      T 13.0000000        S14PB 2.0106D-04        S14AA 0.06348211
    IAA 38.5999985          IPB 982.000000           AA 38.5999985
     PB 982.000000         AA14 2.45040916         PB14 0.19743968
   FDAA 0.38499999         AACD 0.38499999         AAPB 0.21617639
   PBAA 0.21617639           RF 0.31700000           DF 0.31700000
  S14AA 0.06348211        S14PB 2.0106D-04          FSR 0.31544980
  WBFSR 0.31700000

      T 14.0000000        S14PB 2.1488D-04        S14AA 0.06250222
    IAA 38.5999985          IPB 982.000000           AA 38.5999985
     PB 982.000000         AA14 2.41258550         PB14 0.21101186
   FDAA 0.38499999         AACD 0.38499999         AAPB 0.21617639
   PBAA 0.21617639           RF 0.31700000           DF 0.31700000
  S14AA 0.06250222        S14PB 2.1488D-04          FSR 0.31524074
  WBFSR 0.31700000

      T 15.0000000        S14PB 2.2848D-04        S14AA 0.06153755
    IAA 38.5999985          IPB 982.000000           AA 38.5999985
     PB 982.000000         AA14 2.37534929         PB14 0.22437089
   FDAA 0.38499999         AACD 0.38499999         AAPB 0.21617639
   PBAA 0.21617639           RF 0.31700000           DF 0.31700000
  S14AA 0.06153755        S14PB 2.2848D-04          FSR 0.31501908
  WBFSR 0.31700000

      T 16.0000000        S14PB 2.4187D-04        S14AA 0.06058786
    IAA 38.5999985          IPB 982.000000           AA 38.5999985
     PB 982.000000         AA14 2.33869142         PB14 0.23752010
```

```
     FDAA 0.38499999          AACD 0.38499999          AAPB 0.21617639
     PBAA 0.21617639            RF 0.31700000            DF 0.31700000
    S14AA 0.06058786         S14PB 2.4187D-04           FSR 0.31478487
    WBFSR 0.31700000

        T 17.0000000         S14PB 2.5505D-04         S14AA 0.05965293
      IAA 38.5999985           IPB 982.000000            AA 38.5999985
       PB 982.000000          AA14 2.30260291          PB14 0.25046273
     FDAA 0.38499999          AACD 0.38499999          AAPB 0.21617639
     PBAA 0.21617639            RF 0.31700000            DF 0.31700000
    S14AA 0.05965293         S14PB 2.5505D-04           FSR 0.31453818
    WBFSR 0.31700000

        T 18.0000000         S14PB 2.6803D-04         S14AA 0.05873251
      IAA 38.5999985           IPB 982.000000            AA 38.5999985
       PB 982.000000          AA14 2.26707491          PB14 0.26320199
     FDAA 0.38499999          AACD 0.38499999          AAPB 0.21617639
     PBAA 0.21617639            RF 0.31700000            DF 0.31700000
    S14AA 0.05873251         S14PB 2.6803D-04           FSR 0.31427907
    WBFSR 0.31700000

        T 19.0000000         S14PB 2.8080D-04         S14AA 0.05782639
      IAA 38.5999985           IPB 982.000000            AA 38.5999985
       PB 982.000000          AA14 2.23209871          PB14 0.27574106
     FDAA 0.38499999          AACD 0.38499999          AAPB 0.21617639
     PBAA 0.21617639            RF 0.31700000            DF 0.31700000
    S14AA 0.05782639         S14PB 2.8080D-04           FSR 0.31400762
    WBFSR 0.31700000

        T 20.0000000         S14PB 2.9336D-04         S14AA 0.05693435
      IAA 38.5999985           IPB 982.000000            AA 38.5999985
       PB 982.000000          AA14 2.19766575          PB14 0.28808302
     FDAA 0.38499999          AACD 0.38499999          AAPB 0.21617639
     PBAA 0.21617639            RF 0.31700000            DF 0.31700000
    S14AA 0.05693435         S14PB 2.9336D-04           FSR 0.31372390
    WBFSR 0.31700000

        T 21.0000000         S14PB 3.0573D-04         S14AA 0.05605616
      IAA 38.5999985           IPB 982.000000            AA 38.5999985
       PB 982.000000          AA14 2.16376759          PB14 0.30023096
     FDAA 0.38499999          AACD 0.38499999          AAPB 0.21617639
     PBAA 0.21617639            RF 0.31700000            DF 0.31700000
    S14AA 0.05605616         S14PB 3.0573D-04           FSR 0.31342799
    WBFSR 0.31700000

        T 22.0000000         S14PB 3.1791D-04         S14AA 0.05519161
      IAA 38.5999985           IPB 982.000000            AA 38.5999985
       PB 982.000000          AA14 2.13039591          PB14 0.31218787
     FDAA 0.38499999          AACD 0.38499999          AAPB 0.21617639
     PBAA 0.21617639            RF 0.31700000            DF 0.31700000
    S14AA 0.05519161         S14PB 3.1791D-04           FSR 0.31311996
    WBFSR 0.31700000

        T 23.0000000         S14PB 3.2989D-04         S14AA 0.05434048
      IAA 38.5999985           IPB 982.000000            AA 38.5999985
       PB 982.000000          AA14 2.09754255          PB14 0.32395673
     FDAA 0.38499999          AACD 0.38499999          AAPB 0.21617639
     PBAA 0.21617639            RF 0.31700000            DF 0.31700000
    S14AA 0.05434048         S14PB 3.2989D-04           FSR 0.31279990
    WBFSR 0.31700000

        T 24.0000000         S14PB 3.4169D-04         S14AA 0.05350258
      IAA 38.5999985           IPB 982.000000            AA 38.5999985
       PB 982.000000          AA14 2.06519945          PB14 0.33554046
     FDAA 0.38499999          AACD 0.38499999          AAPB 0.21617639
     PBAA 0.21617639            RF 0.31700000            DF 0.31700000
    S14AA 0.05350258         S14PB 3.4169D-04           FSR 0.31246789
    WBFSR 0.31700000

        T 25.0000000         S14PB 3.5330D-04         S14AA 0.05267769
      IAA 38.5999985           IPB 982.000000            AA 38.5999985
       PB 982.000000          AA14 2.03335869          PB14 0.34694194
     FDAA 0.38499999          AACD 0.38499999          AAPB 0.21617639
     PBAA 0.21617639            RF 0.31700000            DF 0.31700000
    S14AA 0.05267769         S14PB 3.5330D-04           FSR 0.31212403
    WBFSR 0.31700000
```

```
        T 26.0000000          S14PB 3.6473D-04          S14AA 0.05186561
      IAA 38.5999985            IPB 982.000000             AA 38.5999985
       PB 982.000000           AA14 2.00201246           PB14 0.35816399
     FDAA 0.38499999           AACD 0.38499999           AAPB 0.21617639
     PBAA 0.21617639             RF 0.31700000             DF 0.31700000
    S14AA 0.05186561          S14PB 3.6473D-04            FSR 0.31176840
    WBFSR 0.31700000

        T 27.0000000          S14PB 3.7598D-04          S14AA 0.05106614
      IAA 38.5999985            IPB 982.000000             AA 38.5999985
       PB 982.000000           AA14 1.97115308           PB14 0.36920941
     FDAA 0.38499999           AACD 0.38499999           AAPB 0.21617639
     PBAA 0.21617639             RF 0.31700000             DF 0.31700000
    S14AA 0.05106614          S14PB 3.7598D-04            FSR 0.31140110
    WBFSR 0.31700000

        T 28.0000000          S14PB 3.8705D-04          S14AA 0.05027910
      IAA 38.5999985            IPB 982.000000             AA 38.5999985
       PB 982.000000           AA14 1.94077299           PB14 0.38008094
     FDAA 0.38499999           AACD 0.38499999           AAPB 0.21617639
     PBAA 0.21617639             RF 0.31700000             DF 0.31700000
    S14AA 0.05027910          S14PB 3.8705D-04            FSR 0.31102224
    WBFSR 0.31700000

        T 29.0000000          S14PB 3.9794D-04          S14AA 0.04950427
      IAA 38.5999985            IPB 982.000000             AA 38.5999985
       PB 982.000000           AA14 1.91086475           PB14 0.39078127
     FDAA 0.38499999           AACD 0.38499999           AAPB 0.21617639
     PBAA 0.21617639             RF 0.31700000             DF 0.31700000
    S14AA 0.04950427          S14PB 3.9794D-04            FSR 0.31063190
    WBFSR 0.31700000

        T 30.0000000          S14PB 4.0867D-04          S14AA 0.04874148
      IAA 38.5999985            IPB 982.000000             AA 38.5999985
       PB 982.000000           AA14 1.88142103           PB14 0.40131307
     FDAA 0.38499999           AACD 0.38499999           AAPB 0.21617639
     PBAA 0.21617639             RF 0.31700000             DF 0.31700000
    S14AA 0.04874148          S14PB 4.0867D-04            FSR 0.31023020
    WBFSR 0.31700000

print /nciprn=1 t,S14Aa,S14Pb,FSR,wbfsr,AaPb,PbAa,AaCd,FdAa,Rf,Df,Aa,Pb
```

Line	T	S14AA	S14PB	FSR	WBFSR	AAPB
0	0.	0.07772020	1.0183E-13	18.8676000	0.31700000	0.21617600
1	1.00000000	0.07651920	1.6975E-05	0.31695900	0.31700000	0.21617600
2	2.00000000	0.07533680	3.3684E-05	0.31690400	0.31700000	0.21617600
3	3.00000000	0.07417280	5.0130E-05	0.31683700	0.31700000	0.21617600
4	4.00000000	0.07302690	6.6320E-05	0.31675600	0.31700000	0.21617600
5	5.00000000	0.07189870	8.2255E-05	0.31666300	0.31700000	0.21617600
6	6.00000000	0.07078810	9.7940E-05	0.31655600	0.31700000	0.21617600
7	7.00000000	0.06969480	1.1338E-04	0.31643700	0.31700000	0.21617600
8	8.00000000	0.06861840	1.2858E-04	0.31630400	0.31700000	0.21617600
9	9.00000000	0.06755870	1.4354E-04	0.31615900	0.31700000	0.21617600
10	10.0000000	0.06651550	1.5826E-04	0.31600100	0.31700000	0.21617600
11	11.0000000	0.06548850	1.7275E-04	0.31583000	0.31700000	0.21617600
12	12.0000000	0.06447750	1.8702E-04	0.31564600	0.31700000	0.21617600
13	13.0000000	0.06348210	2.0106E-04	0.31545000	0.31700000	0.21617600
14	14.0000000	0.06250220	2.1488E-04	0.31524100	0.31700000	0.21617600
15	15.0000000	0.06153750	2.2848E-04	0.31501900	0.31700000	0.21617600
16	16.0000000	0.06058790	2.4187E-04	0.31478500	0.31700000	0.21617600
17	17.0000000	0.05965290	2.5505E-04	0.31453800	0.31700000	0.21617600
18	18.0000000	0.05873250	2.6803E-04	0.31427900	0.31700000	0.21617600
19	19.0000000	0.05782640	2.8080E-04	0.31400800	0.31700000	0.21617600
20	20.0000000	0.05693430	2.9336E-04	0.31372400	0.31700000	0.21617600
21	21.0000000	0.05605620	3.0573E-04	0.31342800	0.31700000	0.21617600
22	22.0000000	0.05519160	3.1791E-04	0.31312000	0.31700000	0.21617600
23	23.0000000	0.05434050	3.2989E-04	0.31280000	0.31700000	0.21617600
24	24.0000000	0.05350260	3.4169E-04	0.31246800	0.31700000	0.21617600
25	25.0000000	0.05267770	3.5330E-04	0.31212400	0.31700000	0.21617600
26	26.0000000	0.05186560	3.6473E-04	0.31176800	0.31700000	0.21617600
27	27.0000000	0.05106610	3.7598E-04	0.31140100	0.31700000	0.21617600
28	28.0000000	0.05027910	3.8705E-04	0.31102200	0.31700000	0.21617600
29	29.0000000	0.04950430	3.9794E-04	0.31063200	0.31700000	0.21617600
30	30.0000000	0.04874150	4.0867E-04	0.31023000	0.31700000	0.21617600

Line	PBAA	AACD	FDAA	RF	DF	AA
0	0.21617600	0.38500000	0.38500000	0.31700000	0.31700000	38.6000000
1	0.21617600	0.38500000	0.38500000	0.31700000	0.31700000	38.6000000
2	0.21617600	0.38500000	0.38500000	0.31700000	0.31700000	38.6000000
3	0.21617600	0.38500000	0.38500000	0.31700000	0.31700000	38.6000000
4	0.21617600	0.38500000	0.38500000	0.31700000	0.31700000	38.6000000
5	0.21617600	0.38500000	0.38500000	0.31700000	0.31700000	38.6000000
6	0.21617600	0.38500000	0.38500000	0.31700000	0.31700000	38.6000000
7	0.21617600	0.38500000	0.38500000	0.31700000	0.31700000	38.6000000
8	0.21617600	0.38500000	0.38500000	0.31700000	0.31700000	38.6000000
9	0.21617600	0.38500000	0.38500000	0.31700000	0.31700000	38.6000000
10	0.21617600	0.38500000	0.38500000	0.31700000	0.31700000	38.6000000
11	0.21617600	0.38500000	0.38500000	0.31700000	0.31700000	38.6000000
12	0.21617600	0.38500000	0.38500000	0.31700000	0.31700000	38.6000000
13	0.21617600	0.38500000	0.38500000	0.31700000	0.31700000	38.6000000
14	0.21617600	0.38500000	0.38500000	0.31700000	0.31700000	38.6000000
15	0.21617600	0.38500000	0.38500000	0.31700000	0.31700000	38.6000000
16	0.21617600	0.38500000	0.38500000	0.31700000	0.31700000	38.6000000
17	0.21617600	0.38500000	0.38500000	0.31700000	0.31700000	38.6000000
18	0.21617600	0.38500000	0.38500000	0.31700000	0.31700000	38.6000000
19	0.21617600	0.38500000	0.38500000	0.31700000	0.31700000	38.6000000
20	0.21617600	0.38500000	0.38500000	0.31700000	0.31700000	38.6000000
21	0.21617600	0.38500000	0.38500000	0.31700000	0.31700000	38.6000000
22	0.21617600	0.38500000	0.38500000	0.31700000	0.31700000	38.6000000
23	0.21617600	0.38500000	0.38500000	0.31700000	0.31700000	38.6000000
24	0.21617600	0.38500000	0.38500000	0.31700000	0.31700000	38.6000000
25	0.21617600	0.38500000	0.38500000	0.31700000	0.31700000	38.6000000
26	0.21617600	0.38500000	0.38500000	0.31700000	0.31700000	38.6000000
27	0.21617600	0.38500000	0.38500000	0.31700000	0.31700000	38.6000000
28	0.21617600	0.38500000	0.38500000	0.31700000	0.31700000	38.6000000
29	0.21617600	0.38500000	0.38500000	0.31700000	0.31700000	38.6000000
30	0.21617600	0.38500000	0.38500000	0.31700000	0.31700000	38.6000000

Line	PB
0	982.000000
1	982.000000
2	982.000000
3	982.000000
4	982.000000
5	982.000000
6	982.000000
7	982.000000
8	982.000000
9	982.000000
10	982.000000
11	982.000000
12	982.000000
13	982.000000
14	982.000000
15	982.000000
16	982.000000
17	982.000000
18	982.000000
19	982.000000
20	982.000000
21	982.000000
22	982.000000
23	982.000000
24	982.000000
25	982.000000
26	982.000000
27	982.000000
28	982.000000
29	982.000000
30	982.000000

quit

Appendix D. Data from Bernier and Calvert (1987)

Time (min)	S14Aa (uCi/umol)	S14Pb (uCi/umol)
2.5000000000000	0.0746060000000	0.0000403000000
5.0000000000000	0.0702030000000	0.0000780000000
10.0000000000000	0.0613960000000	0.0001455000000

15.0000000000000	0.0525900000000	0.0002024000000
20.0000000000000	0.0437840000000	0.0002486000000
30.0000000000000	0.0261710000000	0.0003095000000

Appendix E. ACSL Builder and ACSL Math–Optimize Instructions

TO RUN ACSL BUILDER

Three files are required to run a model:

1. the .CSL file contains the model code
2. the .CMD file configures the model output when the model is run
3. the .PRJ file which tells ACSL what files to use (and where they are located)
   You will create this file. Printouts of files 1 and 2 are in the back of this handout

NOTE: Drive and directory references in instructions refer to the author's system and will need to be changed accordingly.

click on START (in lower left corner)
then PROGRAMS
ACSL OPTIMIZE 1.3
ACSL BUILDER (the top ACSL icon in the group)
an empty acsl builder window will appear

click on PROJECT
then NEW PROJECT
a directory window will appear
double click on the ACSLOPT directory
go to a directory named WG (double click on it)
enter WG.PRJ in the FILENAME window
click OK

The acsl builder project window is now filled with some information
click on the WG.CSL file until it is highlighted
click on the ADD button
the project WG.PRJ has now been created and is ready to run

TO RUN PROJECT WG
click on TOOLS
click on RUN
the program (WG) will compile and then an output window will appear
type OUT and hit RETURN (or ENTER key)
the model will run and you will see the output scroll down the screen
there are graphing options (for graphing S14AA and S14PB) under ANALYSIS
and then PLOT
you can also change model parameters and rerun the model by typing code in the START window. For instance, to change the protein turnover rate from 31.7%/d to 15%/d
type **RF=0.15** (hit ENTER)
type **DF=0.15** (hit ENTER)
the type **OUT** (hit ENTER)
the model will output new information based on the new synthesis and degradation rates

You can also change model parameters by editing the WG.CSL file and then recompiling and rerunning the file.

TO EXIT ACSL BUILDER
click on the x in the upper right corner of the window

TO RUN ACSL MATH

YOU MUST HAVE RUN **ACSL BUILDER** FIRST!
ACSL builder compiles the model so that it is ready to run (fit data, do sensitivity analysis etc.) in ACSL MATH
ACSL math uses the most currently compiled version of your project

click on START (in lower left corner)
then PROGRAMS
ACSL OPTIMIZE 1.3
ACSL MATH

an empty acsl math window will appear
follow instructions under **ACSL MATH**

ACSL MATH
the commands to type in are in **bold**
menu instructions are also in bold but are proceeded with **!!!**

```
MATH> set @format=long
!!!GO TO FILE MENU, LOAD MODEL (WG.PRJ)

MATH> Working directory now "C:\Heidi\wg"
Model loaded from "wng.prx"
Switching CMD unit to 4 to read wg.cmd

!File:WG.cmd
PROCEDURE MTH
SET WEDITG=.FALSE.
SET HVDPRN=.F.
FILE/PRNFILE='NUL'
END

PROCEDURE OUT
SPARE
SET HVDPRN=.T.
OUTPUT t,S14Pb,S14Aa
OUTPUT iAa,iPb,Aa,Pb
OUTPUT Aa14,Pb14
OUTPUT FdAa,AaCd,AaPb,PbAa,Rf,Df
OUTPUT S14Aa,S14Pb,FSR,WbFSR,psaa,pspb
PREPARE t,S14Pb,S14Aa
PREPARE iAa,iPb,Aa,Pb
PREPARE Aa14,Pb14
PREPARE FdAa,AaCd,AaPb,PbAa,Rf,Df
PREPARE S14Aa,S14Pb,FSR,WbFSR,psaa,pspb
set ndbug=1
s pcwprn=120
START
print /nciprn=1 t,S14Aa,S14Pb,psaa,pspb
END

PROCEDURE DP
d Aa Pb RF DF
d FdAa AaCd AaPb PbAa
d S14Aa S14Pb FSR wbFSR
END
End of file found on unit 4
Reverting to logical unit number 5
MATH> load bnc @file=bnc.txt @format=ascii
Loading from "bnc.txt"
MATH> bnc

bnc =
      2.5000000000000       0.0746060000000       0.0000403000000
      5.0000000000000       0.0702030000000       0.0000780000000
     10.0000000000000       0.0613960000000       0.0001455000000
     15.0000000000000       0.0525900000000       0.0002024000000
     20.0000000000000       0.0437840000000       0.0002486000000
     30.0000000000000       0.0261710000000       0.0003095000000
MATH>
!!!GO TO OPTIMIZE ON MENU BAR, PARAMETER FITTING SETTINGS
   ENTER METHOD-GRG
   TARGET S14Aa,S14Pb
   ADJUST AaCd
   CONSTRAIN AaCd>0
   EXPERIMENTS SINGLE
   DATA BNC
   OK

OpAlgorithm =
     3

OpAdjustable =
FDAA RF
```

```
OpStartValues =
      0.3849999904633       0.3170000016689

OpConstrained =
FDAA

OpConstraints =
      0     -1
```

!!!!GO TO MENU BAR-OPTIMIZE
SELECT FIT PARAMETERS

```
Opening logical unit 99 as file C:\Heidi\wng\op.cmd
data Op_Data0(T)
2.5
5
10
15
20
30
end
End of file found on unit 99
Reverting to logical unit number 5
            DESCRIPTION         PARAMETER ESIMATES    STANDARD
            -----------         ------------------    DEVIATION
                              INITIAL        FINAL
LOG LIKELIHOOD FUNCTION        69.692        93.865
                  FDAA          0.385        1.0897     0.06238
                    RF          0.317       0.31432   0.0037613

             S14AA         S14AA
       T     OBSERVED      PREDICTED     PCT ERROR
     2.5     0.074606     0.0714259       4.26257
       5     0.070203     0.0656418       6.49711
      10     0.061396     0.0554425       9.69684
      15      0.05259       0.04683       10.9527
      20     0.043784     0.0395573       9.65358
      30     0.026171     0.0282302      -7.86817
             S14PB         S14PB
       T     OBSERVED      PREDICTED     PCT ERROR
     2.5    4.03e-005  4.06581e-005      -0.88855
       5     7.8e-005  7.80014e-005   -0.00173975
      10    0.0001455   0.000143797       1.17067
      15    0.0002024   0.000199285       1.53905
      20    0.0002486    0.00024607       1.01777
      30    0.0003095   0.000318735       -2.9838

 STATISTICAL SUMMARY

               MAXIMIZED      WT RESID      WEIGHTED    PERCENTAGE
                     LOG        SUM OF      RESIDUAL     VARIATION     WEIGHTING
              LIKELIHOOD       SQUARES           SUM     EXPLAINED     PARAMETER
 ---------- ------------ ------------ ------------ ------------ ------------
     S14AA.        24.01    3.946e-003   1.145e-001        100.00         1.164
      S14PB        69.86    1.407e-003  -3.938e-005        100.00             2
    OVERALL        93.87    5.354e-003   1.145e-001        100.00

 CORRELATION MATRIX
                           FDAA            RF
 FDAA                         1
 RF                     0.84068             1

 VARIANCE-COVARIANCE MATRIX
                           FDAA            RF
 FDAA                 0.0038913
 RF                  0.00019725   1.4147e-005

 Optimization Method:  Generalized Reduced Gradient
 INFORM = 0
 GRG2 concluded successfully.

 Time elapsed:                        35.81 seconds
 Number of function evaluations:      213
 !!dp
 d Aa Pb RF DF
           AA 38.5999985           PB 982.000000           RF 0.31700000
```

```
          DF 0.31431567
d FdAa AaCd AaPb PbAa
        FDAA 0.38499999          AACD 1.08970264          AAPB 0.21434583
        PBAA 0.21434583
d S14Aa S14Pb FSR wbFSR
       S14AA 0.02823018         S14PB 3.1873D-04           FSR 0.28880071
       WBFSR 0.31431567
MATH>
```

TO EXIT ACSL MATH
click on the x in the upper right corner of the window

DEVELOPING AND TESTING INTEGRATED MULTICOMPARTMENT MODELS TO DESCRIBE A SINGLE-INPUT MULTIPLE-OUTPUT STUDY USING THE SAAM II SOFTWARE SYSTEM

David M. Foster

Department of Bioengineering 352255
University of Washington
Seattle, WA 98195-2255

ABSTRACT

As measurement devices become more sophisticated, it is possible to design more complex input-output studies, i.e., studies where data are obtained from several sites in the system under study. To interpret the resulting data requires models which can integrate known information about the system under study while simultaneously describing the data. In this chapter, we will illustrate how to develop and test a model structure for a single-input multiple-output study using the SAAM II software system. This system has been designed to make the use of sound modeling principles easy.

It will be assumed that a known amount of a radiolabeled substance was injected as a bolus into plasma, that this substance can bind to and be taken up by red cells, that its only route of elimination is through the urine, and that external measurements are possible over a target organ. The steps in developing a model structure will make use of SAAM II's forcing function capability to show how the system can be decoupled; this will permit us to postulate model structures for the various subsystems accessible to measurement. We will then show how to use this information to postulate a model describing all the data, and how to test this model structure. This will permit us to comment on those parts of the system not accessible for experimental measurement. We will end with a general discussion of how to test for goodness-of-fit and model order.

INTRODUCTION: THE DATA

In 1979, Foster et al. (1979) postulated an integrated model for zinc metabolism. This model, which was based upon data collected over a 5-day period, accounted for only 10% of total body zinc. This fraction, however, could be rapidly mobilized. This chapter is designed to illustrate how to use forcing functions as a tool to develop and test model structures. We will utilize a subset of the data presented in this earlier work (Foster, 1979). The data are given in the following figures.

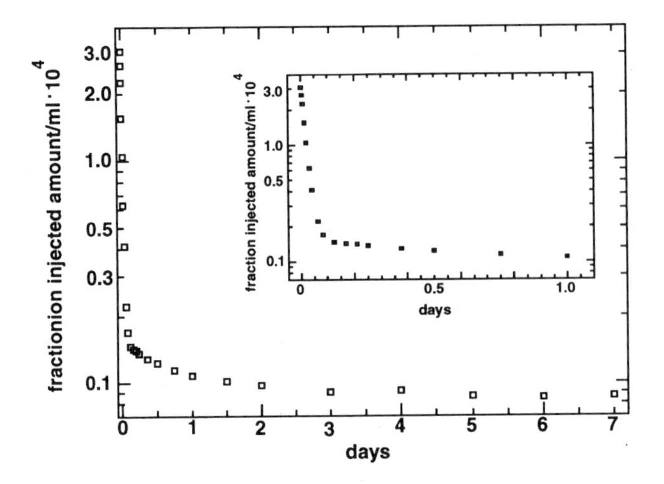

Figure 1. Plasma data collected following a bolus of 69^m-Zn injected into plasma. Data were collected at 2, 5, 10, 20, 30, 40, and 45 minutes, 1, 1.5, 2, 3, 4, 5, 6, 9, 12, and 18 hours, and 1, 2, 3, 4, 5, 6, and 7 days. Data are expressed in terms of fraction of injected amount per ml of plasma. The decay in plasma during the first day is shown in the inset.

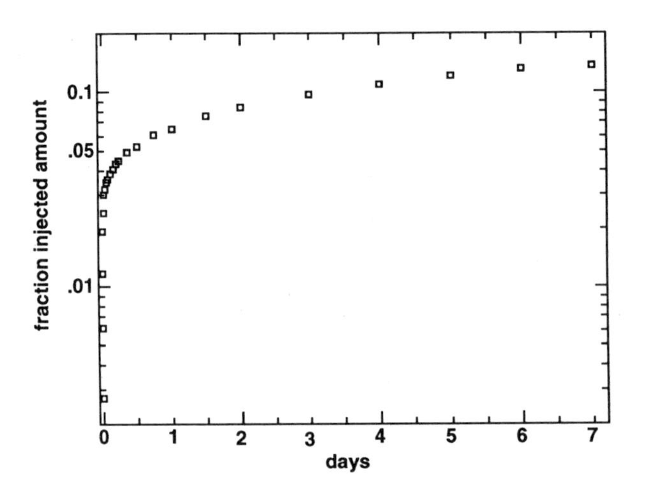

Figure 2. Red cell data collected following a bolus of 69^m-Zn injected into plasma. Data were collected at the same time plasma data were collected. Data are expressed as a fraction of the total amount injected.

It was assumed a known amount of free 69^m-zinc was given as a bolus into plasma. By dividing all samples by this amount, the data were converted either to fraction of injected amount per mL or fraction of injected amount. The problems to be discussed in the next sections relate to the postulation of a multicompartmental model structure which is compatible with these data. This will illustrate how to use many of the modeling tools incorporated in the SAAM II software system (SAAM II, 1997).

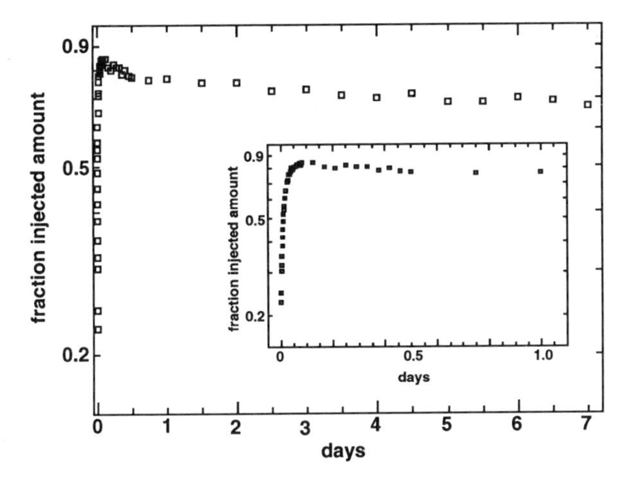

Figure 3. Data obtained from an external counter placed over the liver following an injection of 69^m-Zn into plasma. Data were collected at 1, 2, 4, 5, 8, 10, 12, 14, 16, 18, 20, 25, 30, 35, 40, 45, 50, 55, 60, 65, 70, 75, 80, 85, 90, 95, 100, 105, 110, 115, and 120 minutes, 3, 4, 5, 6, 7, 8, 9, 11, 12, and 18 hours, and 1, 1.5, 2, 2.5, 3, 3.5, 4, 4.5, 5, 5.5, 6, 6.5, and 7 days. Data are expressed in terms of fraction of the total amount injected. The appearance of radioactivity during the first day is shown in the inset.

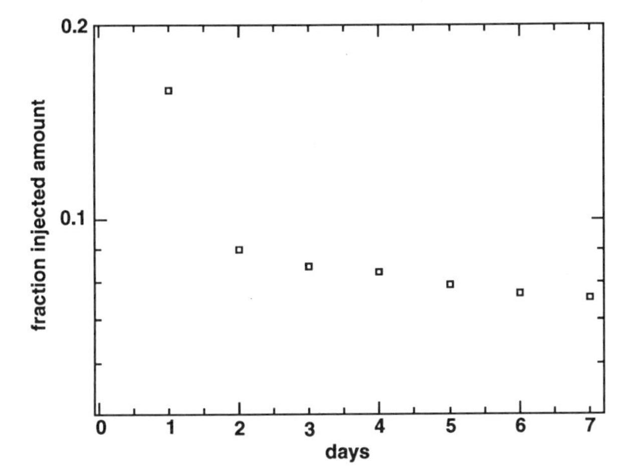

Figure 4. 24-hour urine collections of 69^m-Zn. Data are expressed in terms of fraction of injected amount.

Using Forcing Functions to Decouple the System

What is a forcing function, and what does it mean to decouple the system (Foster, 1979; SAAM II, 1997)? For the data given above, suppose we want to postulate a model structure for the red cell system, and describe the elimination of 69^m-Zn through the urine. Because the only way the 69^m-Zn can interact with the red cells is via the plasma, and the only way it can leave the body is via the plasma through the urine, one can sketch the following relationships among plasma, red cells and urine.

This figure indicates that the 69^m-Zn can exchange with plasma, and that there is a unidirectional loss from plasma into the urine.

Assume that the 69^m-Zn in plasma is *kinetically homogeneous*. This means that the 69^m-Zn exists free in plasma, and does not bind in any significant way to any of the plasma proteins (or if it does, the binding equilibrium is to fast too be "seen" in these data). This assumption thus means that the plasma pool can be described by a single compartment in a multicompartmental model.

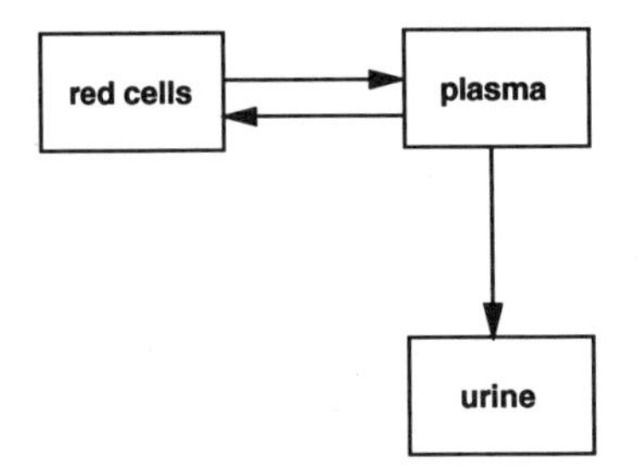

Figure 5. The metabolic relationships of 69^m-Zn among plasma, red cells, and urinary excretion.

Without any *a priori* knowledge of the system, one does not know what kind of a multicompartmental structure will be required to describe red cells or urinary excretion. As a first attempt, one might want to try the following model.

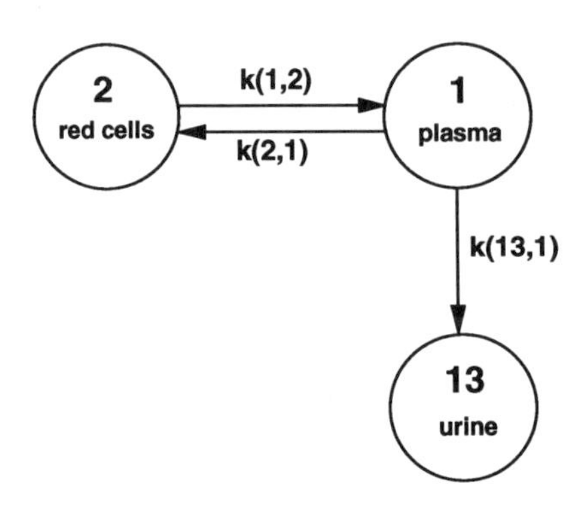

Figure 6. A potential model structure describing the interrelationships of 69^m-Zn among plasma, red cells, and urinary excretion. Compartment numbers are 1, 2, and 13 respectively; the $k_{I,j}$ are the fractional transfer coefficients (see Eqs. 1-3).

The differential equations describing this model are

$$\frac{dq_1}{dt} = -(k_{2,1} + k_{13,1})q_1(t) + k_{1,2}q_2(t) \qquad q_1(0) = 1 \tag{1}$$

$$\frac{dq_2}{dt} = k_{2,1}q_1(t) - k_{1,2}q_2(t) \qquad q_2(t) = 0 \tag{2}$$

$$\frac{dq_1}{dt} = k_{13,1}q_{13}(t) \qquad q_{13}(0) = 0 \tag{3}$$

where $q_i(t)$ is the amount of material in compartment i at time t, and the k_{ij} are the *fractional transfer coefficients*, i.e., the fraction of material in compartment j transferred to compartment i per unit time. The $q_i(0)$ are the "initial conditions" in the system, and reflect the unit bolus injection into compartment 1 with no *a priori* existing material in compartments 2 and 13.

Obviously this model will not fully describe the system because the organ data are not represented. In particular, $q_1(t)$ will not describe the plasma data. However, if the plasma data were known, then the rate constants $k_{2,1}$, $k_{1,2}$, $k_{13,1}$ could be determined. That is, it is necessary that these systems "see" what is actually in plasma to determine the rate constants and to determine if compartments 2 and 13 can describe red cell and urine data, respectively.

To do this, one uses the forcing function machinery in SAAM II to substitute a function which has the characteristics of the plasma decay shown in Figure 1 for $q_1(t)$ in (2 and 3). In SAAM II, such a forcing function is denoted $q1 \cdot FF$. With the forcing function, (2 and 3) become:

$$\frac{dq_1}{dt} = -(k_{2,1} + k_{13,1})q_1(t) + k_{1,2}q_2(t) \qquad q_1(0) = 1 \tag{4}$$

$$\frac{dq_2}{dt} = k_{2,1}q1 \cdot FF(t) - k_{1,2}q_2(t) \qquad q_2(t) = 0 \tag{5}$$

$$\frac{dq_1}{dt} = k_{13,1}q1 \cdot FF \qquad q_{13}(0) = 0 \tag{6}$$

Notice that (1) and (4) are the same; the forcing function has no affect on $dq_1(t)/dt$. However, in (5 and 6), $q_1(t)$ is replaced with $q1 \cdot FF$. The system has been decoupled in the sense that we can now deal independently with the red cell and urinary subsystems. Once we have a proposed model structure for these two subsystems, we will deal with the organ subsystem. At that point, we will have known model structures for the three subsystem models (red cell, urine, organ), and once we can reconstruct the plasma curve, we know these structures will be valid.

First, we must postulate a functional description for $q1 \cdot FF$. We will then develop and test model structures for the red cell and urinary subsystems. With the red cell model, we can postulate a structure for the organ subsystem, estimating at the same time the blood volume of the organ.

Defining the Forcing Function in SAAM II

The model is built in SAAM II using the "point and click" method to drag and drop icons on the drawing canvas. To create the forcing function $q1 \cdot FF$, we must create a functional description of the data shown in Figure 1. There are two ways this can be done in SAAM II; both methods are invoked by opening the attribute box associated with compartment 1.

The first option is to internally create $q1 \cdot FF$ by letting SAAM II perform a linear interpolation between sequential pairs of data. Clearly this will recreate the shape of the input curve, but if the data are noisy, there can be problems. Alternatively, one can write an expression directly for $q1 \cdot FF$, this normally will smooth out the noise in the data. For the example here, we will describe the data shown in Fig. 1 by a sum of four exponentials:

$$\begin{gathered} \text{plasma} = \frac{1}{V}(A_1 e^{-a_1 t} + A_2 e^{-a_{21} t} + A_3 e^{-a_3 t} + A_4 e^{-a_4 t}) \\ A_1 + A_2 + A_3 + A_4 = 1 \end{gathered} \tag{7}$$

At this point, we need to make some observations about (7). First, because the units of the plasma data are fraction of injected amount per mL, the units of (7) are in these terms. However, the units of the forcing function must be in terms of fraction of injected amount. Second, our constraint (the sum of the coefficients A_1, A_2, A_3, and A_4 equals 1) is equivalent to the bolus injection of 1 into the system. Dividing the sum by the unknown parameter V will permit an estimate the plasma volume and thus assure that the units of (7) are fraction of injected amount per mL. When a fit has been obtained, the forcing function $q1 \cdot FF$ can be written:

$$q1 \cdot FF = A_1 e^{-a_1 t} + A_2 e^{-a_2 t} + A_3 e^{-a_3 t} + A_4 e^{-a_4 t} \tag{8}$$

A fit of (7) to the data given in Figure 1 is shown in Figure 7.

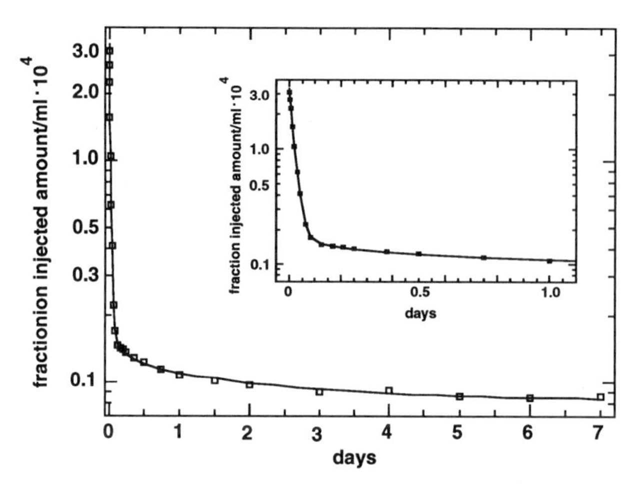

Figure 7. A fit of the plasma data given in Figure 1 to the sum of exponentials given by (7). See text for additional explanation.

It is important to emphasize that (7) is used to provide a functional description of the data given in Figure 1. It is not important that one know the statistics of the fit; all one is interested in is an accurate description of the shape of the plasma curve! In fact, if one tried to fit (7) to these data using SAAM II, one will obtain a best fit but learn that the model is overdefined.

For this system, an estimate for the volume V is 2975 mL, and (8) can be written for this example:

$$q1 \cdot FF = 0.95e^{-62t} + 0.013e^{-5t} + 0.013e^{-0.5t} + 0.025e^{-0.0001t} \tag{9}$$

(Equation 9) will be entered as the forcing function in the attribute dialog box associated with compartment 1, and the individual subsystem models will be determined as described below.

The Red Blood Cell System

To determine the model for the red blood cell system, we will first try the single compartment model suggested in Figure 6. An example of the model constructed in SAAM II is shown in Figure 8. In SAAM II, bullets labeled s1, s2, etc., represent samples. In the dialog box associated with the sample, the user writes the measurement equation, and associates the sample with the data. When a sample is associated with a single compartment, there is a single line to that compartment (thus, in Figure 8, the measurement equation is s2 = q2). When the sample is associated with several compartments, there are lines from the bullet to each compartment involved (thus, in Figure 9, the sample equation is s2 = q2 + q3). When a sample is not associated with a compartment, there is no line from the bullet (thus, in Figure 8, $s1 = q1 \cdot FF$, i.e., $s1$ is not associated with $q1$).

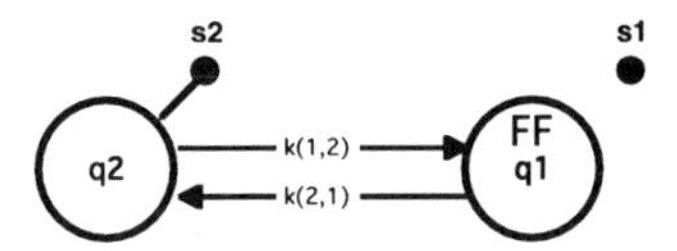

Figure 8. A model to determine the structure of the red blood cell system. Compartment q1 is a forcing function with the equation given by (9). Compartment q2 represents the red blood cell system. The bullets labeled s1 and s2 are samples. Sample s2 is associated with the red blood cell data. Sample s1 associated the forcing function with the plasma data, and is used to be sure the forcing function has been correctly specified; s1 is not connected to compartment 1 because the sample equation is written in terms of $q1 \cdot FF$ and not $q1(t)$ (see SAAM II, 1997).}

The differential equation for the red blood cell system represented by this model is

$$\begin{aligned} \frac{dq_2}{dt} &= k_{2,1} q1 \cdot FF(t) - k_{1,2} q_2(t) \\ &= k_{2,1}(0.95e^{-62t} + 0.013e^{-5t} + 0.013e^{-0.5t} + 0.025e^{-0.0001t}) - k_{1,2} q_2(t) \end{aligned} \tag{10}$$

If this model is fitted to the red blood cell data, one will see the single compartment system for the red blood cells is not sufficient, and that a second compartment must be added. At this point, however, one has a choice of model structures, and to select the most appropriate, one must rely on what is known about 69^m-Zn and its interaction with red blood cells. The two different model structures are shown in Figure 9.

The two models shown in Figure 9 have different physiological interpretations. It is essential to realize that both will produce the same fit to the data. Model A is consistent with the 69^m-Zn interacting with the red blood cell membrane (q2) as indicated by the rate constant $k_{2,1}$. The 69^m-Zn can then be internalized (q3), represented by $k_{3,2}$, or return to plasma (q1) indicated by $k_{1,2}$. The internalized 69^m-Zn can also bind to the interior of the membrane, $k_{2,3}$. Model B is consistent with there being two different classes of red blood cells, and that the 69^m-Zn has different kinetic interactions with each class. In general, it is difficult to support this hypothesis, and so we will choose Model A.

It is important to note that derived parameters of interest can also be estimated from the primary parameters, the $k_{I,j}$. For example, if one wanted to know the fraction of 69^m-Zn bound to the red blood cell membrane that was internalized, one could write

$$\text{frac\_in} = \frac{k_{3,2}}{k_{3,2} + k_{1,2}} \tag{11}$$

In this expression, the denominator $k_{3,2} + k_{1,2}$ is the total transfer from compartment 2, so frac_in is the fraction contributed by $k_{3,2}$.

If we assume that 69^m-Zn can bind to the red cell membrane and subsequently be internalized, then Model A is the model of choice. Figure 10 shows the best fit of this model to the red blood cell data, remembering that it is the shape of the plasma curve as represented by the forcing function that is driving the system.

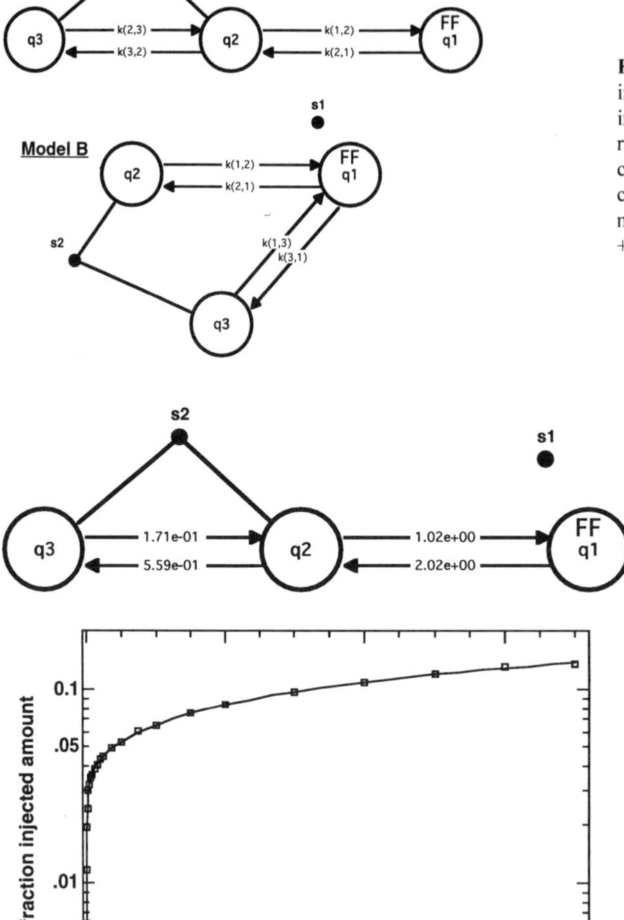

Figure 9. Two different models incorporating a second compartment into the red blood cell system. The red cell system is represented by compartments q2 and q3, and the red cell sample s2 is the sum of the material in the compartments, $s2 = q3 + q3$. $s1$ was described in Figure 8.

Figure 10. The best fit of Model A shown in Figure 9 to the red blood cell data. The model is shown at the top with the rate constants producing the best fit. The data (squares) and model prediction (continuous line) are shown in the bottom panel. Samples s1 and s2 are described in Figure 9.

It thus appears that the two-compartment system represented by compartments 2 and 3 will produce a good description of the red blood cell system. Technically, one should try a third compartment, a compartment 4 exchanging with compartment 3, for example, to determine if yet another component is required. In this situation, such a compartment is not justified; this can be tested using the tests for goodness-of-fit and model order discussed in a later section.

The Urinary System

As noted in Figure 4, the urine data represent 24-hour urine collections, and the data are expressed in terms of fraction of injected amount excreted during a 24-hour period. To simulate such an experimental protocol in SAAM II, we take advantage of its "change condition" capability. Basically, this permits us to reset the value for the urine compartment equal to zero at 24-hour intervals. The single compartment urine system described in Figure 6 can be implemented in SAAM II as indicated by Figure 11.

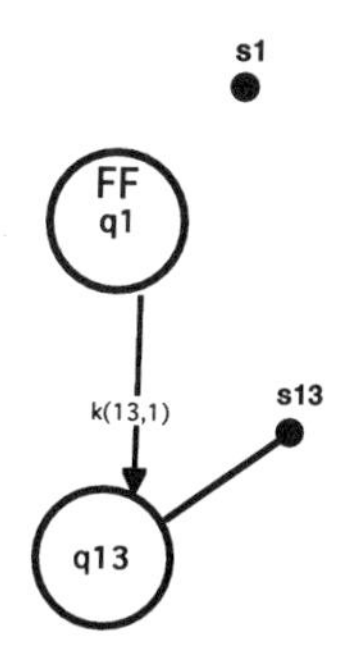

Figure 11. A single compartment urinary excretion system being driven by the plasma forcing function. The model indicates there is no delay in appearance of 69^m-Zn in the urine. The sample s1 has been described previously; the sample s13 is the 24-hour excretion of 69^m-Zn.

It is important to note that this model is consistent with some urinary excretion happening immediately. If this were not the case, then an intermediary compartment could be placed between the forcing function and compartment 13, or if needed, the delay element in SAAM II could be used. The best fit of the model shown in Figure 11 to the urinary excretion data shown in Figure 4 is shown in Figure 12.

The Liver System

We are now in a position to postulate a model structure to describe the liver data. Remember that these data were collected using an external counter, and this counter will quantitate material both in the organ cells and in the blood. That is, a fraction of the material "seen" by the counter will be due to the 69^m-Zn in blood and the rest presumably to the 69^m-Zn in the liver's cells.

Suppose a two-compartment system is required to account for the radioactivity seen by the external counter. Then we are in a situation similar to that described above for the red blood cells. Specifically, there can be one compartment which exchanges with plasma and a second exchanging with this compartment; as before, this model would be consistent with the 69^m-Zn interacting with the organ cell's membrane with the possibility for intracellular uptake. A second possibility would be two compartments exchanging with plasma. This would be consistent with the organ containing two different cell types with different 69^m-Zn kinetics for each type. In complex organ systems such as the liver, this is not an unreasonable situation, and care should be taken in deciding which structure to use.

These two situations are illustrated in models A and B in Figure 13. Notice that the red blood cell subsystem is included; this is because it is assumed that plasma, represented by the forcing function $q1 \cdot FF$ and the red cells, represented by the sample s2 = q2 + q3, can account for all the 69^m-Zn in the blood. The urinary subsystem is not included because it plays no part in quantitating the 69^m-Zn in the organ.

For both models A and B shown in Figure 13, the measurement equation for the 69^m-Zn, as seen by the external counter in the organ, is

$$s3 = \text{fraction}(q1 \cdot FF(t) + s2(t)) + q4 + a5 \tag{12}$$

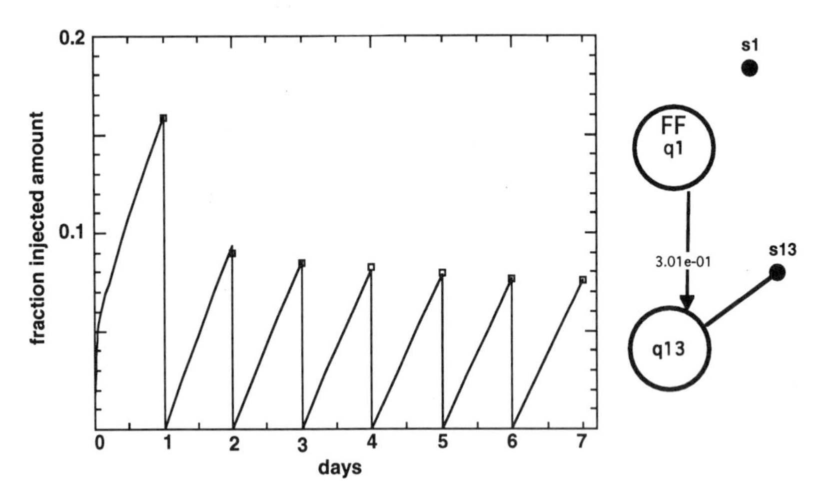

Figure 12. The best fit of the model shown in Figure 11 to the urinary excretion data. The model is shown to the right hand side of the fit, and indicates $k_{13,1} = 0.301/day$. Notice the saw-tooth nature of the model-predicted values; this reflects the fact that the data are 24-hour collections, hence the contents of compartment 13 are set equal to zero each day after the last collection. The samples s1 are s13 are described in Figure 11.

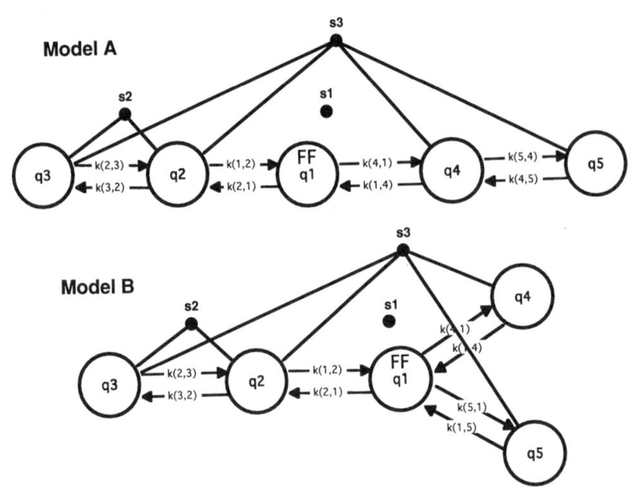

Figure 13. Two different multicompartmental models which can describe the organ kinetics of 69^m-Zn. See text for additional explanation. Samples s1 and s2 have been described previously; sample s3 accounts for the 69^m-Zn data seen by the external counter over the organ. Note from (12): s3 includes $q1 \cdot FF$ and not $q_1(t)$, thus the sample is not connected to compartment 1.

The parameter "fraction" can be estimated from the data as the fraction of blood in the organ. This is important information because the blood volume of many organs is known. The estimate of the parameter "fraction" can be checked against the known blood volume as a step in validating the model structure.

The best fit of Model A to the organ data shown in Figure 3 is given in Figure 14, and indicates a that a two-compartment subsystem is sufficient to describe the organ data. As with the red blood cell subsystem, one can add a third compartment to test if this improves the fit, and use the tests for goodness-of-fit and model order discussed below to determine the best model. However, as noted, one cannot distinguish with these data any difference between Models A and B. To do this will require additional experiments.

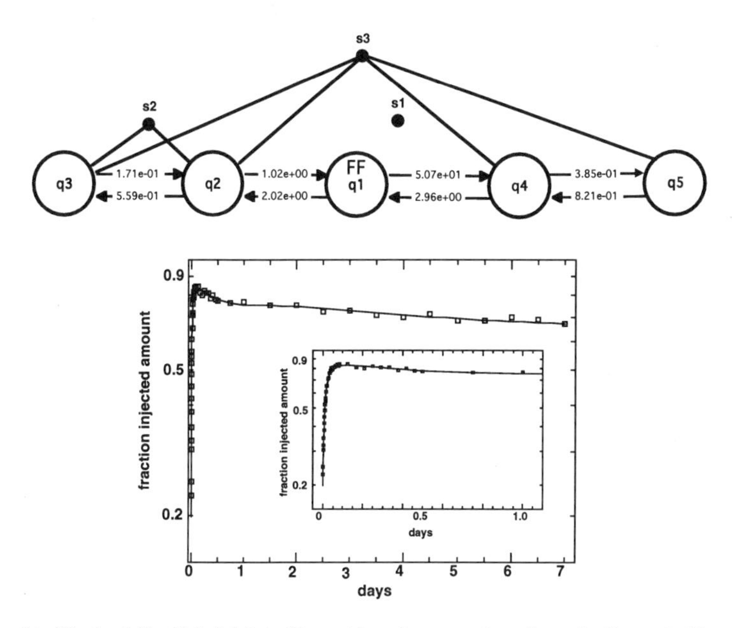

Figure 14. The best fit of Model A in Figure 13 to the organ data shown in Figure 3. The top panel shows the model and the parameter values for the exchange rate constants. The lower panel shows the fit with the inset giving the details for the first day. Samples s1 and s2 have been described previously; sample s3 accounts for the 69^{m}-Zn data seen by the external counter over the organ. Note from (12): s3 includes $q1 \cdot FF(t)$ and not $q_1(t)$, thus the sample is not connected to compartment 1.

Postulating the System Model Structure

What is known at this stage of the model development process is that, given the correct shape of the plasma input function, we have subsystem models which can describe the data. The system has thus been truly decoupled, in that we have examined the subsystems first. We must now build the final structure to account for the totality of the data. This is a two-step process. In the first step, we postulate, given the known structure and parameter values for the individual subsystems, the plasma model. In the second step, all parameters are permitted to be adjustable so that we can generate the best estimates for the model's parameters and their uncertainties. This is the first stage towards evaluating the model; the last stages, which are discussed below, deal with tests for goodness-of-fit and model order.

The Plasma Model

In determining the plasma model, we first create a model in which all subsystems are incorporated with their rate constants fixed, remove the forcing function on plasma, and ask whether or not the plasma data can be fitted. Figure 15 describes this situation.

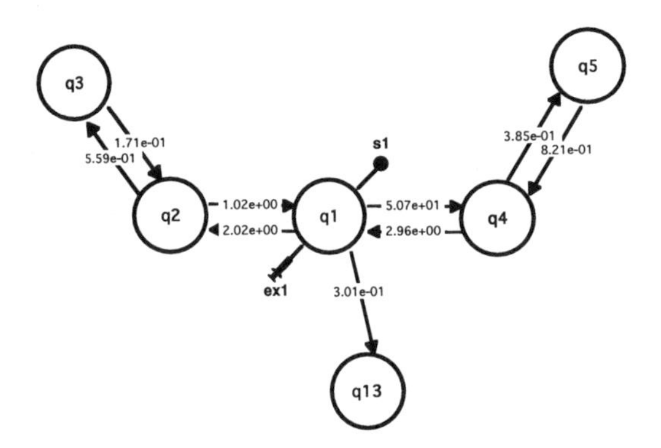

Figure 15. A model incorporating all known subsystems with rate constants fixed. Notice that s1 is now attached to compartment 1; this is because the measurement equation is now given by (12). The input ex1 represents the bolus input into compartment 1. See text for additional explanation.

In removing the plasma forcing function, the differential equation for the compartments other than compartment 1 change from the form given in (5 and 6) back to the form given by (2 and 3). Now, however, since there is no function driving the system, one must specify $q_1(0)$ in (1). In Figure 15, this is represented by ex1, and is a bolus equal to 1. Because the units of the plasma data are fraction of injected amount per mL, the measurement equation s1 becomes

$$s1 = q1 / vol \tag{13}$$

The parameter *vol* provides an estimate for the plasma volume, the only adjustable parameter in the model shown (Figure 15). This model will either fit the plasma data or not. If it does, then the subsystems for which data are available completely specify the total system, and the structure given in Figure 15 would be the proposed model structure necessary to describe all the data.

If, on the other hand, this model does not fit the plasma data, then additional structures must be proposed. Such a situation is illustrated in the model shown in Figure 16. In this model, an additional compartment that exchanges with plasma, compartment 7, has been added. In this case, in addition to the adjustable parameter *vol*, there are two adjustable rate constants, $k_{7,1}$ and $k_{1,7}$. It is important to realize that the rate constants characterizing the subsystems remain fixed. This is because we are trying to build a structure necessary to describe the plasma data, and if the subsystem rate constants were allowed to adjust, we would lose the subsystem fits. This is an important point behind the decoupling process, and the process of reconstructing the plasma input function.

In the example we are discussing here, the model shown in Figure 16 can describe the plasma data, and hence it is proposed as the system model structure compatible with the data.

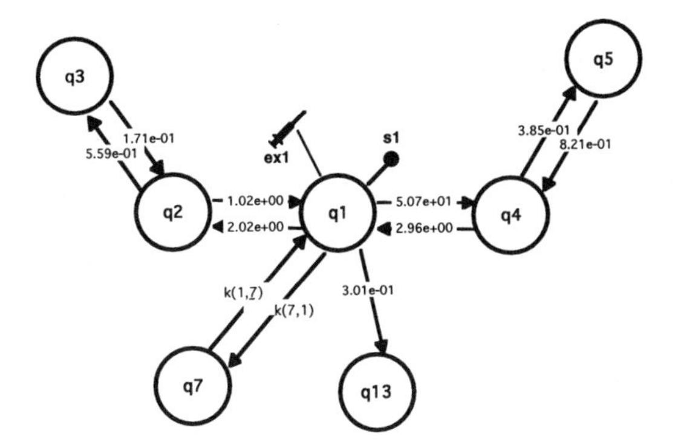

Figure 16. An expanded model to describe the plasma data. See text for additional explanation.

The System Model

With the model structure proposed, it remains only to let all parameters of this model adjust. In the previous step, the subsystem parameters were fixed because we wanted to know the structure needed to recreate the plasma curve. In Figure 17, we show the model which indicates that all parameters are adjustable. We do not show any of the fits because they are essentially identical to those shown for the individual subsystem fits, and for the plasma forcing function. It should be noted that some of the numerical values for the rate constants are slightly different from those derived for the subsystem models; this is because we are now fitting the totality of the data.

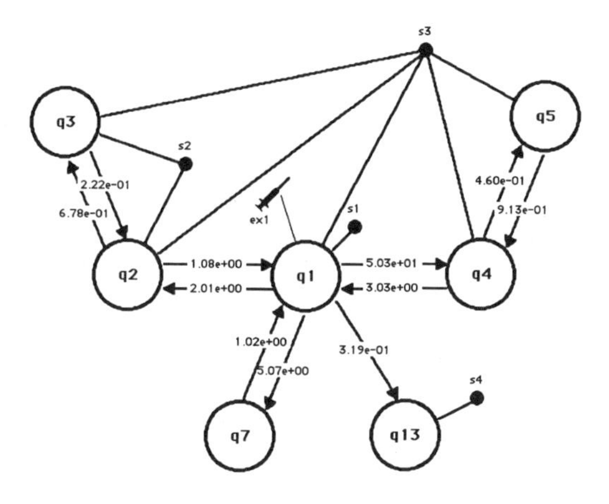

Figure 17. The model with the best parameter estimates for the rate constants $k_{l,j}$. See text for additional explanation.

The last two figures provide statistical information about the fit. Figure 18 is a screen snapshot of the statistical information available from SAAM II, and lists the individual parameters, their values, the standard deviations and coefficients of variation of the parameter estimates, and the 95% confidence limits. Figure 19 shows the correlation matrix. The optimization routine used in SAAM II is described elsewhere (Bell et al., in press). More information in the theory and interpretation can be found in standard texts (Neter and Wasserman, 1974; Bates and Watts, 1988). We will summarize briefly how this information can be used.

Parameter/Variable	Value	Std.Dev.	Coef.of Var.	95% Confidence Interval	
frac	0.19927	1.67576e-03	8.40960e-01	0.19594	0.20260
k(1,2)	1.07781	3.66678e-02	3.40207e+00	1.00494	1.15068
k(1,4)	3.02546	2.75663e-02	9.11143e-01	2.97068	3.08025
k(1,7)	1.02191	5.87676e-02	5.75075e+00	0.90512	1.13871
k(13,1)	0.31893	4.58261e-02	1.43688e+01	0.22785	0.41000
k(2,1)	2.01372	8.22416e-03	4.08406e-01	1.99738	2.03007
k(2,3)	0.22203	2.55172e-02	1.14929e+01	0.17131	0.27274
k(3,2)	0.67838	5.89215e-02	8.68563e+00	0.56128	0.79548
k(4,1)	50.27794	2.37546e-01	4.72466e-01	49.80584	50.75004
k(4,5)	0.91296	5.61649e-02	6.15195e+00	0.80134	1.02458
k(5,4)	0.46018	3.16151e-02	6.87014e+00	0.39735	0.52301
k(7,1)	5.07315	1.19808e-01	2.36162e+00	4.83504	5.31126
vol	2992.27541	1.62575e+01	5.43315e-01	2959.96532	3024.58549

○ Correlation Matrix ○ Covariance Matrix ◉ Objective

	Objective	Scaled Data Variance
s4 : urine	-5.413718e-01	8.578328e+01
s3 : organ	-4.362497e+00	1.484460e-02
s2 : rbc	-3.154522e+00	1.012920e-02
s1 : plasma	-6.390931e+00	9.391255e-03
Total objective	-1.444932e+01	
AIC	-1.626089e+01	
BIC	-1.605341e+01	

Figure 18. A screen snapshop showing the statistical information available on the adjustable parameters from SAAM II following a successful fit.

The fractional standard deviation is the quotient of a parameter's estimated standard deviation and the parameter value. When multiplied by 100, it is a percent fractional deviation or a coefficient of variation (CV). The smaller the CV value, i.e., the closer it is to zero, the better the estimated value of the parameter. Thus a parameter such as $k_{4,1}$, with a CV of 0.47, is very well defined.

As can be seen in Figure 18, the highest coefficient of variation for this model is less than 20%, so one can feel confident that the parameters have been estimated with very good precision. Depending upon the situation (mainly noise in the data), it is reasonable to expect that the coefficients of variations of all parameters be less than 50%. In the range between 50% and 90%, one should be concerned about model accuracy. Above 90%, one can probably conclude that the model contains more adjustable parameters than can be supported by the data. Those parameters with high coefficients of variation need to be fixed at some reasonable value, or a constraint such as a functional relationship among parameters needs to be introduced in order to reduce the number of adjustable parameters.

The correlation matrix in Figure 19 provides information on the relationship among the parameters in the model. It is especially useful when one is having difficulty optimizing the model to the data because some parameters have high coefficients of variation. This usually indicates a high correlation among some parameters in the model. When two parameters that are highly correlated, such as when the correlation coefficient is between 0.9 and 1, or between -1 and -0.9, one parameter can change and the other will also change in a way such that the fit of the data doesn't change significantly. This indicates which parameters are causing problems in the fitting process.

Ideally, the correlation coefficients should be close to zero. In our experience in biomedical research, this rarely happens, and often one is forced to live with correlation coefficients that are near 1 or –1. Because of the nature of the optimization problem,

	frac	k(1,2)	k(1,4)	k(1,7)	k(13,1)	k(2,1)	k(2,3)	k(3,2)	k(4,1)	k(4,5)	k(5,4)	k(7,1)	vol
frac	1	0.06808	-0.10567	0.11422	0.04246	-0.13542	0.02206	0.0415	-0.40976	-0.0371	-0.04406	0.1664	0.18425
k(1,2)	0.06808	1	0.1202	-0.06618	-0.06022	0.39435	0.6112	0.87455	-0.21635	-0.07239	-0.04695	-0.14238	0.368
k(1,4)	-0.10567	0.1202	1	0.15584	0.10214	0.1422	0.08456	0.09481	0.37856	0.48271	0.67813	0.1326	0.11577
k(1,7)	0.11422	-0.06618	0.15584	1	0.83576	0.08681	0.00304	-0.05161	-0.08237	0.46525	0.49973	0.36931	-0.10876
k(13,1)	0.04246	-0.06022	0.10214	0.83576	1	0.0548	0.01953	-0.03934	0.0153	0.56427	0.48615	-0.05822	-0.10474
k(2,1)	-0.13542	0.39435	0.1422	0.08681	0.0548	1	0.17456	0.28443	0.41971	0.05336	0.09988	0.18602	-0.34245
k(2,3)	0.02206	0.6112	0.08456	0.00304	0.01953	0.17456	1	0.89167	-0.06317	0.13441	0.14002	-0.04204	0.07035
k(3,2)	0.0415	0.87455	0.09481	-0.05161	-0.03934	0.28443	0.89167	1	-0.13023	0.02575	0.06086	-0.08101	0.16763
k(4,1)	-0.40976	-0.21635	0.37856	-0.08237	0.0153	0.41971	-0.06317	-0.13023	1	0.17246	0.22454	-0.03096	-0.58274
k(4,5)	-0.0371	-0.07239	0.48271	0.46525	0.56427	0.05336	0.13441	0.02575	0.17246	1	0.9215	-0.0502	-0.09676
k(5,4)	-0.04406	-0.04695	0.67813	0.49973	0.48615	0.09988	0.14002	0.06086	0.22454	0.9215	1	0.12085	-0.15246
k(7,1)	0.1664	-0.14238	0.1326	0.36931	-0.05822	0.18602	-0.04204	-0.08101	-0.03096	-0.0502	0.12085	1	-0.34425
vol	0.18425	0.368	0.11577	-0.10876	-0.10474	-0.34245	0.07035	0.16763	-0.58274	-0.09676	-0.15246	-0.34425	1

Figure 19. The correlation matrix associated with the fit described in Figure 18.

however, this situation can occur and yet the coefficients of variations on the parameters are still acceptable.

SUMMARY: MODEL DEVELOPMENT

The development and testing of an integrated model for a single-input multiple-output experiment requires a number of steps. This chapter has illustrated how to use the forcing function machinery to postulate a model structure for such an experiment. In summary, the steps are:

1. Provide a functional description of the input function as a forcing function.
2. Use the forcing function to develop and test model structures for the subsystems.
3. Fix these subsystem structures and postulate a most for the input function.
4. Optimize on all model parameters to obtain best estimates of the parameters and their uncertainties.

MODEL IDENTIFIABILITY AND GOODNESS-OF-FIT

In the above example, we have concentrated on the steps needed to develop a model structure which is consistent with the experimental data. We assume in this discussion that the reader is aware of the principles of designing sound kinetic experiments, and is aware of tools, for example, to help in determining optimal sample selection times (see, for example, Carson, 1983).

When one has postulated a model structure consistent with the data, a number of questions arise:

- Given an ideal set of data and the correct model of the system, can I estimate uniquely the parameters of the model?
- Given the model of the system, can I estimate the parameters of the model with some statistical precision?
- How do I know when the model *fits* the data?
- How do I know if I need more compartments in the model?

The first question is known as the *a priori* identifiability question (Carson, 1983). It is set in the context of knowing the correct model structure and having noise-free, continuous time measurements (ideal data). The problem is that while some model structures can have a finite number of solutions which will produce a best fit to the data, others will have an infinite number of solutions. What is desired is a situation where there is a unique global minimum, i.e. a unique set of parameter values which will produce a best fit of the data. This question is very important because physiological conclusions are drawn from the model, and different sets of parameter values will lead to different conclusions. A general solution to the problem is difficult, but a software package, GLOBI (Audoly, in press) has recently been developed for linear compartmental models.

The second question relates to the *a posteriori* identifiability problem. Given a model structure and an initial set of parameter estimates, can a best fit be obtained where the error estimates of the parameters are acceptable?

To obtain statistical information about the precision of a model's fit to a set of data such as that presented above, requires the assignment of weights to the data. Suppose $y(\vec{\mathbf{p}},t)$ is the model being tested. Here t is the independent variable time, and

$\vec{\mathbf{p}} = (p_1, \cdots, p_n)$ are the parameters characterizing the model. For a linear compartmental model, these would be the rate constants $k_{i,j}$ and, for example, a volume term. Then for each time t_i for which there is a datum $y_{obs}(t_i)$, there is a model predicted value $y(\vec{\mathbf{p}}, t_i)$. There are two common, flexible error models that are used to calculate weights w_i associated with data:

$$w_i = \frac{1}{A + B \cdot y_{obs}(t_i)^C} \tag{14}$$

or

$$w_i = \frac{1}{A + B \cdot y(\vec{\mathbf{p}}, t_i)^C} \tag{15}$$

In these equations, A, B, and C are parameters that can, in principal, be estimated from the data. SAAM II currently supports both options, assuming A, B, and C are known. For example, if B is zero, one has constant weights.

There is a relationship between the weights assigned to a datum and the variance of the datum. One case, known as *absolute weights*, assumes that the weights are known. The relationship (for data as opposed to model weighting) is given by :

$$\text{var}(y_{obs}(t_i)) = \frac{1}{w_i} \tag{16}$$

An alternate case is *relative weights*. This assumes the error structure in the data is known up to a proportionality constant; this constant is estimated by the software. The advantage of this scheme is that SAAM II can deal with data sets that can differ over several orders of magnitude. In the example we have just discussed, there would be separate estimates of this proportionality constant, the variance factor of the data set, and for the plasma, red blood cell, urine, and organ data. In this case, the relationship between the variance and weight is given by:

$$\text{var}(y_{obs}(t_i)) = \frac{\upsilon_j}{w_i} \tag{17}$$

where υ_j is the variance factor for the *j*th data set. The default setting in SAAM II is relative weights based upon experimental data, and a constant coefficient of variation.

When the weights have been assigned to the data and a best fit has been achieved, one can then investigate the statistical output. An example is given in Figures 18 and 19. As noted above, one should be sure the estimated fractional standard deviations or coefficients of variation are "reasonable." We provided guidelines for this previously. Finally, one should check the correlation coefficients.

When one has obtained a best fit and is satisfied with the statistical estimates, and has addressed the question of *a priori* identifiability, then it is possible to test for goodness-of-fit and model order.

A common test for goodness-of-fit is the runs test. The runs test is based upon the residual or weighted residual where, if $\hat{\mathbf{p}}$ is the parameter vector producing the best fit to the data, then the *i*th residual or weighted residual is given respectively by

$$res(t_i) = y(\hat{\mathbf{p}},t_i) - y_{obs}(t_i) \tag{18}$$

or

$$wres(t_i) = w_i \cdot y(\hat{\mathbf{p}},t_i) - y_{obs}(t_i) \tag{19}$$

A run is a subset of sequential residuals having the same sign, plus or minus. The runs test is described in standard statistical text books (eg. Dixon, 1969 or Carson, 1983). It is also discussed in Cobelli et al. (this volume).

A final question deals with the number of compartments which are required to fit the data. Normally, one can determine the number by examining the statistics of a fit and performing a runs test. Sometimes, however, different model structures will give similar fits. The question is whether or not there is a test to determine if there is a "best" structure.

Tests exist that are based on the principal of parsimony (see Carson, 1983). These are the Akaike Information criterion (AIC) and the Schwarz criterion (SC). Suppose WRSS is the value for the weighted residual sum of squares following a model fit to a set of data. For the absolute weight case, the AIC and SC are respectively

$$\begin{aligned} AIC &= WRSS + 2P \\ SC &= WRSS + P \cdot ln(N) \end{aligned} \tag{20}$$

where P is the number of adjustable parameters and N is the number of data. For the relative weight case, these equations change to

$$\begin{aligned} AIC &= N \cdot ln(WRSS) + 2P \\ SC &= N \cdot ln(WRSS) + P \cdot ln(N) \end{aligned} \tag{21}$$

The AIC and SC can be calculated for any number of model structures, and that model with the lowest criterion is said to be the best.

For the example given here, the total number of adjustable parameters is 13, the number of data points is 111 (plasma 24, rbc 24, urine 7, organ 56), and the WRSS is 602. Because relative weighting has been used, (21) gives the proper formulas to calculate AIC and SC as 736 and 81, respectively. If a second model structure were proposed that had a different number of parameters, and if that model also gave a reasonable fit with reasonable statistical information on the parameters, one would calculate the AIC and SC for that model, and choose as the best model the one with the lower criterion. This is explained in more detail elsewhere (Carson, 1983).

The combination of tests for goodness-of-fit and model order can provide one with confidence that the model selected is the most appropriate for the specific set of experimental data.

CONCLUSION

The purpose of this chapter has been to discuss the development and testing of integrated multicompartmental models to describe a single-input multiple-output experiment, and to illustrate the application of modeling strategies using the SAAM II software system (SAAM II, 1997). A fundamental point is that by going through this process, one will actually learn the information content in the data set being analyzed, and thus extract the maximum amount of information, and at the same time gain a greater confidence in the model that is developed. Very often, one will design an experiment and then try to apply an existing model to analyze the data. This can be a dangerous approach as the prior model may often not fit the data for a variety of reasons, largely due to differences in the experimental design. While the original model can be taken as a starting point, one should always subject it to the same scrutiny we have described here.

In the first part of this chapter, we went through the steps to postulate a model which is compatible with a set of data, and summarized the main steps. The second part of the chapter dealt with issues related to model identifiability and assessing the goodness-of-fit. These are the steps that give one confidence in the validity of the model and its application.

We saw that the identifiability issue has two parts. *A priori* identifiability addresses the question as to whether the model's parameters can be estimated from the proposed experimental data. If the answer is no, then the experimental design must be changed. The other part is *a posteriori* identifiability. This relates to assessing the statistical information that is returned following optimization. If there are problems in the optimization process, i.e., some parameters are not well-determined, then there is information in the statistics that can help identify which parameters are causing the difficulties.

Finally, we discussed standard tests for goodness-of-fit and model order. These are the tests that give one confidence in the model and its order, i.e., the number of compartments and interconnections that are required.

The SAAM II software system has been designed and is constantly being upgraded to make all of these processes easier so that it can be a useful research tool as well as a tool for data analysis. We have described how the existing SAAM II tools can help make the modeling process easier. New tools are being designed to help in the model development and testing process.

ACKNOWLEDGEMENTS

This work was supported by a grant from NIH/NCRR/BTP RR-02176.

CORRESPONDING AUTHOR

Please address all correspondence to:
David M. Foster
Department of Bioengineering 352255
University of Washington
Seattle, WA 98195-2255

REFERENCES

Audoly S; D'Angio' L; Saccomani MP; Cobelli C. Global identifiability of linear compartmental models: A computer algebra algorithm. *IEEE Trans Biomed Engr,* 1998, 45(1):36-47.

Bates DM; Watts DG. *Nonlinear Regression and its Application.* John Wiley and Sons: New York. 1988.

Bell BM; Burke JV; Schumitzky A. A relative weighting method for estimating parameters and variances in multiple data sets. *Comput Stat and Data Anal,* (in press)

Carson ER; Cobelli C; Finkelstein L. *The Mathematical Modeling of Metabolic and Endocrine Systems.* John Wiley and Sons: New York. 1983.

Cobelli C; Foster DM. Compartmental models: Theory and practices using the SAAM II software system. (this volume)

Dixon WJ; Massey Jr FJ. *Introduction to Statistical Analysis.* McGraw-Hill: New York. 1969.

Foster DM; Aamodt RL; Henkin RI; Berman M. Zinc metabolism in humans: A kinetic model. *Am J Phys,* 1979, 237:R340-R349.
Foster DM; Boston RC. The use of computers in compartmental analysis: The SAAM and CONSAM programs, in*: Compartmental Distribution of Radiotracers.* Robertson JS; Ed. CRC Press: Boca Raton, FL. 1983.
Neter J; Wasserman W. *Applied Linear Statistical Models.* R. D. Irwin, Inc.: Homewood, IL. 1974.
SAAM II User Guide. SAAM Institute: Seattle, WA. 1997.

COMPARTMENTAL MODELS:
THEORY AND PRACTICE USING THE SAAM II SOFTWARE SYSTEM

Claudio Cobelli[1] and David M. Foster[2]

[1]Department of Electronics and Informatics
University of Padua
35131 Padua, Italy
[2]Department of Bioengineering
University of Washington
Seattle, WA 98195-2255

ABSTRACT

Understanding *in vivo* the functioning of metabolic systems at the whole-body or regional level requires one to make some assumptions on how the system works and to describe them mathematically, that is, to postulate a model of the system. Models of systems can have different characteristics depending on the properties of the system and the database available for their study; they can be deterministic or stochastic, dynamic or static, with lumped or distributed parameters. Metabolic systems are dynamic systems and we focus here on the most widely used class of dynamic (differential equation) models: compartmental models. This is a class of models for which the governing law is conservation of mass. It is a very attractive class to users because it formalizes physical intuition in a simple and reasonable way.

Compartmental models are lumped parameter models, in that the events in the system are described by a finite number of changing variables, and are thus described by ordinary differential equations. While stochastic compartment models can also be defined, we discuss here the deterministic versions—those that can work with exact relationships between model variables. These are the models most widely used in discussions of endocrinology and metabolism. In this chapter, we will discuss the theory of compartmental models, and then discuss how the SAAM II software system, a system designed specifically to aid in the development and testing of multicompartmental models, can be used.

INTRODUCTION

To understand the *in vivo* functioning of metabolic systems at the whole-body or regional levels requires one to make certain assumptions about how these systems work and to describe them mathematically, that is, to postulate a mathematical *model of the system.* Additionally, one often finds that there is a mismatch between the information content of the accessible pool measurements and the complexity of the model necessary to adequately describe the system.

To meet these challenges and to develop better models of the system, dynamic data are necessary. The tool of choice to generate such data is the tracer probe, which is normally a radioactive or a stable isotope. Why do tracers help to augment the information content of the accessible pool measurements? Tracer data contain information on the system because the tracer travels in the system the same way that the original substance, the tracee, does—thanks to the former's indistinguishability and negligible perturbation properties. Tracer data have a definite advantage with respect to tracee data. Because the tracer inputs are known—and controlled by the researcher—tracer data can be related to a simpler model of system. The inputs of the tracee model are more complex because endogenous productions are present. Thus, one can resolve first the tracer model of the system (which reflects exchange and disposal processes only) and subsequently use it, in conjunction with the accessible pool tracee measurements, to arrive at the tracee model of the system. This applies both to the steady and nonsteady state. For instance, if glucose fluxes in the body need to be quantitated in the basal steady state, one can obtain parameters such as endogenous glucose production (which should equal disposal because the system is in steady state), pool sizes, and fluxes in the body (Cobelli, 1984a; Ferrannini, 1985).

The nonsteady state situation is more difficult to tackle. For instance, if one wants to quantitate endogenous glucose production and glucose disposal after a meal or a test like an oral glucose tolerance test (OGTT), there is the need to use two different glucose tracers, one given orally and one intravenously (Taylor, 1996). Another interesting example of how a tracer can increase the information content of a clinical test is provided by the intravenous glucose tolerance test (IVGTT). When interpreted with the so-called minimal model of glucose kinetics (Bergman, 1979), this test is very informative and provides parameters like whole-body insulin sensitivity, which is the effect of insulin in stimulating glucose utilization in peripheral tissue and inhibiting its production. This model does not allow one to separately describe the insulin effect on the liver and on the peripheral tissues. However, if a tracer is added to the glucose bolus and its time course is measured in plasma, a richer model of the system can be resolved (Cobelli, 1986; Caumo, 1993; Vicini, 1997). This expanded model allows a segregation of the insulin effect on peripheral tissues from that on the liver.

At this point it is clear that, to quantitatively answer questions about both the accessible and nonaccessible system properties from the accessible pool, we need a sufficient model of the system and an adequate database. Ideally, the latter would be obtained by a suitably designed tracer experiment. Space considerations force us to concentrate here on the model development process only. We are not going to discuss issues related to tracer experiment design and measurement, such as sites for tracer input and measurement (Cobelli, 1980; Saccomani, 1992; Audoly, 1998), or how to express stable isotope measurements and test for their possible perturbation (Cobelli, 1992). The reader is urged to read about other treatments of these topics in this volume and in the references cited here.

System models can have different characteristics depending on the properties of the system in question and the database available for their study. Models can be deterministic or stochastic, dynamic or static, with lumped or distributed parameters. Metabolic systems are dynamic systems and we will focus here on the most widely used class of dynamic (differential equation) models, the compartmental models. This is a class of models for which the governing law is conservation of mass. They are very attractive to users because they formalize physical intuition in a simple and reasonable way. Compartmental models are lumped parameter models, in that the events in the system are described by a finite number of changing variables, and are thus described by ordinary differential equations. We will discuss here their deterministic version, although stochastic compartmental models can also be defined. Using exact relationships between variables, these models are by and large the form most widely used in endocrinology and metabolism studies.

Compartmental models have been widely employed to solve a broad spectrum of physiologic problems related to the distribution of materials in living systems. Their usefulness in research, diagnosis, and therapy has been demonstrated at whole-body, organ, and cellular levels. Examples and references can be found in books (Berman, 1982; Jacquez, 1985) and reviews (Berman, 1982; Carson, 1979; Cobelli, 1984b; Cobelli, 1987a; Cobelli, 1990; Radziuk, 1982). The purposes for which compartmental models have been developed include the following.

- *Identification of system structure.* Such models examine different hypotheses regarding the nature of specific physiologic mechanisms.
- *Estimation of unmeasurable quantities.* These quantities might include the estimation of internal parameters and other variables of physiologic interest.
- *Simulation of the intact system behavior* where ethical or technical reasons would not allow direct experimentation on the system itself.
- *Prediction and control of physiologic variables by the administration of therapeutic agents.* Such models might predict an optimal administration of a drug while maintaining one or more physiologic variables within desirable limits.
- *Optimization of the cost effectiveness of dynamic clinical tests.* These models would help to obtain the maximal information possible from the minimum number of blood samples withdrawn from a patient.
- *Diagnostic uses.* Such models could help augment quantitative information from laboratory tests and clinical symptoms, thus improving the reliability of diagnosis.
- *Teaching.* Models could aid in the teaching of many aspects of physiology, clinical medicine, and pharmacokinetics.

In this chapter, we will discuss the theory of compartmental models, and then discuss how the SAAM II software system, a system designed specifically to aid in the development and testing of multicompartmental models, can be used.

THE THEORY OF COMPARTMENTAL MODELS

Before discussing the theory of compartmental models, we first need to give some definitions. A *compartment* is an amount of material that acts as though it is well mixed and kinetically homogeneous. A *compartmental model* consists of a finite number of compartments with specified interconnections among them. These interconnections represent a flux of material which physiologically represents transport from one location to another or a chemical transformation or both. An example is shown in Figure 1. Control signals arising in endocrine-metabolic control systems can also be described. In this case, one can have two separate compartmental models: one for the hormone and one for the substrate, which interact via control signals. An example is shown in Figure 2.

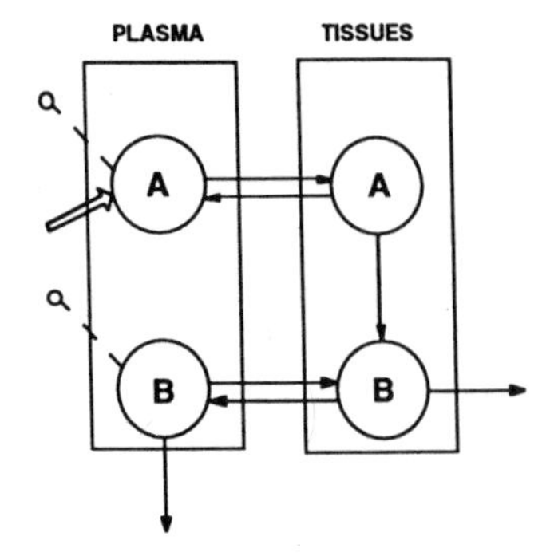

Figure 1. The compartmental system model showing the interconnections among compartments. The administration of material into and sampling from the accessible pools are indicated by the input arrow and measurement symbols (dotted line with bullet) respectively. The solid arrows represent the flux of material from one compartment to another.

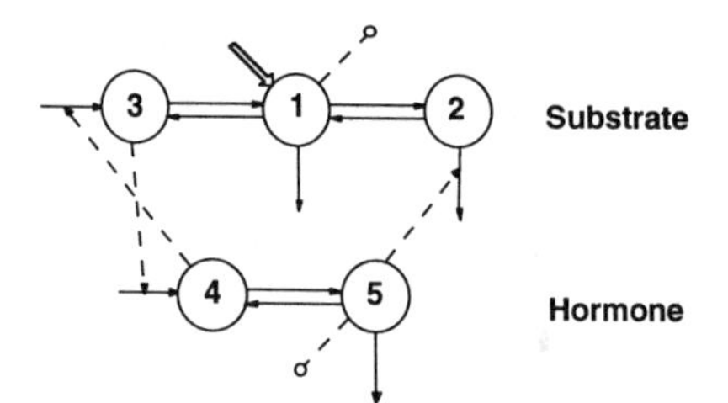

Figure 2. An example of a multicompartmental model of an endocrine-metabolic control system. The top and bottom multicompartmental models describe the metabolism of the substrate and hormone respectively. The dotted arrows represent control signals [examples are given in (8,10, and 11)]. For example, the dotted arrow from compartment 3 to the input arrow into compartment 4 indicates that the amount of material in compartment 3 controls the input of material into compartment 4.

Given our introductory definitions, it would be useful to discuss possible candidates for compartments before explaining what we mean by a compartment's *well-mixed and kinetic homogeneity*. Consider the notion of a compartment as a physical space. Plasma is a candidate for a compartment, a substance such as plasma glucose could be a compartment, zinc in bone could be a compartment also, as could insulin in β-cells. In some experiments, several different substances in plasma can be followed, such as glucose, lactate, and alanine. Thus, there can be more than one plasma compartment in the same experiment, one for each of the substances being studied. This notion extends beyond plasma. Glucose and glucose-6-phosphate might need to be shown by two different compartments, depending on whether they are found in liver or muscle tissue. Thus a single physical space or substance may actually be represented by more than one compartment, depending on the components measured or their location.

In addition, one must distinguish between compartments that are accessible and nonaccessible for measurement. Researchers often try to assign physical spaces to the nonaccessible compartments. This is a very difficult problem which is best addressed be the recognition that a compartment is actually a *theoretical* construct, one which may in fact combine material from several different *physical* spaces. To equate a compartment with a physical space depends upon the system under study and the assumptions made about the particular model.

With these notions of what might constitute a compartment in mind, it is easier to define the concepts of well-mixed and kinetic homogeneity. *Well-mixed* means that any two samples taken from a compartment at the same time would have the same concentration of the substance being studied and therefore be equally representative. Thus the concept of well-mixed relates to the uniformity of the information contained in a single compartment.

Kinetic homogeneity means that every particle in a compartment has the same probability of taking any pathways leaving the compartment. When a particle leaves a compartment, it does so because of metabolic events related to transport and utilization, and all particles in the compartment have the same probability of leaving due to one of these events.

This process of combining material with similar characteristics into collections that are homogeneous and behave identically is what allows one to reduce a complex physiologic system into a finite number of compartments and pathways. The number of compartments required depends both on the system being studied and on the richness of the experimental configuration. A compartmental model is clearly unique for each system studied, because it incorporates known and hypothesized physiology and biochemistry specific to that system. It provides the investigator with insights into the system's structure and is only as good as the assumptions that are incorporated into its structure.

The Tracee Model

In this section, we will discuss the definition of the tracee model using Figure 3. This is a typical compartment, the *i*th compartment. The tracee model can be formalized by defining precisely the flux of tracee material into and out from this compartment, and be establishing the measurement equation if this compartment is accessible for sampling. Once this is understood, the process of connecting several such compartments together into a multicompartmental model and writing the corresponding equations is easy.

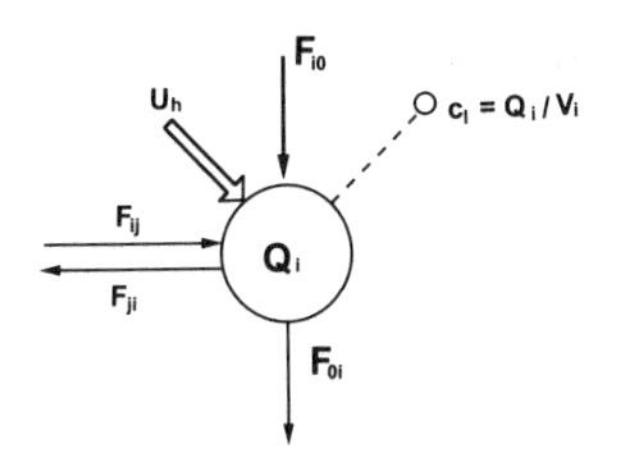

Figure 3. The *i*th compartment of an *n*-compartmental model showing fluxes into and out from the compartment, inputs, and measurements. See text for explanation.

Let Figure 3 represent the *i*th compartment of an *n*-compartment model of the tracee system with Q_i denoting the mass of the compartment. The arrows represent fluxes into and out of the compartment. The input flux into the compartment from outside the system, the *de novo* synthesis of material, is represented by F_{i0}; the flux to the environment and therefore out of the system by F_{0i}; the flux to and from compartment *j* by F_{ji} and F_{if}, respectively; and finally, U_h (h = 1, …, r) denotes an exogenous input. All fluxes F_{ij} (i = 0,1,…, n; j = 0,1,…, n; $i \neq j$)and masses Q_i (i = 1,2,…, n) are ≥ 0. The dashed arrow with a bullet indicates that the compartment is accessible to measurement. This measurement is denoted by C_l (l = 1, …, m) where we assume it is a concentration, $C_l = Q_i / V_i$ where V_i is the volume of compartment *i*. As already noted, usually only a small number of compartments are accessible to test inputs and measurements.

By using the mass balance principle, one can write for each compartment

$$\dot{Q}_i(t) = -\sum_{\substack{j=0 \\ j\neq i}}^{n} F_{ji}(Q_1(t),\cdots,Q_n(t)) + \sum_{\substack{j=1 \\ j\neq i}}^{n} F_{ji}(Q_1(t),\cdots,Q_n(t)) + F_{i0}(Q_1(t),\cdots,Q_n(t)) + U_h(t) \tag{1}$$

$$C_l(t) = \frac{Q_1(t)}{V_i} \qquad Q_i(0) = Q_{i0}$$

where $\dot{Q}_i(t) = \frac{dQ_i(t)}{dt}$ and $t > 0$ is time, the independent variable. All the fluxes F_{ij}, F_{i0}, and F_{0i} are assumed to be functions of the compartmental masses Q_i.

If one writes the generic flux F_{ji} ($j = 0,1,\ldots,n$; $i = 1,2,\ldots,n$; $j \neq i$) as

$$F_{ji}(Q_1(t),\cdots,Q_n(t)) = k_{ji}(Q_1(t),\cdots,Q_n(t))(Q_i(t) \tag{2}$$

where $k_{ji}(\geq 0)$ denotes the fractional transfer coefficient between compartment i and j, (1) can be rewritten as

$$\begin{aligned}\dot{Q}_i(t) &= -\sum_{\substack{j=0\\ j\neq i}}^{n} k_{ji}(Q_1(t),\cdots,Q_n(t))Q_i(t) + \sum_{\substack{j=1\\ j\neq i}}^{n} k_{ji}(Q_1(t),\cdots,Q_n(t))Q_j(t)\\ &\quad + F_{i0}(Q_1(t),\cdots,Q_n(t)) + U_h(t)\\ C_l(t) &= \frac{Q_1(t)}{V_i} \qquad Q_i(0) = Q_{i0}\end{aligned} \tag{3}$$

Equation (3) describes the *nonlinear* compartmental model of the tracee system.

To make the model operative, one has to specify how the k_{ij} and F_{i0} depend upon the Q_i. This obviously depends upon the system being studied. Usually the k_{ij} and F_{i0} are functions of one or a few of the Q_i. Some possible examples include the following.

- k_{ij} are constant, and thus do not depend upon any Q_i.

$$k_{ij}(Q_1(t),\cdots,Q_n(t)) = k_{ij} = \text{constant} \tag{4}$$

- k_{ij} are described by a saturative relationship such as Michaelis-Menten.

$$k_{ij}(Q_j(t)) = \frac{V_M}{K_m + Q_j(t)} \tag{5}$$

or the Hill equation

$$k_{ij}(Q_j(t)) = \frac{V_M Q_j^{m-1}(t)}{K_m + Q_j^m(t)} \tag{6}$$

Note that when $m = 1$ in the above, (6) becomes (5).

- k_{ij} is controlled by the arrival compartment, such as by a Langmuir relationship.

$$k_{ij}(Q_i(t)) = \alpha(1 - \frac{Q_i(t)}{\beta}) \tag{7}$$

- k_{ij} is controlled by a remote compartment different from the source (Q_j) or arrival (Q_i) compartments. For example, using the model shown in Figure 2, one could have

$$k_{02}(Q_5(t)) = \gamma + Q_5(t) \tag{8}$$

or a more complex description such as

$$k_{02}(Q_2(t), Q_5(t)) = \frac{V_m(Q_5(t))}{K_m(Q_5(t)) + Q_2(t)} \tag{9}$$

where now one has to further specify how V_m and K_m depend on the controlling compartment, Q_5.

The input F_{i0} can also be controlled by remote compartments. For example for the model shown in Figure 2, one can have

$$F_{30}(Q_4(t)) = \frac{\delta}{\varepsilon + Q_4(t)} \tag{10}$$

$$F_{40}(Q_3(t)) = \eta + \lambda Q_3(t) + \mu \dot{Q}_3(t) \tag{11}$$

The nonlinear compartmental model given in (3) permits the description of an endocrine-metabolic system in nonsteady state under very general circumstances. Having specified the number of compartments and the functional dependencies, there is now the problem of assigning a numerical value to the unknown parameters which describe them. Some of them may be assumed to be known, but some need to be tuned to the particular subject studied. In terms of model construction, it is now time to turn to parameter estimation, that is, how we must use our observations on the system to fit the model properties to those of the system. Often, however, the data are not enough to arrive at the unknown parameters of the model and a tracer is employed to enhance the information content of the data.

The Tracer Model

In this section, we will formalize the definition of the tracer model using Figure 4. This parallels exactly the notions introduced above, except now we follow the tracer, denoted by lowercase letters, instead of the tracee. The link between the two, the tracee and tracer models, is given in the following section.

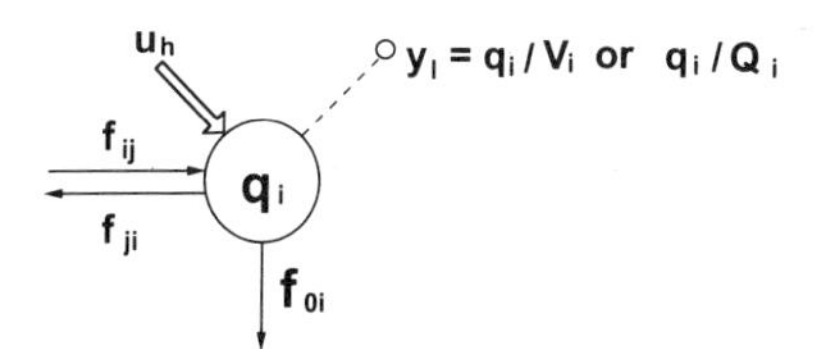

Figure 4. The *i*th compartment of an *n*-compartmental tracer model showing fluxes into and out from the compartment, inputs, and measurements. See text for explanation.

Suppose an isotopic (radioactive or stable) tracer is injected (denoted by u_h) into the *i*th compartment and denote $q_i(t)$ its tracer mass at time t (Figure 4). Assuming an ideal tracer, the tracer-tracee indistinguishability ensures that the tracee rate constants k_{ij} also apply to the tracer. Again as with the tracee, the measurement is usually a concentration; for a radioactive tracer, it can be written $y_l(t) = q_i(t)/V_i$. For the stable isotope tracer, the measurement is the tracer-to-tracee ratio, $y_l(t) = q_i(t)/Q_i(t)$.

The tracer model, given the tracee model (3), is

$$\dot{q}_i(t) = -\sum_{\substack{j=0 \\ j\neq 1}}^{n} k_{ji}(Q_1(t),\cdots,Q_n(t)q_i(t) + \sum_{\substack{j=0 \\ j\neq 1}}^{n} k_{ij}(Q_1(t),\cdots,Q_n(t)q_j(t) + u_h(t)$$

$$q_i(0) = 0 \qquad y_l(t) = \frac{q_i(t)}{V_i} \quad \text{or} \quad y_l(t) = \frac{q_i(t)}{Q_i(t)} \tag{12}$$

Note that the endogenous production term F_{i0} in (3) does not appear in (12); this is because this term applies only to the tracee.

Linking the Tracer and Tracee Models: The Tracer-Tracee Model

The model of system necessary to arrive at the nonaccessible system properties is obtained by linking the tracee and tracer models to form the tracer-tracee model; this model is described by (3 and 12). The problem one wishes to solve is how to use the tracee data $C_l(t)$ and the tracer data $y_l(t)$ to obtain the unknown parameters of the model. This is accomplished by using the parameter estimation principle described below.

In the general setting, the problem is complex. Before going into the details of this problem, it is useful to consider an important special situation, the tracee steady-state case. This situation is the experimental protocol most frequently encountered in nutritional studies.

The Tracee Steady State

If the tracee is in a constant steady state, the exogenous input U_h is zero, all the fluxes F_{ij} and masses Q_i (t) in the tracee model (1) are constant, and the derivatives $\dot{Q}_i(t)$ are zero. As a result, all the fractional transfer coefficients k_{ij} [see (2)] are constant.

The tracee and tracer models given in (3 and 12), respectively, thus become

$$0 = -\sum_{\substack{j=0 \\ j\neq i}}^{n} k_{ji}Q_i + \sum_{\substack{j=1 \\ j\neq i}}^{n} k_{ij}Q_j + F_{i0} \qquad Q_i(0) = Q_{i0} \qquad C_l = \frac{Q_i}{V_i} \tag{13}$$

$$\dot{q}_i(t) = -\sum_{\substack{j=1 \\ j\neq i}}^{n} k_{ji}q_i(t) + \sum_{\substack{j=1 \\ j\neq i}}^{n} k_{ij}q_j(t) + u_h(t) \qquad q_i(0) = 0$$

$$y_l(t) = \frac{q_i(t)}{V_i} \quad \text{or} \quad y_l(t) = \frac{q_i(t)}{Q_i} \tag{14}$$

This is an important result: the tracer compartmental model is linear if the tracee is in a constant steady state, irrespective of whether it is linear or nonlinear. The modeling machinery for (13 and 14) is greatly simplified with respect to the nonlinear models shown in (3 and 12). The strategy is to use the tracer data to arrive at the k_{ij} and the accessible pool volume V_i or mass Q_i of (14), and subsequently use the steady-state tracee model of (13) to solve for the unknown parameters F_{i0} and the remaining Q_i.

As one can expect, there is a price to be paid for this simpler modeling machinery. While it gives a mechanistic picture of the steady-state operating point, it cannot describe the tracee system in its full nonlinear operation, including saturation kinetics and control signals. To make this point clear, it is worth considering the simple example shown in

Figure 5. The tracee system is in steady state but it is nonlinear: both its production and utilization are nonlinearly controlled by Q_1. An ideal, (stable isotope) tracer experiment can provide information on the parameters k_{01} (a constant) and Q_1, and thus utilization and production $F_{01} = k_{01}Q_1 = F_{10}$. Even this experiment will not allow us to arrive at α, β, V_m, or K_m. This simple example was deliberately chosen, but one can readily extend the reasoning to real endocrine-metabolic system models such as the one in Figure 2.

TRACEE MODEL

$C_1 = Q_1/V_1$

$F_{10} = \dfrac{\alpha}{\beta + Q_1}$ Q_1 $F_{01} = \dfrac{V_M}{K_m + Q_1} Q_1$

TRACER MODEL

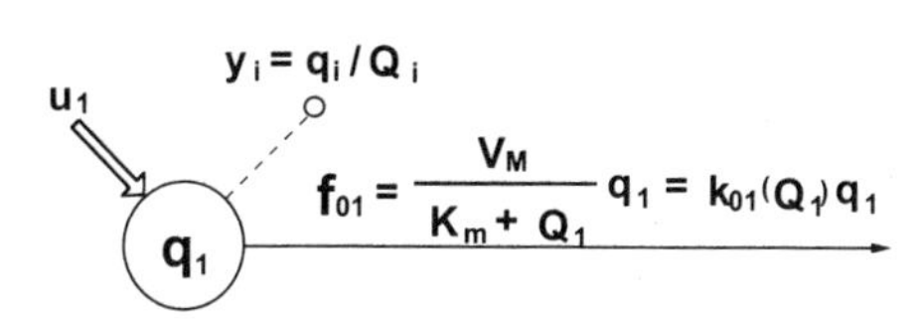

Figure 5. An example of a system where the tracee is in steady state but nonlinear. See text for explanation.

To describe the intimate behavior of a tracee system, there is thus the need to have more information than just a single steady-state tracer experiment. One possibility is obviously to use the more demanding nonsteady-state models of (3 and 12) discussed previously. An appealing alternative is to use multiple steady-state tracer modeling, that is, to repeat the tracer experiment in various steady states and use linear tracer modeling to describe the system in each. In this way the information we can obtain increases—the k_{ij} for each of the various steady states become available, for example—and we can arrive at the parameters describing the functional dependencies of the tracee model.

MODEL IDENTIFICATION

Introduction

With the tracer-tracee model described by (3 and 12) or (13 and 14) in steady-state conditions), we can now proceed to model identification, the process by which we can arrive at a numerical value of the unknown model parameters from the measurements. The measurement configuration for the tracee and tracer models has been described through the variables C_l and y_l in (3 and 12) respectively, or (13 and 14). However, real data are noisy, or affected by errors, and are usually collected at discrete time instants $t_1,\ldots,t_N$.

In this chapter, we will assume that measurement error is additive and the tracee and tracer actual measurements are thus described at sample time t_j respectively by:

$$C_l^{obs}(t_j) = C_l(t_j) + E_l(t_j) \tag{15}$$

$$y_l^{obs}(t_j) = y_l(t_j) + e_l(t_j) \tag{16}$$

where E_l and e_l are the measurement error of tracee and tracer respectively. The errors are usually given a probabilistic description, in that they are assumed to be independent and often Gaussian. With (15 and 16), and the model equations (3 and 12) or (13 and 14), the compartmental *model identification* problem can now be defined: we can begin to estimate the unknown model parameters from the noisy data contained in C_l^{obs} and y_l^{obs}.

Before solving this problem, however, we must deal with a prerequisite issue for the well-posedness of our parameter estimations. This is the issue of *a priori* identifiability. As seen in the next section, this requires reasoning which uses ideal noise-free data, such as that from (3 and 12) or (13 and 14).

A Priori Identifiability

The following question arises before performing the experiment to collect the data to be analyzed using the model or, if the experiment has already been completed, before using the model to estimate the unknown parameters from the data: Do the data contain enough information to estimate all the unknown parameters of the postulated model structure? This question is usually referred to as the *identifiability* problem. It is set in the ideal context of an error-free model structure and noise-free (*a priori*) measurements, and is an obvious prerequisite for determining if later parameter estimation from real data is well-posed. If the postulated model structure is too complex for the particular set of ideal data—if some model parameters are not identifiable from the data, for example—it is highly unlikely that parameters could accurately be identified in a real situation when there is error in the model structure and noise in the data. *A priori* identifiability is also crucial in experimental design (Cobelli, 1980) and helps to establish the input-output configuration necessary to ensure optimal estimation of the unknown parameters. Particular questions, such as which is the minimal input-output configuration, can thus be answered (Saccomani, 1993).

A priori identifiability thus examines whether, given ideal noise-free data and assuming an error-free compartmental model structure as described by (3 and 12) or (13 and 14), it is possible to make unique estimates of all the unknown model parameters. A model can be *uniquely* (globally) identifiable or *nonuniquely* (locally) identifiable (one or more of the parameters has more than one but a finite number of possible values) or nonidentifiable (one or more of the parameters has an infinite number of solutions).

A simple example of the linear two-compartment tracer model shown in Figure 6 can serve to better focus the problem. The upper model is *a priori* uniquely identifiable from the ideal data y_1, in that V_i, k_{21}, k_{12}, and k_{01} can be uniquely estimated, while the lower model is nonidentifiable (Carson et al., 1983; Jacquez, 1985). Here V_1 can be uniquely estimated but the rate constants k_{21}, k_{12}, k_{01}, and k_{02} have an infinite number of solutions. This simply means that one cannot extract too much from the data, because the model is too complex for the designed experiment. Clearly, there is no way that the actual measurements y_1^{obs} which are affected by errors and are finite in number, can improve the identifiability properties of this model. Knowing in advance the identifiability properties of a model is thus an obvious prerequisite to its success.

The identifiability problem is in general a large, nonlinear, algebraic one, and thus difficult to solve because one would need to solve a system of nonlinear algebraic equations which are increasing in the number of their terms and nonlinearity degree with the number of compartments in the model (the model order). Various methods for testing identifiability of linear and nonlinear compartmental models are available (Cobelli, 1980; Carson, 1983; Jacquez, 1985; Godfrey, 1987). For linear compartmental models, various specific methods based have been developed, such as those based on the transfer function. Explicit identifiability results on catenary and mamillary compartmental models (Cobelli,

1979) and on two- (Carson, 1983) and three-compartmental models (Norton, 1982) are also available. A parameter-bound strategy for dealing with nonidentifiable compartmental models has also been developed (DiStefano, 1983). The problem is more difficult for nonlinear compartmental models. For small models, the output series expansion method (Pohjanpalo, 1978) is usually employed.

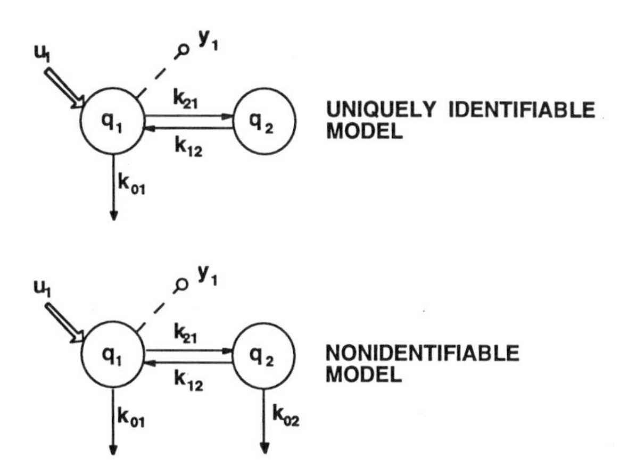

Figure 6. Two different two-compartmental models. The upper model is uniquely identifiable while the lower model is not. See text for explanation.

However, all the proposed methods apply to models of relatively low dimension; when applied to large models, these methods involve nonlinear algebraic equations too difficult to be solved even by resorting to symbolic algebraic manipulative languages such as Reduce or Maple. These difficulties have stimulated new approaches for the study of global identifiability based on computer algebra. In particular, nonlinear model approaches based on differential algebra have been investigated (D'Angio', 1994; Ljung, 1994), while for linear compartmental models, a method based on the transfer function and Gröbner basis algorithm has been proposed (Audoly, 1998).

If a model is *a priori* uniquely or nonuniquely identifiable, then identification techniques such as least squares can be used to estimate the numerical values of the unknown parameters from the noisy data. If a model is *a priori* nonidentifiable, then not all its parameters can be estimated using identification techniques. Various alternate strategies can be used, such as the derivation of bounds for nonidentifiable parameters, model reparametrization (parameter aggregation), the incorporation of additional knowledge, or the design of a more informative experiment. An example on the use of some of these approaches for dealing with a nonidentifiable model of glucose kinetics is given by Cobelli (1987b).

From the above considerations, it follows that *a priori* unique identifiability is a prerequisite for well-posedness of parameter estimation and for the reconstructability of stated variables in nonaccessible compartments. It is a necessary step, that, because of the ideal context where it is posed, does not guarantee a successful estimation of model parameters from real input-output data.

Parameter Estimation

Model identification deals with estimating the values of the unknown parameters of the model (3 and 13), or (13 and 14) from the set of noisy real data as described in (15 and 16). This is the parameter estimation problem, and in what follows, we will discuss some of its underlying principles.

Basically one needs a method for providing a set of numerical estimates of the parameters which best fit the data, such as $\hat{k}_{ij}$, and a measure of their precision, such as their standard deviation, $SD(\hat{k}_{ij})$ or their percent fractional standard deviation or coefficient of variation $CV(\hat{k}_{ij}) = SD(\hat{k}_{ij})/(\hat{k}_{ij}) \cdot 100$. The parameter estimation problem for both linear and nonlinear compartmental models is nonlinear in that parameters do not appear in the model linearly as a straight line or a polynomial model. This makes it a difficult problem. For details on parameter estimation of physiologic system models, we refer to Carson (1983).

Weighted nonlinear least squares is the most commonly used parameter estimation technique. To better grasp the method, consider the linear compartmental model of (13 and 14), and suppose that we want to estimate its parameters from a set of radioactive noisy data [as in (16)]. Weighted nonlinear least squares defines the weighted residual sum of squares (*WRSS*) as a cost function:

$$WRSS = \sum_{j=1}^{N} \frac{1}{w_l(t_j)} (y_l^{obs}(t_j) - y_l(t_j, k_{ij}, V_i))^2 \tag{17}$$

and the desired parameter estimates, $\hat{k}_{ij}$ and $\hat{V}_i$, are those which minimize the *WRSS*.

As already pointed out, models are nonlinear in their parameters and thus there is no closed form analytical solution to the *WRSS* minimization, as is the case with linear parameter problems such as a linear regression for estimating a straight line or polynomial parameters. This requires an iterative solution to the *WRSS* minimization. Iteration can be thought of as a series of steps where the nonlinear problem is handled in a linear fashion at each step in such a way that the *WRSS* is minimized in passing from one step to next. The interested reader can refer to Carson (1983) for more details. The process is illustrated schematically in Figure 7.

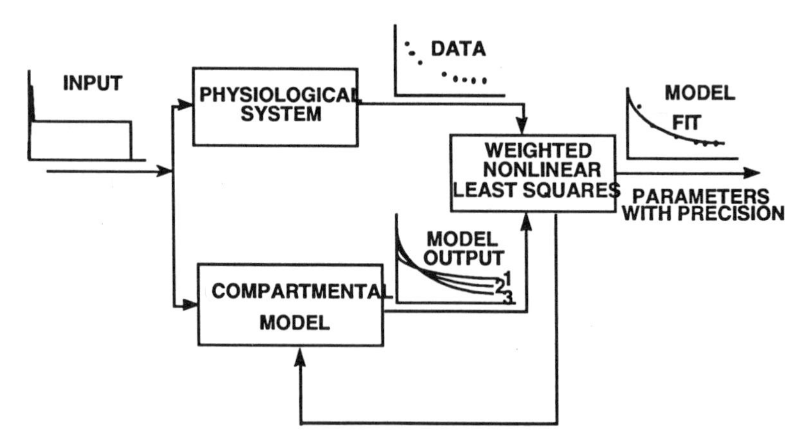

Figure 7. A schematic of the *WRSS* minimization. The goal in iterated weighted least squares is to obtain a best fit to the data. At each iteration step, a model output is generated, as indicated in the figure, and compared with the data. Model outputs 1, 2, and 3 could be the results of three successive iterations. See text for explanation.

Various robust techniques are available to carry out this iterative process (Carson, 1983). Some are based on successive linearizations of the model, so that at each iteration a linear least square problem needs to be solved. It is also clear that to start the iterative scheme with the initial model linearization, one needs an initial value for the unknown parameters and this adds to the difficulties of the problem.

Another important but critical ingredient of the weighted nonlinear least squares process is the choice of the weights [the $w_i(t_j)$ in (17)]. In fact, in order to have a correct summary of the statistical properties of the estimates, the weights should be chosen optimally, such that they are equal to the inverse of measurement error variances if they are known, or of their relative values, if variances are known up to a proportionality constant (Carson, 1983). Under these circumstances, one can obtain an approximation of the covariance matrix of the parameter estimates which can then be used to evaluate their precision, or their *a posteriori* or *numerical* identifiability. The diagonal of the covariance matrix contains the variances of the parameter estimates. Precision is usually expressed in terms of percent fractional standard deviation or coefficient of variation. For instance, as noted previously, if $\hat{k}_{ij}$ denotes the estimate for k_{ij}, one calculates the percent fractional standard deviation or coefficient of variation as

$$\frac{\sqrt{\operatorname{var}(\hat{k}_{ij})}}{\hat{k}_{ij}} \cdot 100 = \frac{SD(\hat{k}_{ij})}{\hat{k}_{ij}} \cdot 100 \tag{18}$$

A scheme which summarizes the various ingredients of weighted nonlinear least squares from the user's point of view is shown in Figure 8. One has obviously to supply not only the data and the model, but also the measurement error of the data which will be used to assign the weights and an initial estimate for each of the unknown parameters. From the weighted nonlinear least squares, one can obtain the model fit to the data and the residuals (the difference between the datum and the model predicted value at specific sample times), as well as the estimated parameter values and their precision.

Assessing Model Quality

The quantitative assessment of model quality from parameter estimation is crucial as it is also an essential component of model validation. One has first to examine the quality of model predictions to observed data. In addition to visual inspection, various statistical tests on weighted residuals, such as the weighted difference between data and model prediction, are available to check for the presence of model misfitting. In fact, because the sequence of the residuals can be viewed as an approximation of the measurement error sequence, one would expect them to be approximately independent; any nonrandomness in the residuals would indicate that the model is too simple for the data, and that one may need more compartments than those postulated.

Clearly if the model does not fit the data there is a reason: it is too simple. But how does one know whether the model is too complex, and that the system is *over*modeled instead of *under*modeled? Suppose for instance, one has some bolus decay tracer data obtained in a steady-state system. A linear compartmental model would be appropriate and one can model the data with one-, two-, three-, or four-compartment models. How does one know the right model order? Residuals and indices like the *WRSS* are not of help because they will always improve as model order increases. One has to look instead at the covariance matrix of parameter estimates: precision of parameter estimates will clearly worsen as model order increases, and parameters estimated with poor precision are a symptom of a model too complex for the data.

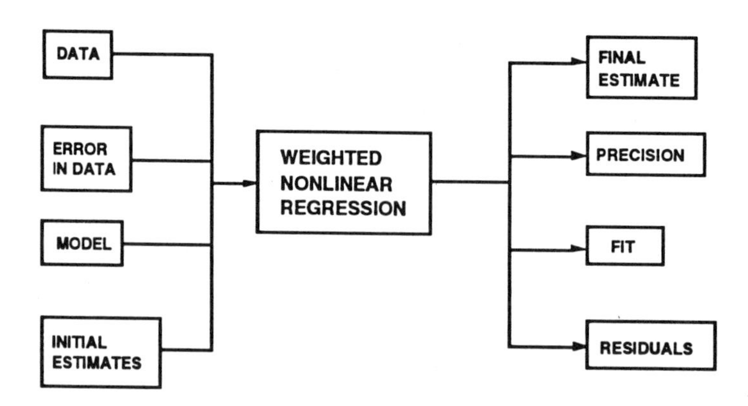

Figure 8. The ingredients of weighted nonlinear least squares from the user point of view showing input and output.

One speaks of parsimonious modeling in terms of finding a happy compromise between good model fit to the data and a high precision of the parameter estimates. For linear compartmental models there are guides as the F-test, and the Akaike and the Schwarz criteria, which can be used if measurement errors are independent and Gaussian. For instance, for the case of known measurement error and optimally chosen weights, the Akaike (AIC) and Schwarz (SC) criteria are given by Carson (1983) as:

$$AIC = WRSS + 2P \tag{19}$$

$$SC = WRSS + P\ln N \tag{20}$$

where P is the number of unknown model parameters and N is the number of data. One would choose as the most parsimonious model the one which has the smallest value of AIC or SC.

Model Validation

It is not difficult to build models of systems—the difficulty lies in making them accurate and reliable in answering the question asked. For the model to be useful, one has to have confidence in the results and the predictions that are inferred from it. Such confidence can be obtained by *model validation*. Validation involves the assessment of whether a compartmental model is adequate for its purpose. This is a difficult and highly subjective task when modeling physiologic systems, because intuition and an understanding of the system, among other factors, play an important role in this assessment. It is also difficult to formalize related issues such as model credibility, or the use of the model outside its established validity range. Some efforts have been made, however, to provide some formal aids for assessing the value of models of physiologic systems. Validity criteria and validation strategies for models of physiological systems are available (Cobelli, 1984c) which take into account both the complexity of model structure and the extent of available experimental data. Of particular importance is the ability to validate a model-based measurement of a system parameter by an independent experimental technique. A model that is valid is not necessarily a true one; all models have a limited domain of validity and it is hazardous to use a model outside the area for which it has been validated.

However, a valid model can be used not only for measuring nonaccessible parameters but also for simulation.

Suppose one wishes to see how the system behaves under certain stimuli, but it is inappropriate or impossible to carry out the required experiment. If a valid model of the system is available, one can perform an experiment on the model by using a computer to see how the system would have reacted. This is called *simulation*. Simulation is thus an inexpensive and safe way to experiment with the system. Clearly, the value of the simulation results depend completely on the quality or the validity of the model of the system.

SOFTWARE FOR COMPARTMENTAL MODELING

There are a number of software packages that are used in the development and testing of multicompartmental models. While it is not our intent to list all, those frequently used in nutritional studies include the SAAM/CONSAM system (Boston, this volume), ACSL (Johnson, this volume), and SAAM II (SAAM Institute, 1997).

The SAAM II software system is a totally reengineered software system based upon Berman's SAAM (Simulation, Analysis, and Modeling) (Boston, this volume). The original SAAM was a powerful research tool to aid in the design of experiment and the analysis of data. Its dictionary permitted compartmental models to be developed without formally specifying the differential equations. However, SAAM was developed before the era of modern computer software design and implementation techniques, and as a result, is not user-friendly. In addition, the computational kernel was not well documented, which made validation, maintenance, and further enhancements difficult.

It was the recognition of these problems that led to the development of SAAM II. SAAM II deals easily with compartmental and noncompartmental models in order to help researchers create models, design and simulate experiments, and analyze data quickly, easily, and accurately. Compartmental models are constructed graphically using drag and drop icons. The models can be either linear or nonlinear.

SAAM II creates a system of ordinary differential equations from the model and experiment created using the graphical user interface. Powerful integrators and a new optimization technique developed specifically for the software (Bell, 1996) are employed. The statistical information it provides following a successful fit helps the user to assess the *a posteriori* identifiability, goodness-of-fit, and model order. Examples are provided later in this chapter. When computational problems are detected either in the integrator or optimizer, SAAM II displays useful warnings or error messages.

It is important to point out that the objective (cost) function used in SAAM II is not the *WRSS* as given in (17), but a modification of that function which permits different weighting schemes and deals more accurately with situations where a set of experimental data might include subsets (from different measurement sites, for example) whose magnitude may be significantly different. The interested reader is referred for more details (SAAM Institute, 1997; Bell, 1996).

EXAMPLES

Example 1

To illustrate the points we have been making, consider first the following set of data which will be described by a sum of exponentials. In our previous discussions, we have focused on compartmental models. However, when one is starting "from scratch," it is often wise to fit the data to a sum of exponentials, because this gives a clue as to how many compartments will be required in the model.

Table 1. Plasma data from a tracer experiment.

Time	Plasma	σ	Time	Plasma	σ
2	3993.50	99.87	28	2252.00	65.04
4	3316.50	86.33	31	2169.50	63.39
5	3409.50	88.19	34	2128.50	62.57
6	3177.50	83.55	37	2085.00	61.70
7	3218.50	84.37	40	2004.00	60.08
8	3145.00	82.90	50	1879.00	57.58
9	3105.00	82.10	60	1670.00	53.40
10	3117.00	82.34	70	1416.50	48.33
11	2984.50	79.69	80	1333.50	46.67
13	2890.00	77.80	90	1152.00	43.04
14	2692.00	73.84	100	1080.50	41.61
15	2603.00	72.06	110	1043.00	40.86
17	2533.50	70.67	120	883.50	37.67
19	2536.00	70.72	130	832.50	36.65
21	2545.50	70.91	140	776.00	35.52
23	2374.00	67.48	150	707.00	34.14
25	2379.00	67.58			

Consider the data given in Table 1; these data are radioactive tracer glucose concentrations measured in plasma following an injection of tracer at time zero. The time measurements are minutes, and the plasma measurements are dpm/mL. The experiment was performed in a normal subject in the basal state (Cobelli, 1984a). In order to select the order of the multi-exponential model which is best able to describe these data, one-, two- and three-exponential models can be considered:

$$\begin{aligned} y(t) &= A_1 e^{-\lambda_1 t} \\ y(t) &= A_1 e^{-\lambda_1 t} + A_2 e^{-\lambda_2 t} \\ y(t) &= A_1 e^{-\lambda_1 t} + A_2 e^{-\lambda_2 t} + A_3 e^{-\lambda_3 t} \end{aligned} \tag{21}$$

The measurement error is assumed, as discussed above, to be additive:

$$y^{obs}(t_i) = y(t_i) + e(t_i)$$

where the errors $e(t_i)$ are assumed to be independent, Gaussian with a mean of zero, and an experimentally-determined standard deviation of:

$$SD(e(t_i)) = 0.02 y^{obs}(t_i) + 20$$

These values are shown associated with each datum in Table 1. The three models are to be fitted to the data by applying weighted nonlinear regression with the weights chosen equal to the inverse of the variance. The plots of the data and the model predictions together with the corresponding weighted residuals are shown in Figure 9; the model parameters are given in Table 2.

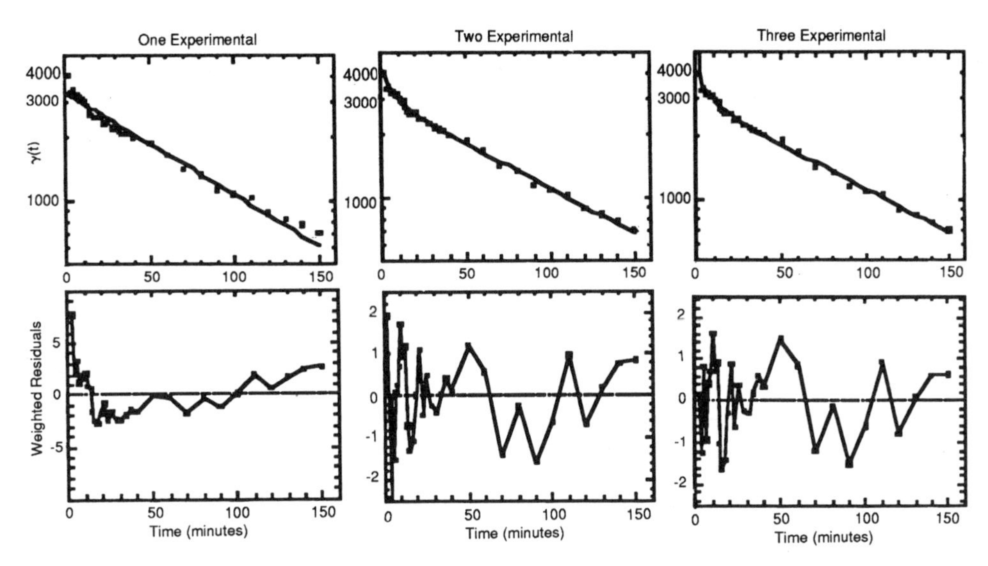

Figure 9. The best fit of the data given in Table 1 to a single exponential model and a sum of two- and three-exponential models, together with a plot of the weighted residuals for each case. The coefficients and exponentials (eigenvalues) for each are given in Table 2.

Examining Table 2, we can see that all parameters can be estimated with acceptable precision in the one- and two-exponential models, while some parameters of the three-exponential model have very high values for the fractional standard deviation. This means that the three-exponential model cannot be resolved with precision from the data. In fact, the first exponential is so very rapid, $\lambda_1 = 4.6\ \text{min}^1$, it has practically vanished by the time of the first available datum at 2 minutes. The other two exponential terms have values similar to those obtained for the two-exponential model. In addition, the final estimates of A_1 and λ_1 are also dependent upon the initial estimates; that is, starting from different initial points in parameter space, the nonlinear regression procedure yields different final estimates while producing similar values of the *WRSS*. Therefore the three-exponential model is not identifiable *a posteriori*, and can be rejected at this stage.

One can now compare fit of the one- and two-exponential models. Nonrandomness of the residuals for the one-exponential model is evident because the plot reveals long runs of consecutive residuals of the same sign. The run test allows one to check the independence formally, and from the values of Z, one can conclude that the residuals of the two-exponential model is consistent with the hypothesis of independence because the Z value lies within the 5% region of acceptance (-1.96,1.96), or equivalently the P-value is high. Conversely, the Z value for the one-exponential model indicates that the hypothesis of independence is to be rejected, with a P value less than 0.5.

Most residuals for the two-exponential model lie between -1 and 1, which indicates they are compatible with the assumptions on the variance of the measurement error. On the other hand, only a few of the residuals of the one-exponential model fall in this range. To test formally if the weighted residuals have unit variance, as expected if the model and/or assumptions on the variance of the measurement error are correct, the X^2-test can be applied. For the one-exponential model, the degrees of freedom $df = N - P = 31$, and for the level of significance equal to 5%, the region of acceptance is (16.8,47.0). Because the *WRSS* is greater than the upper bound 47.0, the assumption of unit variance of the residuals has to be rejected with a P-value less than 0.5%. For the two-exponential model, the P-values are higher, and the *WRSS* lies within the 5% region (16.05,45.72), indicating that the residuals are consistent with the unit variance assumption. The *WRSS* decreases, as expected, when the number of parameters in the model increases.

The *F* test indicates that the two-exponential model reduces the *WRSS* significantly when compared with the one-exponential model because the *F* value is greater than F_{max} = 3.33 (evaluated for a 5% level of significance from the $F_{92,29}$ distribution). Similar conclusions can be derived from the *AIC* and *SC*, which assume their lower values for the two-exponential model.

Table 2. Precision of the parameter estimates expressed as percent fractional standard deviation is shown in parentheses.

		1 Exponential	2 Exponentials	3 Exponentials
A_1		3288 (1%)	1202 (10%)	72486 (535789%)
λ_1		0.0111 (1%)	0.1383 (17%)	4.5540 (7131%)
A_2			2950 (2%)	1195 (14%)
λ_2			0.0098 (3%)	0.1290 (22%)
A_3				2925 (2%)
λ_3				0.0097 (3%)
Run test:	Z value	-5.13	-1.51	
	5% region	[-1.96,1.96]	[-1.96,1.96]	
	P value	< 0.5%	> 6%	
χ^2 test	WRSS	167.10	32.98	
	5% region	[16.8,47.0]	[16.0,45.7]	
	P value	< 0.5%	>20%	
F test	F ratio	2 vs 1:	29,59	
		5% region	[0,3.33]	
		P value	<0.5%	
AIC		171.10	40.98	
SC		174.09	46.97	

Example 2

Example 1 was a glucose turnover study in humans in which 4.3037x10^7 dpm of labeled glucose was injected as a bolus into an individual weighing 50.85 kg, with the measured steady-state glucose concentration in plasma of 100 mg/mL. It was found in the two-exponential model that the following provided the best fit to the data.

$$y(t) = 1202e^{-0.138t} + 2950e^{-0.0098t} \tag{24}$$

These data will be fitted using the two-compartment models shown below in Figure 10. In these models, compartment 1 is assumed to be plasma and compartment 2 an extravascular compartment which equilibrates with plasma.

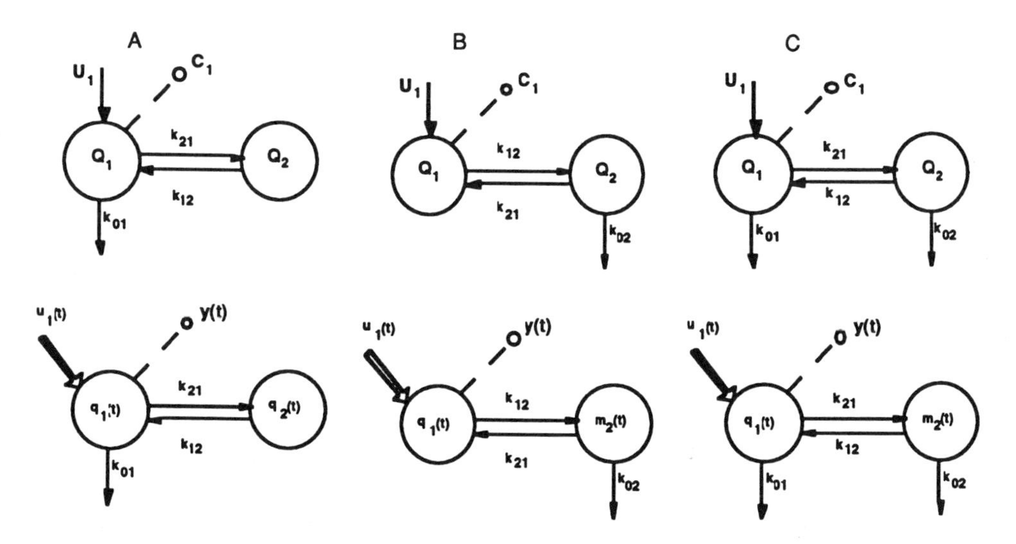

Figure 10. Three distinct two-compartment models. The upper row shows the tracee models and the lower row the tracer models. Panel A: The two-compartment model with no loss from compartment 2. Panel B: The two-compartment model with no loss from compartment 1. Panel C: The two-compartment model with losses from both compartments 1 and 2.

Models A and B are uniquely identifiable, which means that the rate constants, K_{ij}, and the volume of the accessible pool, V_1,can be identified from the data. The results of fitting these two models to the data are shown in Figure 11.

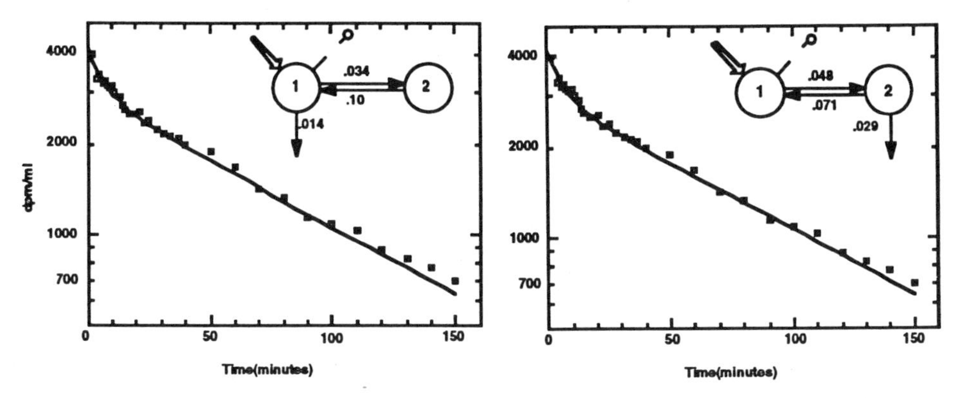

Figure 11. Left figure: Plot of data (labeled plasma glucose concentration dpm/mL) and results of the best fit of model A to the data. Right figure: Plot of the data and results of the best fit of model B to the data. The model parameters are shown in the inset.

The fit for both models with residuals, shown in Figure 12, is identical to that shown in Figure 9 for the two-exponential model. Then the mass Q_1 of compartment 1 can be calculated $Q_1 = C_1 \cdot V_1$, and other kinetic parameters of the tracee system, such as the tracee mass Q_2 in the nonaccessible compartment 2, and tracee production U_1, can be estimated by solving the tracee steady-state system. The results are summarized in Table 3.

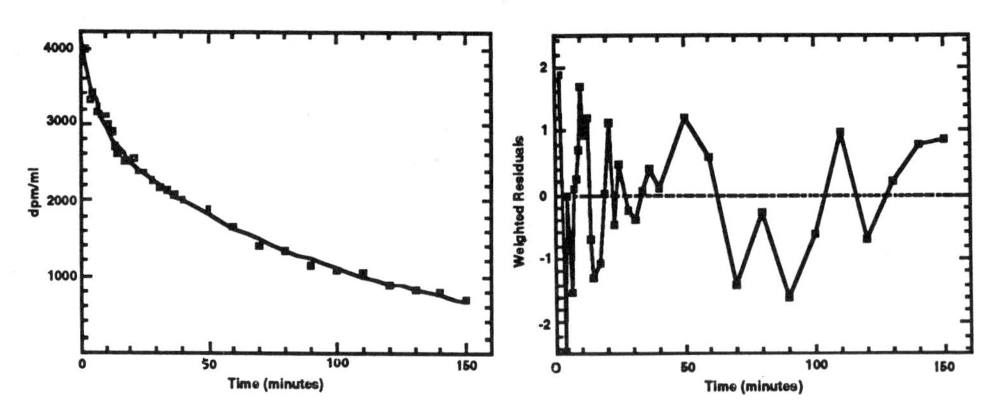

Figure 12. The fit with residuals of Models A, B, and C of Figure 10 to the data. All fits are identical, and are identical to that shown in Figure 9 for a sum of two exponentials.

Table 3. Precision of the parameter estimates expressed as percent fractional standard deviation is shown in parentheses.

	Model A	Model B
k_{21} (min^{-1})	0.0336 (24)	0.0469 (18)
k_{12} (min^{-1})	0.1011 (16)	0.0721 (21)
k_{01} (min^{-1})	0.0134 (4)	0
k_{02} (min^{-1})	0	0.0288 (7)
V_1 (mL)	10367 (3)	10367 (3)
Q_1 (mg)	10367 (3)	10367 (3)
Q_2 (mg)	3445 (9)	4818 (6)
Q_{tot} (mg)	13812 (1)	15185 (2)
U_1 (mg/min)	139 (1)	139 (2)

While models A and B are both *a priori* uniquely identifiable, they are, in general, not physiologically plausible. From a physiological point of view, model C with losses from both compartments is more plausible, but it is not *a priori* identifiable. There are two alternatives in dealing with model C. One is to estimate the bounds of the rate constants. For this example, using models A and B, these are:

$$0 \leq k_{01} \leq 0.0134 \qquad 0 \leq k_{01} \leq 0.0288$$
$$0.0336 \leq k_{21} \leq 0.0470 \qquad 0.0723 \leq k_{12} \leq 0.1011 \tag{25}$$

The other alternative is to incorporate *a priori* knowledge into the model by introducing a constraint among the rate constants k_{ij}. To illustrate this approach, consider the two-compartment model shown as Model C in Figure 10. If the rapidly equilibrating compartment 1 is hypothesized to be responsible for the insulin-independent tissues which, in the normal state, utilize about 75% of the total glucose disposal, one can write the following constraint:

$$k_{01}Q_1 = 3k_{02}Q_2 \tag{26}$$

From the steady-state equations, one can write $k_{21}Q_1 = (k_{12} + k_{02})\, Q_2$which can be used to develop the following relationship:

$$k_{21} = \frac{3k_{02}k_{21}}{k_{12} + k_{02}} \tag{27}$$

This constraint allows Model C to become *a priori* uniquely identifiable because it provides an additional independent relation among the unknown parameters. The results of fitting this constrained model are summarized in Table 4. The fit and weighted residuals are identical to those for Models A and B. One can calculate the system parameters in this case, and compare them with the corresponding parameters from Models A and B. The fact that some are different illustrates a crucial point: If one is using a model to make predictions about nonaccessible compartments in the system, the results are model-dependent and it is thus essential that the model structure reflect the physiology of the system.

Table 4. Numbers in parentheses are the precision of the parameter estimates expressed as a percent fractional standard deviation.

	Model C
k_{21} (min^{-1})	0.0370 (22)
k_{12} (min^{-1})	0.0920 (18)
k_{01} (min^{-1})	0.0101 (4)
k_{02} (min^{-1})	0.0092 (8)
V_1 (mL)	10367 (3)
Q_1 (mg)	10367 (3)
Q_2 (mg)	3788 (8)
Q_{tot} (mg)	14155 (2)
U_1 (mg/min)	139 (1)

CONCLUSION

In this chapter, we have summarized the underlying theory that is required to develop and test sound models of physiological systems. When one has a grasp of the points we have made, one can have greater confidence in the quality of the models that are postulated and in their utilization to calculate events in the accessible pool and predict events in the nonaccessible pools.

The SAAM II software system was designed with these considerations in mind. SAAM II makes the postulation of multicompartmental models easy using the icons available in its graphical user interface. This ease of use has drawbacks, especially for the novice user, because it is easy to postulate models whose complexity cannot be supported by the available data. For this reason, SAAM II has incorporated in its computational kernel a number of messages, warnings, and errors, which point out potential problems and their possible sources.

Development of the software, which is supported by a grant from BTP/NCRR/NIH, continues. New features are constantly being added, both at the request of users and in response to technological changes. These include a number of 'tips' for use during the model development phase.

The overall goal of this chapter has been to present a balance between the basic theory of compartmental models and a discussion of the SAAM II software package as one means by which individuals involved in nutritional studies can use modeling as a powerful research tool with confidence.

In summary, once a model structure is postulated, one must test the *a priori* identifiability of the model. While SAAM II permits the easy construction of such models, testing the *a priori* identifiability of such structures is not always easy. As models become larger and more complex, this issue becomes even more critical. The next step is to be aware of the error structure in the data. Again, SAAM II permits many ways to specify the error, but this cannot be done blindly: the quality of the results, both in terms of the estimated parameter values and their estimated errors, depends upon the error structure of the data. This, in turn, impacts the *a posteriori* identifiability issue.

Lastly, one must address the quality of the model and the model validity. SAAM II returns information which can and should be used to assess goodness-of-fit and model order. These are the obvious and necessary steps to take before one can discuss the results of a given modeling effort with any confidence.

CORRESPONDING AUTHOR

Please address all correspondence to:
Claudio Cobelli
Department of Electronics and Informatics
Via Gradenigo, 6/A
35131 Padova
ITALY

REFERENCES

Audoly S; D'Angio' L; Saccomani MP. Global identifiability of linear compartmental models. A computer algebra algorithm. *IEEE Trans Biomed Eng,* 1998, 45(1):36-47.

Bell BM; Burke JV; Schumitzky A. A relative weighting method for estimating parameters and variances in multiple data sets. *Comp Stat & Data Anal,* 1996, 22:119-135.

Bergman RN; Ider YZ; Bowden CR; Cobelli C. Quantitative estimation of insulin sensitivity. *Am J Phys,* 1979, 236:E667-E677.

Berman M; Grundy SM; Howard BV; Eds. *Lipoprotein Kinetics and Modeling.* Academic Press: New York. 1982.

Berman M. Kinetic analysis and modeling: Theory and applications to lipoproteins, in: *Lipoprotein Kinetics and Modeling.* Berman M; Grundy SM; Howard BV; Eds. Academic Press: New York. 1982. pp. 3-36.

Boston RC. Balancing needs, efficiency, and functionality in the provision of modeling software: A perspective of the NIH WinSAAM Project. (this volume).

Carson ER; Jones EA. The use of kinetic analysis and mathematical modeling in the quantitation of metabolic pathways in vivo. Application to hepatic anion metabolism. *NEJM,* 1979, 300:1016-1027 and 1078-1086.

Carson ER; Cobelli C; Finkelstein L. *The Mathematical Modeling of Metabolic and Endocrine Systems.* John Wiley: New York. 1983.

Caumo A; Cobelli C. Hepatic glucose production during the labeled IVGTT: Estimation by deconvolution with a new minimal model. *Am J Phys,* 1993, 264:E829-E841.

Cobelli C; Lepschy A; Romanin Jacur G. Identifiability results on some constrained compartmental systems. *Math Biosci,* 1979, 47:173-196.

Cobelli C; DiStefano JJ, III. Parameter and structural identifiability concepts and ambiguities: A critical review and analysis. *Am J Phys,* 1980, 239:R7-R24.
Cobelli C; Bergman RN; Eds. *Carbohydrate Metabolism.* John Wiley: Chichester. 1981.
Cobelli C; Toffolo G; Ferrannini E. A model of glucose kinetics and their control by insulin. Compartmental and noncompartmental approaches. *Math Biosci,* 1984a, 72:291-315.
Cobelli C. Modeling and identification of endocrine-metabolic systems: Theoretical aspects and their importance in practice. *Math Biosci,* 1984b, 72:263-289.
Cobelli C; Carson ER; Finkelstein L. Validation of simple and complex models in physiology and medicine. *Am J Phys,* 1984c, 246:R259-R266.
Cobelli C; Pacini G; Toffolo G; Sacca' L. Estimation of insulin sensitivity and glucose clearance from a minimal model: New insights from labelled IVGTT. *Am J Physiol,* 1986, 250:E591-E598.
Cobelli C. Identification of endocrine-metabolic and pharmacokinetic systems, in: *Control Aspects of Biomedical Engineering.* Nalecz N; Ed. Pergamon Press: Oxford. 1987a. pp. 235-249.
Cobelli C; Toffolo G. Theoretical aspects and practical strategies for the identification of unidentifiable compartmental systems, in: *Identifiability of Parametric Models.* Walter E; Ed. Pergamon: Oxford. 1987b. pp. 85-91.
Cobelli C; Mariani L; Eds. *Modeling and Control of Biomedical Systems.* Pergamon Press: Oxford. 1989.
Cobelli C; Bier DM; Ferrannini E. Modeling glucose metabolism in man: Theory and practice, in: *Computers and Quantitative Approaches to Diabetes.* Schretzenmeir J; Kraegen EW; Beyer J; Eds. G. Thieme Verlag: Stuttgart (Horm Metab Res Suppl 24) 1990. pp. 1-10.
Cobelli C; Toffolo G; Foster D. Tracer-to-tracee ratio for analysis of stable isotope tracer data: link with radio-active kinetic formalism. *Am J Phys,* 1992, 263:E968-E975.
Cramp DG; Ed. *Quantitative Approaches to Metabolism.* John Wiley: Chichester. 1982.
D'Angio' L; Audoly S; Bellu G; Saccomani MP; Cobelli C. Structural identifiability of nonlinear systems: Algorithms based on differential ideals, in: *Proceedings of the 10th IFAC Symposium on System Identification, Vol. 3.* Blanke M; Söderström T; Eds. Danish Automation Society: Copenhagen. 1994. pp. 13-18.
DiStefano JJ, III. Complete parameter bounds and quasi-identifiability conditions for a class of unidentifiable linear systems. *Math Biosci,* 1983, 65:51-68.
Ferrannini E; Smith DJ; Cobelli C; Toffolo G; Pilo A; DeFronzo RA. Effect of insulin on the distribution and disposition of glucose in man. *J Clin Invest,* 1985, 76:357-364.
Gibaldi M; Perrier D. *Pharmacokinetics, 2nd Ed.* Marcel Dekker: New York. 1982.
Godfrey KR; DiStefano JJ, III. Identifiability of model parameters, in: *Identifiability of Parametric Models.* Walter E; Ed. Pergamon Press: Oxford. 1987.
Jacquez JA; Ed. Berman memorial issue. *Math Biosci,* 1984, 72(2).
Jacquez JA. *Compartmental Analysis in Biology and Medicine, 2nd Ed.* University of Michigan Press: Ann Arbor. 1985.
Johnson H. Modeling protein turnover. (this volume).
Ljung L; Glad T. On global identifiability for arbitrary model parametrizations. *Automatica,* 1994, 30:265-276.
Norton JP. An investigation of the sources of non-uniqueness in deterministic identifiability. *Math Biosci,* 1982, 60:89-108.
Pohjanpalo H. System identifiability based on the power series expansion of the solution. *Math Biosci,* 1978, 41:21-34.
Radziuk J; Hetenyi G, Jr. Modeling and the use of tracers in the analysis of exogenous control of glucose homeostasis, in: *Quantitative Approaches to Metabolism.* Cramp DG; Ed. John Wiley: Chichester. 1982. pp. 73-142.
Rowland M; Tozer T. *Clinical Pharmacokinetics: Concepts and Applications, 3rd Ed.* Williams and Wilkins: Baltimore. 1995.
SAAM II User Guide. SAAM Institute: Seattle. 1997.
Saccomani MP; Cobelli C. Qualitative experiment design in physiological system identification. *IEEE Control Systems Magazine,* 1992, 12:18-23.
Saccomani MP; Cobelli C. A minimal input-output configuration for a priori identifiability of a compartmental model of leucine metabolism. *IEEE Trans Biomed Eng,* 1993, 40:797-803.
Taylor R; Magnusson I; Rothman DL; Cline GW; Cline GW; Caumo A; Cobelli C; Shulman GI. Direct assessment of liver glycogen storage by 13C nuclear magnetic resonance spectroscopy and regulation of glucose homeostasis after a mixed meal in normal subjects. *J Clin Invest,* 1996, 97:126-132.
Vicini P; Caumo A; Cobelli C. The hot IVGTT two-compartment minimal model: Indices of glucose effectiveness and insulin sensitivity. *Am J Phys,* 1997, 273:E1024-E1031.

APPROACHES TO POPULATION KINETIC ANALYSIS WITH APPLICATION TO METABOLIC STUDIES

Paolo Vicini,[1,2] P. Hugh R. Barrett,[1] Claudio Cobelli,[2] David M. Foster,[1] and Alan Schumitzky[1,3]

[1]Department of Bioengineering
University of Washington, Seattle, WA
[2]Department of Electronics and Informatics
University of Padova, Padova, Italy
[3]Department of Mathematics
University of Southern California, Los Angeles, CA

ABSTRACT

Population kinetic analysis is the methodology traditionally used to quantify inter-subject variability in pharmacokinetic studies. In the statistics literature, it is also called analysis of repeated measurement data or analysis of longitudinal data.

In this work, we will state the population kinetics problem and give some historical background to its significance. Then we will describe and apply to case studies in intermediary metabolism various two-stage and other parametric methods for nonlinear mixed effects models. We will then briefly review the software available for population kinetic analysis.

INTRODUCTION: THE POPULATION KINETICS QUESTION

Definition of population kinetics

Population kinetic analysis is the methodology traditionally used to quantify inter-subject variability in pharmacokinetic studies. In the statistics literature, it is also called analysis of repeated measurement data or analysis of longitudinal data.

In kinetic analysis, the focus of interest is the relationship between a given experimental protocol and the resulting effect. This cause/effect relationship usually differs markedly among individual subjects (here subject can mean human, lab animal, or lab preparation). Population kinetics is the discipline that models this inter-subject variability.

Mathematical Modeling in Experimental Nutrition
Edited by Clifford and Müller, Plenum Press, New York, 1998

Population kinetic analysis is the methodology used to quantify this inter-subject variability relative to a given population kinetic model.

It is assumed that each subject's cause/effect relationship has been defined in terms of a model which depends on certain known quantities: e.g., the dosage regimen, sampling schedule design, and measured covariates; and on certain unknown quantities: e.g., the unknown model and noise parameters. These unknown parameters are assumed to be random variables with a common but unknown distribution (the population distribution). The population kinetic analysis problem is then the estimation of this population distribution based on a collection of individual subject data (the population data).

Metabolic applications of population kinetics: an example

Let us begin with a simple model used in metabolic applications and demonstrate how we can apply population kinetics. Suppose one would like to determine the clearance of a substance from a system in the steady state, and the kinetics of the system can be described by a single compartment model. One can then perform a tracer experiment, such as injecting a tracer bolus into the system and monitoring (i.e. sample at discrete times) the disappearing tracer concentration y(t) for a certain length of time. In this case, the model for the tracer data is a simple one-compartment model, and is described by the equation:

$$\begin{aligned} Q(t) &= e^{-Kt} \\ S(t) &= Q(t)/V \end{aligned} \qquad (1)$$

where $Q(t)$ is the mass of the tracer (with $Q(0) = 1$); K is a rate constant with units of time^{-1}; $S(t)$ the concentration of the tracer; and V is the volume of distribution of the tracer. The model-predicted response can then be fitted (e.g. by least squares techniques) to the data $y(t)$ obtained in a single subject by adjusting the model parameters K and V. Usually, to obtain more robust information, this experiment is repeated in a number of individuals, with a similar experimental setup.

When applying population kinetics to this particular situation, several questions might come to mind:

- Is it possible to precisely estimate the model parameters K and V in a cohort of subjects with a "small" number of samples?
- Is it possible to make efficient use of the knowledge (when available) that all the subjects belong to the same population?
- Is it possible to reliably estimate the characteristics of the population (i.e. the value of the parameters and their distributions) from a limited number of subjects?

These questions address potentially crucial aspects of the clinical and research applications of models used in intermediary metabolism. This example has helped us to set the stage for the remainder of this chapter, where we will formalize the population kinetics problem and then describe and apply it to case studies in intermediary metabolism using various two-stage methods and other parametric methods for nonlinear, mixed-effects models.

PROBLEM DEFINITION–MIXED EFFECTS MODELS

Let us define the structure of the kinetic model for the generic jth subject, $j=1, \ldots, N$, where N is the number of subjects. For each subject, we have the model:

$$y_{j,i} = f\,(t_{j,i}\, \beta_j) + e_{j,i}) \tag{2}$$

where $y_{j,i}$ is the ith measurement for the jth subject at time t_{ji}; f the model function, β_j the vector of random parameters of the kinetic model; and $e_{j,i}$ is the measurement error. The β_j are a known function of parameters γ which do not vary across the population (*fixed effects*) and of parameters z_j which change across the population (*random effects*):

$$\beta_j = g\,(\gamma, z_j) \tag{3}$$

For this reason, these models are called *mixed effects models*. Moreover, in the parametric approach, the assumption is usually made that the random effects z_j are normally distributed with mean m and covariance D:

$$z_j \sim N\,(m, D) \tag{4}$$

In this case, the population kinetic analysis problem is to estimate m and D based on the population data.

The method most often used for this estimation is maximum likelihood (Davidian and Giltinan, 1995), based on the maximization of the likelihood function. This method requires integration over high-dimensional spaces, which is a formidable numerical task, even for simple models. Consequently, other ways have been considered to approximate the nonlinear model so as to lessen the complication of this integration. Different approximations, based mostly on first-order linearization, have given rise to alternate approaches to population kinetics. For example, linearizing the kinetic model $[f(t,\beta\,)]$ around the mean of the population of the random effects β is at the basis of the most commonly used method for population analysis, NONMEM (Beal and Sheiner, 1982). A different linearization, more accurate but more computationally expensive, is performed around the value of the random effects for each individual, and gives the Lindstrom-Bates iterative algorithm, discussed later (Lindstrom and Bates, 1990).

BACKGROUND AND HISTORY

Before the 1970s, the two approaches for estimating population means and covariances were based on *pooling* and *averaging*. The *pooling approach*, also called naïve pooling in the literature (Steimer et al., 1984), assumes that all population data come from the same individual. The model parameters estimated by this approach give an estimate of the population mean but no estimate of the population covariance.

The *averaging approach* gives estimates of both population means and covariances in those situations where there are enough data available for each subject in the population to accurately estimate that subject's unknown parameters. The sample average of these individual estimates is then the estimate of the population mean, and the sample covariance of these individual estimates is then the estimate of the population covariance. This approach is also known as the Standard Two-Stage (STS) method. The main drawbacks of the averaging approach are that:

1. equal weight is given to each individual's parameter estimates, even though some are more accurate than others; and
2. the method requires sufficient data for each individual to obtain that individual's parameter estimates.

The first method for population kinetic analysis that took into account the accuracy of individual estimates appears to be the work described in the 1964 thesis of K. Pettigrew (Lyne et al., 1992). The Pettigrew method is similar to what is now known as the Global Two-Stage method. Independently of Pettigrew, the Global Two-Stage (GTS) method was developed in the 1977 technical report of G. Prevost (Steimer et al., 1984). Both Pettigrew's and Prevost's methods required sufficient data for each individual to obtain not only that individual's parameter estimates but also the covariance matrix of those estimates. This covariance matrix estimate provided the information needed to properly weight the individual estimates.

Two-stage methods can generally be described as follows. In the first stage, each subject's parameters are estimated; in the second stage, the corresponding population parameters are estimated. These methods can be nonparametric, as in the Standard Two-Stage approach; or parametric, as in the Global Two-Stage and Iterative Two-Stage approaches. The distinction between the two classes is at times blurred: for example, the Standard Two-Stage could be regarded also as parametric (if the assumption of a normal distribution is implicitly made), and the Global Two-Stage is likely to perform well even if the (required) assumption of normality is not entirely met.

The Standard Two-Stage method is as follows: the model parameters are identified in each subject separately, and the mean and covariance of the population are then established as the sample mean and covariance. As mentioned earlier, this method does not take into account the reliability of the individual estimates and has been shown to overestimate the population covariance (Davidian and Giltinan, 1995).

The Global Two-Stage (Steimer et al., 1984), a more refined approach, iteratively optimizes on the individual estimates and their covariances, again identifying all the subjects separately. In contrast to the Standard Two-Stage, however, this method pools together the individual estimates and their precision in a theoretically sound fashion. Intuitively, taking this information into account is qualitatively better than not considering it al all, as the Standard Two-Stage does.

The Iterative Two-Stage (Steimer et al., 1984) is another iterative method, whose most attractive feature is that, at each iteration, the information from the sample mean and covariance of the population are used as prior knowledge (in the context of Bayesian estimation) in quantifying the individual parameters from the data. This method results in more accurate estimates of the individual parameters and, at convergence, a more reliable measure of the population parameters. It takes into account all individual contributions but does not improve dramatically upon the Global Two-Stage estimate (Racine-Poon, 1985).

Two-stage methods are not appropriate for routine clinical data which sometimes contain only one data point per subject. Nor do they take into account how population parameters depend on covariate information such as age, sex, creatinine clearance, etc. In the early 1980s, these shortcomings led Sheiner and coworkers to develop the first true population kinetic analysis method, NONMEM (Beal and Sheiner, 1982).

Parametric methods, such as the Global and Iterative Two-Stage, and NONMEM essentially assume that the distribution of the population parameters is multivariate normal. Variations on the NONMEM method were later developed by Lindstrom and Bates (1990) and Vonesh and Carter (1992). By contrast, nonparametric approaches, such as the Standard Two-Stage, make no assumptions about the form of the underlying population distribution. The first population kinetics analysis alternative using a nonparametric approach was developed by Mallet (1986). Later variations on this method, based on the EM algorithm, were developed by DerSimonian (1986) and Schumitzky (1991). There is also an approach that lies between the parametric and nonparametric, proposed by Davidian and Gallant (1993).

All of the above approaches use the method of maximum likelihood as the basis for estimation. A Bayesian methodology was introduced for population kinetic analysis by Wakefield et al. (1994), which requires the calculation of very high dimensional integrals. The relatively new method of Gibbs sampling was used for this purpose.

An excellent survey of the subject of population kinetic analysis is given in Davidian and Giltinan (1995).

AVAILABLE POPULATION KINETICS SOFTWARE

At the time of this writing, there are only three "commercial-grade" population kinetic analysis programs with general model-building capabilities: NONMEM (Beal and Sheiner, 1982), P-PHARM (Gomeni et al., 1994), and BESTFIT (http://www.bestfit.lu/npml.htm). The first two programs are parametric, while BESTFIT can be both parametric and nonparametric. NONMEM performs a first-order linearization of the random effects model and uses maximum likelihood as the method of estimation; P-PHARM uses the Iterative Two-Stage approach with a diagonal population covariance matrix; and BESTFIT is based on the NPML algorithm described by Mallet (1986). In addition to these programs, there are many "in-house" programs for population kinetic analysis in various stages of development.

Clearly, this is a very active area of theoretical research and software development activities although, at present, most efforts are in the pharmaceutics area, due to the technological demands made by the analysis of phase 1, 2, and 3 clinical trials.

The development of user-friendly and comprehensive population kinetic software is, in principle, a major contribution to both the pharmacokinetic and metabolic arenas. In this latter research area, there is a growing interest in easy-to-use software tools that can be used to measure the contribution of specific covariates (e.g., sex, height, weight, lipid levels) to population kinetics parameters. Such tools will open up new kinds of metabolic kinetic experimental designs.

POP3CM is a prototype population kinetics software tool developed by the Resource Facility for Kinetic Analysis (RFKA) at the University of Washington. It features a graphical user interface to population kinetic analysis based on the Lindstrom-Bates algorithm. Recall that the Lindstrom-Bates algorithm is one that applies maximum likelihood estimation and linearizes the kinetic model around the individual subjects' parameter estimate. It supports data which can be analyzed using the model shown in Figure 1, one widely used in the interpretation of pharmacokinetic data.

In this figure, compartment c is the central or plasma compartment. It exchanges with an extravascular (peripheral) compartment, compartment p. Bolus inputs, B_a, B_c, or B_p can be made into any of the compartments. A constant infusion, I_c, can be made into compartment C. A sample, denoted by the bullet with dotted lines, can be taken from compartment C, P, or a linear combination of the two.

The program accommodates this open two-compartment model with multiple inputs. The adjustable parameters for this model are K_a, K_e, K_{pc}, K_{cp} and V and various one-to-one transformations of same, such as CL and 1/V. A general analytic solution for the multiple dosing case is employed. Further, the partial derivatives of the solution with respect to the microparameters are also calculated analytically. These analytic derivatives are used to provide analytic gradients for the optimization algorithms. A prototype version of the parametric program has been written in the language O-Matrix (Bell et al., 1994). This prototype program can be downloaded from the RFKA homepage: http://weber.u.washington.edu/~rfka/. In Figure 2, we show two screen shots from the homepage describing POP3CM.

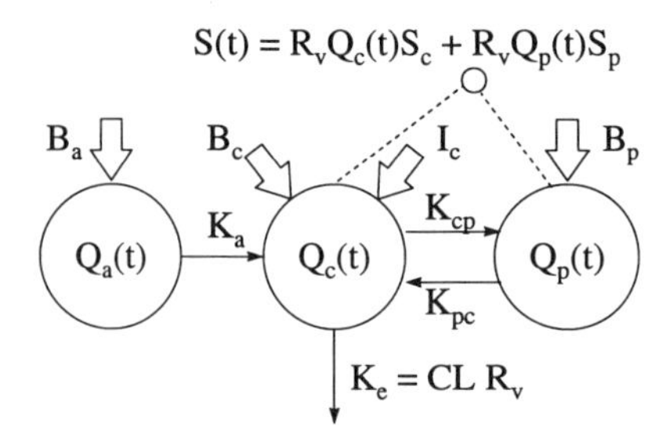

Figure 1. The three-compartment model implemented in POP3CM. Subscripts indicate: a, absorption compartment; c, central compartment; p, peripheral compartment. Explanation of symbols: S(t) is the measurement function; $Q_a(t)$ is the mass in the absorption compartment; $Q_c(t)$, the mass in the central compartment; $Q_p(t)$, the mass in the peripheral compartment; B_a, the amount of mass injected into the absorption compartment; B_c, the amount of mass injected into the central compartment; B_p, the amount of mass injected into the peripheral compartment; S_c, a scale factor for the central compartment; S_p, a scale factor for the peripheral compartment; I_c, the rate of mass infused into the central compartment; K_a, the transfer rate from the absorption to the central compartments; K_{cp}, the transfer rate from the central to the peripheral compartments; K_{pc}, the transfer rate from the peripheral to the central compartments; K_e, the transfer rate from the central compartment to the outside world; CL, the clearance from the system to the outside world; $R_v = 1/V$, the reciprocal of volume.

CASE STUDIES

Estimates of Mean Population Parameters from Limited Data Sets

We present here results from a simulation study where we defined a one-compartment kinetic model with two parameters (1) of given population mean and covariance, and then sampled its response for 20 simulated subjects, extracting only one data point per subject. In this case, the standard method of analysis (identify the model parameters in each subject and then take their average, i.e., the Standard Two-Stage method) is not applicable. If you have two parameters and only one data point per subject, you will not be able to obtain parameter estimates for each individual separately. Given this limitation, it then follows that you will not be able to obtain population parameter estimates.

We have used the prototype software, POP3COM (see above), to apply the Lindstrom-Bates algorithm (1990) to this database. What the Lindstrom-Bates and other, similar, algorithms do is, basically, pool the available information from all the individual subjects together in a statistically meaningful sense. The final product of this pooling is the population mean and covariance, which can then be used as additional information to extract individual parameter estimates even from data sets where there is only one data point. In fact, in this case, the information you obtain is much more meaningful: there is the population mean and covariance (which should reflect also the parameter values of an individual subject), and there is the single data point, which helps in 'individualizing' the estimate. This is also sometimes called a 'post hoc' estimate.

The true population mean was K=1.5 and 1/V=5, and the population covariance was diagonal with Var[1/V]=1 and Var[K]=0.01. The measurement error structure was constant CV. Twenty subjects were simulated and one data point per subject was taken at $t = 30/i$, $i=1,\ldots, 20$, adding different levels of noise to the data. The estimates of the population mean for 1/V and K are reported in Table 1 for different levels of noise in the data.

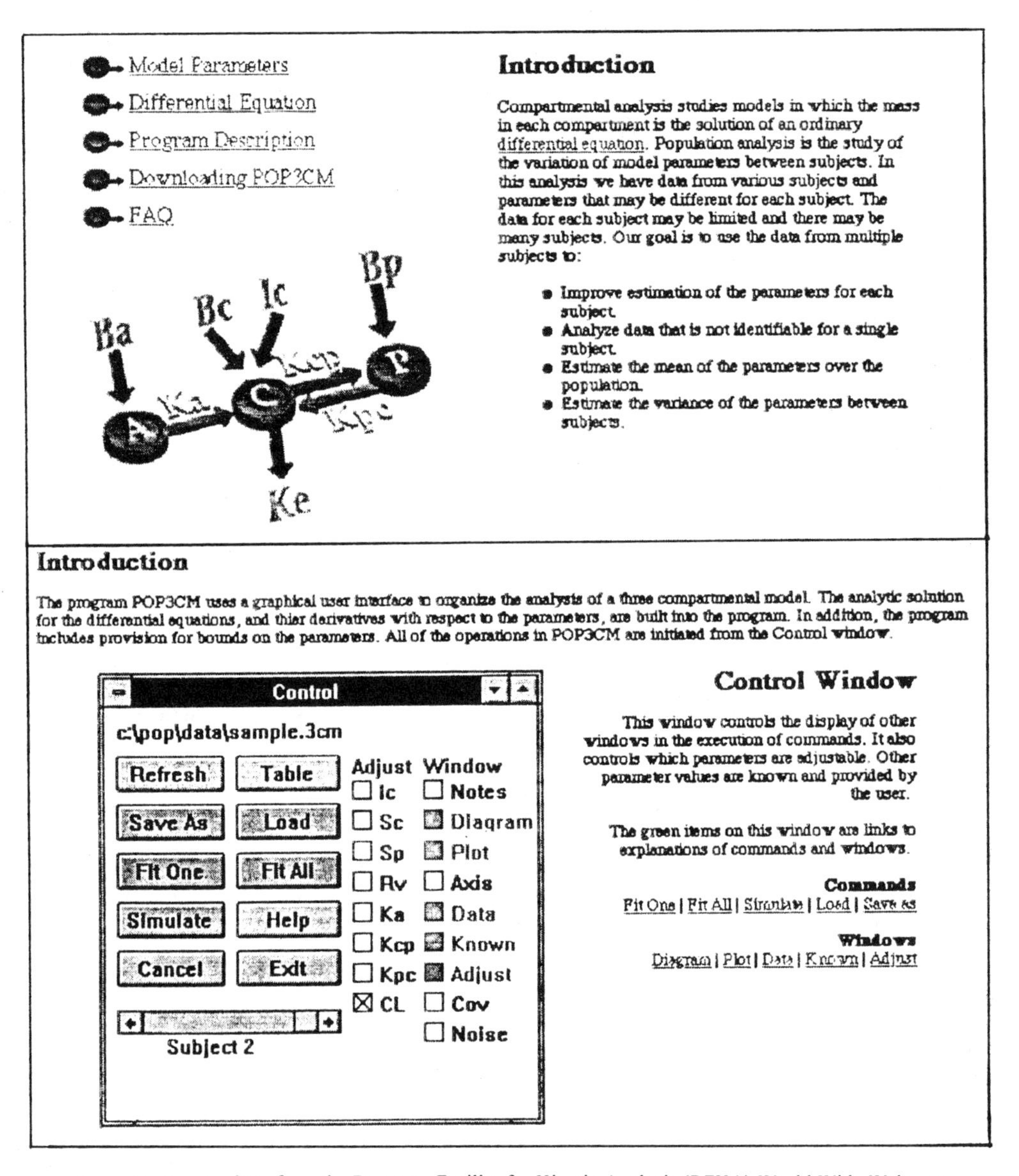

Figure 2. Screen shots from the Resource Facility for Kinetic Analysis (RFKA) World Wide Web page at http://weber.u.washington.edu/~rfka/, showing POP3CM control windows.

We can see that for small noise levels, the estimated means are virtually identical to the true population mean, while the estimates of the covariance are generally different, with the appearance of an off-diagonal term. The accuracy of the covariance is strongly influenced by the small database considered, and reflects the difficulty intrinsic to estimating higher-order moments from limited data (Davidian and Giltinan, 1995). However, these results show that an estimate of the population mean was possible in this sparse-data situation, even if the single subjects parameters were not resolvable from the data.

However, as mentioned earlier, the Lindstrom-Bates algorithm, as implemented in this case by POP3CM, allows one to not only obtain the population parameters, but also the individual parameters. At each iteration, a Bayesian fit is performed on the subjects data using as a prior the current estimate of the population distribution. In this way, it is possible to obtain meaningful estimates of the single subjects parameters, thus "individualizing" the population distribution despite a very limited data set.

Table 1. Simulated population study with a one-compartment open model. See text for details.

Error level	1/V	K	Var[1/V]	Var[K]	Cov[1/V,K]
True Values	5.00	1.50	1.00	0.010	0.000
Data Error = 2%	4.81	1.50	0.50	0.012	-0.030
Data Error = 5%	4.82	1.39	0.38	0.005	-0.046
Data Error = 10%	4.82	1.55	0.41	0.002	-0.030
Data Error = 20%	3.73	1.26	1.85	0.106	0.410

Precise Individual Parameters from Limited Data Sets

The classic minimal (hereafter cold) model of glucose kinetics (Bergman et al., 1979) during an intravenous glucose tolerance test (IVGTT) is commonly used to investigate glucose metabolism *in vivo* in pathophysiological and epidemiological studies in humans. Its application is very widespread; as early as 1994, it had appeared in more than 240 papers in the medical and physiological literature. The model is described by the following system of differential equations:

$$\begin{aligned} \frac{Q(t)}{dt} &= -[S_G + X(t)]Q(t) + S_G Q_b \qquad Q(0) = D + Q_b \\ \frac{X(t)}{dt} &= -p_2\{X(t) - S_I[I(t) - I_b]\} \qquad X(0) = 0 \\ G(t) &= \frac{Q(t)}{V} \end{aligned} \tag{5}$$

where $Q(t)$ (mg) is the glucose mass in plasma; D, the glucose dose (mg kg^{-1} body weight); $G(t)$ (mg dL^{-1}), the plasma glucose concentration; $I(t)$ (μU mL^{-1}), the plasma insulin concentration; and $G_b = Q_b/V$ and I_b, the glucose and insulin concentration basal values. V (dL kg^{-1}) is the distribution volume of glucose; p_2 (min^{-1}), a parameter related to insulin action. S_G (min^{-1}) and S_I (min^{-1} per mU mL^{-1}) are the minimal model metabolic indices of glucose effectiveness and insulin sensitivity respectively, and reflect the effect of both glucose and insulin on the glucose rate of disappearance and endogenous production. Insulin concentration acts as a known (assumed without error) input (forcing function) in the second equation, and model parameters are identified from fitting the model to glucose concentration only. $X(t)$ represents insulin action on glucose disappearance from a compartment remote from plasma.

The usual protocol for the standard IVGTT is as follows. A glucose bolus (between 300 and 330 mg kg^{-1}) is administered at the beginning of the experiment. Blood samples for the assessment of the changing plasma glucose and insulin concentrations are then drawn for four hours after the bolus administration, according to a schedule comprising 30 samples (usually, 0, 2, 3, 4, 5, 8, 10, 12, 14, 16, 18, 20, 24, 28, 32, 40, 45, 50, 60, 70, 80,

90, 100, 110, 120, 140, 160, 180, 210, and 240 minutes). The minimal model parameters are then identified from these glucose and insulin kinetic data, gathered in a certain number of subjects.

We will now apply the minimal model to clinical data from a standard IVGTT performed in a population of 15 normal subjects (Figure 3). The glucose bolus was administered at time 0. Experiments were performed at the Washington University School of Medicine (Saint Louis, MO) and the Department of Metabolic Diseases, University of Padova, Italy. Samples were drawn according to the full schedule (30 samples); a reduced schedule of only 12 samples was also extracted from the data (2, 3, 4, 5, 8, 20, 30, 60, 70, 120, 180, and 240 minutes). The glucose measurement error was experimentally determined to be 2% of the datum.

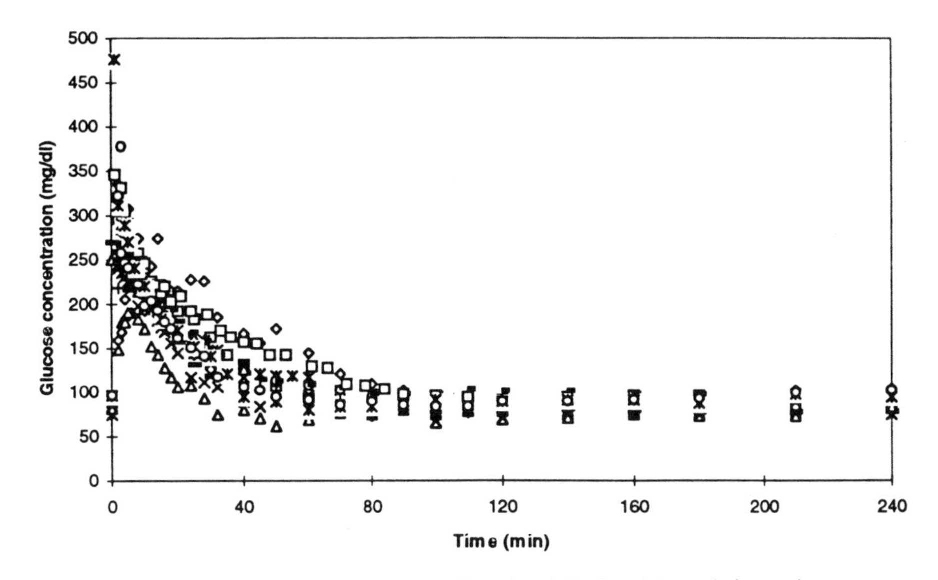

Figure 3. Sample glucose curves from the minimal model population study.

We applied the Standard and Iterative Two-Stage methods to both the full and the reduced sampling schedule; the results are shown in Table 2. For the estimated value of each parameter (S_G, S_I, $p2$, and V), there are numbers between parentheses which represent the mean of the 'precision of the estimate,' expressed as percent fractional standard deviation (%FSD) of the estimate, and numbers below the estimate, which represent the 'population spread.'

To make a practical example, if we look at the first line: S_G is equal to 2.63, with a mean precision of 63%, and a corresponding population spread of 41%. Precision is a quantitative measure of the reliability of the estimate in the individuals (thus, a smaller value would then represent a more precise estimate), while the population spread is a measure of how much the parameter varies in the population. Both are expressed as percentages. With this in mind, it is not difficult to understand why the mean precision of S_G with the Standard Two-Stage and the reduced sampling schedule (90%) is worse than the mean precision of the estimate with the full sampling schedule and the Standard Two-Stage (63%), and why the mean precision of S_G with the reduced sampling schedule and the Iterated Two-Stage algorithm (28%) is better than both of them, despite having been obtained with a reduced sampling schedule.

While one can obtain very similar values of the population mean and spread with all three methods, the Iterative Two-Stage approach allows us to recover much more precise individual parameter estimates, such as glucose effectiveness, S_G, and insulin sensitivity, S_I, even with a reduced sampling schedule.

Table 2. Minimal model population study. Numbers between parenthesis are the mean precision of the estimate across the individuals expressed as percent fractional standard deviation of the estimate. %CV is the population spread, expressed as the ratio between the population standard deviation and mean. See text for details.

	STS, Full Schedule (N=30 samples)			
	S_G x10^2	S_I x10^4	p_2	V
Mean	2.63 (63)	6.15 (42)	5.00 (82)	1.65 (8)
%CV	41	104	44	11
	STS, Reduced Schedule (N=12 samples)			
	S_G x10^2	S_I x10^4	p_2	V
Mean	2.53 (90)	5.58 (95)	5.79 (209)	1.69 (12)
%CV	40	104	73	14
	ITS, Reduced Schedule (N=12 samples)			
	S_G x10^2	S_I x10^4	p_2	V
Mean	2.58 (28)	5.46 (33)	4.85 (40)	1.69 (5)
%CV	39	100	35	13

CONCLUDING REMARKS

Of central interest in kinetic analysis is the relationship between a given experimental protocol and the resulting effect. This cause/effect relationship usually differs markedly between individual subjects. By definition, population kinetic analysis is the methodology used to quantify this inter-subject variability.

Population kinetic analysis is widely used in pharmacokinetic/pharmacodynamic (PK/PD) studies because it is the key to understanding how drugs behave in humans and animals. More specifically, it provides the foundation for the intelligent design of dosage regimens to treat disease processes. In metabolic studies, it is used to identify which parameters in a model change when a population of normal subjects is compared to a population of subjects with a pathological condition. In this way it can be used to identify potential aberrant pathways, which is a first step in planning an intervention. It is also used to analyze the action of a specific intervention in a baseline and treated state. Finally, population kinetic analysis is necessary in situations where there is not sufficient data on each subject to estimate the individual subject parameters.

In this work, we have reviewed existing population kinetic theory and put it into the perspective of metabolic studies. We have described some of the software options currently available, and we also reported some work in progress based on the application of population analysis to classical metabolic problems and studies. We showed how some of these methods, besides providing reliable estimates of the population parameters and covariance matrix, also allow the estimation of individual parameters with very good precision, even in a sparse data situation.

ACKNOWLEDGMENTS

This work has been partially supported by NIH/NCRR grant RR-02176, and by NIH grant GM-53930. The authors would like to acknowledge the invaluable support of Dr. Bradley M. Bell in the preparation of this paper.

CORRESPONDING AUTHOR

Please address all correspondence to:
Prof. David M. Foster
Resource Facility for Kinetic Analysis
Box 352255
Department of Bioengineering
University of Washington
Seattle, WA 98195-2255
ph: 206-685-2009
fax: 206-543-3081
foster@saam.washington.edu

REFERENCES

Beal SL; Sheiner LB. Estimating population kinetics. *CRC Crit Rev Bioeng,* 1982, 8:195-222.

Bell BM; Paisley B; Trippel D. *O-Matrix for Windows Users Guide.* Harmonic Software, Inc.: Seattle. 1994.

Bergman RN; Ider YZ; Bowden CR; Cobelli C. Quantitative estimation of insulin sensitivity. *Am J Physiol,* 1979, 236:E667-E677.

Berman M; Weiss MF. *The SAAM Manua.* USPHS(NIH) Publication No. 78-180, US Government Printing Office: Washington DC. 1978.

Davidian M; Giltinan D. *Nonlinear Models for Repeated Measurement Data.* Chapman and Hall: New York. 1995.

Davidian M: Gallant A. The nonlinear mixed effects model with a smooth random effects density. *Biometrika,* 1993, 80:475-488.

DerSimonian R. Maximum likelihood estimation of a mixing distribution. *Appl Stat,* 1986, 35:302-309.

Gomeni R; Pineau G; Mentre F. Population kinetics and conditional assessment of the optimal dosage regimen using the P-PHARM software package. *Anticancer Res,* 1994, 14:2321-2326.

Lindstrom MJ; Bates DM. Nonlinear mixed effects models for repeated measures data. *Biometrics,* 1990, 46:673-687.

Lyne A; Boston R; Pettigrew K; Zech L. EMSA: A SAAM service for the estimation of population parameters based on model fits to identically replicated experiments. *Comp Meth Prog Biomed,* 1992, 38:117-151.

Mallet A. A maximum likelihood estimation method for random coefficient regression models. *Biometrika,* 1986, 73:645-656.

Racine-Poon A. A Bayesian approach to nonlinear random effects models. *Biometrics*, 1985, 41:1015-1023.

Schumitzky A. Nonparametric EM algorithms for estimating prior distributions. *App Math Comp,* 1991, 45:141-157.

Steimer JL; Mallet A; Golmard J-L; Boisvieux J-F. Alternative approaches to estimation of population pharmacokinetic parameters: Comparison with the nonlinear mixed-effect model. *Drug Metab Rev,* 1984, 15:265 - 292.

Vonesh EF; Carter RL. Mixed-effects nonlinear regressions for unbalanced repeated measures. *Biometrics,* 1992, 48:1-17.

Wakefield J; Smith AFM, Racine-Poon A; Gelfand A. Bayesian analysis of linear and nonlinear population models. *Appl Stat,* 1994, 43:201-222.

THE MATHEMATICS BEHIND MODELING

Judah Rosenblatt

University of Texas Medical Branch
Shriners Burns Institute
Galveston, Texas

INTRODUCTION

Scientific research has changed quite a bit since the early 1970s, in large part because computers and computer software packages have been dropping in price and have become progressively easier to use and more powerful. But while the packages have gotten easier to invoke, many researchers are still uncomfortable with them, because they don't have much understanding of what this software is really doing, and they can check neither the validity of the assumptions used to justify the data processing, nor the resulting conclusions.

Even as long as 40 years ago, enzyme kinetic models relating glucose production, G, to time, t, of the form

$$G(t) = ae^{-At} + be^{-Bt} \tag{1}$$

were considered. Then, just as now, many individuals using such a model had little idea of its origin; even more, they had almost no reasonable means of estimating the unknown constants, a, A, b, B from observed data. In contrast, today we can implement the fitting of much more complex models to data, but many individuals using these models have little more understanding of them than those researchers of years ago. And even more of them don't know what the computer is doing to estimate the unknowns from the data.
The purpose of this chapter is to provide the essence of the theoretical background that I feel is needed for effective and comfortable use of available software packages, such as SAAM and MLAB.

To begin, even at the risk of presenting material familiar to many, I think it is worthwhile to talk a bit about models and their uses (Rosenblatt and Bell, 1998). From my perspective as an applied mathematician/statistician, *a model is a description of measurements*, often embodying how the measurements are related to one another. There can be a variety of different models to describe a given system, depending on the use to be made of the model. These uses usually involve prediction and/or control. For example, we may want to predict the future weight of an individual based on a prescribed diet, or based on a prescribed diet and exercise regimen.

Predictions made from a model may be *deterministic* or *statistical.* Statistical predictions are used when factors other than those being explicitly accounted for are important influences on the quantity to be predicted. For instance, when basal metabolic rate is an important influence on future weight, but it is not being explicitly used in future weight prediction.

It is critical to choose the appropriate model for a given use. One example might be to determine the clearance of a given substance from the body. Frequently, we may need to concern ourselves solely with modeling how much of this substance remains in the body as time passes, and, for example, consider it as effectively eliminated when its level falls below a specified threshold in the bloodstream. On the other hand, we would want to use a more refined model if the substance is radioactive with a long half-life. In this case, even a small amount remaining in fatty tissue for an extended period could cause serious damage. We can thus see that a model used for determining the clearance of some medication might very well be substantially different from that used to describe the clearance of a radioactive tracer.

COMPARTMENTAL MODELS

One important class of descriptions used in metabolism and nutrition studies consists of what we call *compartmental models* (Jacquez, 1988). A model of an organic system is compartmental if it views this system as a collection of related compartments with well-defined rules for the flow of some specified substance between compartments. (The compartments need not correspond to physically or even physiologically distinct regions.) The flow of glucose produced by the liver (one compartment), to the bloodstream (a second compartment), to the intersticial compartment, and to the cells, and back again, might be analyzed by means of some compartmental model. Washout of a substance from the body can sometimes be adequately described by a one-compartment model, but a model with at least two compartments is likely to be needed for adequate description of radioactive substance washout.

We often represent a compartmental model by means of a diagram of the type shown in Figure 1. The arrows indicate flow of the substance between the compartments they connect, or to the outside.

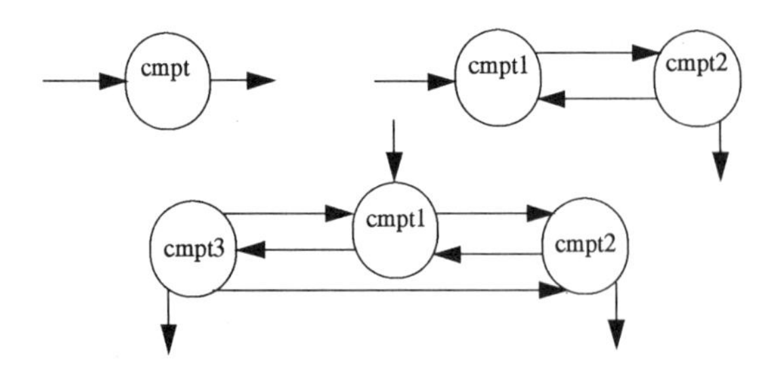

Figure 1. Diagrams representing various one-, two-, and three-compartment models.

We now need to get specific about the way that flow between compartments can be specified. To motivate our choice, let us think of the substance of interest as a collection of randomly moving spheres, and the compartments as physical enclosures, as illustrated in Figure 2, with the compartments separated by a wall with smaller doors on springs. Some of these doors open one way, others open the other way, with the springs on the doors in one direction being possibly stiffer than the others.

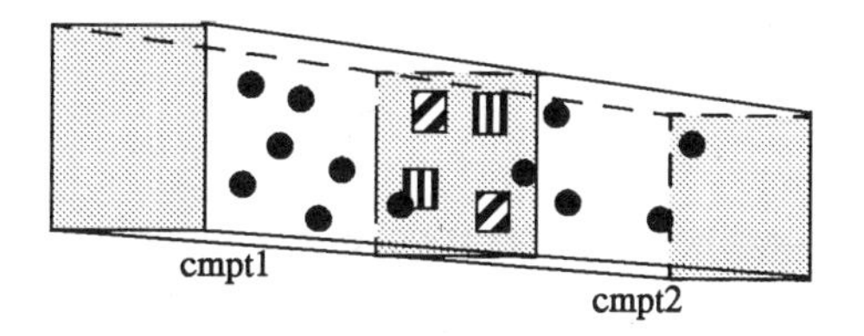

Figure 2. A simplified representation of a two-compartment model.

Now assume that these spheres are bouncing around, rarely hitting each other (which is reasonable if they represent molecules, that aren't highly concentrated). We can then expect, at least for a short time period, that the number of spheres per second moving from compartment 1 to compartment 2 would be proportional to the current number of spheres in compartment 1. That is, with for example, 5000 spheres in compartment 1, you would expect about five times as many transfers to compartment 2 than if there were only 1000 spheres in compartment 1. This relationship would hold for as long as the numbers of spheres in the compartments remain relatively close to the originally specified values.

Some notation is needed to make these ideas precise. So we let $X_i(t)$ denote the number of molecules of the substance of interest in compartment i at time t. We denote the number of molecules per (small) unit time exiting compartment i in any short time interval including the time value t by $-X_i'(t)$. If this number is *negative*, it indicates the number of molecules entering compartment 1. More precisely, this symbol represents the value this rate is approaching as the time interval shrinks to being arbitrarily small. For practical purposes, we may as well think of it as the number exiting per small unit of time.

It is essential that the time interval referred to in defining $-X_i'(t)$ be short, so that in this interval only a small proportion of the molecules in compartment i will exit. Then our description that the number moving out of compartment i per unit time isn't changed substantially for the duration of this small interval, because there are about the same number of opportunites for *exit* for any short time sub-interval of given duration.

If there are only two compartments as shown in Figure 2, (with no exits to the outside), then we would specify the rate of exit of molecules from compartment 1 to compartment 2 by $k_{21}X_1(t)$ to indicate that this exit rate is proportional to the number of molecules, $X_1(t)$, at time t. The quantity k_{21} is called the compartment 21 *rate constant*, and is the constant of proportionality referred to in our verbal description. Assuming that molecules can also flow from compartment 2 to compartment 1, we can write $k_{12}X_2(t)$ for the rate of exit of molecules from compartment 2 to compartment 1. Now denote the net *change* in the number of molecules in compartment 2 per unit time in a short interval of time including the point t by $X_2'(t)$. We find

$$\begin{aligned} X_1'(t) &= -k_{21}X_1(t) + k_{12}X_2(t) \\ X_2'(t) &= +k_{21}X_1(t) - k_{12}X_2(t) \end{aligned} \tag{2}$$

because molecules exiting compartment 1 enter compartment 2 and those exiting compartment 2 enter compartment 1. Again, if the net change is negative, more molecules are exiting than entering compartment 2. Looking at the first half of (Eq. 2), this occurs if its first term has bigger magnitude than its second one; i.e. if more molecules exit compartment 1 than enter it. This description is indicated graphically in Figure 3.

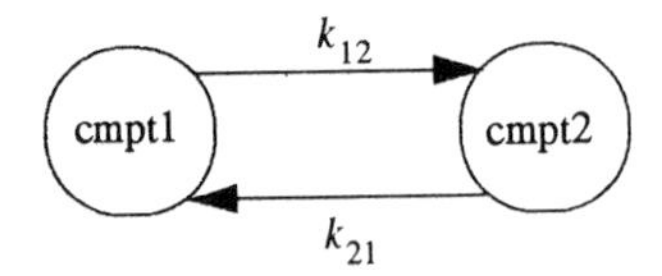

Figure 3. A simple two-compartment model, including rate constants k_{21} and k_{12}.

The simplest situation of this type is where $k_{12} = 0$. This is called *irreversible loss*, and is pictured in Figure 4. (Here, the notation is slightly changed, with the area outside the model considered as compartment 0, which is not pictured via a circular enclosure.)

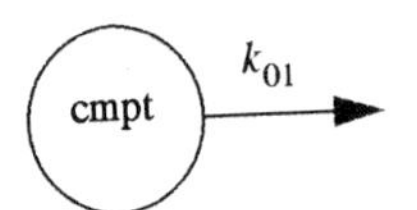

Figure 4. A simple one-compartment model, demonstrating irreversible loss to the outside.

To describe this situation we need only write

$$X_1'(t) = -k_{01}X_1(t) \tag{3}$$

In mathematical words, $X_1'(t)$ is called the *derivative* of X_1 at time t and (Eq. 3) is a called a *differential equation* (Golomb and Shanks, 1965). This particular differential equation represents the simplest pure washout model. We will see that even this simplest of models can provide useful information in certain situations. For example, we examine how we might calculate the amount of material remaining in the compartment at time t, if we knew the rate constant, k_{01}, and the amount of substance in this compartment at some initial time, which we'll choose to be $t = 0$.

Although we originally thought of $X_1(t)$ as the *number* of molecules in compartment 1 at time t, for mathematical purposes we can modify this to be the amount of substance, (e.g., in gm) present at time t. Then $X_1'(t)$ represents the change of amount of substance per small unit time in compartment 1 at time t.

If h stands for a short duration of time, the change in amount of substance in the compartment in the time interval from t to $t + h$ is

$$X_1(t+h) - X_1(t) \tag{4}$$

so that the change in amount per small unit of time is

$$\frac{X_1(t+h) - X_1(t)}{h} \cong X_1'(t) \tag{5}$$

Substituting into (Eq. 3) from (Eq. 5), we find

$$\frac{X_1(t+h) - X_1(t)}{h} \cong -k_{01}X_1(t) \tag{6}$$

or,

$$X_1(t+h) \cong X_1(t)(1 - k_{01}h) \tag{7}$$

These equations allow the computation of the amount of material in the compartment at time $t + h$ from the amount of material in the compartment at the earlier time, t. Suppose that the initial amount of material in the compartment was 10 gm, and suppose k_{01} = 0.0001/sec. To make use of (Eq. 7), we choose a suitably small value of h, say h = 1 sec. For this particular choice, we rewrite (Eq. 7) as

$$X_1(t+1) \cong X_1(t)(1-0.0001) = X_1(t)(0.9999) \tag{8}$$

To make use of the condition that (initially) $X_1(0) = 10$ gm, in (Eq. 8) we choose t = 0. Substituting this choice into (Eq. 8) yields

$$X_1(1) \cong 10(0.09999) = 9.9999 \tag{9}$$

Now that we have a good estimate of the amount, $X_1(1)$, left at 1 second, we can use (Eq. 8) again, this time with t chosen to be 1. When we do this, we find

$$X_1(2) \cong X_1(1)(0.9999) = 10(0.9999)^2 \tag{10}$$

We can continue this process, obtaining

$$X_1(T) \cong 10(0.9999)^T \tag{11}$$

Via hand calculator, we find that after 10 hours (36,000 seconds), the amount remaining is approximately

$$X_1(36{,}000) \cong 10(0.9999)^{36{,}000} = 10(0.0273) = 0.273 \tag{12}$$

Thus, this particular model predicts that only about 0.273 gm, or 3%, of the original substance will remain in the compartment after 10 hours .

In this case, we have developed a fairly simple formula to use to calculate the amount of material remaining in the compartment after T seconds of washout (Eq. 11), so that this value can be found using a hand calculator. Unfortunately, simple formulas are very hard to come by for multi-compartment models. However the type of tedious computation illustrated by (Eq. 7) does extend to more complicated compartmental models. Forty years ago, this observation wouldn't have been of much real value. Today, such computations are very feasible on inexpensive and readily available personal computers. To illustrate this, we choose a value for h, such as $h = 5$ sec, and actually carry out the computation

$$X_1(t+h) \cong X_1(t)(1-k_{01}h) \tag{13}$$

for the given choice of k_{01} and $h = 5$, successively for t = 0, h, 2h, ..., 36,000h. To show how easily this can be implemented, let us look at an example of a *For* loop using the Mathematica software program (Wolfram, 1991):

```
For[start,test,increment counter,computation]
```

The commas separate the different parts of the *For* loop. The counter is a symbol whose value reflects how many times the computation in the body has been done. At any stage, we'll let c stand for the counter's value, t the corresponding time, and *Temp* the value of the current amount of substance in the compartment. The actual program is

$$\text{For}[\underbrace{\text{Temp} = 10; \text{h} = 5\text{c} = 0\text{t} = 0}_{\text{start}}, \underbrace{\text{c} < 36000/\text{h}}_{\text{test}}, \underbrace{\text{c}++}_{\text{counter}},$$
$$\underbrace{\text{x} = \text{Temp}(1 - 0.0001\text{h}); \text{Temp} = \text{x}; \text{t} = \text{t} + \text{h}}_{\text{(body) computation}}]$$

The way the program runs is as follows:

- First (and only once) the starting values are set.
- Then the computations in the body are carried out in order (once).
- Then the counter is incremented (the first time it gets raised from 0–its starting value–to 1, the next time from 1 to 2, and so forth).
- Then the counter value, *c,* is evaluated to see if the test passes or fails. If $c < 36{,}000/h$ (pass), the body computations are carried out again, the counter is incremented, and the test is repeated. The process stops the first time the test fails

After running this program, entering x yields the desired value for the approximation to $X_1(36{,}000)$: 0.272991 (as expected).

We can also vary h and see whether the same answer could have been obtained with a larger value of h. With $h = 10$, we get an answer of 0.272746, which is pretty close to our original estimate. The fact that it was derived after half the original number of computations indicates, but doesn't guarantee, that this new estimate is adequate. Our conclusion is based on the widely-used, but not completely trustworthy rule of thumb, that holds that the desired value is well-approximated when further computational refinement yields essentially no change (Ralston, 1965).

Let's now see how simply this type of computation extends to a two-compartment model as shown by Figure 5.

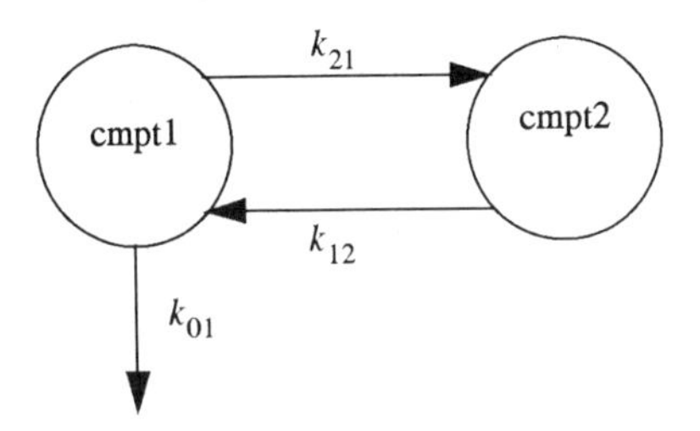

Figure 5. A simple two-compartment model, including rate constants and irreversible losses from compartment 1.

The equations are almost the same as (Eq. 2), which were missing the irretrievable loss indicated by k_{01}. They are

$$X_1'(t) = -(k_{21} + k_{01})X_1(t) + k_{12}X_2(t)$$
$$X_2'(t) = k_{21}X_1(t) - k_{12}X_2(t) \tag{14}$$

Suppose here that at $t = 0$ seconds, a bolus of 10 grams is injected into compartment 1, and mixes almost instantly throughout the entire compartment. We will assume that none of this material is in compartment 2 at time $t = 0$, but that some will appear at any time $t > 0$. Let us further assume that our rate constants were such that

$$k_{21} = 0.0001/\text{sec}, \quad k_{12} = 0.000001/\text{sec}, \quad k_{01} = 0.0001/\text{sec} \tag{15}$$

For quite a while, material flowing into compartment 2 will tend to build up, because until compartments 1 and 2 have near the same amount, much more will be entering than leaving compartment 2. Roughly speaking, you would expect that after 5 hours, compartment 1 will be pretty empty, and compartment 2 will have about 5 grams.

To see if our intuition is correct, let's approximate differential equations (Eq. 14) just the way we did before, and replace

$$X_1'(t) \text{ by } \frac{X_1(t+h)-X_1(t)}{h} \text{ and } X_2'(t) \text{ by } \frac{X_2(t+h)-X_2(t)}{h} \tag{16}$$

We can then use some algebra to get $X_1(t+h)$ and $X_2(t+h)$ written in terms of $X_1(t)$ *and* $X_2(t)$. We find

$$\begin{aligned} X_1(t+h) &\cong (1-0.0002h)X_1(t)+0.000001hX_2(t) \\ X_2(t+h) &\cong 0.0001hX_1(t)+(1-0.000001h)X_2(t) \end{aligned} \tag{17}$$

For the Mathematica program to approximate X_1 and X_2 at $t = 5$ hours (18,000 seconds), we let *Temp1* and *Temp2* be used to store the (current) values of X_1 and X_2 at t, and let $x1$ and $x2$ store the corresponding values at time $t + h$. For $h = 5$ seconds, the program is

```
For[Temp1 = 10; Temp2 = 0; h = 5; c = 0; t = 0, c < 18000/h, c++,
    x1 = (1 - .0002 h) Temp1 + .000001 h Temp2; x2 = .0001 h Temp1 +
    (1 - .000001 h) Temp2; Temp1 = x1; Temp2 = x2; t = t + h]
```

We can see that the Mathematica code for this model is not that much more complicated than our earlier version.

Our results indicate that at 5 hours, x1 = 0.294454 and x2 = 4.82011, as we predicted. At 50 hours from the start, we find x1 = 0.02291 and x2 = 4.5706. So it seems that in the final 40 hours, about 5% of what was in compartment 2 has disappeared. To see how long it would take for 98% of the contents to be eliminated from compartment 2, when compartment 1 is nearly empty (based on the result just derived concerning the final 40 hours, when compartment 1 is almost empty), we will assume that 5% of compartment 2 empties out every 40 hours. That means under compartment 1's condition of near emptiness, at the end of $40k$ more hours, the proportion of the initial amount remaining in compartment 2 is 0.95^k. Hence, if we want to determine how many hours are needed before only 2% remains, the answer is 40k, where $(0.95)^k = 0.02$, from which, using the well-known property of the logarithm function, that $ln(y^j) = j\, ln(y)$, we find

$$k = ln\,(0.02)\,/\,ln\,(0.95) \cong 76. \tag{18}$$

So about 76 steps of 40 hours (or 3040 hours, or 126 days) are needed.

Two more points are worth making in regard to differential equation modeling. First, we have only touched on the simplest methods of numerical solution. There are numerous refinements which allow much greater efficiency. Nevertheless, the basic ideas underlying numerical solution are quite well illustrated by the method that we have examined. Second, the models that we have looked at both assumed constant rate parameters. It may be necessary to use different models for different input ranges; or alternatively, to build models which take saturation into account. One such model is that of Michaelis-Menten kinetics, which seems to describe a clogging effect – so that at high concentrations, a given increase in concentration has less effect than this same increase at low concentrations. This model is given by the differential equation

$$X'(t) = -\frac{VX(t)}{K + X(t)} \tag{19}$$

Note that the outflow rate approaches V for large amounts of material, $X(t)$, in the compartment, while for very small amounts of material the outflow rate is close to $(V/K)X(t)$. For specified initial amounts and values of V, K, determination of the solution, $X(t)$, proceeds in much the same way as in the cases already treated.

ESTIMATION IN COMPARTMENTAL MODELS

The next question that arises is that of using observed data to learn something about the model (Venables and Ripley, 1996). For example, we might have assumed a model of the form

$$\begin{aligned} X_1'(t) &= -(k_{21} + k_{01})X_1(t) + k_{12}X_2(t) \\ X_2'(t) &= k_{21}X_1(t) - k_{12}X_2(t) \end{aligned} \tag{20}$$

By the phrase "of the form," we mean that at least one (and probably all) of the flow rate parameters, k_{01}, k_{21}, k_{12}, are unknown, and we want to determine something about them from experimental data. Maybe we want to know whether $k_{21} = 0$; if so, then we are dealing with a one-compartment model; if $k_{21} \neq 0$, then perhaps it is a model with two compartments, (a determination, as we discussed earlier, which may have important medical significance). Or maybe we want good estimates of the k_{ij}.

In order to simplify handling general problems of this type, we will convert to a less cumbersome vector notation. We let the symbol $\boldsymbol{K}$ stand for any arbitrary guess for the vector $(K_1, \ldots, K_m)$ of rate constants for whatever system we are considering. In this example, the system governed by (Eq. 20), $m = 3$, and $\boldsymbol{K} = (K_1, K_2, K_3)$ would stand for (k_{01}, k_{21}, k_{12})

As in our discussion of methods of solving systems like (Eq. 20), we'll only examine the simplest general approach, although most software packages use refined methods which are usually more efficient. While the details can get messy, the underlying ideas are not hard to comprehend. In some cases, in particular for the second model we examined, there is a simple formula involving the rate parameters to be used for the solution of the system of differential equations. Though this situation is one of the most common and thoroughly investigated, there seems to be no advantage to investigating such cases separately from the general case.

We begin by gathering data described by some system of differential equations; for illustrative purposes we'll use (Eq. 20) above and assume all of its rate parameters to be unknown.

Just about all approaches start out with initial guesses for the vector of rate parameters, and successively improve on them until satisfactory values for these parameters have been obtained. It's best if these initial guesses have some reasonable relation to reality, because almost all software packages used for estimating these rate constants can misbehave badly if this is not the case.

In order to determine when we are obtaining an improvement, we must choose a criterion to evaluate the merit of our first and succeeding guesses, and use this criterion to lead us to a satisfactory choice of the parameter vector $\boldsymbol{K}$ (Venables and Ripley, 1996). As the criterion for the quality of an estimate $\boldsymbol{K}$ of the rate constant vector, we will use the *total squared deviation* Ψ, which is almost equivalent to the standard deviation, but is mathematically easier to handle.

$$\psi(\boldsymbol{K}) = \sum_i [\{X_1(t_1,\boldsymbol{K}) - x_{1,i}\}^2 + \{X_2(t_i,\boldsymbol{K}) - x_{2,i}\}^2 + \cdots] \tag{21}$$

Here, the $X_j(t_j,\boldsymbol{K})$ are the calculated values of the solution to the model equations (such as Eq. 20), at times t_i, when the vector of rate constants is $\boldsymbol{K}$. All that need be kept in mind regarding the $X_j(t_j,\boldsymbol{K})$ is that there are readily available and very effective software programs (such as SAAM, Mlab, Mathematica, etc.) of the type we examined earlier, for their determination. The values, $x_{1,i}$, $x_{2,i}$, ..., represent actual experimentally-generated data, subject to experimental error, for $X_1(t_i, \boldsymbol{K}_{true})$, $X_2(t_i, \boldsymbol{K}_{true})$, ..., where $\boldsymbol{K}_{true}$ is the (unknown) vector of actual rate constants.

Notice that $\Psi(\boldsymbol{K})$ is never negative, and if we found a value of the vector $\boldsymbol{K}$ for which $\Psi(\boldsymbol{K}) = 0$, the fit would be perfect. This is something that almost never occurs, due to factors not accounted for by the model. These factors may not be of primary importance, but they do have an effect on the measurements..

More important, the more the calculated values $X_j(t_i, \boldsymbol{K})$, tend to differ from the observed data, $x_{j,i}$, the larger $\Psi(\boldsymbol{K})$ tends to be. So it seems reasonable to seek the value of $\boldsymbol{K}$ which makes the quantity $\Psi(\boldsymbol{K})$ smallest. This *minimizing value*, $\boldsymbol{K}_{min}$ of $\boldsymbol{K}$, can reasonably be considered to yield the *best fit* to the observed data, for a model of the type we're assuming, .

To get a better handle on the approach we'll be taking, it helps to visualize the criterion function, Ψ, as a surface lying above a base, (x, y), plane. The base plane is at sea level; and the surface may be thought of as a fairly smooth mountain. The points in the base plane may be thought of as possible values of the rate constant vector $\boldsymbol{K}$, and the value $\Psi(\boldsymbol{K})$ as the height of this surface above the point $\boldsymbol{K}$. Now unless $\boldsymbol{K}$ has only two coordinates, this picture is mainly symbolic. But even the usual picture of a three-dimensional figure on a flat page, (which would be used when $\boldsymbol{K}$ has only two coordinates) suffers from this difficulty. So think of the x, y base plane in the example given by (Eq. 20) as including three axes, one for each of the rate constants, $K_1 = k_{01}$, $K_2 = k_{12}$, $K_3 = k_{21}$ in $\boldsymbol{K}$, as illustrated in Figure 6.

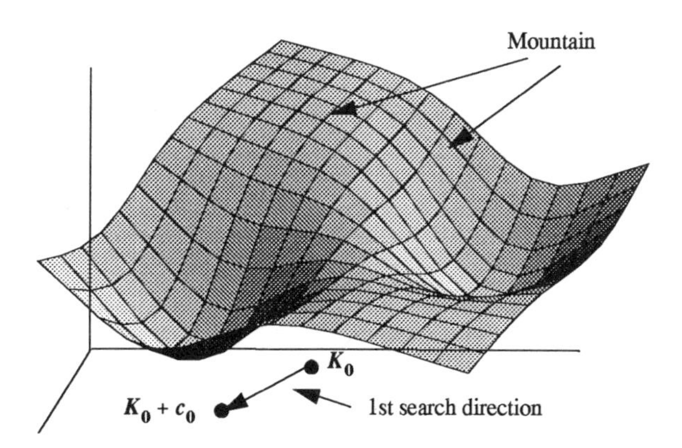

Figure 6. Geometrical illustration of the search for the low point on a surface whose height can be computed for any base-plane value.

In terms of this geometrical description of the criterion function, Ψ, our object is to find the value of $\boldsymbol{K}$ in the base plane whose corresponding surface height, $\Psi(\boldsymbol{K})$, is a minimum. The difficulty with this problem is that the only values available to you are the observed data, $x_{1,i}$, $x_{2,i}$, ..., and the corresponding computed solutions $X_1(t_i, \boldsymbol{K})$, $X_2(t_i, \boldsymbol{K})$, ..., for any finite number of given values of $\boldsymbol{K}$, from which you can compute $\Psi(\boldsymbol{K})$ for this set of $\boldsymbol{K}$ values – somewhat like operating with instruments which can provide you with mountain heights at any finite number of longitudes and latitudes you specify.

The method to be examined for finding the $\boldsymbol{K}$ minimizing $\Psi(\boldsymbol{K})$ is called the *steepest descent* (Luenberger, 1973). This designation arises from a very natural method of working our way down a mountain. The idea behind steepest descent is to start off with an initial guess, $\boldsymbol{K}_0$. Now from your position $P_0 = (\boldsymbol{K}_0, \Psi(\boldsymbol{K}_0))$ on the mountain, find the direction in which to proceed from $\boldsymbol{K}_0$, that gives the steepest looking dropoff, *close* to $\boldsymbol{K}_0$ – i.e. find a very short displacement, c_0 from $\boldsymbol{K}_0$, for which the slope

$$\frac{\psi(\boldsymbol{K}_0 + c_0) - \psi(\boldsymbol{K}_0)}{\|c_0\|} \tag{22}$$

is most negative. Here $\|c_0\|$, the length of $(c_{0,1}, c_{0,2}, \ldots, c_{0,m})$ is given by

$$\|c_0\| = \sqrt{c_{0,1}^2 + c_{0,2}^2 + \cdots + c_{0,m}^2} \tag{23}$$

where m is the number of rate constants in $\boldsymbol{K}$.

Restricting ourselves to values of $\boldsymbol{K}$ along this direction of steepest descent in the base plane, we can find one yielding a lower value than $\Psi(\boldsymbol{K}_0)$, and call it $\boldsymbol{K}_1$. At $\boldsymbol{K}_1$, find *its* direction of steepest descent . Continue on with this process, generating successive vectors $\boldsymbol{K}_2$, $\boldsymbol{K}_3$, ..., with each successive height $\Psi(\boldsymbol{K}_{i+1})$ smaller than the previous one, $\Psi(\boldsymbol{K}_i)$. Choose some reasonable way of terminating this process when no substantial improvements are being made.

Under very favorable circumstances, the path through the successive values $\boldsymbol{K}_1$, $\boldsymbol{K}_2$, $\boldsymbol{K}_3$,..., might look like the paths in Figure 7. Note that in the illustration, the successive heights are decreasing.

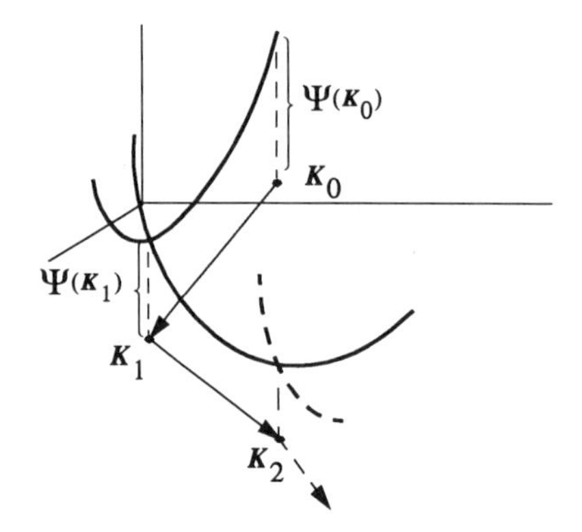

Figure 7. Illustration of paths taken on the surface when implementing steepest descent.

There are three fundamental problems which need solution to carry out this procedure:

1. Determining the steepest descent directions (the directions of the line segments emanating from each $\boldsymbol{K}_i$ toward $\boldsymbol{K}_{i+1}$),
2. Determining a point $\boldsymbol{K}_{i+1}$ on the line segment emanating from $\boldsymbol{K}_i$ with a reasonably lower surface height $\Psi(\boldsymbol{K}_{i+1})$ than $\Psi(\boldsymbol{K}_i)$ If possible, try for the lowest such surface height.
3. Deciding when to terminate this process, using the last computed $\boldsymbol{K}_i$ as the sought-for value, $\boldsymbol{K}_{min}$.

To present the solution to the first of these problems, we must introduce the concept of partial derivatives. Recall that the derivative, $X'(t)$, of a function, X, of time, is geometrically, just the slope of the tangent to the curve X, at the point $(t,X(t))$, as indicated in Figure 8.

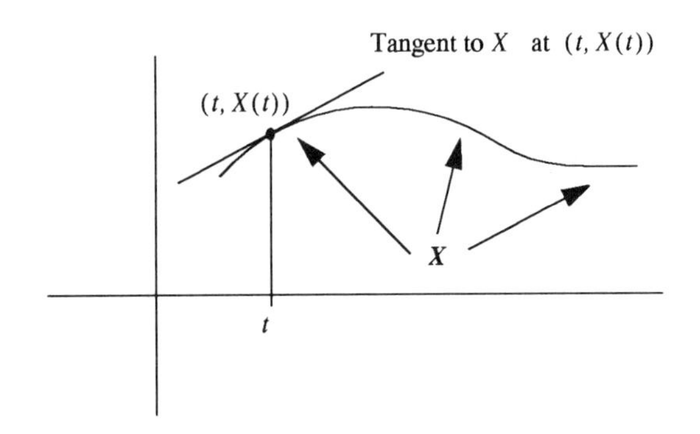

Figure 8. Tangent line illustration, for extension to one-dimensional steepest descent searches.

The slope

$$X'(t) \cong \frac{X(t+h) - X(t)}{h} \tag{24}$$

where h is a small, nonzero number. The partial derivative is a simple extension of this concept to a function of several variables, whose graph may be thought of as a surface, rather than a curve. The partial derivatives are just slopes of curves obtained by restricting to that portion of the surface lying above lines parallel to the coordinate axes. From an algebraic view, the partial derivative of the function Ψ with respect to the first coordinate of the rate constant vector, at its value $\boldsymbol{K} = (K_1, K_2, \ldots, K_m)$ and denoted by $D_1 \Psi(\boldsymbol{K})$ is computable from the equation

$$D_1\psi(\boldsymbol{K}) \cong \frac{\psi(K_1 + h, K_2, \cdots, K_m) - \psi(K_1, K_2, \cdots, K_m)}{h} \tag{25}$$

for h, a very small nonzero number.

The partial derivative of Ψ with respect to its jth coordinate, $D_j \Psi(\boldsymbol{K})$ is defined similarly, (with $K_j + h$ in the jth coordinate, instead of $K_1 + h$ in the first coordinate.)

The vector $grad\,\Psi(\boldsymbol{K}) = (D_1 \Psi(\boldsymbol{K}), \ldots, D_m \Psi(\boldsymbol{K}))$, of partial derivatives is called the gradient of the function Ψ at $\boldsymbol{K}$.

A result is provided in the following theorem. The proof for this result can be found in Rosenblatt and Bell (1998).

The direction of steepest descent of Ψ at $\boldsymbol{K}$ is that of $-grad\,\Psi(\boldsymbol{K})$. That is, (22) is most negative for small c_0 when c_0 is a small negative multiple of $grad\,\Psi(\boldsymbol{K})$, (presuming that this gradient is not a vector of all 0s).

This solves the first of the problems we discussed earlier. It shows that when we try to determine $\boldsymbol{K}_{i+1}$ from $\boldsymbol{K}_i$, we need only examine values of the form $\Psi(\boldsymbol{K}_i + c_i)$, where c_i is a negative multiple of $grad\,\Psi(\boldsymbol{K}_i)$. So, steepest descent reduces the problem of finding the lowest height of the (m-dimensional) surface Ψ, to a sequence of one-dimensional search problems.

But even though one-dimensional searches tend to be simpler than multidimensional ones, searching for the minimum height of a curve, with the tools available to us, can end in failure, as might occur in Figure 9. Here, because we can only compute the heights at a finite number of points, we could miss a low point entirely and be lead to P rather than Q.

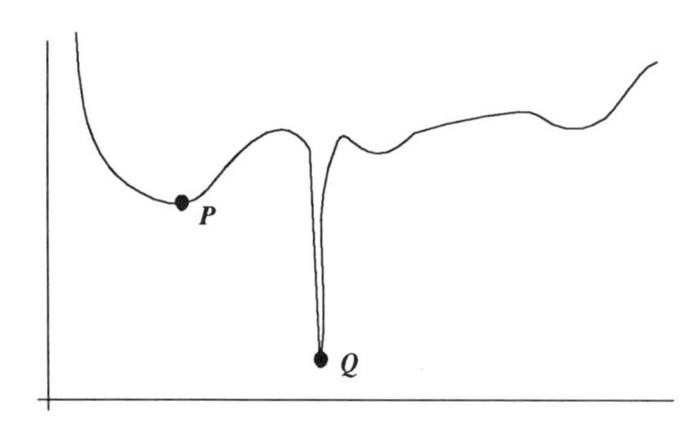

Figure 9. Illustration of curve whose absolute minimum is easy for programs to miss.

We try our best to get the first guess, $\boldsymbol{K}_0$ close enough to the value $\boldsymbol{K}_{min}$ that we want, so that we don't encounter such behavior in the regions being searched. The following one-dimensional search method will find the desired low point, when the curve is concave up, with the low point inside, as illustrated in Figure 10. Here we let $u_i = c_i / \|c_i\|$ be the unit length vector in the steepest descent direction emanating from $\boldsymbol{K}_i$. For some positive value d that we choose, we compute $\Psi(\boldsymbol{K}_i + ndu_i)$ for $n = 0{,}1$, and if $\Psi(\boldsymbol{K}_i) \leq \Psi(\boldsymbol{K}_i + du_i)$, then we reduce the magnitude of the positive value d until we achieve the condition $\Psi(\boldsymbol{K}_i) > \Psi(\boldsymbol{K}_i + du_i)$. Next, for $n = 2, ..., N$, we successively compute $\Psi(\boldsymbol{K}_i + du_i)$, where N is the first value exceeding 2, for which $\Psi(\boldsymbol{K}_i + [\,N-1]\,du_i) < \Psi(\boldsymbol{K}_i + Ndu_i)$. (In Fig. 10, it looks as if N = 3.) The midpoint of the interval $[\,\Psi(\boldsymbol{K}_i + [\,N-1]\,du_i),\ \Psi(\boldsymbol{K}_i + Ndu_i)]$ is within $d/2$ of the low point, $\boldsymbol{K}_{i+1}$ that we are looking for. By choosing d small enough, any desired accuracy in determination of the low point can be achieved.

Under the specified conditions, this method solves the second of the three fundamental estimation problems. One-dimensional searches have been extensively studied for centuries, and there are many more efficient methods available for this purpose. Nonetheless, this method is often used preliminary to switching to a more refined one.

Assuming success in the one-dimensional searches, we only need consider the solution of the final fundamental estimation problem – when to terminate. The rule of thumb is to stop when $\boldsymbol{K}_{i+1}$ is sufficiently close to $\boldsymbol{K}_i$.

A final word regarding estimation procedures: even an apparently successful procedure may only lead to a relative minimum and not necessarily the smallest value of $\Psi(\boldsymbol{K})$, and I know of no package which always guarantees success. But success comes much more often to those whose first guess, $\boldsymbol{K}_0$ is reasonably close to the desired value $\boldsymbol{K}_{min}$.

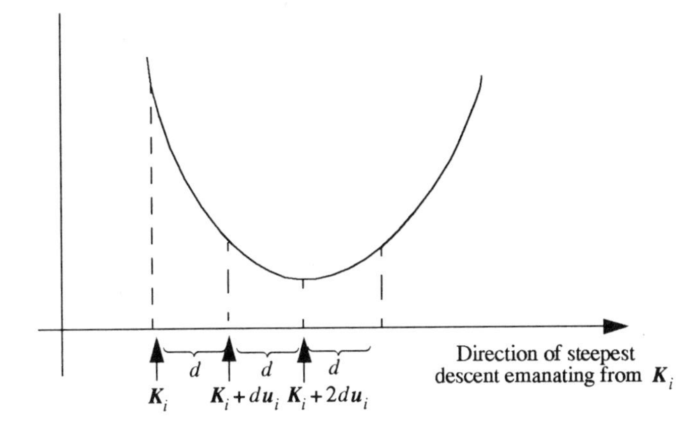

Figure 10. Illustration of simple method for finding minimum of function of one variable, for a one-dimensional search.

ACCURACY OF ESTIMATION IN COMPARTMENTAL MODELS

The final task is determining how well the final computed estimate of the differential equations solution fits the data, and how accurate are the estimates of the rate constants. For determining how well the final solution estimate fits the data, use is made of the root mean square (RMS), deviation s, given by (26).

$$s = \sqrt{\frac{1}{n-m}\psi(\boldsymbol{K}_{final})} \tag{26}$$

where m is the number of rate parameters estimated and n is the number of data values. The RMS deviation is often close to the true raw data standard deviation. This RMS deviation can usually be interpreted as a reasonable average measure of deviation of data from the theoretical value – deviations due to factors not explicitly taken account of by the model. If s is too large, the model may be of little use to you for predicting or controlling future data. You will have to be the judge. A rule of thumb which has some basis in theory, is that for a large enough sample size, approximately 95% of the raw data will be no further than $2s$ from the model that was derived by minimizing $\Psi(\boldsymbol{K})$.

The problem of determining the accuracy of rate parameter estimates, k_1, k_2, ..., k_m is often handled by the so-called asymptotic theory, a theory which is valid provided the size of the data set is sufficiently large. Because there is usually no agreed-upon meaning for the phrase 'sufficiently large,' the asymptotic theory may give misleading answers. An alternative, more reliable, approach is via Monte Carlo Simulation (Bratley et al., 1983).

To get an idea of this approach in a very simple situation, imagine that we have a coin that we suspect may not be a fair one. So we toss it 200 times, and obtain 120 heads. Our estimate of the probability of a head is 0.6. But how much variability is there in this estimate? In this situation, there is a simple answer based on theory, and our best estimate of the standard deviation of the estimate is $\sqrt{0.6 \times 0.4 / 200}$.

But if there were no simple theoretical answer, here's what you might do. Find a coin whose probability of heads is known to be 0.6. Let E200 be the experiment of tossing *this* coin 200 times. Now repeat E200 many times, and let $\hat{p}_k$ be the observed proportion of heads on the kth run of experiment E200. Use the following computable sample standard deviation as an estimate of your desire standard deviation,

$$s_{p_k} = \sqrt{\frac{1}{N-1}\sum_{k=1}^{N}(\hat{p}_k - \bar{\hat{p}})^2} \tag{27}$$

where $\bar{\hat{p}}$ is the average of the values $\hat{p}_k$. Carrying this procedure out in the Minitab program was a matter of typing three short lines. The theoretical estimate in the case we ran was 0.0346, while the Monte-Carlo result yielded 0.0343 (when E200 was run 2000 times).

In this situation, let us assume that the obtained model seems a close fit to the data–as judged by either a graph of the solutions of the differential equations using the final estimate of $\boldsymbol{K}$ on which the data are superimposed, or by a table of the residuals – the differences between the observed data values and the chosen model's predictions, or by a sufficiently small value of s. Here is how we extend the simple procedure to the illustration above.

First: Generate some theoretical data from the model just obtained, at the same time as points of the data used for developing the model.

Second: Add random error with standard deviation s to each data point. This can be accomplished using random number generators, which are available in most statistical computer packages

Third: Using this newly generated data, find new estimates of the parameters k_1, k_2, ..., k_m (using whatever method is available)

Repeat the second and third steps many r times, obtaining estimates $k_1^{(i)}$, $k_2^{(i)}$, ..., $k_m^{(i)}$ where $i = 1, 2, \ldots, r$.

Define $\bar{k}_j$ to be the average over $i = 1, \ldots,$ r of the $k_j^{(i)}$, and let

$$s_j = \sqrt{\frac{1}{r-1}\sum_{i=1}^{r}(k_j^{(i)} - \bar{k}_j)^2} \tag{28}$$

The coefficient of variation, (CV), s_j / k_j is a reasonable measure of relative error for the estimate of the jth rate parameter, k_j. In my own experience, a CV below 0.1, (or an average relative error of 10%), is about as good as is seen in human studies; a much smaller CV is often observed in the engineering field, which of course, permits a tighter control of experimental conditions.

SUMMARY

The purpose of this chapter has been to furnish insight into the theoretical background on which compartmental modeling software packages are based. To accomplish this goal, only the basic ideas were stressed, avoiding discussion of the intricacies required for efficiency. The object was to remove the mystery from these powerful programs by examining the fundamental ideas which make them tick.

The first section was concerned with how compartmental models are built up and how to obtain information concerning the system behavior described by these models. The cornerstone here is to describe a system by determining how it behaves over (typically) very short time periods. This leads to a differential equation description of a compartmental system. Information can be extracted from these equations by returning to their basic meaning, illustrated in their derivation. Computers are ideal for obtaining this information by piecing together the results obtained over short time periods, to find the behavior over long time periods.

In an actual situation governed by a compartmental model, it's often the case that we may know only the form of the model, but not the values of the rate constants which must be known for its effective use. The second section of the chapter was devoted to the practical problem of determining the rate constants of the model, based on observed data. This is a matter of searching for those values which, in some sense, best fit the data. To attack this problem we need a reasonable criterion to judge how well a proposed model fits the data. We chose to use the *total squared deviation*, Ψ, which is the most common such criterion—but not the only reasonable one. The search technique we examined—*steepest descent*—is based on a simple idea: looking at the total squared deviation criterion geometrically. In graphical terms, the *best fit* corresponds to finding the low point on a surface, whose height above any point at sea level is computable. If we could imagine the view of someone trying to find the low point from some arbitrarily chosen initial position on this mountain-like surface, we would look around and find the direction where (*close by*) the mountain drops off most steeply. We would go in that specific direction until we reach a low point, moving only along this initially chosen direction. At this new low point, we could change again to a direction of steepest descent, and keep repeating this process until we make no further effective downward progress. No guarantee in general is made for this process, but it often works.

Finally, having found the *best values* for the rate constants, we must recognize that if the data is affected by factors not explicitly taken into account in the model, the variability induced by these factors precludes a perfect fit. For this reason, it is finally necessary to determine how good the model is, as a description of the data, and how accurate are the fitted rate constants. For the model fit, the sample RMS error (simply related to the total squared deviation, and often the same as the sample standard deviation) may be used. For determining the accuracy of the fitted rate constants, practical methods based on computer software simulations are recommended.

CORRESPONDING AUTHOR

Please address all correspondence to:
Judah Rosenblatt
University of Texas Medical Branch
301 University Blvd.
Galveston, TX 775555-1220

REFERENCES

Bratley, P; Bennett LF; Schrage LE. *A Guide to Simulation.* Springer-Verlag: New York. 1983.

Golomb M; Shanks M; *Elements of Ordinary Differential Equations, 2nd Ed.* McGraw-Hill: New York. 1965.

Jacquez JA. *Compartmental Analysis in Biology and Medicine, 2nd Ed.* University of Michigan Press: Ann Arbor. 1988.

Luenberger DG. *Introduction to Linear and Nonlinear Programming.* Addison-Wesley: Reading, MA. 1973.

Ralston A. *A First Course in Numerical Analysis.* McGraw-Hill: New York. 1965.

Rosenblatt J; Bell S. *Mathematical Analysis for Modeling.* CRC Press: Boca Raton. 1998.

Venables WN; Ripley BD. *Modern Applied Statistics with S-Plus.* Springer-Verlag: New York. 1996.

Wolfram S. *Mathematica, 2nd Ed.* Addison-Wesley: Reading, MA. 1991.

DISTRIBUTING WORKING VERSIONS OF PUBLISHED MATHEMATICAL MODELS FOR BIOLOGICAL SYSTEMS VIA THE INTERNET

ME Wastney,[1] DC Yang,[2] DF Andretta,[1] J Blumenthal,[3] J Hylton,[3] N Canolty,[4] JC Collins,[5] and RC Boston[6]

[1]Department of Pediatrics
[2]Department of Chemistry
[3]Dahlgren Memorial Library
Georgetown University Medical Center, Washington, D.C.
[4]Department of Foods and Nutrition
The University of Georgia, Athens, GA
[5]Department of Medicine
Vanderbilt University School of Medicine, Nashville, TN
[6]Biomathematics Unit, School of Veterinary Medicine
University of Pennsylvania, Kennett Square, PA

ABSTRACT

Mathematical models are useful tools for investigating complex systems. By representing physiological systems as models, theories can be tested quantitatively against data from the system. Models can be used to explore new theories prior to experimentation and to design studies to optimize experimental resources. They can also be used as teaching tools to illustrate physiochemical principles. In spite of their usefulness and the time invested in developing models, published models are often underused due to the difficulty in obtaining working versions of the model. To address this problem we have designed a library for mathematical models of biological systems on the Internet. The library contains published models of biological systems in formats compatible with several modeling packages, from the fields of physiology, metabolism, endocrinology, biochemistry, and chemistry. The models can be viewed graphically, model solutions can be viewed as plots against data, and models can be downloaded to be run with software on the user's own system. The address of the library is: http://biomodel.georgetown.edu/model/

Investigators are invited to submit working versions of published models to the library. Models can be submitted electronically at the time a manuscript is accepted for publication. As journals go online, articles containing models can be linked to working versions of the models in the library. By increasing access to working versions of models, more of the investment in kinetic studies and model development can be realized.

INTRODUCTION

Engineers, economists, and physicists dealing with complex systems use mathematical models routinely in the study of these systems. While bioengineers use models to study biological systems, they are not used by most biologists. There are a number of reasons why modeling is not used routinely by biologists and they relate to a lack of expertise and the availability of tools. Firstly, it is difficult for biologists to think about systems quantitatively if they have not been trained in quantitative sciences or mathematics. Even with this training, modeling biological systems differs sufficiently from modeling physical systems (where the structure of the system is usually known) so that different tools and approaches need to be learned. Furthermore, modeling software has often been developed by individual groups for a specific area. Translating and applying the software to other systems has been difficult because of the low-level language of the software and incomplete user documentation. Unfortunately, models that are developed are often not described fully or made available to the biological community.

The use of modeling in the biological community is increasing, however. Software tools for modeling are now readily available. Educating biology students in the concepts of engineering and training investigators in the use of modeling is beginning with the emergence of modeling courses at various institutions. This paper describes a mechanism for increasing accessibility to models by making published models available via the Internet. Facilitating access to models ensures that the models can be obtained, tested and used by other investigators. In this way models of biological systems can be used to design new studies, and proposed models can be expanded or replaced by new models when new data are obtained. Using models in this manner enables limited research resources to be used more efficiently.

USING MODELS TO UNDERSTAND BIOLOGICAL SYSTEMS

Biological systems are complex because many processes occur within the organism. These processes are linked and the systems are dynamic; many behave nonlinearly, and in a manner that cannot always be predicted from intuition. Because of this complexity and the rapid accumulation of knowledge about this complexity, models may be the best way to integrate and understand information from biological systems. Simplified descriptions of the real system, models are used to identify system features, such as the number of processes or the value of a parameter of a particular process, and to describe the behavior of the system under different conditions. Mechanistic models incorporate features such as the structure of a system. For example, a compartmental model may contain separate compartments for a compound of interest that is located in different tissues. The connections between the compartments would then represent movement of the compound between the tissues. Once a model solution fits observed data and is consistent with other known information about a system, the model can be used to calculate parameter values for the system (such as rates of transfer between specific tissues).

Once a model has been developed to fit data from a particular study, it can be used to make predictions about the behavior of a system under other conditions. For example, a model of whole-body metabolism can be used to predict changes in plasma concentration when intake of a compound increases, or a model for cellular uptake can be used to predict intracellular levels of a compound if an inhibitor is added. These predictions can then be tested experimentally. The information in models can be used therefore to design more efficient experiments.

ACCESSING MODELS THROUGH A MODEL LIBRARY

In the past, published models could be used by other investigators only if the authors were willing to supply an electronic version of the model or if investigators had the time and expertise to recreate the model from the paper. Unfortunately, often neither situation existed and the information residing within models was unused. Setting up models in working format from papers is sometimes difficult because important information may be omitted or printed incorrectly in the publication. Furthermore, software and technical difficulties may also occur.

To expedite access to models we, and others, have proposed a library of models (Summers and Montani, 1994; Wastney et al., 1995) and we have developed a pilot library of mathematical models accessible on the Internet (Wastney et al., 1997; Wastney et al., 1998). The library also acts as a forum for information on new publications, modeling software, demonstrations, and modeling resources. The *Mathematical Models of Biological Systems Library* can be accessed at: http://biomodel.georgetown.edu/model/

From the home page, users can access information on the purpose of the library, the personnel involved in the development of the library, and a list of publications which contain models. The publication list is updated regularly from literature searches on terms relevant to modeling such as biological model, computer simulation, kinetic, dynamic, and compartmental analysis. Users also have the opportunity of subscribing to a "Who's Who," a list of people with expertise in modeling in various areas, that is being collated so that investigators interested in specific areas can easily contact people with appropriate modeling expertise.

Users can search for models and submit models to the library (as described below). Models can be located in various ways (for example, by entering a search term such as a compound of interest or a type of model). Alternatively, all models in the library can be listed, or the models can be listed in terms of hierarchy (e.g., molecular, cellular, whole body, or population level model), species studied, type of model, and software used. If data are supplied with a model, these are also made available through the model library.

OBTAINING MODELS AND DEPOSITING MODELS IN THE LIBRARY

The library currently contains about forty models, including some commonly used chemical models. The chemical models differ in that a tutorial on use of the model and some exercises are included. Specifically, there is a model of enzyme kinetics (Hensley et al., 1990; Hensley et al., 1992), a cellular model of fluid-phase endocytosis (Blomhoff et al., 1989), several at the tissue and organ level including ruminant digestion (Dijkstra et al., 1992; France et al., 1982; Beever et al., 1981) and glucose-insulin metabolism (Weber et al., 1989), models of whole body metabolism of calcium, lithium, magnesium, selenium, and zinc (Ramberg et al., 1973; Everts et al., 1996; Neer et al., 1967; Avioli and Berman, 1966; Patterson et al., 1989; Wastney et al., 1986), a model of pharmacokinetic dosing regimens (Jackson and Zech, 1991) and a population model of AIDS (Anderson et al., 1986).

The chemical models section includes models of chemical reactions and enzyme kinetics. The models are presented in several formats: as chemical equations, as rate equations in the form of differential equations, as a graphic, and as a working version of the model. In addition, simulated plots are shown using varying parameter values, and examples of experimental results are incorporated for illustration. Because these models are simple and well-established, researchers from different fields can learn the basics of simulating reactions by modeling through this section of the *Model Library*.

Once a model of interest has been located, a working version of the model can be accessed in a format compatible with one or more commonly used software packages. These include SAAM (Berman and Weiss, 1978; Berman et al., 1983; Boston et al., 1981), SAAM II (Foster et al., 1994), ACSL (Mitchell and Gauthier, 1986), and in the future, Scientist (Bogen, 1989) and MLAB (Knott, 1992). Using the menu on the browser, models can be downloaded to the user's system to be used with their own software. Although it might be helpful to be able to run a model on the library server, due to software license restrictions, SAAM (Berman and Weiss, 1978) is the only software package of those mentioned above that could potentially be run in this way.

Criteria for submitting models to the library are that they have been published in a peer-review journal and that they have been fitted to data. Investigators are encouraged to check the web site or contact the library developers for the latest instructions prior to submitting a model.

MODELING AND NUTRITION RESEARCH

Providing access to working versions of models on the Internet will likely have a direct effect on nutrition research by helping investigators to identify areas for research, test hypotheses prior to experimentation, and design new studies. Furthermore, models of various nutrients will be able to be linked to explore sites of interaction. These may include specific interactions between nutrients, such as the trace elements zinc and copper, or broad interactions, such as trace elements with vitamins. Endocrine models may be linked to nutrient models, to describe interactions quantitatively, such as those of calcitropic hormones on calcium metabolism (Jung, 1982). This concept of linking and integrating nutrition models could form the basis of a 'Nutriome Project' for relating genetics, nutrition, and physiology.

ACKNOWLEDGEMENTS

This article was supported by NSF-BIR 950-3872 and by NIH NCRR RR-00095.

CORRESPONDING AUTHOR

Please address all correspondence to:
Meryl E. Wastney, PhD
Division of Neonatology
Georgetown University Medical Center
3800 Reservoir Road, NW
Washington DC 20007
ph: 202-687-5004
fax: 202-784-4747
wastneym@gunet.georgetown.edu

REFERENCES

Anderson RM; Medley GF; May RM; Johnson AM. A preliminary study of the transmission dynamics of the human immunodeficiency virus (HIV), the causative agent of AIDS. *IMA J Math Appl Med Biol,* 1986, 3:229-263.

Avioli LV; Berman M. Mg^{28} kinetics in man. *J Appl Physiol,* 1966, 21:1688-1694.
Beever DE; Osbourn DF; Cammell SB; and Terry RA. The effect of grinding and pelleting on the digestion of Italian ryegrass and timothy by sheep. *Br J Nutr*, 1981, 46:357-370.
Berman M; Weiss MF. *SAAM Manual.* DHEW Publication No. NIH 78-180. US Printing Office, Washington, DC. 1978.
Berman M; Beltz WF; Greif PC; Chabay R; and Boston RC. *CONSAM User's Guide.* DHEW Publication No. 1983-421-123:3279. US Government Printing Office, Washington, DC. 1983.
Blomhoff R; Nenseter MS; Green MH; Berg T. A multicompartmental model of fluid-phase endocytosis in rabbit liver parenchymal cells. *Biochem J,* 1989, 262:605-610.
Bogen DK. Simulation software for the Macintosh. *Science*, 1989, 246:138-142.
Boston RC; Greif PC; Berman M. Conversational SAAM – An interactive program for kinetic analysis of biological systems. *Computer Programs in Biomedicine*, 1981, 13:111-119.
Dijkstra J; Neal HDSC; Beever DE; France J. Simulation of nutrient digestion, absorption and outflow in the rumen: Model description. *J Nutr,* 1992, 122:2239-2256.
Everts HB; Jang HY; Boston RC; Canolty NL. A compartmental model predicts that dietary potassium affects lithium dynamics in rats. *J Nutr,* 1996, 126:1445-1454.
France J; Thornley HM; Beever DE. A mathematical model of the rumen. *J Agric Sci Camb,* 1982, 99:343-353.
Foster DM; Barrett HR; Bell BM; Beltz WF; Cobelli C; Golde H; Jacquez JA; Phair RD. SAAM II: simulation, analysis and modeling software. *BMES Bulletin*, 1994, 18:19-21.
Hensley P; Nardone G; Chirikjian JG; Wastney ME. The time-resolved kinetics of superhelical DNA cleavage by BamHI restriction endonuclease. *J Biol Chem,* 1990, 265 (25):15300-15307.
Hensley P; Nardone G; Wastney ME. Compartmental analysis of enzyme-catalyzed reactions. *Meth Enzymol,* 1992, 210:391-405.
Jackson AJ; Zech LA. Easy and practical utilization of CONSAM for simulation, analysis, and optimization of complex dosing regimens. *J Pharm Sci*, 1991, 80:317-320.
Jung A. Endocrine control of calcium metabolism, in: *Quantitative Approaches to Metabolism.* Cramp DG; Ed. John Wiley & Sons Ltd., 1982. pp. 219-245.
Knott G. MLAB Interactive computer program for mathematical modeling. *J NIH Res,* 1992, 4:93.
Mitchell; Gauthier A. *Advanced Continuous Simulation Language (ACSL) User Guide/Reference Manual.* Concord, MA. 1986.
Neer R; Berman M; Fisher L; Rosenberg LE. Multicompartmental analysis of calcium kinetics in normal adult males. *J Clin Invest*, 1967, 46 (8):1364-1379.
Patterson BH; Levander OA: Helzlsouer K; McAdam PA; Lewis SA; Taylor PR; Veillon C; Zech LA. Human selenite metabolism: A kinetic model. *Am J Physiol,* 1989, 257 (Reg. Integr Comp Physiol 26):R556-R567.
Ramberg CF Jr; Delivoria-Papadopoulos M; Crandall ED; Kronfeld DS. Kinetic analysis of calcium transport across the placenta. *J Appl Physiol,* 1973, 35:682-688.
Summers RL; Montani J-P. Interface for the documentation and compilation of a library of computer models in physiology. *JAMIA*, 1994, Symposium Supplement:86-89.
Wastney ME; Aamodt RL; Rumble WF; Henkin RI. Kinetic analysis of zinc metabolism and its regulation in normal humans. *Am J Physiol*, 1986, 251 (Reg Integ Comp Physiol 20):R398-R408.
Wastney ME; Broering N; Ramberg CFR Jr;. Zech LA;, Canolty N; Boston RC. World-wide access to computer models of biological systems. *Info Serv & Use*, 1995, 15:185-191.
Wastney ME; Siva Subramanian KN; Broering N; Boston R. Using models to explore whole-body metabolism and accessing models through a model library. *Metabolism*, 1997, 46:330-332.
Wastney ME; Wang XQ; Boston RC. Publishing, interpreting and accessing models. *J Franklin Instit,* 1998, 335B:281-301.
Weber KM; Martin IK; Best JD; Alford FP; Boston RC. Alternative method for minimal model analysis of intravenous glucose tolerance data. *Am J Physiol*, 1989, 256 (Endocrinol Metab 19):E524-E535.

Part II

STATISTICAL MODELING IN NUTRITION

MEASUREMENT ERROR AND DIETARY INTAKE

Raymond J. Carroll,[1] Laurence S. Freedman,[2] and Victor Kipnis[2]

[1]Department of Statistics
Texas A&M University
College Station, TX 77843
[2]Biometry Branch, DCP
National Cancer Institute
Bethesda, MD 20892

ABSTRACT

This chapter reviews work of Carroll, Freedman, Kipnis, and Li (1998) on the statistical analysis of the relationship between dietary intake and health outcomes. In the area of nutritional epidemiology, there is some evidence from biomarker studies that the usual statistical model for dietary measurements may break down due to two causes: (a) systematic biases depending on a person's body mass index; and (b) an additional random component of bias, so that the error structure is the same as a one-way random effects model. We investigate this problem, in the context of (1) the estimation of the distribution of usual nutrient intake; (2) estimating the correlation between a nutrient instrument and usual nutrient intake; and (3) estimating the true relative risk from an estimated relative risk using the error–prone covariate. While systematic bias due to body mass index appears to have little effect, the additional random effect in the variance structure is shown to have a potentially important impact on overall results, both on corrections for relative risk estimates and in estimating the distribution usual of nutrient intake. Our results point to a need for new experiments aimed at estimation of a crucial parameter.

INTRODUCTION

This paper reviews work of Carroll, Freedman, Kipnis, and Li (1998) on the statistical analysis of the relationship between nutrient intake data and health outcomes. The quantification of an individual's usual diet is difficult, and so various dietary assessment instruments have been devised, of which three main types are most commonly used in contemporary nutritional research. The one that is most convenient and inexpensive to use is the Food Frequency Questionnaire (FFQ), which is the instrument of choice in large nutritional epidemiology studies. However, while dietary intake levels reported from FFQs are correlated with true usual intake, they are thought to involve a systematic bias (i.e.

under- or over–reporting at the level of the individual). The other two instruments that are commonly used are the 24-hour Food Recall and the multiple-day Food Record. Each of these is more work-intensive and more costly, but is thought to involve less bias than a FFQ. However, the large daily variation in a western diet makes a single Food Record/ Recall an imprecise measure of true usual intake.

The major difference between these instruments is easily explained. The FFQ is exactly what it seems. Study participants are asked to fill out a questionnaire concerning their typical eating patterns over a recent period of time (often 6 months to a year), including typical foods and portion sizes. Similarly, Food Records are typically determined by participants keeping a diary of what was eaten per day, including portion sizes, usually for a 3–7 day period. Food Recalls are generally interviews in which the study participant's report their food consumption in the previous 24-hour period. The results of all three instruments are then converted to quantitative dietary amounts, e.g., total energy, total fat, etc. Thus, the data available for analysis are numeric, e.g., measures of Percent Calories from Fat (%CF) in the diet, over a longer period (FFQ), a shorter period (Food Records) or a single day (Food Recalls).

Hunter et al. (1996) reported on a pooled analysis of seven cohort studies of dietary fat and breast cancer, with nutrient intakes measured by a FFQ. They observed a relative risk of only 1.07 for the highest as opposed to the lowest quintile of energy–adjusted total fat intake, with a similar trend across %CF. This suggests that cohort studies yield an observed relative risk for these quintile medians of approximately 1.10 or less when measuring %CF from a FFQ. The essential controversy arises over the meaning of these figures. It is well known (Freedman et al., 1991) that FFQs contain substantial measurement error, and that this error may have a systematic component, and that bias and measurement error have impacts on relative risks (Rosner et al., 1989; Carroll et al., 1995). To take into account the effects of measurement error in these instruments, they used an analysis due to Rosner et al. (1989). Their analysis is closely related to a model we define below as (1–2), and is given as follows. It is typical to give a FFQ to all study participants. In a subset of the main study, often called a calibration study, a selected group of study participants are also asked to fill out (typically multiple) Food Records or 24-hour Food Recalls. In the calibration study, the mean of these Food Record/Recalls is regressed against the value of the FFQ, here called Q, forming a prediction equation of the form $a + b \times Q$. This prediction equation is then calculated for all study participants, and a logistic regression is run with cancer incidence as the outcome. More motivation for this procedure is given below.

Hunter et al. examined the impact that bias and measurement errors in FFQs have on observed relative risk. From their analysis, they conclude that even accounting for errors in FFQs, the impact of %CF on breast cancer relative risk is unlikely to be large. Prentice (1996), using data from the Women's Health Trial Vanguard Study (WHTVS) (Henderson et al., 1990), uses a different model from Hunter et al., and concludes exactly the opposite, namely that in fact the cohort data are consistent with a potentially large and biologically important effect of %CF. Prentice's model is complex, but it is closely related to ours in spirit.

In this chapter, we re-examine this controversy. We consider the errors–in–variables model of Freedman et al. (1991) which is essentially the same as that of Hunter et al., and we develop a new linear errors–in–variables model which has important similarities to Prentice's model, one of which is explained below. We apply the old and the new model not only to the Women's Health Trial Vanguard Study, but also to data from an American Cancer Society Study. Analysis of the former study yields similar results to those of Prentice, but the latter study yields results somewhat more in line with the Hunter et al. conclusions. Thus, the estimated impact of dietary bias and measurement error may depend on the model for the bias and measurement error, but it also may depend on the data used to examine the impact.

This chapter is organized as follows. In the first section, we describe the usual measurement error model for dietary intake, apply it to the two data sets mentioned above, and discuss some of its potential shortcomings. In the next section, we introduce a new measurement error model, and apply it as well to the two data sets. The final section has concluding remarks.

THE USUAL MODEL

In seven of their studies, Hunter et al. set out to understand the bias and measurement error in FFQs. The investigators undertook calibration studies in which Food Record/Recalls were measured on a subset of the study participants. A model, which is similar in many respects to theirs, is the linear errors–in–variables model used by Freedman et al. (1991) and defined as follows. In the calibration studies, the individual reports diet using a FFQ on m_1 occasions ($m_1 \geq 1$) and using a Food Record/Recall on m_2 occasions ($m_2 \geq 2$). The model related intake of some nutrient (e.g., %CF) reported on FFQs (denoted by Q) and intake reported on Food Record/Recalls (denoted by F) to long-term usual intake (denoted by T) is a standard linear errors--in--variables model, namely

$$Q_j = \beta_0 + \beta_1 T + r + \varepsilon_j \qquad j = 1, \cdots, m_1 \tag{1}$$

$$F_j = T + U_j \qquad j = 1, \cdots, m_2 \tag{2}$$

In model (1), r is called the *equation error* (Fuller, 1987). The terms ε_j represent the within individual variation in FFQs, while the U_j are the within individual variation in Food Record/Recalls. Among these random variables, T has mean μ_t and variance σ_t^2, U_j has mean zero and variance $\sigma_u^2, \varepsilon_j$ has mean zero and variance σ_ε^2, and r has mean zero and variance σ_r^2. For j = 1,..., $m_3 \leq \min(m_1, m_2)$, the FFQs and corresponding Food Record/Recalls were recorded close together in time, and the covariance between ε_j and U_j is denoted by $\sigma_{\varepsilon u}$. Otherwise, all random variables are uncorrelated. We must have $m_3 \geq 1$ in order to estimate $\sigma_{\varepsilon u}$, and $m_1 \geq 2$ in order to estimate σ_r^2.

The model used by Hunter et al. assumes (2). In addition, it assumes that conditional on usual intake T, Q, and F are independent, which in our context means that $\sigma_{\varepsilon u} = 0$. There are some data sets for which this assumption is questionable, in particular the Women's Health Trial Vanguard Study. Equation error lies at the heart of our discussion. Its interpretation is that even if one could observe an infinite number of repeated FFQs ($m_1 = \infty$) *on the same individual*, then the resulting mean would not be *exactly linearly related* to usual intake, because (by what statisticians call the law of large numbers) its value would equal $\beta_0 + \beta_1 \mathrm{T} + r$. By exactly linearly related, we mean that if we were to plot the mean of many FFQs against usual intake, the plot would not fall exactly on a straight line because the equation errors r vary from individual to individual. There is ample evidence that such within–individual equation errors occur in nutrition, see Freedman et al. (1991).

We illustrate the method using the variable %CF, with data from the Women's Health Trial Vanguard Study, which has n=86 women measured with $m_1 = m_2 = m_3 = 2$. For the WHTVS, each Food Record/Recall was the average of results from a 4–day food record. The results of the analysis are summarized in Table 1.

Of major concern is the attenuation of the estimated fat effect caused by error and bias in the FFQ, namely the slope γ in the regression of usual intake on FFQ and B (i.e., body mass index: (weight in kilograms) / (height in meters)^2). In the case that usual intake is independent of B, $\gamma = \beta_1 \sigma_t^2 / \sigma_q^2$. An auxiliary quantity is what we called deattenuated relative risk, which is calculated as follows. In a logistic regression of a binary response Y on T and B, so that pr(Y=1|T) = $H(\alpha_0 + \alpha_1 T + \alpha_2 B)$ where $H(\cdot)$ is the logistic distribution function, the regression calibration approximation (Rosner et al., 1989; Carroll et al., 1995) states that the observed data also follow a logistic model in Q and B, but with the slope for Q being $\alpha_{1\gamma}$. Thus, if the observed relative risk for a disease outcome is 1.10, then the estimated relative risk due to T is $1.10^{1/\gamma}$. The results from this are in basic agreement with the Hunter et al. conclusion, namely that with models (1-2), one does not expect the relative risk due to true %CF intake to be particularly large.

Table 1. Results for % Calories from Fat using the linear errors–in–variables model of Freedman et al. (1991). Here WHTVS is the Women's Health Trial Vanguard Study. The term *attenuation coefficient* means the slope in the regression of usual intake T on the Food Frequency Questionnaire Q. *Relative Risk* is the estimated relative risk due to usual intake T assuming that the estimate from the error–prone Food Frequency Questionnaires was 1.10.

Variable	WHTVS
Attenuation Coefficient	0.34
Relative Risk	1.32

We want to emphasize the first basic conclusion. After taking into account the bias and measurement error in FFQs, the standard model (1)–(2) suggests that the estimated relative risk for true %CF is less than 1.35. This is the basic conclusion of Hunter et al.

A NEW MODEL FOR NUTRITIONAL MEASUREMENT ERROR

It is interesting to reflect carefully on the meaning of (2). Consider the mean of a large number of Food Record/Recalls *within the same individual*. Again, by the statistical law of large numbers, model (2) says that this mean would exactly equal usual intake T. Thus, in effect, the usual linear measurement error model is *defining* the usual intake T on an individual to be the average of many Food Record/Recalls for that individual, if such a large number of Food Record/Recalls could be obtained. In most fields, this is usually a perfectly reasonable way to proceed, because such an average would usually represent the best measure that one could possibly obtain. In other words, measurement error models with additive structure such as (2) implicitly define the "true" value of the latent, unobservable value as the average of a large number of independent error–prone replicates of an instrument at hand.

In nutritional epidemiology, however, it has recently been shown by comparing reports from food records with precise measurements of energy expenditure determined by the "doubly–labeled water" method, that many individuals tend to under–report energy (caloric) intake (Livingston et al., 1990; Martin et al., 1996; Sawaya et al., 1996). Moreover, each individual may misreport their intake by a different average amount. In other words, Food Record/Recalls may be subject to individual–specific bias. In addition, it appears that the extent of under–reporting may be related to an individual's body size (Mertz et al., 1991).

Based on this evidence, we are led to consider alternatives to the standard model (1)–(2). The simplest such model adds an effect due to BMI as well as a random effect to the Food Record/Recalls, so that if B denotes BMI, we have

$$Q_j = \beta_0 + \beta_1 T + \beta_2 B + r + \varepsilon_j \qquad j = 1, \cdots, m_1 \tag{3}$$

$$F_j = T + s + \beta_3 B + U_j \qquad j = 1, \cdots, m_2 \tag{4}$$

The main difference between (2) and (4) is the presence of the random variable s, which represents within–individual random bias. In model (3)–(4), the random variable s is assumed to have mean zero, variance σ_s^2 and to be independent of all other random variables except possibly r, in which case we write the covariance and correlation between r and s as σ_{rs} and $\rho_{r,s}$, respectively. The model of Prentice (1996) is similar, but also allows the correlation between r and s to depend on B (BMI).

Unfortunately, all the parameters in model (3)–(4) cannot be identified in general. To see this, suppose that BMI is also normally distributed with mean μ_b and variance σ_b^2. Then there are 15 parameters, namely β_0, β_1, β_2, β_3, μ_t, σ_t^2, σ_s^2, σ_ε^2, σ_u^2, $\sigma_{\varepsilon u}$, $\sigma_{r,s}$, $\mathrm{cov}(T,B)$, μ_b, and σ_b^2. However, there are only 12 uniquely identified moments, namely $E(Q)$, $E(F)$, $E(B)$, $\mathrm{var}(Q)$, $\mathrm{var}(F)$, $\mathrm{var}(B)$, $\mathrm{cov}(Q_1,Q_2)$, $\mathrm{cov}(F_1,F_2)$, $\mathrm{cov}(Q_1,F_1)$, $\mathrm{cov}(Q_1,F_2)$, $\mathrm{cov}(Q,B)$ and $\mathrm{cov}(F,B)$. Clearly then, identifiability under normality assumptions requires that we restrict three of the parameters, or that we have additional information from other experiments, such as biomarker studies.

Our approach is a mixture of the two, which we illustrate for the variable % Calories from Fat. In the absence of a direct biomarker data satisfying model (1)–(2), we must use indirect information and some reasonable intuition to achieve the three restrictions. We summarize one such approach as follows.

- Evidence (Mertz et al., 1991) suggests that food records under–report true energy intake by as much as 20% in those with higher BMI. One might reasonably assume that the same is true of fat intake, in which case the %CF would have an under–report essentially independent of BMI, and thus $\beta_3 = 0$. We took this value as fixed in our analysis.
- One might reasonably expect that consistent under– or over–reporting of %CF is fairly highly correlated in all instruments, i.e., that $\rho_{r,s} > 0$. To allow for possibly very strong correlations, we set $\rho_{r,s} = 0.3$ and $\rho_{r,s} = 0.6$. These correspond in broad outline to cases considered by Prentice.
- The variability in the error components r and r should be of roughly the same order of magnitude. Evidence for a fixed value for the ratio of σ_s^2 to σ_r^2 has been reported for neither %CF nor total energy intake. Lacking any real information on this point, we set $\sigma_s^2 = 0.5\sigma_r^2$. However, we did perform sensitivity analyses. If this ratio is below 0.25, then the results are qualitatively in line with the standard model (1)–(2). For a ratio greater than 0.5 up to 1.5 (the range we studied), the results are qualitatively in agreement with model (3)–(4).

  The results from the new model (3)–(4) are strikingly different from those of the standard model (1)–(2), when $\rho_{r,s} = 0.6$, see Table 2. An examination of Table 2 reveals the following conclusion, namely that the attenuation and hence the corrected relative risk estimates change in important ways.
- In the WHTVS, instead of an estimated corrected relative risk of 1.32, the new model suggests a corrected relative risk of 1.46 when $\rho_{r,s} = 0.3$ and 3.77 when $\rho_{r,s} = 0.6$. This is a practically important difference, well in line with that observed by Prentice

(1996). Our reanalysis of the WHTVS suggests that a small observed relative risk, such as observed in the Hunter et al. study, could be masking a large and important effect if the errors r and s are highly correlated. There is, of course, no direct information available to verify that this correlation is indeed high enough to matter.

- Carroll et al. (1998) report results from another study done by the American Cancer Society that has somewhat different conclusions. They found that correlations between r and s of less that 0.60 do not suggest large deattenuated relative risks. However, note that if the correlation between r and s is changed to $\rho_{r,s} = 0.8$ or greater, they found a large inflation in the corrected relative risk.

Table 2. Comparison of data sets for the variable % Calories from Fat using the linear errors–in–variables model of Freedman et al. (1991). Here WHTVS is the Women's Health Trial Vanguard Study. Attenuation is the slope in Q of the regression of usual intake on the Food Frequency Questionnaire Q and BMI, and *Relative Risk* is the deattenuated relative risk assuming that the estimated relative risk from the error–prone Food Frequency Questionnaires was 1.10.

Variable	WHTVS Unbiased	WHTVS $\rho_{r,s} = 0.3$	WHTVS $\rho_{r,s} = 0.6$
Attenuation Coefficient	0.34	0.25	0.07
Relative Risk	1.32	1.46	3.77

DISCUSSION

In this paper, we have suggested that evidence points to the possibility that when measuring nutrient intake, the standard linear errors–in–variables model may be missing an important component of variability. Our new model (3)–(4) includes this missing error component, as well as an allowance for possible systematic biases. Unfortunately, the model is unidentified, and we were forced to make estimates of enough of the parameters to force identifiability. Whether our estimates are in fact a fair approximation of the truth is unknown, and can only be tested in future experiments.

We have shown that in one data set (the WHTVS), under certain conditions, the impact of dietary measurement error depends strongly on the measurement error model used. However, when Carroll et al. (1998) applied these models to the ACS study, major differences between the two models arise only when $\rho_{r,s}$ becomes 0.80 or larger. Thus, in assessing the potential impact of dietary measurement error, not only is the measurement error important, but potentially as important is the data set used to perform the assessment. It is not *a priori* clear whether the results of the new model, when applied to the seven studies in Hunter et al., will mirror the WHTVS or the ACS experience. Clearly, our results suggest that the calibration studies of Hunter et al. should be reanalyzed.

We point out once again that in this example that the major difference between the old model (1)–(2) and the new model (3)–(4) is the possibility of correlated error components r and s in the latter model. The effect of body mass index is minimal, and in addition, it can be shown that if these two error components have correlation $\rho_{r,s} = 0$, then the old model and the new model lead to the same deattenuated relative risk estimates. At this stage, there are no data available known to us in which this correlation could be estimated for % Calories from Fat or Total Fat.

We reemphasize that the new model has two components which are not estimable from that data currently available. One of the two components, which we have called $\rho_{r,s}$, is the correlation of the within–individual equation errors. Setting this correlation equal to 0.0 and applying our model would result in an analysis similar to that of Hunter et al. Setting this correlation to be larger suggests a much greater impact of measurement error than expected in the Hunter et al. analysis. There is no data to which we have access which can be used to assess $\rho_{r,s}$.

Our overall conclusion is that until further understanding of dietary measurement error is available, measurement error corrections must be done on a study–to–study basis, together with sensitivity analyses. Even then, the results of nutritional epidemiology studies which relate diet to disease should be interpreted cautiously. The lack of knowledge about a crucial parameter is disconcerting, and points to a need for new experiments aimed at its estimation.

ACKNOWLEDGMENTS

Dr. Carroll's research was supported by a grant from the National Cancer Institute (CA-57030). The authors thank Drs. Elaine W. Flagg, Ralph J. Coates, Eugenia Calle, and Michael Thun (the American Cancer Society CPS II Nutrition Survey Validation Study) for providing us with the data used in this paper.

CORRESPONDING AUTHOR

Please address all correspondence to:
Raymond J. Carroll
Department of Statistics
Texas A&M University
College Station, TX 77843-3143

REFERENCES

Carroll RJ; Freedman LS; Kipnis V; Li L. A new class of measurement error models, with application to dietary intake data. *Can J Stat,* 1998.

Freedman LS; Carroll RJ; Wax Y. Estimating the relationship between dietary intake obtained from a food frequency questionnaire and true average intake. *Am J Epi,* 1991, 134:510-520.

Fuller WA. *Measurement Error Models.* John Wiley & Sons: New York. 1987.

Henderson MM; Kushi LH; Thompson DJ; et al. Feasibility of a randomized trial of a low-fat diet for the prevention of breast cancer: Dietary compliance in the Women's Health Trial Vanguard Study. *Prev Med,* 1990, 19:115-133.

Hunter DJ; Spiegelman D; Adami H-O; Beeson L; van der Brandt PA; Folsom AR; Fraser GE; Goldbohm A; Graham S; Howe GR; Kushi LH; Marshall JR; McDermott A; Miller AB; Speizer FE; Wolk A; Yaun SS; Willett W. Cohort studies of fat intake and the risk of breast cancer – A pooled analysis. NEJM, 1996, 334:356-361.

Livingston ME; Prentice AM; Strain JJ; Coward WA; Barker AE; McKenna PG; Whitehead RJ. Accuracy of weighted dietary records in studies of diet and health. *Brit Med J,* 1990, 300:708-712.

Martin LJ; Su W; Jones PJ; Lockwood GA; Tritchler DL; Boyd NF. Comparison of energy intakes determined by food records and doubly-labeled water in women participating in a dietary intervention trial. *Am J Clin Nut,* 1996, 63:483-490.

Mertz W; Tsui JC; Judd JT; Reiser S; Hallfrisch J; Morris ER; Steele PD; Lashley E. What are people really eating: The relation between energy intake derived from estimated diet records and intake determined to maintain body weight. *Am J Clin Nut,* 1991, 54:291-295.

Prentice RL. Dietary fat and breast cancer: Measurement error and results from analytic epidemiology. *Nat Canc Inst,* 1996, 88:1738-1747.

Rosner BA; Willett WC; Spiegelman D. Correction of logistic regression relative risk estimates and confidence intervals for systematic within—person measurement error. *Statistics,* 1989, 8:1051-1070.

Sawaya AL; Tucker K; Tsay R; Willett W; Saltzman E; Dallal GE; Roberts SB. Evaluation of four methods for determining energy intake in young and older women: Comparison with doubly-labeled water measurements of total energy expenditure. *Am J Clin Nut,* 1996, 63-491-499.

STATISTICAL MODELS FOR QUANTITATIVE BIOASSAY

Matthew R. Facer,[1] Hans-Georg Müller,[1] and Andrew J. Clifford[2]

[1]Division of Statistics
[2]Department of Nutrition
University of California, Davis
Davis, CA 95616

ABSTRACT

We discuss various statistical approaches useful in the analysis of nutritional dose-response data with a continuous response. The emphasis is on the multivariate case with several predictors. The methods which will be discussed can be classified into parametric models, including change-point models, and nonparametric models, which rely on smoothing methods such as weighted local linear fitting. The methods will be illustrated with the analysis of data generated from a folate depletion-repletion bioassay experiment conducted on rats, where the measured growth rate of the rats is the response variable. We also discuss the biological conclusions that can be drawn from applying various statistical methods to this data set.

INTRODUCTION

Modeling a continuous response on several predictors has received much attention and is of interest in nutrition as well as several other scientific fields. Parametric models assume that the response can be expressed explicitly as a function of the predictors by way of an equation acting over the full observed range of the predictors. This equation is defined by the values of parameters which are assumed fixed but unknown, and must be estimated from the data. The most commonly used parametric method, multiple linear or polynomial regression (Neter et al., 1997), assumes the response to be a polynomial function of the predictors, often linear for each predictor (the simplest case). The parameters are usually estimated using the least squares criterion, which produces maximum likelihood estimates if the observed data are normally distributed. One generalization of linear regression is the generalized linear model, or GLM (McCullagh and Nelder, 1989). This technique assumes that the random error associated with the response comes from a distribution which is a member of the exponential family of distributions and that the response is related to a linear combination of the predictors (the "linear predictor") by way of a link function.

Nonlinear regression (Bates, 1988) extends linear least squares regression by modeling the response on the predictors by way of equations that involve the parameters in nonlinear ways, with estimators obtained by the method of nonlinear least squares. A nonlinear regression model of particular interest for applications in nutrition is the change-point or "broken stick" model (Hinkley, 1971; Bhattacharya, 1994). This method, which will be presented here for only one predictor, fits two lines connected continuously at one unknown point (the *change-point*). This technique is of interest when the mechanism driving the response changes abruptly at one point, such as in the sudden onset of a threshold or saturation phenomenon.

Nonparametric regression models correspond to local fitting and allow the equation relating the response to the predictors to change locally as the predictor values vary, leading to an estimated function that is smooth but does not fall into a parametric class. All nonparametric methods have four characteristics:

1. They model data locally, meaning for any fixed value of the predictors, only data points with predictors close to this value matter in estimating the response.
2. The distribution of the random error of the response is not critical as long as certain moment conditions are satisfied.
3. No assumptions beyond smoothness need to be made on the shape of the regression relation.
4. There is a "curse of dimensionality" in that these methods are ineffective for more than three dimensions (predictors).

Two types of nonparametric estimators will be discussed in detail. Convolution-based kernel estimators (see, e.g., Müller, 1988 or Wand and Jones, 1995) form a local weighted average of the observed responses with corresponding predictors close to the predictor value where the regression function is to be computed, with the weighting determined by a kernel function. The second method is locally weighted linear regression (see, e.g., Fan and Gijbels, 1996), which uses similar weighting schemes to the kernel method, but where the weights are used in fitting a line or polynomial locally by weighted least squares. This method contains automatic boundary adaptations when estimating near the limits of the observed predictor values in the data (boundary regions), an important feature for the multivariate case (Ruppert and Wand, 1994). Such boundary adaptations can also be achieved for kernel estimators but require the kernel in this region (the *boundary kernel*) to differ from that in the interior. A disadvantage of local polynomial fitting (as compared to kernel estimation) is that irregular grids of predictors with "holes" in the design can lead to regions where local polynomial fits are undefined, whereas convolution-based kernel estimators are defined for all designs, no matter how irregular.

A DEPLETION-REPLETION FOLATE BIOASSAY

Overview

In a folate depletion-repletion bioassay using rats, Clifford et al. (1993) showed that growth response and diet folic acid concentration in the range of 227 to 680 nmol/kg diet followed a linear relationship. Above 680 nmol/kg diet (the critical concentration), the growth rate remained relatively unchanged. This work further suggested that future rat bioassays following this protocol should use total folate amounts that do not exceed this critical dosage (we will use the terms "concentration" and "dosage" interchangeably).

We will illustrate methods for bioassay with data from an experiment using this depletion-repletion bioassay protocol, but involving variations in the source of the folate as well as in the total folate dosage. Of primary interest was an exploration of how the relationship between growth rate and folate concentration might change with folate source (if at all). Three sources of folate were studied: folic acid (*FA*), folate in cooked refried beans (*BF*), and folate in cooked beef liver (*LF*). The folate-depleted rats were fed an amino acid-based diet supplemented with 11 different concentrations of folate (227, 272, 317, 363, 408, 454, 499, 544, 590, 635, and 680 nmol/kg) from each of 12 different regimens (sources of folate: folic acid, fried beef liver, and cooked pinto beans each provided individually, or as 1/3, 1/1, or 3/1 combinations of folate from folic acid/beans, folic acid/beef liver, and beans/beef liver), for a total of 132 treatments. Each rat was randomly assigned to one of these. After being assigned to a regime, a rat was then randomly assigned to one of eleven folate concentrations. The regime using only folic acid had 2 rats for each of its 11 concentrations, which brought the total number of rats to 132+11=143.

Experimental Design

For a detailed description of the biological protocol which generated these data, we refer to Müller et al. (1996). This protocol was approved by the Animal Use and Care Administrative Advisory Committee of the University of California, Davis. In summary, 143 weanling male Sprague-Dawley rats were depleted of folate by feeding them a folate-free diet (Walzem and Clifford, 1988), supplemented with 1.27 nmol iron citrate/kg diet for 29 days. Each rat was randomly assigned to one of the folate-repleting regimes described above, giving the regime allocation described by the table in the Appendix (which reflects the loss of 19 of the 143 rats as explained below).

Each rat received its diet regime for 26 days, during which time it was weighed daily. These 26 weights were then used to calculate a growth rate for each rat by using the slope coefficient of the least squares line fitted to these weights (with the day of the regime as the predictor). This estimated growth rate served as the response value in the statistical models, while the concentration of each folate source (and functions thereof) served as the predictors.

Our design allowed for the combination of numerous diet folate concentrations with folate source mixtures while also allowing the error variance ($\hat{\sigma}^2$) to be estimated analogously to the more common situation of a design with replication (Müller and Schmitt, 1990). The one subject-per-dose combination design used here is advantageous for analyzing bias and obtaining diagnostics for intricate response surfaces.

We analyzed data for 124 of the original 143 rats. The exclusion of the remaining 19 rats had several well-documented causes (for details, see Müller et al., 1996). The statistical analyses were conducted with the S-PLUS statistical package (1990), v. 3.0, as well as code written in FORTRAN 77, part of which used subroutines from *Numerical Recipes in Fortran* (1992). One plot was made using Harvard Chart XL, v. 2.0 (1995).

Descriptive Plots

Before fitting any model to the data, some basic plots provide an overview of the data. A scatter plot of growth rate against total folate amount (regardless of diet regime) is shown in Figure 1. Despite all total folate concentrations being below the aforementioned critical dosage, the plot clearly shows a leveling-off trend for higher dosages. As no total folate concentration exceeded the critical dosage of the earlier experiment (Clifford et al., 1993), we hypothesized that this trend was due either to interaction between the sources of folate or the folate sources having a different effect on the growth rate.

Further insight is provided in Figure 2. This figure is similar to Figure 1, but is restricted to those points where only one folate source was used in the repletion (thus interactions between folate sources are not possible). This plot shows the same leveling-off trend as the data as a whole, suggesting that much of this trend is due to differences in the three repletion sources rather than, or in addition to, source interaction. A similar plot using only those points with two folate repletion sources (not shown) displayed the leveling-off as well, but not as pronounced as in Figure 2.

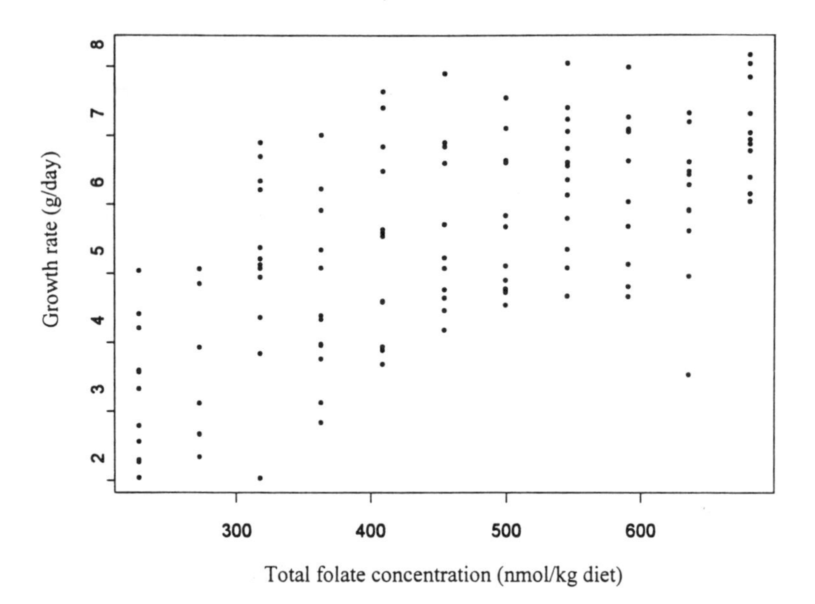

Figure 1. Plot of growth rates vs. total folate concentrations.

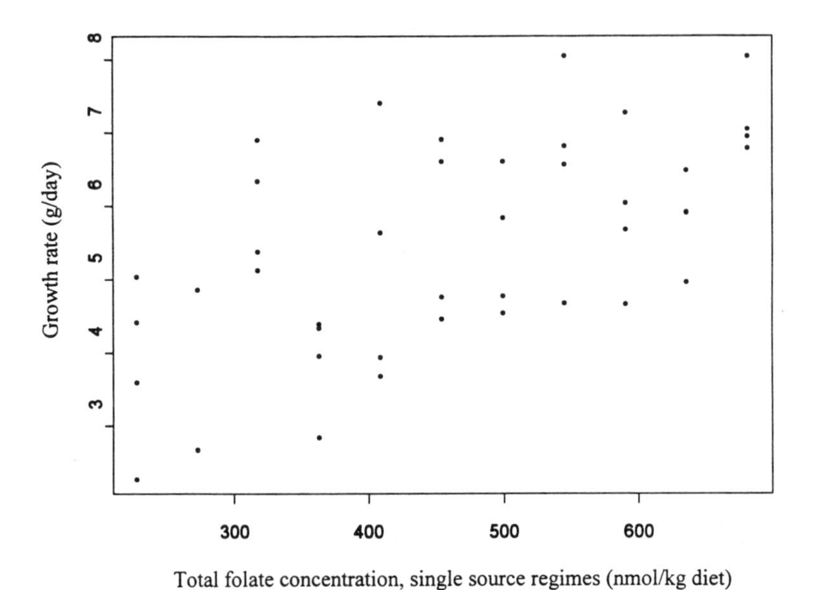

Figure 2. Plot of growth rates vs. total folate concentrations for single source regimes (regime numbers 1, 2, and 6, as shown in the table in the Appendix).

PARAMETRIC APPROACH

Henceforth, the data are assumed to be of the form ($\mathbf{x}_1$, Y_1), ..., ($\mathbf{x}_n$, Y_n), with $\mathbf{x}_1 = [x_{i1} \cdot\cdot x_{ip}] \in \Re^p$ (a p-variate vector of real numbers), and $Y_i \in \Re$ for $i = 1, \ldots, n$. Of interest is modeling the response as a function of the predictors plus random error:

$$Y_i = m(\mathbf{x}_i) + \varepsilon_i \tag{1}$$

where the errors ε_i are independent with $E[\varepsilon_i] = 0$.

We distinguish parametric models, where $m(\cdot)$ belongs to a finitely parameterized class of functions, from nonparametric models where $m(\cdot)$ is only assumed to be smooth. In the parametric case, there are two models of particular interest:

- The *Generalized Linear Model*, or GLM, where the distribution of the errors ε is in the exponential family, and

$$E[Y_i] = g(\beta^T \mathbf{x}_i), \qquad \mathrm{var}(Y_i) = \sigma^2 V(E[y_i]), \qquad i = 1, \ldots, n. \tag{2}$$

  Here, $\beta \in \Re^p$ is a vector of unknown parameters, $g(\cdot)$ is a known link function, and $V(\cdot)$ a known variance function.

- The *Multiple Linear Regression* model with interactions, where

$$m(\mathbf{x}_i) = \beta^T \mathbf{x}_i + \mathbf{x}_i^T A \mathbf{x}_i, \qquad \mathrm{var}(\varepsilon_i) = \sigma^2 = \text{constant}, \qquad i = 1, \ldots, n, \tag{3}$$

  with an upper triangular matrix A, which indicates product interactions among the predictors. If no interactions exist, $A = 0$; usually only a few interaction terms at most are included. If $A = 0$ and the errors ε_i are normally distributed, this is a special case of the GLM with identity link and constant variance function.

The quality of a parametric model depends crucially on how many predictors are included in the assumed model. Too many and too few are both bad as compared to the actual model. Inclusion of too many predictors corresponds to overfitting, in which case the variance of the parameters will get too large; inclusion of too few predictors leads to biases and a systematic lack of fit. A measure of how far away the fitted model is from the observed data is important. Common measures are the *deviance* in GLM models and the *sum of (weighted) squares* in multiple regression models (the basis of the least squares criterion). However, it should be noted that minimizing these deviation measures by including more predictors leads to overfitting.

Determining the Distribution of the Data

Before any statistical tests or confidence intervals can be computed in the parametric case, we need to determine the best distributional fit for the data (that is, for the errors ε in the model). Given that the responses were continuous and positive, the two distribution possibilities we explored for the folate bioassay data were the Gaussian and Gamma distributions.

While a *Q-Q* plot for the Gaussian distribution is standard practice, that for the Gamma distribution differs in that this distribution has associated with it a shape parameter, and so varying this parameter will produce varying *Q-Q* plots. We looked at *Q-Q* plots for the Gamma distribution with several values of this shape parameter. The Gamma *Q-Q* plot with the "best" shape parameter was no better than the Gaussian *Q-Q* plot. These plots showed that the two distributions fit the data about the same, and we thus opted to use the Gaussian distribution due to its well-known properties and common usage.

We further decided to use the multiple regression model rather than the GLM, i.e., to choose the identity link function. Such a model is easier to fit and interactions, which we knew existed from previous work with this data (Müller et al., 1996), are often easier to interpret.

Modeling Single-Source Regimes Separately

To counter any effect of folate source interaction, we analyzed the subset of the data consisting only of those repletion regimes having exactly one folate source (regimes 1, 2, and 6 from the table in the Appendix). First, we considered only models linear in their predictors. The best such model fit a linear term for each of the three folate sources. This model yielded residuals showing a faint quadratic trend when plotted against its fitted values, suggesting that the linear fits could be improved upon by adding one or more quadratic terms.

Upon fitting single-source quadratic terms to the linear model, the only one that was significant was $(LF)^2$. This yielded

$$Growth = 1.825 + 0.0073(FA) + 0.0074(BF) + 0.0169(LF) - 0.000015(LF)^2 \tag{4}$$

with $R^2 = 0.492$, SSE = 46.4, $p < .001$ for each linear term, and $p < .03$ for the quadratic term.

In conclusion, the best model looking at this single-source subset of these data is (4), a regression model with separate linear terms for each folate source and a quadratic term for folate from liver (LF). Thus folate from liver has a different effect on growth rate than folate from the other two sources. Looking at this model further, one can see that the linear coefficients for (FA) and (BF) are very close to each other. We decided to test this notion only with the full data set, along with interaction effects.

Modeling Over All Regimes

In light of the information gathered by looking at the three single-source regimes, we fit a series of models to the whole data set, of which we show the better ones in increasing order of quality:

Separate source effects; linear:

$$Growth = 2.3345 + 0.0066(FA) + 0.0062(BF) + 0.0080(LF) \tag{5}$$

Source effect for (FA) and (BF) same, different for (LF); linear:

$$Growth = 2.3347 + 0.0064(FA+BF) + 0.0080(LF) \tag{6}$$

Testing model (5) against its nested model (6) yielded an F-statistic having p-value > 0.5, and we thus concluded that the effect of (FA) and (BF) was exchangeable for these data. A three-dimensional scatterplot of the data with this adjustment is shown in Figure 3.

We then considered modeling interaction among folate sources, but doing so with the above results included in the model-building process. Multiplying the fraction of total folate coming from a specific source with the absolute level of folate coming from that same source, and then including these newly created variables among the predictors was a successful approach to incorporate interaction terms. For example, doing this for liver folate gave

$$(\%LF) = \frac{(LF)}{(LF) + (BF) + (FA)}, \tag{7}$$

with one possible interaction term, $(\%LF) \times (LF)$.

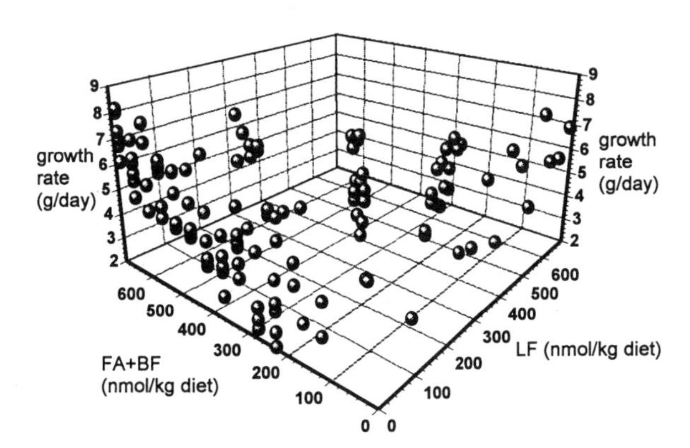

Figure 3. Scatter plot of folate data assuming an additive effect for (*FA*) and (*BF*).

If one or more of these terms were significant, it would imply that the slope for that source changes with the percentage of total folate coming from that source. Note that the percentages (expressed as fractions) themselves could not be used as predictors as this would be interpreted as a change in intercept, which is biologically implausible (all specimens with no folate repletion should have the same "baseline" growth rate). The best model without interactions described above was used as a starting model to which significant interaction terms were added. This led to the following model with interactions:

$$Growth = 1.618 + .0092(LF) + .0079(FA{+}BF) + .0107(\%LF)(LF) - .00002(LF)^2 \tag{8}$$

with $R^2 = 0.494$, SSE = 143.3, and $p < .0001$ for all terms except $(\%LF) \times (LF)$, which had $p < .005$.

The adequacy of this model was verified in several ways. First, it was checked whether the model describes the existing interaction between the various folate sources adequately. This was done by comparing (8) with an alternate, more complicated model that allowed the coefficients for (LF), $(FA{+}BF)$, and $(LF)^2$ to vary with regime. While this regime-based model more accurately describes the source interactions than (8), it does so with far less efficiency. The regime-based model was tested against the simpler model (8) by means of an F-test, which yielded a *p*-value of about 0.178, suggesting the simpler model does not fail to adequately describe the interaction between various sources of folate.

Two residual plots — standardized residuals vs. fitted values (Figure 4) and a Gaussian *Q-Q* plot for standardized residuals (Figure 5) — show that (8) provides a reasonable fit to these data.

Alternatively, we also fitted the model using the more standard product interaction $(FA{+}BF) \times (LF)$ as the interaction term rather than $(\%LF) \times (LF)$. This gave the model

$$Growth = 1.346 + .0202(LF) + .0084(FA{+}BF) - .00002(FA{+}BF)(LF) - .00002(LF)^2 \tag{9}$$

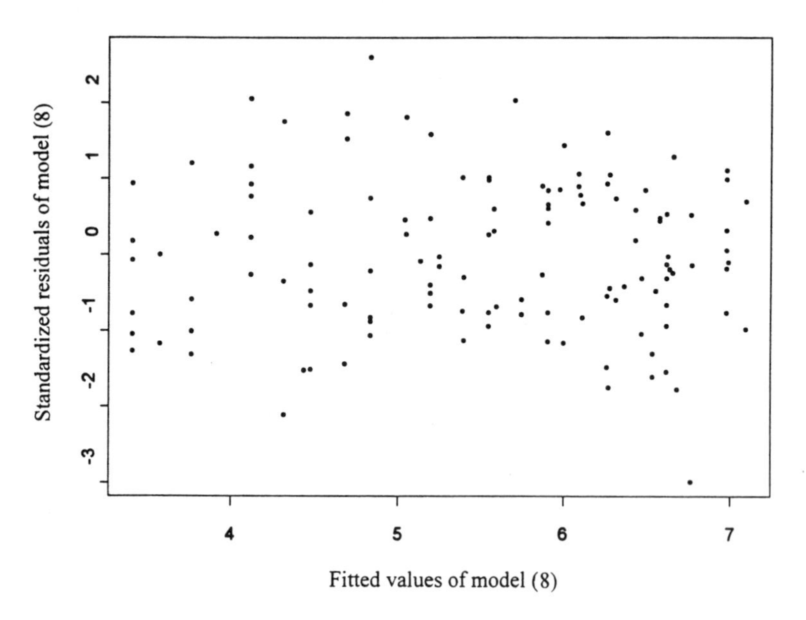

Figure 4. Plot of fitted values vs. standardized residuals from model (8). See text for details.

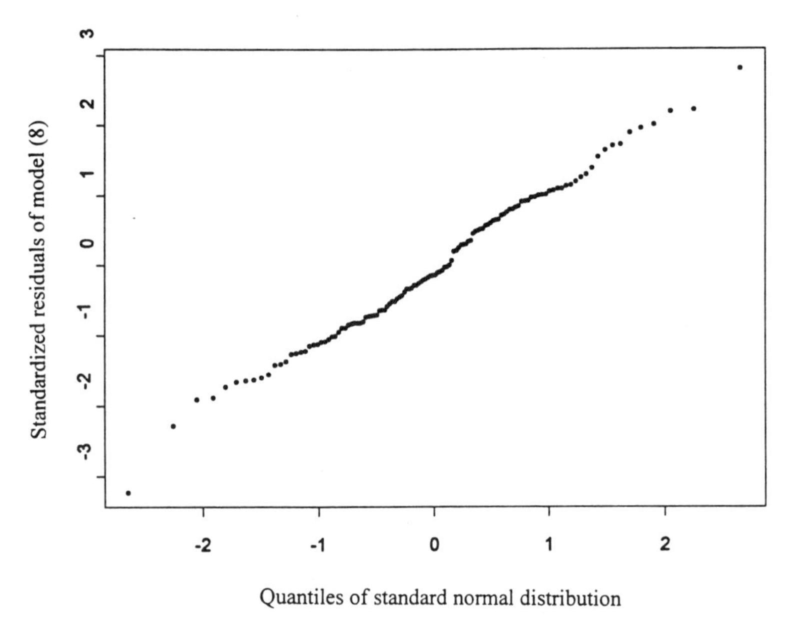

Figure 5. Gaussian *Q-Q* plot for standardized residuals from model (8). See text for details.

with $R^2 = 0.505$, SSE = 129.1, and $p < .0001$ for (*LF*) and (*FA*+*BF*) coefficients and $p < .005$ for the other terms. While this model provided a very similar fit to (8) above, we decided to use (8) as the final parametric model since it expresses the antagonistic interaction between (*FA*+*BF*) and (*LF*) more clearly.

Through its interaction term, model (8) provides information on how (*LF*) and (*FA*+ *BF*) interact when combined in a repletion diet regime. Written in its full form with the parameter estimate (recall this parameter has p-value < .0001), the interaction term is written

$$0.0107\left[\frac{(LF)^2}{LF+(FA+BF)}\right]$$

Through this expression, one sees that the interaction between (LF) and ($FA+BF$) is antagonistic. For example, when (LF) = ($FA+BF$) = 200 nmol/kg diet, the term above equals 1.07; when (LF) and ($FA+BF$) are 100 and 300 nmol/kg diet, respectively, this interaction term equals 0.27. In both cases, the total concentration of folate given is 400 nmol/kg diet, illustrating that the presence of more ($FA+BF$) reduces the effect of the (LF).

The surface plot produced by model (8) is shown in Figure 6, which reveals much information about the relation between folate source and growth rate. Of immediate interest is how the two sources of repletion folate affected growth rate differently. For ($FA+BF$) by itself, the effect is linearly increasing, while for (LF) it is quadratic. Hence growth rate increases with increasing (LF) up to a point, and then levels off (it even appears to decrease, which will be discussed later). It can also be seen in this estimated surface plot that there is an interactive effect between the two folate sources, which will be discussed further below. The other major point of the model is that (FA) and (BF) are equivalent and exchangeable in promoting growth.

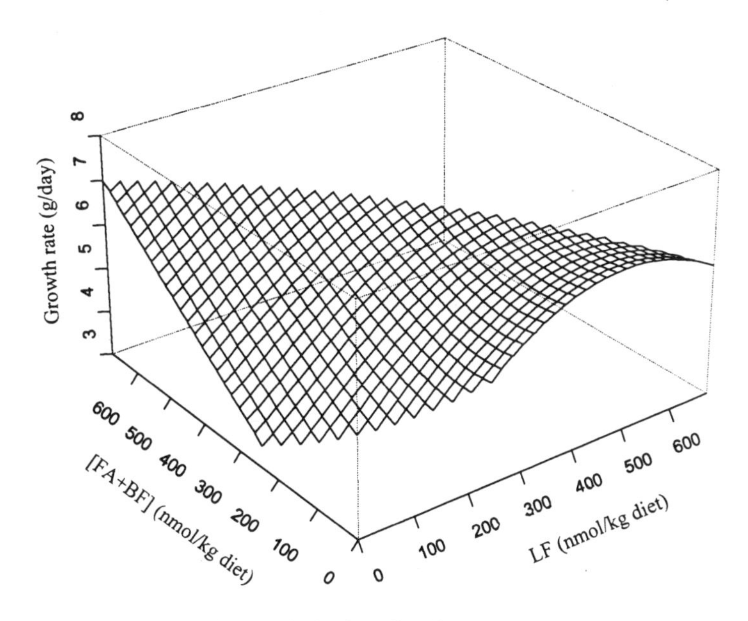

Figure 6. Surface plot of model (8).

A plot of some contours at various growth rate levels for the surface generated from model (8) is shown in Figure 7. The difference between the folate sources are clearly evident, expanding on the nature of the interaction between ($FA+BF$) and (LF). In particular, Figure 7 shows the interaction between the two sources to be antagonistic, as the small positive (linear) effect of additional ($FA+BF$) stays the same when in combination with any fixed dose of LF, but the presence of a fixed dose of ($FA+BF$) lowers the impact additional (LF) has on promoting growth. This can be seen by looking at the contours near the two axes: the distance between the contours is constant near the (*FA+BF*) axis, but increases with increasing growth rate near the (*LF*) axis. The convexity of the contours for larger growth rates near the (*LF*) axis further demonstrates this antagonistic interaction.

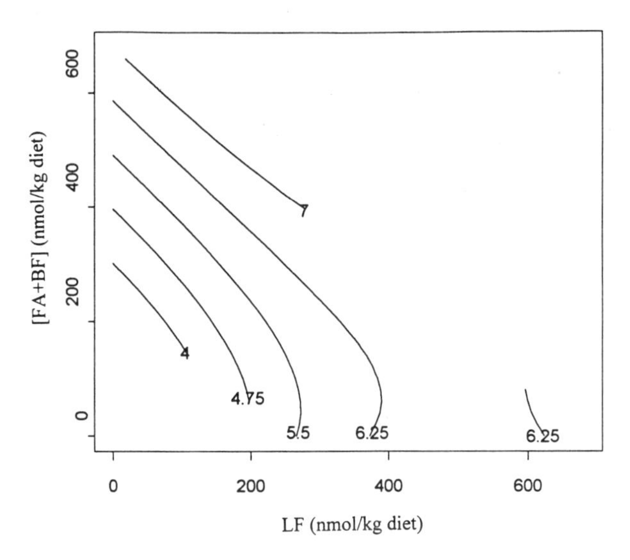

Figure 7. Contours of surface given by model (8). See text for details.

Nonlinear Modeling of Folate Source Interaction

Classical (Gaussian) regression analysis revealed that (*LF*) and (*FA+BF*) had an antagonistic interactive effect on growth rate. We thus conducted a separate analysis of this interaction. To do this, we considered the subset of data consisting of those double-source folate repletion regimes containing (*LF*) and either (*FA*) or (*BF*) (that is, regimes 7-12 from the Appendix). A simple scatter plot of total folate concentration against growth rate for this subset (comparable to Figure 1, but using the subset described above) revealed a saturation effect more pronounced than in Figure 1.

This saturation effect is due mainly to the aforementioned antagonistic interaction between (*LF*) and (*FA+BF*). An issue then arises as how best to model such an effect univariately without specifying folate source; that is, by using total folate concentration as the one predictor as opposed to (*LF*) and (*FA+BF*) as two separate predictors. Such a univariate model is of interest due to its simplicity, particularly when the number of interacting predictors is large, although it could not be considered as a final model.

Given the above parametric response surface analysis, a natural starting point was a quadratic model, which yielded

$$Growth = -.6638 + .0233[LF+(FA+BF)] - .00002[LF+(FA+BF)]^2 \qquad (10)$$

with $R^2 = 0.340$. For this model and all that follow, we let (*FA+BF*) have separate parentheses to denote the exchangeability of (*FA*) and (*BF*) and to clarify that only one or the other is given. As with the multivariate polynomial regression analysis, we have a negative quadratic term that "penalizes" for very large concentrations of folate (in fact, the parameter estimates are nearly equal). However, from a biological viewpoint, this leads to a weakness in interpretation. Using elementary calculus, the above model predicts that for dosages of folate exceeding 613.9 nmol/kg diet, the growth rate starts *decreasing*, which is biologically implausible, as excessive folate dosage does not stunt growth.

Of interest was modeling a saturation phenomenon, meaning that at most, the effect of exceeding a critical dose of folate would lead to little or no increased growth, but with no decrease of growth. This type of effect in a univariate case but with only one folate repletion source was modeled in Clifford et al. (1993) using several nonlinear regression techniques. There, a simple nonlinear regression model providing a good fit to their data was the simple inverse model, which we applied to our data to get:

$$Growth = 8.2053 - \frac{1052.043}{(LF)+(FA+BF)} \tag{11}$$

with R^2 = 0.350. While the fit is at best marginally improved compared to the quadratic model (10), this model has two advantages. First, growth rate will not decrease for large concentrations of folate. Second, it is a simpler model, fitting one less parameter than the quadratic model yet still fitting the data as well. The main disadvantage of this model is that it is somewhat difficult to interpret the biological meaning of the parameter estimates.

The final model we fit to this subset of the data is a change-point model or two-phase regression (broken line regression). These models can be written as:

$$y = \begin{cases} \alpha_0 + \beta_0 x, & \text{if } x \le x_c \\ \alpha_1 + \beta_1 x, & \text{if } x \ge x_c \\ y_c = \alpha_0 + \beta_0 x_c = \alpha_1 + \beta_1 x_c, & \text{if } x = x_c \end{cases} \tag{12}$$

The point (x_c, y_c) where the two regression lines meet to form a "bend" is known as the change-point (see also Clifford et al., 1993). We fit this model by varying x_c over an interval, fitting the two lines joined at x_c via least squares, calculating R^2 for each model fit, and then choosing the one with the largest R^2 as the final model. Thus the parameter x_c was found numerically as a maximizing argument while estimates of α_0, α_1, β_0, and β_1 were calculated explicitly, given x_c. In this particular case, we expected the least squares estimates for the slope coefficients β_0 and β_1 to both be nonnegative for biological plausibility. The estimated model for our data was:

$$Growth = \begin{cases} -1.921 + 0.02265x, & \text{if } 227 \le x \le 321 \\ 4.170 + 0.003675x, & \text{if } 321 \le x \le 680 \\ 5.35, & \text{if } x = 321 \end{cases} \tag{13}$$

with the estimated change-point at $(x_c, y_c) = (321, 5.35)$ and $R^2 = 0.362$.

This model seems to provide the best fit to this subset of the data among the models considered, but at the cost of fitting more parameters. However, the interpretation of the parameters is quite easy: for folate concentrations between 227 and 321 nmol/kg diet, an increase of 1 nmol/kg diet increases the growth rate of rats by 0.02265 g/day, while for concentrations between 321 and 680 nmol/kg diet, this increase is only 0.003675 g/day, a drop of nearly 85%. Again, this interpretation applies only to folate repletion diet regimes containing two sources, one of which is folate derived from beef liver.

A plot of the fitted change-point model and quadratic model placed on top of the original points is shown below in Figure 8. This figure suggests that the change-point model fits the data better as it more effectively models the burst in growth for smaller concentrations of folate. Further, the fitted quadratic model starts to show a decrease in growth near its right tail, which as mentioned above is biologically implausible. Looking at the points themselves, one clearly sees a saturation effect.

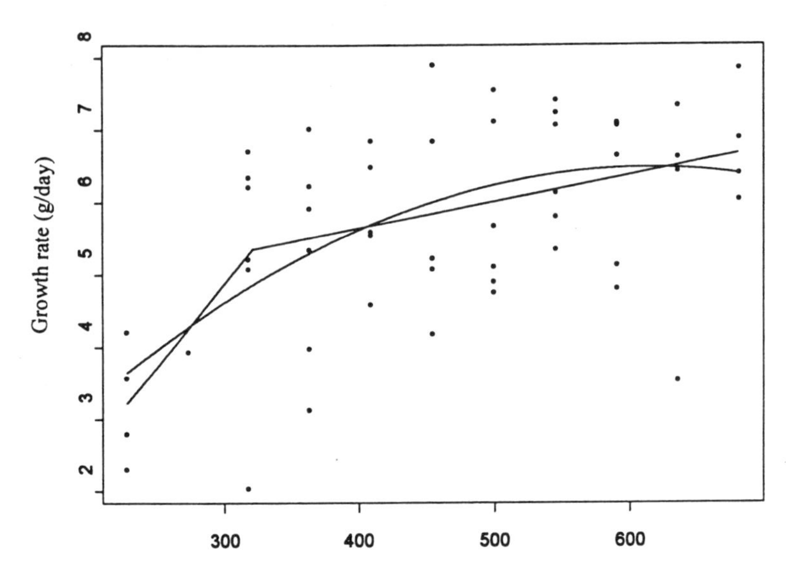

Figure 8. Plot of total folate concentration vs. growth rate for double-source regimes containing (*LF*) (regimes 7–12), along with the fitted curves of models (10) and (13).

By comparing model (13) to the change-point model fit by Clifford et al. (1993) for the case of (*FA*) being the only folate repletion source with dose range 227 to 2270 nmol/kg diet, the difference between using only (*FA*) as the folate repletion source and multiple repletion sources (with (*LF*) as one of the sources) is quite evident. The most obvious difference is in the position of the change-points themselves. With (*FA*) as the sole repletion source, the change-point was estimated to occur at 680 nmol/kg diet (Clifford et al., 1993), while for diet regimes containing (*LF*) in combination with either (*FA*) or (*BF*) (which are exchangeable), the change-point was estimated to occur at 321 nmol/kg diet, less than half as much. Furthermore, when (*FA*) was the only repletion source, the slope before the change-point was estimated to be 0.0079 (Clifford et al., 1993), while for our case [(*LF*) in combination with either (*FA*) or (*BF*)], the slope before the change-point was estimated to be 0.02265, a nearly threefold increase. By the parametric response surface analysis done earlier, we know this to be due to the (*LF*); that is, for smaller concentrations (say below 321 nmol/kg diet), (*LF*) has a more dramatic effect on growth rate than either (*FA*) or (*BF*).

There are other plausible nonlinear regression models for this subset of the data. Two such models fit by Clifford et al. (1993) used natural logarithm and exponential terms. Nonlinear regression models seem to be better suited to model this subset of the data than linear regression. A full scope of such models is presented in Bates (1988). For the particular case of change-point models, various references can be found in a collection of papers edited by Carlstein, Müller, and Siegmund (1994), with the one by Bhattacharya (1994) of particular interest.

NONPARAMETRIC APPROACH

Again consider (1), the general form of the model of study

$$Y_i = m(\mathbf{x}_i) + \varepsilon_i, \qquad i = 1,\ldots, n.$$

Earlier it was assumed that both the (parametric) form of $m(\cdot)$ and the distribution of ε were known. We will now assume that both of these are unknown, defining what is known as the nonparametric regression model. These methods can also be used to estimate the νth derivative of $m(\cdot)$ as well, for $\nu = 1,2,\ldots$, as long as these derivatives or partial derivatives in higher dimensions exist. The focus here will be on the case $\nu = 0$, estimating the function $m(\cdot)$ itself.

There are two design types to consider. If the predictors $\mathbf{x}_i$, $i = 1,\ldots,n$ are fixed *a priori*, we have a *fixed design*. The rat folate depletion bioassay data is of this design type, which will be assumed unless otherwise noted. The case of a *random design*, where the $\mathbf{x}_i$ are randomly sampled, is also of interest for some applications.

Local Versus Global Fit

The parametric models discussed in the previous section are *global* models and the corresponding parameter estimates are global estimators. That is, the fitted model determines the behavior of the fit over the entire range, and any data point has influence on the fit at points far away. To see the problems that can arise from this, we can look to Figure 8. In the parametric case, we considered fitting a quadratic regression to these data (10). However, it is quite evident that for points where the total folate dosage is small, a quadratic trend does not exist. Since the points where the total concentration of folate is large behave differently than those where the concentration is small, why should both areas be modeled in the same way? This question is especially important when trying to predict the growth rate given a total folate amount. We alleviated this problem somewhat by fitting a change-point model (13), but such models are constrained in the number of regions that can be separately modeled, and require additional parameters.

Often, data collected from different parts of the domain where data were sampled will not follow the same shape of regression relation, and then local modeling of the data is warranted. To do this, we make each input point the center of an interval or window whose length or size is determined by a *bandwidth*, and then fit a model to those data whose predictors fall within that interval or window. Hence for any point, only those data having predictor values nearby have an influence on the local model fit. The value of the fitted model at the input point is the estimate of the function $m(\cdot)$ at this point. By taking a continuum of input points, the resulting plot of the estimates against the inputs will be a smooth curve (the *curve estimate*) or surface (the *surface estimate*).

Assessing a Curve Estimator

Not only do several different curve estimators exist, but each one can be made slightly different by altering the bandwidth, weighting scheme (the kernel function), or the type of local model fit. There arises, then, a need for a criterion by which we could compare the performances of various curve estimators. Establishing a criterion depends also on whether the main interest is to assess the performance at one value of $\mathbf{x}$ or over the whole range of input predictors. For the former case, we use the *mean squared error* (MSE), defined as

$$MSE(\mathbf{x}) = E[(\hat{m}(\mathbf{x}) - m(\mathbf{x}))^2]$$
$$= [E(\hat{m}(\mathbf{x}) - m(\mathbf{x}))]^2 + Var[\hat{m}(\mathbf{x})] \tag{14}$$
$$= [bias]^2 + variance$$

Clearly, optimization at **x** refers to making MSE(**x**) as small as possible. The decomposition of MSE into bias and variance is fundamental. Two estimators may have the same bias but different variances or vice versa, allowing one to compare the statistical behavior of two estimators at **x**.

When the goal is a best estimator over the entire range of **x**, a possible criterion is the *mean integrated squared error:*

$$MISE = \int MSE(\mathbf{x}) w(\mathbf{x})\, d\mathbf{x} \tag{15}$$

where $w(\mathbf{x}) \geq 0$ is a weight function (not the same as the kernel function used in the definition of the estimator). Often, $w(\cdot)$ is chosen in such a way as to de-emphasize the boundary regions, where the behavior of curve estimators often deteriorates.

Bandwidth Selection

All of the estimators mentioned will rely on modeling the data locally, controlled by the bandwidth h. Choosing a good value of h is vital to successful nonparametric modeling since it effectively controls model complexity, with $h = 0$ and $h = \infty$ corresponding to exact interpolation and the fitting of one global model, respectively. Recalling the relation $MSE = (bias)^2 + variance$, we can relate how the choice of h affects the performance of an estimator (for simplicity, we will temporarily assume that we are in the univariate case). For very small h, the bias of these estimates is typically small, since $m(\cdot)$ should not change much in $[x-h, x+h]$; however, with a small h, few x_i from the data will be captured in this interval, resulting in $\hat{m}(x)$ having a large variance. If there are too few data in a window, certain estimators such as local polynomials are not even defined. By making h larger, we can capture more x_is in the window $[x-h, x+h]$ and decrease the variance of $\hat{m}(x)$; however, the local model now has to cover a larger range, resulting in a larger bias. An optimal choice of h, then, balances the variance incurred for h being too small with the bias incurred for h being too large.

In the multivariate case (dimension $p > 1$), the analog to the bandwidth is a $p \times p$ bandwidth matrix, which we denote by H. Usually this matrix is diagonal, meaning the directions of the smoothing are the same as the coordinate axes (though this need not be the case). One important issue in the multivariate case is whether to smooth the same amount in each covariate direction by using the same bandwidth in each direction (assuming each covariate has the same units, which can always be done by standardizing). Covariates showing more variability would tend to need larger bandwidths in their direction. In discussing bandwidths below, we use the notation h, with the understanding that we mean a matrix H in the multivariate case.

A further consideration in bandwidth selection is a local/global issue. We may want to use the same h for estimation at all predictors, or allow h to vary with x. In particular, we may need to alter h in boundary regions. For either case, however, the optimal h (in terms of minimizing MSE or MISE) involves $m(\cdot)$, which is unknown. Several ways to estimate the optimal h, denoted by h_{opt}, have been proposed.

The simplest way to estimate h_{opt} is to calculate $\hat{m}(\cdot)$ for various values of h and to choose the h giving the most aesthetically pleasing curve or surface. This is clearly subjective, abandoning the notion of a standard criterion. A more objective way would involve using the data to estimate h_{opt}. To do this, we must first have a criterion to measure the effectiveness of any h in modeling the data.

A widely used technique incorporating the data to estimate h_{opt} is the *least squares cross-validation* (see, e.g., Müller, 1988, or Fan and Gijbels, 1996). For any estimator $\hat{m}(\cdot)$, the goal is to find $h = \hat{h}_{opt}$ which minimizes the cross-validation sum of squares (CVSS),

$$CVSS(h) = \frac{1}{n}\sum_{i=1}^{n}\left[Y_i - \hat{m}_{h,-i}(\mathbf{x}_i)\right]^2 w(\mathbf{x}_i) \tag{16}$$

where $w(\mathbf{x}_i)$ is a weight function and $\hat{m}_{h,-i}(\cdot)$ is the estimator $\hat{m}(\cdot)$ calculated using h with the data point (X_i,Y_i) removed from the data. This criterion is comparable to the MSE/ MISE criterion discussed previously. The resulting *least squares cross-validation bandwidth* is not hard to find numerically.

Several other methods to estimate h_{opt} have been proposed but will not be discussed here. Among these are plug-in methods and pilot methods (see, e.g., Müller, 1988, or Fan and Gijbels, 1996). Although subjective, choosing h at least in part by trial and error is still a good idea since more objective methods may yield an h that produces a surface that looks oversmoothed or undersmoothed. For the rat folate depletion bioassay data, global bandwidths were chosen both subjectively and by the cross-validation method.

Kernel Smoothing

An intuitive way to locally model data is to take a (weighted) average of the responses corresponding to data points in the local window. A weighting scheme should give more weight to data located closer to the point where the estimate is desired. Usually, the weighting scheme employed is a symmetric probability density evaluated at a standardized Euclidean distance (componentwise in the multivariate case) from $\mathbf{x}$. This kernel function is usually denoted by $K_h(\cdot) = K(\cdot/h)/h$. Note the estimate is influenced by $\mathbf{x}_i$ only by how close $\mathbf{x}_i$ is to $\mathbf{x}$ according to the weighting scheme. An often-used form of the kernel estimate (see, e.g., Müller, 1993) is given by

$$\hat{m}(\mathbf{x}) = \sum_{i=1}^{n}\left[\int_{A_i} K_h(\mathbf{u}-\mathbf{x})\,d\mathbf{u}\right]Y_i \tag{17}$$

where $\bigcup_{i=1}^{n} A_i = A$ is a (pairwise disjoint) partition of the domain of the data, and $K_h(\cdot)$ is the kernel weighting function as defined above. In the univariate case, the most common form of this is written as (Gasser and Müller, 1979),

$$\hat{m}(x) = \sum_{i=1}^{n}\left[\int_{s_{i-1}}^{s_i} K_h(u-x)du\right]Y_i \tag{18}$$

with $s_i = \dfrac{x_i + x_{i+1}}{2}$, where it is assumed that the x_i are ordered.

For the p-variate case, let $\mathbf{x} = (x_1,\ldots,x_p)$. As mentioned previously, a bandwidth is needed in each coordinate direction, but its size may vary with direction. Thus for this case, the term $1/h$ means $\prod_{i=1}^{p} h_i^{-1}$, where h_i is the bandwidth in the direction of x_i, $i = 1,\ldots,p$.

A special set-up often used to simplify the multivariate case employs product kernels. Here, the p-variate kernel is written as a product of p-univariate kernels acting on $\mathbf{x}$ componentwise, $K(\mathbf{x}) = \prod_{j=1}^{p} K_j(x_j)$. For a certain type of design (rectangular design), a further simplification is that by using product kernels, it is possible to do the smoothing componentwise, if the design is on a regular grid. That is, instead of one p-variate smoothing, one can do p univariate smoothings, which allows for a simple implementation of this algorithm. This saves a substantial amount of computation (for details see, e.g., Müller, 1993).

Locally Weighted Least Squares (LWLS) Regression

Locally weighted least squares regression, as mentioned above, fits regression models to the data locally. Presented here will be the case of local linear regression, but the results can be generalized to any order of polynomial regression. In the multivariate framework, the boundary region takes on added importance, and LWLS models have built-in boundary adjustments that alleviate boundary effects.

Of interest is the local linear least squares kernel estimator of $m(\cdot)$, which is $\hat{m}(\cdot) = \hat{\alpha}$, where $\hat{\alpha}$ is the solution for α in the weighted least squares problem

$$\text{Minimize} \sum_{i=1}^{n} \left[Y_i - \alpha - \beta^T (\mathbf{X}_i - \mathbf{x})\right]^2 K_H(\mathbf{X}_i - \mathbf{x}), \tag{19}$$

where H is a $p \times p$ diagonal matrix depending on n, $K(\cdot)$ is a p-variate kernel such that $\int K(u)du = 1$ and $K_H(u) = |H|^{-1/2} K(H^{-1/2}u)$. Here $|H|$ is the determinant of the matrix H. The solution is easily obtained once (19) is recognized as a weighted least squares problem, with weights determined by the kernel function.

One can visualize how the data are being used to estimate $m(\cdot)$ at $\mathbf{x}$, by fitting a plane to the data in a neighborhood of $\mathbf{x}$ using weighted least squares with weights given by the kernel function. In particular, this kernel function is usually a p-dimensional probability density function, which gives less weight to points further away from $\mathbf{x}$. The size of the neighborhood around $\mathbf{x}$ is determined by H.

Applying LWLS Regression to Folate Bioassay Data

In order to look at how parametric and nonparametric estimates model the same data, we applied the locally weighted least squares method to some of the data modeled in the parametric section. First, we used univariate LWLS regression to model the subset of data consisting of double-source regimes containing (*LF*) (regimes 7-12). Recall that these data were modeled parametrically in an earlier section using quadratic and change-point models (Figure 8). We first ran the model with a global cross-validation bandwidth (our FORTRAN program had this as an option), and then reran it with various global bandwidths, selecting the one giving the "nicest" curve (trial-and-error method). Both of these curves are shown in Figure 9, where they are overlaid on the original data of regimes 7-12 (as in Figure 8).

The selected bandwidths differed considerably, but for these data, the resulting curves did not. The global cross-validation bandwidth was calculated to be $h = 263.2$, while the one we chose by eye was 150.0, nearly half as much. Although normally not the case, it appears here that the global bandwidth calculated by cross-validation is too large, and the resulting estimated curve is a bit oversmoothed.

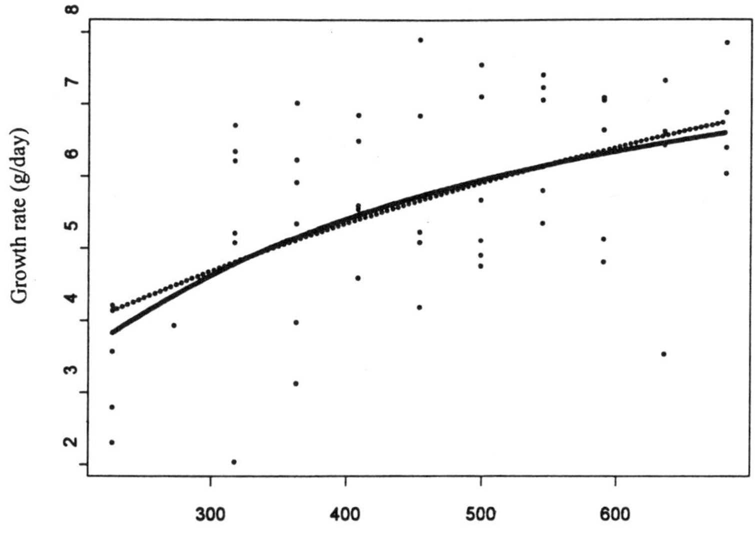

Figure 9. Curves produced by LWLS linear regression for regimes 7–12, using two different bandwidths. The dashed curve is constructed with $h_{cv} = 263.2$ (cross-validation bandwidth), and the solid curve with $h = 150$.

We then extended the application of LWLS regression to the two-dimensional case by modeling the response surface, using growth rate as the response and (*LF*) and (*FA+BF*) as the predictors, in analogy to the parametric model (8), displayed in Figure 6. As in the univariate application above, Gaussian kernel weighting was used, and global bandwidths were chosen both by cross-validation and by visual trial and error. A smoothed surface is shown in Figure 10, the estimated surface using the bandwidth chosen by cross-validation, which is 385 in the (*LF*) direction and 152 in the (*FA+BF*) direction (we chose H to be diagonal). The bandwidth in the (*LF*) direction was bigger since the values of (*LF*) are more variable than those of (*FA+BF*).

Looking at Figure 10, we see that the general shape of this nonparametric surface estimate is quite similar to that of the fitted parametric model (8). In particular, the surface shows linearity for (*FA+BF*), saturation in the (*LF*) direction, and moreover, that (*LF*) yields more growth than (*FA+BF*) when the amount of total folate is small (say between 227 and 500 nmol/kg diet). One advantage of the nonparametric surface over the parametric surface shown in Figure 6 via model (8), is that growth rate does not decrease when the dosage of (*LF*) is large, as it does for the parametric model.

Figure 11 presents a contour plot of this nonparametric surface. The response values where we took contours are the same as those used in the parametric case (based on model (8)), shown in Figure 7. The parametric surface contours (Figure 7) and the nonparametric ones exhibit both similarities and differences. Both show a saturation effect for (*LF*) in that the distance between the contours increases with increasing growth response, as the dosage of (*LF*) gets higher. In particular, looking at Figure 11 near the (*LF*) axis, the distance between the contours for *growth* = 4.75 and *growth* = 5.5 g/day is much smaller than that between those for *growth* = 5.5 and *growth* = 6.25 g/day, despite both pairs of contours being 0.75 g/day apart. Thus, the amount of (*LF*) required to increase growth rate by 0.75 g/day is much higher if the current diet regime in place is mostly (*LF*) at a growth rate of 5.5 g/day than if it were at a level of growth rate of 4.75 g/day; this is clearly a saturation effect. Near the (*FA+BF*) axis, the distance between the contours is roughly constant, in accordance with the parametric assumption of linearity of the effect of (*FA+BF*) on the growth rate. On the other hand, the two sets of contour curves do differ in shape, with the nonparametric contours having less curvature. Furthermore, these nonparametric contours generally have larger (*LF*) amounts than the parametric ones. For example, for *growth* = 6.25 g/day, the (*LF*) intercept is less than 400 nmol/kg diet for the parametric situation, but nearly 500 nmol/kg for the nonparametric situation.

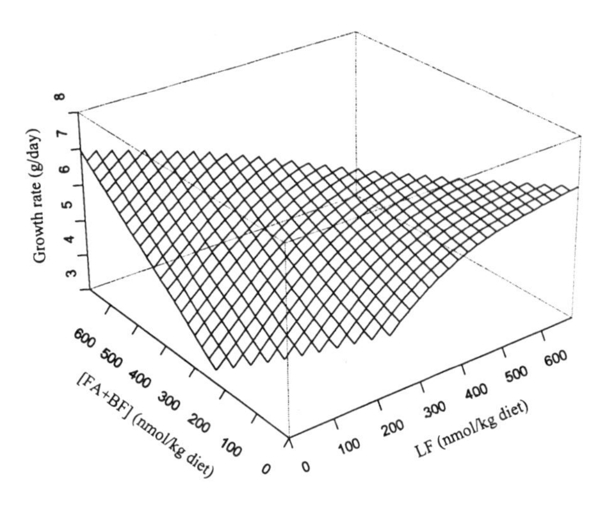

Figure 10. Estimated surface using two-dimensional LWLS, using global bandwidth H_{cv}= *diag*(*LF*,[*FA+BF*]) = (385,152) found by cross-validation.

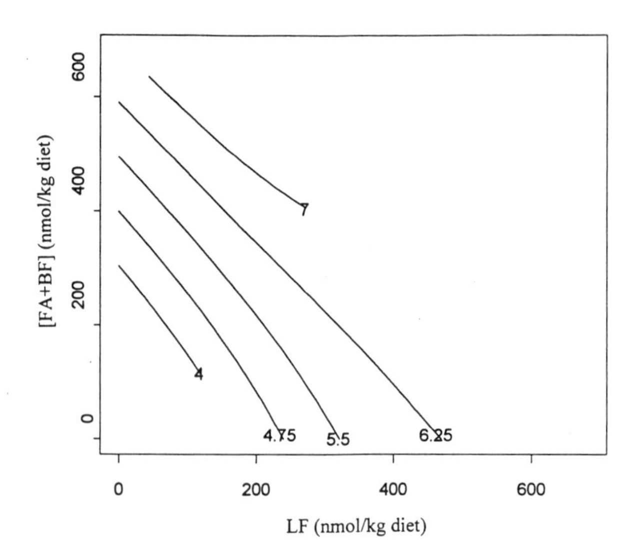

Figure 11. Contours of nonparametric surface estimate shown above in Figure 12, two-dimensional LWLS using H_{cv}.

An alternative surface fit is shown in Figure 12, produced by using the same bandwidth of smaller magnitude in each direction, namely 80.0 for both (*LF*) and (*FA*+*BF*). A comparison of this surface with both the parametric case (Figure 6) and the previous nonparametric case (Figure 10) provides some interesting observations. First, it has the same general shape as the other fits, and again better models the saturation effect than in the parametric case, since growth does not decrease with increasing (*LF*) when (*LF*) is large and (*FA*+*BF*) small. It also reveals more detail than the previous two surface estimates; in particular there appears to be a small "hump" in the middle of this surface not seen in the other fits. This may point to a particularly efficient folate combination in this area. Here we see how bandwidth choice can affect the estimator considerably, with smaller bandwidth choice often revealing additional structure.

The contour plot corresponding to this second nonparametric surface is shown in Figure 13. While these contours exhibit more curvature than those using H_{cv} in Figure 11, the nature of this curvature differs from that for the parametric estimate in Figure 7. In particular, the curvature in this case seems to be drastic in the middle of each contour, but minimal at both ends. For the previous two contour plots (Figures 7 and 11), the curvature tends to increase as each contour approaches the (*LF*) axis. Figure 13 is much more similar to Figure 11 than to Figure 7, indicating that the nonparametric procedures give roughly the same picture.

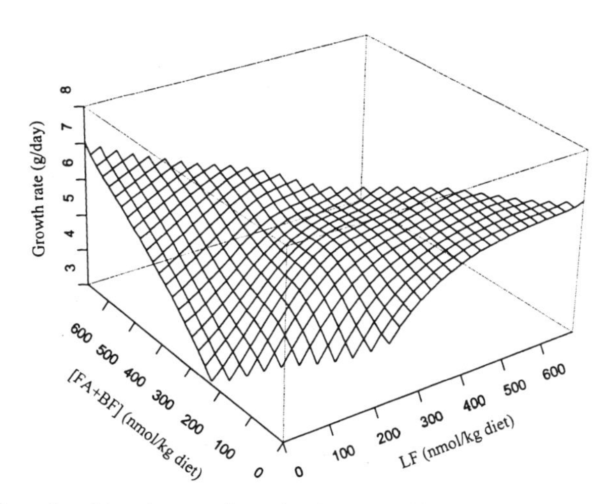

Figure 12. Estimated surface using two-dimensional LWLS with global bandwidth H = diag (LF, $[FA+BF]$) = (80,80).

Comparison of Nonparametric and Parametric Fits

As mentioned above, the plot of the estimated surface for our parametric model (8) shown in Figure 6 is similar in shape to that using nonparametric LWLS regression with bandwidth H_{cv}, shown in Figure 10. We further compared these two approaches graphically. For this, we calculated the fitted value given by each of the two methods calculated on the same two-dimensional grid of predictor variables, and then compared these two sets of response values (paired by input values).

Figure 14 is a simple scatter plot of the non-parametric fits obtained by LWLS using H_{cv} vs. the parametric fits obtained by model (8). The points for the most part fall on a line

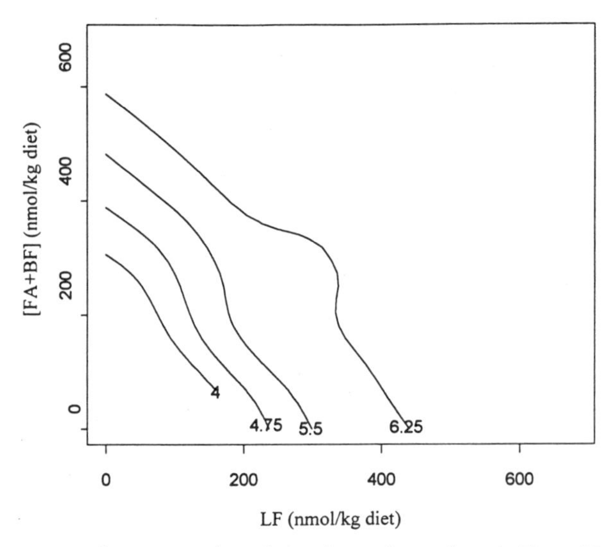

Figure 13. Contours of nonparametric surface estimate shown above in Figure 12, two-dimensional LWLS regression using H= diag(LF,[FA+BF])=(80,80).

nearly described by the equation $y = x$, indicating that the two methods generally agree in fit. However, one also sees that for predicted values in the middle of the distribution, the parametric model yields higher estimated response values than the nonparametric one at the same predictor points. This is then followed by a small area where the nonparametric method yields estimated responses that exceed those of the parametric one. We hypothesize that these trends are due to how the two methods model the effect (LF) has on growth rate. The parametric model (8) predicts that (LF) has a large positive effect on growth rate for smaller doses of (LF), but then eventually a decrease in growth rate occurs when the dosage of (LF) gets large.

A closer look at how the two estimators differ can be seen in the surface plot of Figure 15. Here we plotted the predictor values using the same axes representation as in Figures 6 and 10, but now the response is defined as

$$Y^*(\mathbf{x}) = \hat{m}_1(\mathbf{x}) - \hat{m}_2(\mathbf{x}) \tag{20}$$

where $\mathbf{x} = [LF,(FA+BF)]$, $\hat{m}_1(\mathbf{x})$ is the curve estimate at $\mathbf{x}$ using the nonparametric LWLS regression surface with bandwidth H_{cv}, and $\hat{m}_2(\mathbf{x})$ is the parametric surface estimate at $\mathbf{x}$ using model (8). Here we indeed see that much of the difference between the two estimators is explained by how they model the effect of (LF) dosage on growth rate.

Looking at the surface near the (LF) axis (so that (FA+BF) is near 0), we see that for small (LF), the parametric fit is larger than the nonparametric one (that is, Y^* is negative). However, as (LF) increases, Y^* increases quadratically, until finally it is positive with a large magnitude, meaning that the nonparametric fit has larger estimated growth rates than the parametric one for larger (LF) values. This is consistent when observing the individual fits of the two models in Figures 6 and 10. Recall that we found that when (FA+BF) is near 0 and (LF) is large, model (8) predicts that growth rate decreases as (LF) increases, while the LWLS model predicts that it only levels off. This would also explain the discrepancy between the estimators seen in Figure 14.

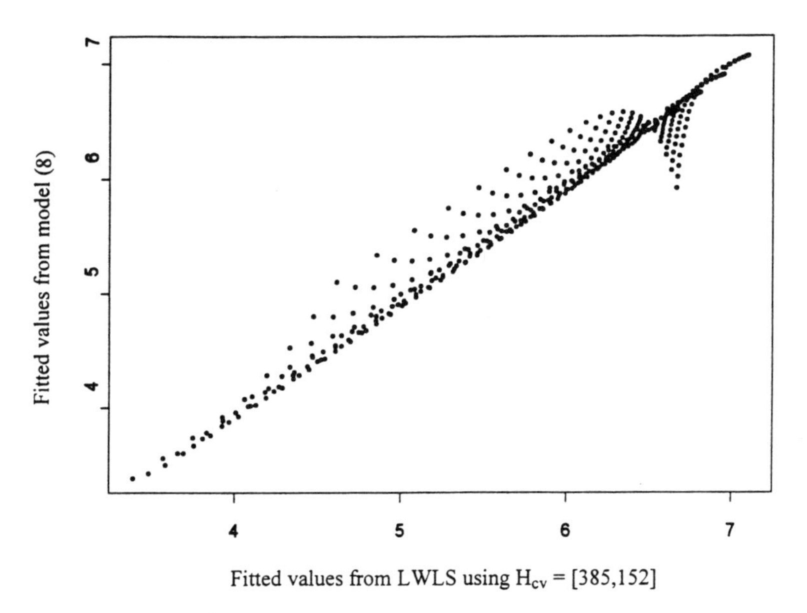

Figure 14. Scatter plot of predicted response values of the LWLS surface using H_{cv} against those using model (8) for the same set of inputs.

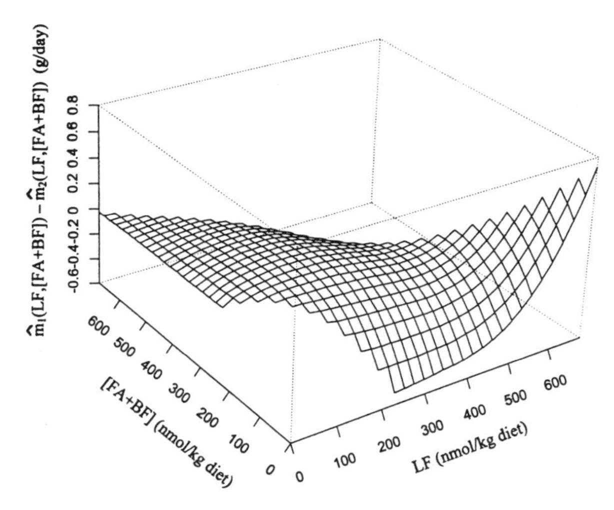

Figure 15. Surface plot of the difference in predicted response values between the LWLS surface using H_{cv}, denoted by $\hat{m}_1(\cdot)$, and the parametric surface corresponding to model (8), denoted by $\hat{m}_2(\cdot)$, for the same set of inputs.

Figure 15 also shows that when (*LF*) is near 0 and (*FA*+*BF*) is allowed to vary, the parametric model yields estimated growth responses greater than those of the non-parametric model, by about 0.2 g/day (i.e., Y^* = -0.2). This difference, unlike that for the case above, is relatively constant as (*FA*+*BF*) varies. This shows that the two estimators are stable with respect to each other in areas where the relation between folate and growth was simple (that is, when (*LF*) is much smaller than (*FA*+*BF*); recall that model (8) shows that when (*LF*) = 0, the relationship between growth rate and (*FA*+*BF*) is linear).

CONCLUSION

We have attempted to provide a broad overview of how folic acid affects the growth rate of rats, and in the process we used this nutritional investigation to illustrate parametric and nonparametric modeling techniques. Both techniques showed that (*LF*) behaved differently than (*BF*) or (*FA*), which seemed to be exchangeable. In particular, for small amounts of total folate, (*LF*) promoted growth more effectively than (*BF*) or (*FA*). For higher amounts, it seemed optimal to combine (*LF*) with (*BF*) or (*FA*), with the percentage of (*LF*) decreasing with increasing folate amount.

It is remarkable how close the parametric and nonparametric fits turned out in this application. This validates the parametric model in this particular instance. Generally, however, it is more prudent to rely on a nonparametric analysis to the largest extent possible, due to the advantages it has over parametric approaches. First, nonparametric techniques do not need distributional assumptions. More importantly, nonparametric techniques model data locally, allowing them to capture the relation between predictors and responses more accurately than parametric techniques, which force a global model across the whole spectrum of the data. This can lead to severe biases, invalidating statistical inferences. In particular, for the folate bioassay data presented, it was of interest to consider how (*LF*) interacted with (*BF*) or (*FA*). Parametric models use one mathematical expression to describe this interaction, but nonparametric techniques automatically change the description of this interaction as the predictor values are changing, which usually provides a more accurate picture. If one is lucky, as in the present example, the two fits will lead to similar scientific conclusions and thus validate each other.

A drawback of the nonparametric approach is that the fitted model cannot be summarized in a parameter vector: how the model behaves across the domain of the data can only be seen in a surface plot. Surfaces with more than two predictors cannot be well visualized. Furthermore, methods for statistical inference are difficult at best to apply in the nonparametric case. Thus there is a definite place for parametric modeling, and an optimal strategy would be to apply both types of techniques in the analysis of data in experimental nutrition.

CORRESPONDING AUTHOR

Please address all correspondence to:
Matthew Facer
Division of Statistics
University of California, Davis
One Shields Ave
Davis, CA 95616

ACKNOWLEDGEMENTS

The authors wish to thank Peg Hardaway for assisting us in several ways during the process of writing this paper. We also wish to thank our referee who provided useful comments.

REFERENCES

Bhattacharya, P.K. Some aspects of change-point analysis, in: *Change-point Problems.* Carlstein E; Müller HG; Siegmund D. Eds. Lecture Notes Monograph Series, Vol. 23. Institute of Mathematical Statistics: Hayward, CA. 1994. pp 28-56.

Bates DM. *Nonlinear Regression Analysis and its Applications.* John Wiley and Sons: New York. 1988.

Carlstein E; Müller HG; Siegmund D; Eds. *Change-point Problems.* Lecture Notes Monograph Series, Vol. 23. Institute of Mathematical Statistics: Hayward, CA. 1994.

Clifford AJ; Bills NA; Peerson JM; Müller HG; Burk GE; Rich KD. A depletion-repletion folate bioassay based on growth and tissue folate concentrations of rats. *J Nutr,* 1993, 123:926-932.

Fan J; Gijbels I. *Local Polynomial Modeling and Its Applications.* Chapman and Hall: New York. 1996.

Gasser T; Müller HG. Kernel estimation of regression functions. in: *Smoothing Techniques for Curve Estimation,* Lecture Notes in Mathematics, 757. Springer-Verlag: New York. 1979. pp 23-68.

Harvard Chart XL Reference Manual. Software Publishing Corporation: Santa Clara, CA. 1995.

Hinkley DV. Inference in two-phase regression. *J Am Stat Assoc,* 1971, 66:736-743.

McCullagh P; Nelder JA. *Generalized Linear Models, 2nd Ed.* Chapman and Hall: New York. 1992.

Müller HG. *Nonparametric Regression Analysis of Longitudinal Data.* Lecture Notes in Statistics, Vol. 46. Springer-Verlag: Berlin. 1988.

Müller HG; Schmitt T. Choice of number of doses for maximum likelihood estimation for quantal dose-response data. *Biometrics,* 1990, 46:117-130.

Müller HG. Surface and function approximation with nonparametric regression. *Rendiconti del Seminario Matematico e Fisico di Milano,* 1993, 63:171-211.

Müller HG; Facer MR; Bills NA; Clifford AJ. Statistical interaction model for exchangeability of food folates in a rat growth bioassay. *J Nutr,* 1996, 126:2585-2592.

Neter J; Kutner MH; Nachtsheim CJ; Wasserman W. *Applied Linear Statistical Models, 4th Ed.* R.D. Irwin: Boston. 1997.

Numerical Recipes in FORTRAN, 2nd Ed. Cambridge University Press: New York. 1992.

Ruppert D; Wand MP. Multivariate weighted least squares regression. *Ann Statist,* 1994, 22:1346-1370.

S-PLUS Reference Manual. Statistical Sciences, Inc.: Seattle. 1990.

Walzem RL; Clifford AJ. Folate deficiency in rats fed diets containing free amino acids or intact proteins. *J Nutr,* 1988, 118:1089-1096.

Wand MP; Jones MC. *Kernel Smoothing.* Chapman and Hall: New York. 1995.

APPENDIX

Table of rat folate depletion-repletion data (N=124)

Amount of folate (nmol/kg diet)			Regime number	Growth rate (g/day)
Chemical source (*FA*)	Bean source (*BF*)	Beef liver source (*LF*)		
227.00	0.00	0.00	1	2.30758
227.00	0.00	0.00	1	3.63096
272.40	0.00	0.00	1	2.70862
272.40	0.00	0.00	1	2.71413
317.80	0.00	0.00	1	5.40885
317.80	0.00	0.00	1	6.37438
363.20	0.00	0.00	1	4.36953
363.20	0.00	0.00	1	3.99358
408.60	0.00	0.00	1	3.71899
408.60	0.00	0.00	1	3.97384
454.00	0.00	0.00	1	4.79681
454.00	0.00	0.00	1	4.49622
499.40	0.00	0.00	1	6.64199
499.40	0.00	0.00	1	5.86936
544.80	0.00	0.00	1	6.85176
544.80	0.00	0.00	1	4.70673
590.20	0.00	0.00	1	5.71150
590.20	0.00	0.00	1	7.30619
635.60	0.00	0.00	1	5.93987
635.60	0.00	0.00	1	4.99342

681.00	0.00	0.00	1	7.07939
681.00	0.00	0.00	1	8.08029
0.00	227.00	0.00	2	4.45270
0.00	317.80	0.00	2	5.15745
0.00	363.20	0.00	2	2.88154
0.00	408.60	0.00	2	5.67045
0.00	454.00	0.00	2	6.93764
0.00	499.40	0.00	2	4.57215
0.00	544.80	0.00	2	6.59765
0.00	590.20	0.00	2	4.69842
0.00	635.60	0.00	2	6.51958
0.00	681.00	0.00	2	6.81902
56.75	170.25	0.00	3	2.07511
68.10	204.30	0.00	3	5.09888
79.45	238.35	0.00	3	4.39684
90.80	272.40	0.00	3	5.11764
102.15	306.45	0.00	3	4.63220
136.20	408.60	0.00	3	6.64972
158.90	476.70	0.00	3	6.31811
170.25	510.75	0.00	3	6.18633
113.50	113.50	0.00	4	2.60258
136.20	136.20	0.00	4	2.38294
158.90	158.90	0.00	4	3.87200
204.30	204.30	0.00	4	7.68098
227.00	227.00	0.00	4	5.74038
249.70	249.70	0.00	4	4.75872
272.40	272.40	0.00	4	6.38952
295.10	295.10	0.00	4	8.02657
317.80	317.80	0.00	4	5.64602
340.50	340.50	0.00	4	8.21232
170.25	56.75	0.00	5	3.36616
204.30	68.10	0.00	5	3.15819
238.35	79.45	0.00	5	4.97919
272.40	90.80	0.00	5	3.79459
306.45	102.15	0.00	5	3.91774
340.50	113.50	0.00	5	4.67794
374.55	124.85	0.00	5	6.67703
408.60	136.20	0.00	5	5.11714
476.70	158.90	0.00	5	7.23355
510.75	170.25	0.00	5	7.35875
0.00	0.00	227.00	6	5.07708
0.00	0.00	272.40	6	4.89117
0.00	0.00	317.80	6	6.93386
0.00	0.00	363.20	6	4.42415
0.00	0.00	408.60	6	7.44538
0.00	0.00	454.00	6	6.63903
0.00	0.00	499.40	6	4.80783
0.00	0.00	544.80	6	8.08638
0.00	0.00	590.20	6	6.07042
0.00	0.00	635.60	6	5.95146
0.00	0.00	681.00	6	6.97935
79.45	0.00	238.35	7	5.24761
90.80	0.00	272.40	7	5.95031
102.15	0.00	306.45	7	6.87965
113.50	0.00	340.50	7	6.87463
124.85	0.00	374.55	7	5.70747
136.20	0.00	408.60	7	6.16560
147.55	0.00	442.65	7	7.09148
113.50	0.00	113.50	8	4.24531
158.90	0.00	158.90	8	6.37479
181.60	0.00	181.60	8	5.37496
204.30	0.00	204.30	8	6.51810
227.00	0.00	227.00	8	7.93781

249.70	0.00	249.70	8	7.58605
272.40	0.00	272.40	8	5.83358
295.10	0.00	295.10	8	4.83991
317.80	0.00	317.80	8	6.65902
340.50	0.00	340.50	8	6.91955
170.25	0.00	56.75	9	3.60842
238.35	0.00	79.45	9	2.06902
272.40	0.00	90.80	9	3.16181
340.50	0.00	113.50	9	5.11015
374.55	0.00	124.85	9	4.93320
408.60	0.00	136.20	9	7.27196
442.65	0.00	147.55	9	7.10505
476.70	0.00	158.90	9	7.36706
510.75	0.00	170.25	9	7.88771
0.00	56.75	170.25	10	2.83226
0.00	79.45	238.35	10	5.10711
0.00	90.80	272.40	10	6.26143
0.00	102.15	306.45	10	5.62290
0.00	113.50	340.50	10	5.25814
0.00	124.85	374.55	10	7.14462
0.00	136.20	408.60	10	5.38269
0.00	147.55	442.65	10	7.12932
0.00	158.90	476.70	10	6.46504
0.00	170.25	510.75	10	6.43016
0.00	136.20	136.20	11	3.96619
0.00	158.90	158.90	11	6.73725
0.00	181.60	181.60	11	7.04508
0.00	204.30	204.30	11	4.61871
0.00	249.70	249.70	11	4.78653
0.00	272.40	272.40	11	7.44529
0.00	295.10	295.10	11	5.16683
0.00	170.25	56.75	12	2.33983
0.00	238.35	79.45	12	6.24630
0.00	272.40	90.80	12	4.00930
0.00	306.45	102.15	12	5.57280
0.00	340.50	113.50	12	4.21356
0.00	374.55	124.85	12	5.14141
0.00	408.60	136.20	12	7.09633
0.00	442.65	147.55	12	6.67226
0.00	476.70	158.90	12	3.57075
0.00	510.75	170.25	12	6.07560

STATISTICAL ISSUES IN ASSAY DEVELOPMENT AND USE

David Giltinan

Genentech, Inc.
South San Francisco, CA

INTRODUCTION

Technical advances in assay methods over the past decades have made reliable measurement of analytes at extremely low concentrations a reality. The potential utility of this enhancement in assay resolution for clinical and laboratory studies, particularly those focusing on pharmacokinetics and metabolism, is obvious. However, any potential gains may be lost by failure of the investigator to appreciate the nuances and limitations of the specific assay being used. In particular, if the study design does not incorporate appropriate measures to minimize any potential confounding effects of assay limitations, the resulting ambiguity in interpreting the data may be great enough to preclude reaching definitive conclusions. To avoid this particular type of experimental failure, scientists should be sufficiently aware of assay-related issues to be able to function as intelligent 'assay consumers.'

This chapter provides a brief overview of the *statistical* issues in the generation and interpretation of assay data. The emphasis is applied, rather than theoretical, focusing on issues that are important in practice. An understanding of the way in which assay values for an unknown sample are calculated, and of any potential limitations of this method, should help investigators make efficient use of existing assay techniques.

There is a huge array of different assay types, depending on the particular experimental objective, as well as the analyte, and the matrix in which it is to be measured. For instance, a biotechnology company using genetic engineering methods to develop proteins for therapeutic use will need to employ many different assay types during development of a molecule. These include a variety of assays to measure RNA and DNA levels, as well as methods to characterize the biochemical identity of the protein product. Quantitative assays for protein content or activity are needed, and may be based on chemical methods such as chromatography or mass spectrometry, on the techniques of immunoassay (e.g., RIA, RIP, IRMA, ELISA), or on some measure of protein bioactivity. Regulatory requirements stipulate that lot release criteria include assessment of potency using a bioassay, recognizing that assays based solely on binding can be misleading with respect to the protein's functional activity. Bioassays in common use range in complexity from *in*

Mathematical Modeling in Experimental Nutrition
Edited by Clifford and Müller, Plenum Press, New York, 1998

vivo evaluation of activity in animal models to *in vitro* testing based on some measure of functional activity in a particular cell line.

In developing a protein to be marketed, a variety of assays is also needed, with the general goal of ensuring the quality, purity, and safety of the product. These typically include assays for degradation products, variant forms of the protein, and possible impurities or contaminants. Assays to rule out potential environmental contaminants (e.g., bacterial or viral contamination) in the manufacturing facility may also be needed. Following administration of the protein to human subjects, determination of its pharmacokinetic and metabolic fate requires the ability to assay protein and metabolite concentration in a variety of biological fluids such as plasma, serum, or urine. A further requirement is an assay to detect the possible occurrence of antibodies to the protein, which could have adverse clinical consequences.

Some of the assay types listed above may not be fully quantitative, in the sense of yielding a single number as an estimate of the analyte concentration in a given test sample. However, semiquantitative assays may be adequate for some purposes—e.g., if the goal is simply to detect the presence of a particular analyte, or to provide reassurance that it remains below a specified threshold level. In this chapter, we will focus on quantitative assays, for which determination of the concentration of analyte in the matrix of interest is carried out using the following basic approach. For each of a number of known (*standard*) concentrations of the analyte, a signal of some type is generated and the signal level is measured. This allows construction of a curve describing the relationship between the signal level and concentration of analyte (the *standard curve*), which should be monotonic over the range of interest in order to give reliable results. The signal generated by an unknown sample is measured in the same system and the sample concentration is estimated by appropriate interpolation along the standard curve. Often, several dilutions of the unknown sample are prepared, to increase the probability that at least one dilution generates a signal within the range of the standard curve. In that case, a combined estimate is generated by averaging concentration estimates (suitably adjusted for dilution factors) across those dilutions yielding data within range.

This review does not attempt to cover the entire spectrum of assay types. We limit detailed exposition to the case of assays where the shape of the standard curve is sigmoidal. This includes almost all immunoassays, as well as many bioassays. Many of the concepts discussed below apply quite generally, although exact computational details may be different for other assay types. An excellent exposition of the challenges in immunoassay development may be found in the text edited by Price and Newman (1991).

The material is organized as follows. The second section gives a brief summary of common statistical terminology used in the context of assay data. A description of the basic mathematical and statistical framework for calibration is given in the following section. The fourth section describes the construction and use of intra-assay precision profiles, followed by a discussion of inter-assay precision. That section also covers the use and interpretation of appropriate control samples. The chapter concludes with recommendations on how to be an intelligent assay consumer, and a brief list of references for further reading. Use of formulae and formal mathematical notation has been confined to the latter sections, although a few Greek symbols may have escaped to other sections.

COMMON ASSAY-RELATED STATISTICAL TERMINOLOGY

This section gives a brief review of statistical terms used commonly in the assay and clinical chemistry literature. These are italicized upon first occurrence in this section. The intent is to introduce the basic concepts used to characterize assay performance, rather than to give definitions, as terminology is not always used consistently in the assay literature. The consensus paper of Shah and collaborators (1992) provides an excellent overview and discussion, as well as design recommendations for the experimental determination of key assay characteristics.

Systematic error corresponds to the situation where measurements are consistently over- or underestimated, by a fixed amount, or possibly by a fixed factor. Existence of this type of error results in *bias* in the estimation of concentration, i.e., a systematic deviation of the calculated estimate of concentration from the actual value. Such deviations from the correct value may correspond to a *fixed bias* or to a *relative bias*, depending on the nature of the underlying systematic error. Use of the term bias generally implies that all measurements are affected equally. Although the concept is straightforward, experimental determination of bias can be tricky, due to the difficulty in the preparation of samples for which the true concentration is known with absolute certainty.

Random errors in measurement should exhibit no particular tendency towards being positive or negative; we would expect the average of several such errors to be about zero. Thus, random error alone does not generally lead to bias in the assay results. The *precision* associated with a measurement is inversely related to the variability that one would expect in a series of similar determinations differing only by random fluctuations. In statistical terms, it is inversely related to the variability in the distribution of these random components. Information on precision is usually expressed using some measure of this variability, such as the standard deviation of the distribution; there is no unique measure of precision. It is also common to express the precision in relative terms, dividing the standard deviation of a series of measurements by the mean of the series. This yields the *coefficient of variation* (CV), sometimes known as the *relative standard deviation* (RSD). It has been argued that *imprecision* would be a more appropriate term for a quantity invariably described in terms of variability, but this usage is not widespread. Random variation in a series of measurements can arise from a variety of sources. In the final two sections of this chapter, we will illustrate how a good understanding of the sources contributing to the random variation can be important to enable efficient use of an assay.

It is common to distinguish between *intra-assay precision*, reflecting variability among replicate determinations within the same assay run and *inter-assay precision*, which relates to variation in values across different assay runs. Unfortunately, these two terms are sometimes used without adequate clarification of what constitutes an assay run. The term *reproducibility* is not always used consistently, but is usually taken to refer to consistency of results across assay runs. *Ruggedness* is a term used to reflect the ability of an assay to provide reliable results under a variety of different experimental conditions. Thus, ruggedness corresponds to a lack of sensitivity of assay performance to features such as changes in instrument, operator, procedural technique, or in other environmental conditions.

Specificity of an assay describes its ability to measure only the intended analyte, and nothing else. The concept of the *sensitivity* of an assay is concerned with its ability to detect/measure analyte at very low concentrations. This idea can be formalized in a number of ways, all subtly different, and this has led to some confusion about the exact meaning of the term *sensitivity*. Used on its own, the meaning is ambiguous; several terms exist relating to the idea of assay sensitivity, each emphasizing a slightly different aspect, and not to be used interchangeably. Two of the most commonly used terms are the *lower limit of detection* and the *quantitation limit*. The former, also known as the *minimum detectable concentration*, (MDC), refers to the lowest concentration that can be declared different from zero with confidence. It is often defined in terms of statistical significance of the difference from zero; as Ekins (1991) points out, this introduces a dependence on the number of sample determinations that is somewhat unsatisfactory.

The *(lower) quantitation limit* is the lowest concentration which can be assayed with 'acceptable' precision; what is acceptable may depend on the assay user and context. As definitions related to sensitivity usually involve assay precision in some way, a good understanding of precision is critical to characterize assay sensitivity correctly.

A STATISTICAL MODEL FRAMEWORK FOR CALIBRATION

Assay Setup and Data Description

Most commonly used immunoassays fall in one of two categories—radioimmunoassay (RIA), or enzyme-linked immunoabsorbent assay (ELISA). The difference lies in the manner in which the signal is generated after binding equilibrium has been reached. In the former case, radioactive labeling is used to generate a signal in the form of a radioactive count. For ELISAs, labeling is accomplished using an enzyme such as hydrogen peroxidase. Upon addition of its substrate, the labeling enzyme generates a colorimetric signal, which increases with the amount of label present. The strength of the signal can be read conveniently as an optical density measurement. This kind of assay typically employs a 96-well microtiter plate. Reference to an assay 'run' in what follows should taken to mean data obtained from a single microtiter plate. Similarly, 'intra-assay' variability refers to (well-to-well) variation within a plate.

Typically, a number of known concentrations are run in replicate, frequently in duplicate, to generate a standard curve. The recommended number of standard concentrations is between 6 and 8, chosen to span the full range of response, from background to maximum signal levels. Including more than 8 concentrations for the standard generally adds little, provided existing standards are spaced appropriately across the range. Geometric spacing of standard concentrations is usual, i.e., equal spacing on the log scale. Unless response variability is large, the gain realized by running standards in triplicate, rather than in duplicate, is usually minimal. Thus, about 15-20 wells per plate are given over to standards. A further 4-8 wells may be used for control samples, as discussed below. The remaining wells accommodate unknown samples. We assume that each plate has its own standard curve. Omitting standards on some plates (e.g. by generating a standard curve for only one of several plates run on a given day) might seem superficially appealing, as a way of freeing up wells to increase throughput for unknown samples. Based on the author's experience, this type of 'shortcut' should be avoided at all costs. The problems induced by having to calibrate 'across plates' typically affect data quality to such an extent that the resultant need for frequent reassays more than cancels out any increase in throughput. If space on the plate is truly a problem, running the standard curve in singlicate, but continuing to do so on each plate, represents a far better option than omitting standards for some plates.

Unknown samples are also typically run in replicate wells on a particular plate (replication may also occur across plates and/or across different days, see the following section on Assay Control). To increase the probability of obtaining a response within the range spanned by the standard concentrations, several dilutions of an unknown sample may be prepared, possibly with some replication at each dilution. Factors which influence the decision whether or not to assay several dilutions include availability of adequate sample volume to do so, and the amount of prior information about the likely concentration value. For instance, in determining protein concentration for vials filled to meet certain target specifications, there is good prior information on the likely value, so it should be straightforward to obtain values within the range of standards using only a single dilution. In contrast, in an early human pharmacokinetic study involving several dose tiers, very little information may be available on the likely value for a given sample, requiring use of several dilutions to obtain a signal value within the range of standards. If several dilutions yield usable data, it is standard to average the concentration estimates from the different dilutions, paying due attention to correcting for the dilution factors. It is important to remember that, for this type of arithmetic manipulation to give correct results, linearity of dilution must hold across the relevant concentration range, both for standards and unknown samples.

Modeling Standard Curve Data

To fix ideas, consider the data shown in Figure 1. The individual panels illustrate concentration-response data obtained for standards in separate runs of a bioassay for the recombinant hormone human relaxin. The assay is based on the ability of relaxin to increase generation and release of intracellular cAMP by normal human uterine endometrial (NHE) cells, in a dose-dependent manner. For each of the assay runs presented, triplicate measurements of cAMP levels were determined by radioimmunoassay following incubation of a fixed number of NHE cells with one of seven known concentrations of relaxin. A single zero-standard response value was also available for each plate. Figure 1 shows the concentration-response data for the standards where, by convention, response at zero concentration is plotted at two dilutions below the lowest standard.

In developing a model to describe data for standards of this type, consider one particular assay run. Suppose that pairs of data values (x_j, y_j) are generated by the standards for that run, where y_j denotes the observed response at concentrations x_j for the standards, $j = 1,\ldots,m$. Inspection of Figure 1 indicates that the data follow an approximate S-shaped, or sigmoidal, curve for each assay run. Certain features of the interpolated curves, such as the maximum, or the steepness, may differ from run to run, but the approximate sigmoidal shape persists across all runs.

In mathematical terms, the postulated interpolating curve for any given assay run may be specified by writing the response, y, as a monotonic function, f, of the analyte concentration, x. Some flexibility must be incorporated when specifying f, to accommodate the types of differences in standard curves seen in Figure 1. This is usually done by specifying a fixed functional form for f, but allowing the function to depend on a set of unknown coefficients, or parameters, denoted here by the $p \times 1$ vector, β. To describe the kind of S-shaped curves encountered in this assay, and in many others, choosing f to be a logistic function, depending on four freely varying parameters, often gives the required degree of flexibility. Thus, choice of the four parameter logistic function, given by (1),

$$f(x,\beta) = \frac{\beta_1 + (\beta_2 - \beta_1)}{[1 + \exp\{\beta_4(\log x - \beta_3)\}]} \qquad (1)$$

is fairly standard for f in this setting. Conceptually, however, any appropriate function f of a general p-dimensional parameter vector β, that is monotonic in concentration, x, could be substituted in (1).

For the logistic function, the four parameters have the following interpretation. β_1 and β_2 represent the response at zero and infinite doses, respectively, and β_4 is a slope parameter. The parameter β_3 gives the location of the inflection point when response is plotted against log concentration; in more familiar biological terms, it is the log*EC*50, where the *EC*50 is the concentration yielding an expected response halfway between the background and maximal response values.

The function f specifies a model for the *structural* component of the response, y. To complete the data description, a model is needed for the *random* component in measuring the response. This is usually done by adding a term to f to incorporate the random error. A simple approach is to add a term ε_j for the jth measurement, where the ε_j values represent independent random draws from a normal distribution with mean zero and (unknown) standard error, σ. However, this would imply that all response measurements are made with equal precision, which is not usually the case for assay data. A more realistic scenario is that variability in measuring the signal increases, as the mean level of the signal increases, at least in absolute terms. In the relaxin data, this increase in variability is quite pronounced (Figure 1). It is common to add a multiplier to the ε_j values, to reflect the increase in variability as a function of the mean response found in many assays.

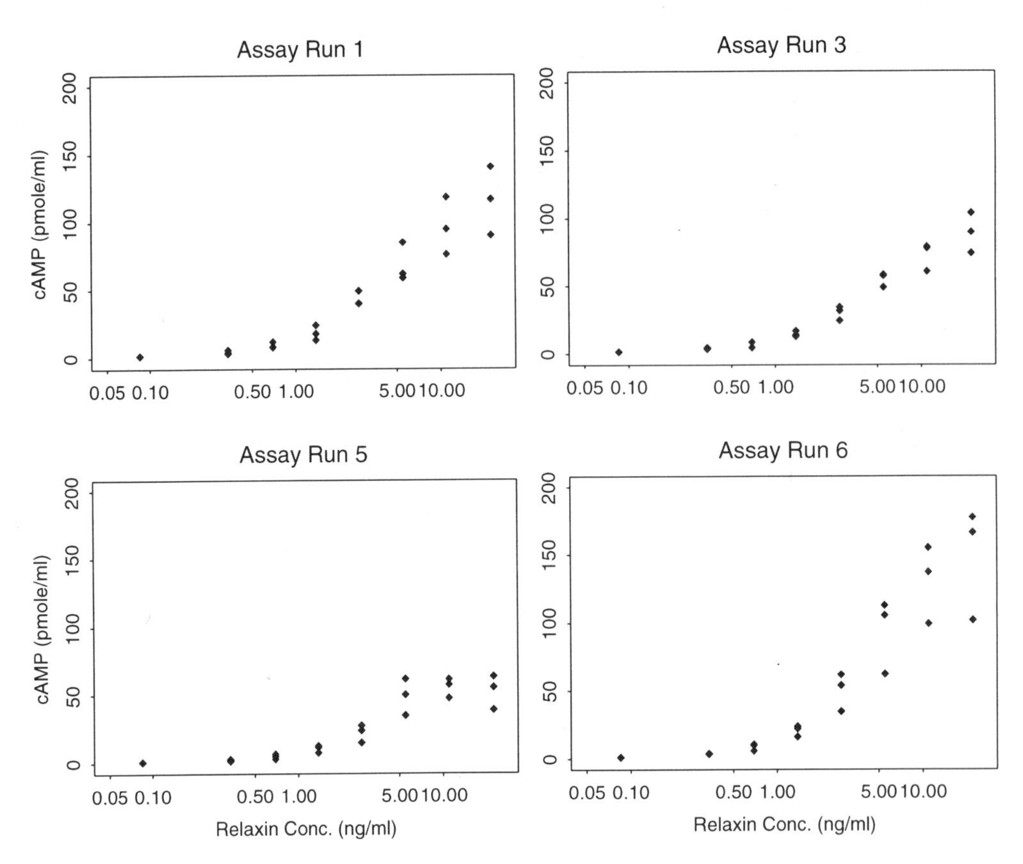

Figure 1. Standard curve data from four runs of a bioassay for relaxin.

There are many possible ways to model the nature of the dependence of variability on the mean; conceptually, one could represent the multiplier by a quite general function g to do so. Statistical terminology refers to the relationship between variance and the mean as the 'variance function'; the phrase 'response error relationship (RER)' appears more commonly in the clinical chemistry literature.

Combining the structural and random components of the model we have:

$$y_j = f(x_j, \beta) + g\left(f(x_j, \beta), \theta\right)\varepsilon_j. \qquad (2)$$

In this equation, the subscript j indexes measurements within the assay run. Replicate measurements are usually taken at some or all of the concentrations for a given run; this is not explicitly highlighted in (2).

A review of some of the more common options for the case of a general variance function g may be found in expository articles by Davidian and Haaland (1990), and O'Connell et al. (1993). In our exposition, we limit attention to one of the simplest possibilities, that where the standard deviation of the response is assumed proportional to some power of the mean. That is, the multiplier used to describe the heterogeneity of variance in response, (in statistical jargon, the *heteroscedasticity*), is taken to be proportional to a power of the mean. It thus has the form $\sigma\mu^{\theta}_{j}$ where μ_j represents the expected response at concentration x_j, which is $f(x_j, \beta)$ by assumption, σ is an unknown scaling constant, and the exponent θ characterizes the severity of the heterogeneity of variance. Typically, one would expect the value of θ to lie between 0, describing homogeneous variance, and 1.0, corresponding to a constant coefficient of variation, which represents quite marked heteroscedasticity. For a radioimmunoassay, the fact that the response variable corresponds to radioactive counts suggests the possibility of Poisson-type variation ($\theta = 0.5$). In the author's experience, other contributions to the variability beyond counting error can result in values of θ that are somewhat larger. Other references in the assay literature (Finney, 1976; Rodbard et al., 1976) offer some corroboration of this phenomenon.

In developing a new assay, the variability may not be well understood; the power model may be reasonable, but common choices, such as θ = 0.5 or 1.0 may be inappropriate. In such cases, a reasonable strategy is to estimate the appropriate value of θ from the data. A discussion of methods for doing so is beyond the scope of this review; further detail may be found in Carroll and Ruppert (1988, chapters 2 and 3), or Davidian and Giltinan (1993; 1995, chapters 5 and 10).

Why is it important to model variability in the response accurately? A common rationale is that statistical theory shows that the quality of fit of the standard curve may be adversely affected if fitting is based on an incorrect assumption about the pattern of response variability. For example, the assumption of homogeneous variability suggests a fit giving equal weight to all data points. However, if larger values are measured less precisely, then a better fit may be obtained by giving such points less weight in the overall fitting criterion. Experience at Genentech for a variety of assays suggests that this concern is largely theoretical when fitting the four-parameter logistic model. In most cases, unweighted and weighted fitting methods give virtually identical fitted curves, thus guaranteeing similar calibration results. This lack of sensitivity to the weighting scheme used does not necessarily extend to other choices of regression function, but the four-parameter logistic function tends to give remarkably stable results, absent extreme data anomalies. However, even though the effect of misspecifying the pattern of response variability or the *accuracy* of calibration of unknown samples may be small, correct specification is critical to obtain a correct assessment of the *precision* associated with the estimated concentrations. It is intuitive that a poor understanding of the variation in the signal results in a poor characterization of variability in estimated concentrations. This is demonstrated explicitly in the following section.

Fitting the Standard Curve

Estimating the coefficient vector β from standards data for a particular assay run is a nonlinear regression problem, as f is a nonlinear function of β. Ordinary least squares fitting is a possibility, but the heterogeneity of variance in response measurements suggests adopting a weighted approach instead, with weights w_j chosen to be inversely proportional to the response variance: $w_j = \sigma^2 \, \mathrm{Var}(y_j)$.

If the weights are considered to be known constants, then the weighted least squares (WLS) estimate of β is the value that minimizes $\sum_{j=1}^{m} w_j \{y_j - f(x_j, \beta)\}^2$. In practice, the correct choice of weights, w_j, will not be known. However, a simple extension of the WLS approach is to use *estimated* weights in a scheme such as the following (assuming θ known). First obtain a preliminary unweighted estimator β^P, e.g. by ordinary least squares. Based on this estimate, form estimated weights $\hat{w}_j = 1/(\hat{\mu}_j^{2\theta})$, where $\hat{\mu}_j = f(x_j, \beta^P)$. Use these weights to reestimate β by weighted least squares. Update the weights based on the new estimate of β and iterate until convergence. This method is known as iteratively reweighted least squares (IRLS).

Nonlinear regression routines with the flexibility to allow iterative reweighting are available in most major statistical packages. These routines generally require explicit specification of the regression function, and possibly its first derivatives with respect to the parameters, and appropriate initial estimates of the coefficients. For the four-parameter logistic model, reasonable starting values may be obtained fairly simply, based on the intuitive interpretation of the four parameters given earlier. Software implementation should also incorporate appropriate criteria for checking the convergence of the parameter estimates to reasonable values.

Measures to evaluate the overall quality of the fit should include a plot of the fitted curve superimposed on the raw data. A plot of weighted residuals against the predicted response values can be useful in assessing the plausibility of the assumed form of the variance function. Agreement between the actual standard concentrations and the calibration estimates obtained from the fitted curve for the standards also provides a useful check. In the author's experience, these checks are more informative than reliance on 'global' measures of fit such as R^2 values, or on standard errors of the estimated coefficients.

Concerns are sometimes raised about the detrimental effects of obvious outliers on the quality of the fit, and of subsequent calibration. Such concerns are often exaggerated, in the author's experience. Usually, such points exert a substantive effect on the fitted curve only if they happen to occur at the extreme concentrations. If the adverse effects on calibration are truly severe, this will generally show up in the behavior of the control samples.

Calibration of Unknown Samples

Let $h(y,\beta)$ denote the inverse of the regression function f specified in (2). In the particular case of the four-parameter logistic function, h has the form

$$x = h(y,\beta) = \exp\left[\beta_3 + \frac{\log\left\{\dfrac{(\beta_2 - y)}{(y - \beta_1)}\right\}}{\beta_4}\right] \tag{3}$$

Then if y_0 represents the average response for r replicates of an unknown sample at concentration x_0, the estimated value of x_0 is given by $\hat{x}_0 = h(y_0, \hat{\beta})$, where $\hat{\beta}$ is an estimate of β. Note that this is simply the algebraic formulation of the intuitive graphical procedure of finding the average response value on the y-axis, moving across horizontally until reaching the fitted curve, and then moving down vertically to find the corresponding concentration value.

INTRA-ASSAY PRECISION

In this section and the next, we consider methods to characterize the variation in estimated concentrations. Knowing which components dominate the variability allows sample submission strategies to be designed that counter the effects of assay variability efficiently. This will be illustrated in the next to last section below. We begin by considering *intra-assay precision*, which relates to consistency of sample results in the same assay run. Definition of an 'assay run' can be ambiguous; in our development here, we take the view that each separate standard curve defines a run. Thus, since we assume that standards are run on each plate, intra-assay precision deals with the within-plate (well-to-well) variation. While variability in the response is immediately evident in the raw data, it is the variability in estimated concentrations that is of primary interest in practice.

Interest in response variation *per se* is minimal; the importance of characterizing it correctly lies in the information it provides about the precision of calibration. There is an assumption in what follows that precision of the reported concentrations plays a major role in determining the *reporting range* of the assay. Specifically, a minimum requirement is assumed on the precision of a result for it to be judged sufficiently reliable to report, which corresponds to an upper threshold on the allowable variability. This threshold should be agreed upon by the assay scientist and the user; an appropriate value may depend on how the results are to be used.

Construction of Intra-assay Precision Profiles

Replication within plate, of standards, controls, or unknown samples, provides some information on within-plate variation. For any one plate, the information is likely to be very limited, as the degree of replication does not usually make it possible to get a good estimate of variability. For instance, if duplicates are run, then the estimate of variability for any sample has only 1 degree of freedom associated with it, meaning that extremely poor estimates are obtained—the estimated standard deviation based on duplicates may be nowhere near the true standard deviation. Controls and standards have an advantage over unknown samples in estimating variability for two reasons: (i) they are more or less guaranteed to span a range of concentrations, and (ii) data from multiple plates can be combined, as discussed in the next section, to augment the degrees of freedom associated with the within-plate variance estimate.

A plot of the variability of a given concentration estimate versus concentration, a so-called *precision profile*, can provide a useful summary of precision across the concentration range. Different variations of the precision profile are possible, depending on the exact measure of precision employed. We shall use the coefficient of variation (CV) in our illustration. The estimation of the CV described below is not based on replication, but rather on the residual variation about the fitted standard curve, and on the assumptions about the nature of the within-plate variation in the response. The calculations involved are approximate, but are nonetheless likely to characterize the precision more accurately than would be expected if one were to base all estimation of variation on duplicate or triplicate observations.

Using the same notation as before, let the inverse of the regression function be denoted by $h(y,\beta)$. Let $h(y,\beta)$ and the p-variate vector-valued function $h_\beta(y,\beta)$ represent the derivatives of h with respect to y and β, respectively. Then if y_0 represents the average response for r replicates of an unknown sample at concentration x_0, the estimated value of x_0 is given by $\hat{x}_0 = h(y_0,\hat{\beta})$, where $\hat{\beta}$ is an estimate of β with estimated covariance matrix V. An estimated approximate variance for $\hat{x}_0$ may be obtained by a standard Taylor series linearization and is given by

$$Var(\hat{x}_0) = \frac{h_y^2(y_0,\hat{\beta})\,\hat{\sigma}^2 g^2(y_0,\hat{\theta})}{r} + h'_\beta(y_0,\hat{\beta})\,V\,h_\beta(y_0,\hat{\beta}), \tag{4}$$

where $(\hat{\sigma},\hat{\theta})$ are estimates of the intra-assay variance parameters. The uncertainty in y_0 propagates to the concentration estimate through the multiplier h_y, the derivative of the inverse regression function. As this multiplier is inversely proportional to the derivative of f, the error in y_0 has least effect when f is steep, but can translate to enormous error in the estimated concentration in the range where the standard curve flattens out. The first term in (4) reflects uncertainty in the measurement of y_0 and may be expected to dominate the second term, which corresponds to uncertainty in the fitted standard curve. For some assays, however, the contribution of the second term may be substantial; in such cases, the widespread practice in the assay and clinical chemistry literature of taking only the first term of (4) may result in unduly optimistic estimates of intra-assay precision.

Use of Intra-assay Precision Profiles

To illustrate the importance of a good understanding of the correct value of the variance parameter θ, consider the profiles shown in Figure 2. These represent precision profiles (CV of estimated concentration versus concentration) based on run #1 of the relaxin bioassay, with the variance being calculated according to (4) under the assumptions $\theta = 0$, 0.5, and 1.0, respectively, and triplicate response measurements for the unknown sample in all cases ($r = 3$). The assessment of intra-assay precision diverges widely, according to our belief about the appropriate value of θ. Not all three can be correct; our point here is simply that accurate knowledge of θ is necessary in order to construct an accurate precision profile.

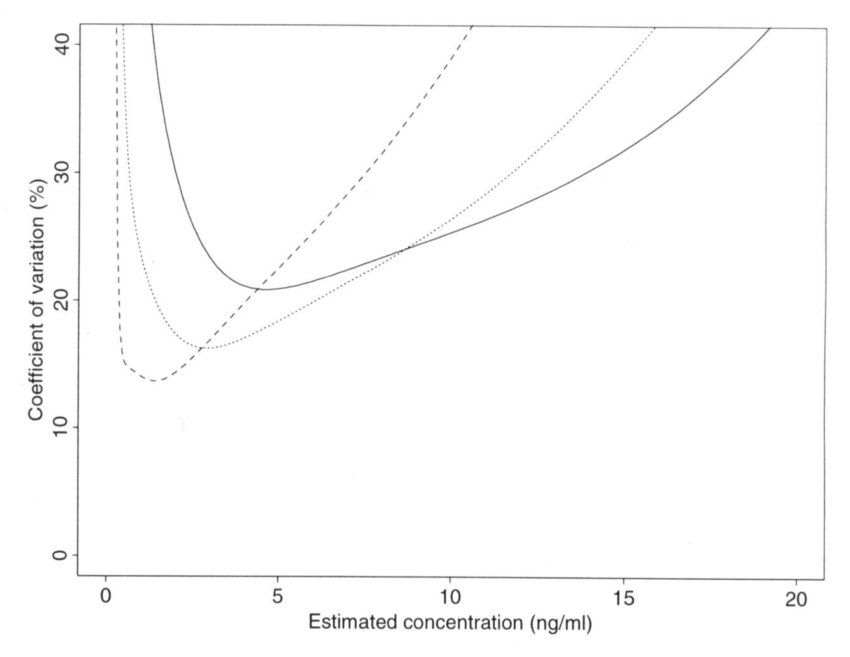

Figure 2. Precision profiles for different θ values, relaxin assay run 1. Solid line, $\theta = 0.0$; dashed line $\theta = 0.5$; dotted line, $\theta = 1.0$.

This type of intra-assay precision profile has two main uses in practice. It provides a useful tool for investigating the effect on precision of changing the degree of replication; for instance, to evaluate the gain in precision by increasing replication to $r = 4$ for this assay, one can simply regenerate the precision profile using a value of $r = 4$ in (4).

A second use for the precision profile is as an objective criterion for determining the 'usable' portion of the assay range. For instance, for the relaxin bioassay, an intra-assay coefficient of variation of 20% or less is considered acceptable by users. The rule of thumb which had been in effect before formal precision analysis was to take the *EC*20 to *EC*80 as the assay range. Inspection of precision profiles for nine separate assay runs showed this rule to be flawed; its symmetry about the inflection point fails to account for the asymmetric nature of the response variability across the response range. The assay ranges derived from the precision profiles, in contrast, were quite asymmetric about the *EC*50 value, typically corresponding to approximate *EC*3 to *EC*55 values; details may be found in Giltinan and Davidian (1994). Given the extreme heteroscedasticity in the response, this represents a more credible assessment of the acceptable range.

ASSAY CONTROLS

Even for assays considered 'routine' by today's standards, the complete sequence of steps required to estimate analyte concentration in an unknown sample is complex. Validity of results generally requires all of the following aspects to be correct: procedures for sample handling and storage, avoidance of operator or instrument error, purity and correct handling of assay reagents, maintenance of acceptable environmental conditions during assay performance, accurate recording of data, and the use of appropriate data reduction techniques. The potential for corruption of results is obvious; furthermore, the ways in which such corruption can arise are many and often subtle. Accordingly, it is essential to implement quality control procedures that maximize the ability to detect any problems that might arise, particularly if an assay is intended for routine long-term use.

This objective is addressed by the inclusion of *control samples* each time an assay is run. These are samples, prepared to a known nominal concentration, in such a way that their expected calibrated value should remain constant. Thus, for instance, the analyte used in preparing controls may be stored at -70°C, to minimize degradation over time. As the expected control concentration is constant, observed differences in estimated concentrations in a series of control values are informative, both in characterizing 'normal' assay variation, and identifying gross anomalies. The first objective of control samples is to gather information which allows the analyst to characterize the normal assay variation, and to facilitate rapid identification of potential problems in running the assay. The second objective is implemented in practice in two ways: (i) control values for a particular run help to judge the acceptability of results for that run, and (ii) monitoring control sample behavior enables the analyst to track assay performance over time, thus allowing early detection (and correction) of any trends indicating deterioration in assay performance. One can view assay controls as a technique to monitor the health of the assay, and to give early warning of any serious problems or adverse performance trends.

General Considerations in Implementing Controls

If judgements about assay performance and the acceptability of results are to be based on controls, then their utility is maximized by ensuring that preparation and handling of controls match procedures for unknown samples as closely as possible. Any differences in the handling of controls and samples will generally increase the potential for flawed sample results to go undetected. For instance, if the samples undergo a particular dilution step that controls do not, an error at this step may introduce error into the sample result that would not necessarily be reflected in the behavior of the controls.

Similarly, the control matrix should ideally mimic that for unknown samples as closely as possible. For example, if the analyte is to be assayed in patient serum, preparation of controls might use serum from a pool obtained from the patient population of interest. Matching may be imperfect: patient-specific differences in potential interfering factors such as binding proteins, or the presence of antibodies, can still complicate interpretation of results. In some cases, data may have been generated to show that use of a non-identical, but similar, matrix (e.g. normal serum in place of patient serum) gives comparable results. In such cases, the control matrix may differ somewhat from that in samples. However, in principle, the closer the match that can be attained, the better.

Controls should be included in each assay run. In the case of assays using microtiter plates, this corresponds to including controls on each plate. As control samples take up space on the plate, there is a tension between obtaining adequate control information and maximizing throughput of unknown samples. It is generally recommended that control samples at each of three concentrations be included. Assaying in duplicate wells at each of the low, medium, and high control concentrations is typical. The control concentrations should provide a reasonable span of the intended range, but high and low concentrations should not be placed too close to the ends of the range, as flatness of the standard curve is likely to affect estimation of control concentrations at the extremes adversely to an extent which prevents the controls from providing information of practical use. Dighe and Adams (1991) suggest the following concentrations for placement of the controls: low control at two- to three-times the limit of quantitation, medium control at 25-50% of the highest standard, high control at approximately 80% of the highest standard. This suggestion could be used to guide initial choice of control concentrations, with changes possible before finalizing placement, once there is sufficient experience with running the assay.

Using Controls to Assess Precision

The list of factors which can contribute to overall assay variation is intimidatingly long. Divergence in handling of the different sample aliquots, possible effects of plate position on signal intensity, differences in response among plates, operators, or due to varying environmental conditions from day to day, are some of the factors that may contribute to the overall variation in concentration estimates for a given sample. We now consider a simple statistical model to describe the different components of variation. We restrict attention to the situation where all contributing factors to the variation are thought to be random, that is, any potential sources of bias have been eliminated. The model can be extended to accommodate systematic factors (*fixed effects*, in statistical terminology); although analysis in such cases is much more complicated, requiring use of inferential techniques for mixed effects models.

To fix ideas, consider the following situation, involving a hierarchy with three levels of variation, and a completely nested design. Suppose for a particular assay, several microtiter plates are run each day, and the assay is run on a regular basis (e.g. three times weekly). Suppose further that control samples at three levels are included each time, in a way that incorporates replication within any one plate, replication across plates within the same day, and replication across days. The following model is appropriate for control samples at one specific level, and would be applied to low, mid, and high controls in three separate analyses.

For the particular control level under consideration, denote the concentration estimate obtained for the kth well, on the jth plate, on the ith assay day by x_{ijk}. The values for i, j, and k, are determined by the degree of replication at each level; in the event of duplicate wells per plate, for instance, k would have 1 and 2 as possible values. If there were no assay variation, this would correspond to a constant value, μ, say. The model which we describe accommodates the different levels of variation by adding random disturbance terms to the 'true value' μ, as appropriate. We may write it as:

$$x_{ijk} = \mu + \delta_i + \pi_{ij} + \varepsilon_{ijk} \tag{5}$$

Here, δ_i is a term representing the (random) deviation of values obtained on the *i*th day from the overall mean, π_{ij} is a random term specific to values on the (i,j)th plate, and the final error term ε_{ijk} represents the random component specific to the particular well indicated by that combination of indices. Each of the three random terms has an associated probability distribution, assumed to be independent of each other. The relative size of the variation from the three contributing factors is expressed in terms of the variances of the three associated distributions. Thus, δ_i is assumed to come from a distribution with variance σ_d^2, π_{ij} represents an draw from a distribution with variance σ_p^2, while the variance of the distribution of the ε term is σ_w^2. The mean of all three distributions is zero in all three cases, reflecting the assumption that systematic bias has been eliminated from the assay, or is small enough to be considered negligible. Note that the values of σ_w^2, σ_p^2, and σ_d^2 are not known and need to be estimated based on existing data.

This model is useful in two ways. It helps develop sensible methods to estimate the different components of variability from control sample data. It also provides insight into the effect of different replication schemes on the precision of an overall determination. Estimating components of variance is tricky, in all but the simplest case, and details are not presented here. The 'simplest case' distinguishes only two components of variability: intra-assay and inter-assay variation. These can be estimated by performing a one-way analysis of variance on the concentration estimates obtained for the control sample in question, with 'assay run' as a factor. Obtaining reasonable estimates of the components of variance in the general case (more than two levels of variability, imbalance in the design, presence of both fixed and random factors in the model, or any combination of the preceding features) is much more difficult, and is best done with help from a statistician.

Even the simple model above can provide useful insight into the effects of replication. Suppose that we are interested in characterizing the 'overall' precision of the final concentration estimate, obtained as the average of all determinations in range, where some replication has been incorporated at each level of the hierarchy. Specifically, suppose that values are obtained for w wells, on each of p plates per day, for each of d days, so that the total number of determinations is $d \times p \times w$. This is a slight oversimplification, but the general message also applies to the case of unequal replication (e.g. different numbers of plates run each day, differences across plates in the numbers of wells giving values in range). The average of the dpw determinations has an associated standard error given by:

$$SE(\bar{y}) = \sqrt{\frac{\sigma_d^2}{d} + \frac{\sigma_p^2}{dp} + \frac{\sigma_w^2}{dpw}}. \qquad (6)$$

This relatively simple formula has important practical implications. Intuitively, we know that the remedy against variability is replication. The average of many determinations gives a more precise estimate of the true concentration than does a single determination. With just one level of variability, and standard deviation σ, this corresponds to the fact that the standard error of the mean of n determinations is $\sigma/\sqrt{n}$. Thus, increasing n always increases precision by a factor of $\sqrt{n}$.

However, (6) shows that the situation is a little more complicated in the presence of several levels of variation. Thus, for instance, increasing the number of days on which determinations are made, d, affects all three terms in the formula, as d appears in the denominator of each. However, if we merely increase the number of wells run on each plate, w, but leave the number of plates per day, and the number of days of assay unaltered, only one term is diminished, that involving σ_w^2, the well-to-well variation. This makes intuitive sense; suppose only a single plate were run—inclusion of more wells on that plate does nothing to counter plate to plate variation. If values for that plate are 'high' (in the sense that the random deviation of results for that plate is positive), then they will all be elevated, and averaging more of them does nothing to alter that. For a nested design of this type, the practical message therefore is: Variability at a particular level can be compensated for only by replication at that level, or higher. Investigators sometimes take comfort because the total number of assay determinations is large. Unfortunately, unless the number is large at the appropriate level of the hierarchy, the sense of security may be misplaced.

This does not mean that replication at the within plate level is worthless. However, the law of diminishing returns takes effect at some point, beyond which the extra gain in precision is minimal. Table 1 gives a concrete illustration of this point, for the following specific values: $\sigma_w = 15$, $\sigma_p = 20$, $\sigma_d = 20$, true concentration = 100, ng/mL. Values in the body of the table represent the precision (% CV) of the final determination for different submission schemes, varying the degree of replication at the different levels of the hierarchy, (# of days, # of plates per day, and # of wells per plate).

By far the biggest payoff comes from replicating across assay days. Thus, for instance, if one has 6 determinations, it is far preferable to have values from 2 wells on each of 3 days (final CV = 17% approx) than from 6 wells on the same plate on a single day (final CV = 29% approx).

Table 1. Precision attained for different replication schemes.

# Days	# Plates per Day	# Wells per Plate	Total # Determinations	Final CV (%)
1	1	1	1	32
1	1	2	2	30
1	1	6	6	29
1	2	2	4	26
1	2	6	12	25
2	1	2	4	21
2	1	6	12	20
2	2	2	8	18
2	2	6	24	17.5
3	1	2	6	17.4
3	1	6	18	16.7
3	2	2	12	14.8
3	2	6	36	14.4

Evaluating the Acceptability of Assay Results

Control samples are also used to make decisions about the acceptability of assay results. Assuming that this judgement is applied to each plate, what would cause one to reject results for a specific plate? Loosely speaking, anomalous behavior of the control samples for the plate would serve as a warning. Identification of anomalous behavior involves an assumption that normal behavior has been adequately characterized. Some idea of the kinds of anomalies that are important practically is also helpful.

The appropriate statistical procedures for tracking performance of a process over time use the ideas underlying the construction of control charts and alert limits, popular in the field of statistical process control. Modifications to suit the assay context have been suggested by Westgard et al. (1981). The reader is referred to that paper, or to chapter 6 of Price and Newman (1991) for further detail, which exceeds the scope of this review.

BEING AN INTELLIGENT ASSAY CONSUMER

Know your Assay

The cliche that the best consumer is an educated consumer certainly holds true when interpreting assay data. The investigator should understand the principles and conduct of the assay in sufficient depth to avoid being misled by some artifact of assay behavior. For example, suppose it is of interest to measure concentrations of the protein IGF-I (insulin-like growth factor I) in subjects with Type I diabetes. Because these subjects will be receiving exogenously administered insulin on a daily basis, it is obviously important to have a clear understanding of how interference by insulin, e.g. due to cross-reactivity, may affect the measurement of IGF-I concentrations. (A trickier question arises if the administration of substantial amounts of exogenous insulin were to induce a downregulation of endogenous IGF-I production, though the question is not strictly an assay issue, if one adopts the view that the function of an assay is merely to provide a reliable measurement of the amount of protein present, not to explain how it got there). Similarly, in this situation, an understanding of the 'typical' milieu of binding proteins to IGF-I that can be expected in this subject group is likely to be important, so that a suitable matrix for control samples can be prepared to match that in the population being assayed as closely as possible.

In addition to knowing basic information such as the usable assay range, estimates of intra- and inter-assay precision, and the recommended degree of replication, the following list of questions may be helpful as a check on one's understanding of a particular assay:

- Does the assay measure free or bound analyte? What binding proteins may be relevant, given the sample matrix?
- What potential sources of interference exist?
- Do matrix effects, or dilution effects, play a role?
- How well does preparation and handling of the control samples mimic that of the unknown samples? If differences occur, what are they, and what are the possible consequences with respect to the ability of the control samples to detect performance problems? (Differences in the matrix and/or dilution procedures are probably the most common source of problems).
- What is known about the sensitivity of assay results to changes such as a new lot or supplier for a key reagent? What about sensitivity to different reading instruments or to different operators?

It is unfortunate that recent trends in the development of assay kits and instrumentation reflect a bias towards 'black box' technology, where involvement by the assay user is deliberately minimized. Estimates of the calibrated concentrations are often generated by internal software accompanying the instrument used to read the signal, with details of the calculations involved being kept from the assay user. This trend, presumably intended to be 'user-friendly,' actually has the undesirable consequence of making it harder for the user to develop intuition for the assay behavior. The problem is compounded by inadequate documentation of the methods used for data reduction, which is common for commercial assay kits and instrumentation. Assuming details of the data reduction methods are available, points to watch out for include the following:

- Is the default choice of standard curve appropriate? What other choices are available?
- Is the fitting technique sensible? Does it reflect the nature of the response error relationship appropriately, by using a suitable weighting scheme?
- If conventions for handling 'outliers' are used, are they sensible? If not, is it possible to change or override them? It is useful to consider this question separately for standards and unknown samples.

Use Appropriate Sample Submission Strategies

Understanding an assay thoroughly allows one to take active steps to use it as efficiently as possible. Two key aspects of intelligent assay use are (i) appropriate batching of unknown samples in submitting them for assay, and (ii) efficient use of replication.

To illustrate the relevant considerations in batching samples for assay, suppose that the samples are from a bioequivalence study to compare the pharmacokinetic behavior of two different formulations of a drug (A and B, say). Suppose further that the study employed a two-period crossover design, so that each subject received both formulations, with serial blood samples taken for assay to establish a concentration time profile over a 72-hour period following dosing of each formulation. Samples are to be assayed by ELISA with an intra-assay (within plate) CV of about 8%. Plate-to-plate variation for assay plates run on the same day is known to be approximately 12%, and day-to-day variation (over and above inter-plate variation) is of the order of 15%. The primary question of interest in the study is a comparison of the pharmacokinetics of both formulations.

In this scenario, comparison of the two formulations can be accomplished most precisely by assaying samples in such a way that, for each subject, samples from both formulations are run on the same plate. This ensures that the substantial inter-plate and inter-assay variability do not compromise the precision of the main comparison of interest, as both formulations can be compared for each subject using measurements obtained on the same plate. This is a simple illustration of a design principle known as *blocking* in the statistical literature. At the other extreme, assaying all samples for formulation *A* on one day, and those for formulation *B* on a different day, would represent an exceptionally poor batching scheme, as it maximizes the extent to which the interassay variability contributes to the imprecision in comparing formulations. Details of how to block efficiently will vary depending on the situation. For instance, in an experiment to compare four formulations, space limitations may preclude running samples for a given subject from all four formulations on one plate; nonetheless, one should try to batch samples so that an acceptable degree of balance in the allocation of formulations across plates and assay days is maintained. Similarly, if samples were to be run in an assay for which two instruments are available, care should be taken to allocate samples to instruments in a sensible manner. Even in the face of manufacturers' assurances about the minimal nature of certain components of assay or instrument variability, batching samples appropriately can protect against worst case scenarios, with little extra effort. Conversely, failure to do so can make it much harder to guarantee unambiguous assay results, and can sometimes result in significant doubts about the validity of experimental conclusions.

Replication is another tool to reduce the uncertainty that results from assay variability. The basic idea is straightforward—the average of several determinations is more precise than is a single measurement, so that precision can be increased by taking more determinations and averaging the result. However, as discussed earlier, to maximize the gain in precision achieved by replication, the variability at different levels of the hierarchy should be taken into consideration, in a manner that takes account of the message illustrated in Table 1.

Report and Interpret Assay Results Sensibly

As with any kind of data analysis, if any points have been omitted, it is the responsibility of the investigator to indicate this when reporting results. At a minimum, the rationale for omitting points, and the exact criteria used to do so should be clearly stated, as well as the number of points omitted as a result, and the likely impact on conclusions. Criteria should be as objective as possible, and should be in place before the analysis. Indicating in a report that certain data points have been omitted based on 'judgement of the investigator' reduces the credibility of the study, and diminishes the strength of the conclusions. Failure to report the exclusion of assay results, believed to be grossly in error, or for any other reason, constitutes an inexcusable omission.

It is important to report results in a manner consistent with the reliability of the assay. Precision should not be overstated, neither should it be sacrificed unnecessarily. A common mistake is to give a misleadingly optimistic impression of precision by retaining too many significant figures in reporting. (Given the nature of the calculations used in assays involving dilutions, it is important to work in terms of significant figures, as opposed to the number of digits after the decimal point). For example, if an assay determination has a 10% CV, it is not only pointless to report results to five or six significant figures, it is somewhat misleading to do so. In this regard, it is useful to bear in mind that the appearance of excessively many significant figures on an instrument panel, or computer readout, confers no particular validity, and does not magically increase the precision of the assay to the implied level. Conversely, one should be cautious about excessive truncation of significant digits; overly aggressive rounding of results can, on occasion, be a source of avoidable error. The 'correct' number of significant figures depends somewhat on the situation, and judgement should be guided by common sense, avoiding a slavish adherence to total consistency (e.g. if results are being reported to three significant figures, and the majority are below 100, writing the few that exceed 100 to four significant digits does not seem unreasonable). In the author's experience, retaining three significant figures is usually about right for immunoassay or bioassay results; note that the maximum possible change induced by rounding using this convention is 0.5%. For assays which attain a considerably greater degree of precision, retention of more significant digits may be appropriate.

Finally, it is important that results be interpreted within the context of the assay precision. For example, suppose one is testing the stability of a particular drug formulation in an accelerated testing protocol, where vials of the drug are stored at high temperatures to accelerate degradation. Following a baseline concentration estimate of 100 ng/mL, determinations after 2 and 4 weeks of storage yield concentrations of 95 and 65 ng/mL, respectively. Is this apparently precipitous loss in potency of between weeks 2 and 4 cause for alarm? The answer depends, of course, on the precision associated with the concentration estimates of 95 and 65 ng/mL. In the case of an HPLC assay, with an overall assay precision associated with a single determination of 5% (combining both intra- and inter-assay variability), then assay variability alone is unlikely to explain the apparent drop between weeks 2 and 4, so the likelihood that the observed difference reflects a true underlying decrease is high. However, if concentrations had been determined in a bioassay with an overall CV of 25%, there might be less cause for alarm, as it is quite plausible that assay variation alone could give rise to an apparent difference of 30 ng/mL between two single determinations made on separate occasions, even in the absence of any real drop in potency. The intention of this illustration is not to comment on the appropriate degree of replication, but to emphasize the importance of accounting correctly for the effects of assay variability and replication when interpreting results.

FURTHER READING

We conclude with a brief list of further references, which contain particularly accessible discussions of assay-related issues. The omission of a particular reference from this list should be interpreted as reflecting the inevitable arbitrariness in selecting such a list; no judgement of merit is implied.

An extremely useful general reference is the *book Principles and Practice of Immunoassay*, a collection of contributed chapters by experts in the field, edited by Price and Newman (1991). Chapters 5-8 discuss questions pertaining to data collection and reduction. The contributed chapter to this volume by Ekins (Immunoassay Design and Optimisation) contains a particularly lucid discussion of important issues in assay design.

Two very helpful tutorial articles on modeling assay data may be found in the journal *Chemometrics and Intelligent Laboratory Systems*. The first of these (Davidian and Haaland, 1990) discusses the case where a straight-line fit is adequate to describe the standard curve, as would be the case for most HPLC assays. Treatment of immunoassay data is considered in the second article (O'Connell et al., 1993). A more detailed development of the material in the sections on statistical model framework and intra-assay precision of this chapter may be found in Giltinan and Davidian (1994), or in chapter 10 of Davidian and Giltinan (1995).

Finally, the book *Statistics for Analytical Chemists*, (Caulcutt and Boddy, 1983), contains a very clear elucidation of general statistical principles in assay development and interpretation, although the exposition is confined to the case where a straight line fit to the standard curve is adequate.

CORRESPONDING AUTHOR

Please address all correspondence to:
David Giltinan
Genentech, Inc.
1 DNA Way
So. San Francisco, CA 94080

REFERENCES

Carroll RJ; Ruppert D. *Transformations and Weighting in Regression*. Chapman and Hall: New York. 1988.

Caulcutt R; Boddy R. *Statistics for Analytical Chemists*. Chapman and Hall: London. 1983.

Davidian M; Giltinan DM. *Nonlinear Models for Repeated Measurement Data*. Chapman and Hall: London. 1995.

Davidian M; Giltinan DM. Some simple methods for estimating intraindividual variability in nonlinear mixed effects models. *Biometrics*, 1993, 49:59-73.

Davidian M; Haaland PD. Regression and calibration with nonconstant error variance. *Chemometrics & Intel Lab Sys*, 1990, 9:231-248.

Dighe SV; Adams WP. Bioequivalence: A United States Regulatory Perspective, in: *Pharmaceutical Bioequivalence*. Welling PG; Tse FLS; Dighe SV; Eds. Marcel Dekker: New York. 1991.

Ekins R. The precision profile: Its use in assay design, assessment and quality control, in: *Immunoassays for Clinical Chemistry*, Hunter WM; Corrie JET; Eds. Churchill-Livingstone: Edinburgh. 1983.

Finney DJ. Radioligand assay. *Biometrics*, 1976, 32: 721-740.

Giltinan DM; Davidian M. Assays for recombinant proteins: A problem in nonlinear calibration. *Stat Med*, 1994, 13:1165-1179.

O'Connell M; Belanger BA; Haaland PD. Calibration and assay development using the four-parameter logistic model. *Chemometrics & Intel Lab Sys*, 1993, 20:97-114.

Price CP; Newman DJ; Eds. *Principles and Practice of Immunoassay*. Stockton Press: New York. 1991.

Rodbard D; Lenox RH; Wray HL; Ramseth D. Statistical characterization of the random errors in the radioimmunoassay response variable. *Clin Chem*, 1976, 22:350-358.

Shah VP; Midha KK; Dighe S; McGilveray IJ; Skelly JP; Yacobi A; Layloff T; Viswanathan CT; Cook CE; McDowall RD; Pittman KA; Spector S. Analytical methods validation: Bioavailability, bioequivalence and pharmacokinetic studies. *Pharm Res*, 1992, 9:588-592.

Westgard JO; Barry PL; Hunt MR. A multi-rule Shewhart chart for quality control in clinical chemistry. *Clin Chem*, 1981, 27:493-501.

STATISTICAL TOOLS FOR THE ANALYSIS OF NUTRITION EFFECTS ON THE SURVIVAL OF COHORTS

Hans-Georg Müller and Jane-Ling Wang

Division of Statistics
University of California
Davis, CA 95616

ABSTRACT

We discuss various methods which can be employed for the comparative analysis of samples of response curves. In the application discussed here, these curves are hazard functions, each generated by the survival data obtained for a cohort of experimental subjects which are fed a specific diet. It is demonstrated how comparisons of the effects of different diets on survival can be carried out by employing statistical techniques for inference on samples of curves. The methods are illustrated with data on the survival of large cohorts of male and female Mediterranean fruit flies under full diet and under protein deprivation. These statistical methods allow one to investigate differences between the samples of hazard functions generated by the four groups defined by combinations of sex and diet.

INTRODUCTION

When analyzing the effects of nutrition on survival and longevity in an experimental setting, typically the lifetimes of subjects which have been fed different diets are recorded. While the lifetimes themselves can be used for such comparisons, there are experiments with many cohorts where the subjects within the same cohort are exposed to cohort-specific influences so that there is a cohort random effect. An example are experiments involving medflies which are kept in cages, such that each cage contains a cohort of flies. In such applications, the cohort survival behavior can be conveniently summarized in the form of an estimated hazard function. The unit of observation then is the entire cohort. The problem of dealing with a sample of cohorts thus is translated into the problem of analyzing a sample of functions. In various types of experiments in nutrition, the recorded outcomes can in fact be viewed as curves. These outcomes may be recorded per subject or per cohort. Examples are pharmacokinetic data, where repeated measurements are recorded per subject

Mathematical Modeling in Experimental Nutrition
Edited by Clifford and Müller, Plenum Press, New York, 1998

(compare Davidian and Giltinan, 1995), or longitudinal weight measurements (see, e.g., Müller and Remschmidt, 1988).

Once it is recognized that such data correspond to a sample of curves, standard statistical techniques such as classical tests or analysis of variance are not directly applicable and generally not efficient, because their power against alternatives is typically rather low. Preferred classical methods include Hotelling's T^2 test to compare the pointwise means of two samples of curves simultaneously, and multivariate analysis of variance (MANOVA) to compare more than two samples of curves, also based on pointwise means. For a description of these methods, we refer the reader to Johnson and Wichern (1992). These methods are problematic when one is dealing with a sample of curves, because they ignore time shifts between curves (the dynamics of individual processes), and they also rely on the fact that the dimension (the number of measurements per curve) is not too high, as well as on normal distribution assumptions. Curve data, however, when viewed as a sample of smooth trajectories which cannot be finitely parameterized, consist of infinite-dimensional objects rather than multivariate vectors. Such data are therefore not covered by the typical methods of statistical multivariate analysis.

Methods for the analysis of curve data have found increasing interest in statistics in recent years. There is one recent monograph (Ramsay and Silverman, 1997) on this topic. Recent articles which discuss its applications to biological data from a statistical perspective include Gasser and Kneip (1995) and Capra and Müller (1997). Further methodological references can be found there. In statistics, this emerging area is also known as *Functional Data Analysis*, or *FDA* in analogy to the classical *Multivariate Data Analysis*, or *MDA*. Related methods have been discussed in applied biological contexts by Guardabasso and Munson (1988), Lindstrom (1995) and Veratta and Sheiner (1992). The reader is urged to also compare Kirkpatrick and Lofsvold (1990).

It is the aim of this chapter to briefly present some methods of statistical inference for curve data and to illustrate these methods with data on the survival of male and female Mediterranean fruit flies (medflies). The experiment, a biologically-oriented curve data analysis, its major biological implications, and their significance are described in Müller et al. (1997a).

The methods of analysis to be discussed in the present chapter include:

1. Displaying samples of estimated curves and their derivatives which can be obtained via curve estimation techniques. Such displays often provide insights into the dynamics of the underlying processes. This is a basic exploratory technique which is recommended as a first step for the analysis of samples of curves data.
2. Comparison of curves via characteristic features (biological parameters). If all curves share a few characteristic features which can be quantified, estimated values for these quantities may then be entered into further statistical analysis such as multivariate analysis of variance (MANOVA).
3. The definition of eigenfunctions and principal components for a sample of curves. The principal components then constitute a projection of each curve onto a low-dimensional vector, and can be entered into further statistical analysis such as MANOVA in the same way as the characteristic features obtained in (2).

We will demonstrate the usefulness of these methods when comparing samples of curves by means of an application example in the following sections.

SAMPLES OF HAZARD RATE FUNCTIONS AND THE MALE-FEMALE LIFE EXPECTANCY REVERSAL FOR MEDFLIES UNDER PROTEIN DEPRIVATION

In an experiment conducted with a total of 416,285 medflies, the survival of cohorts of medflies (*Ceratitis capitata*) was studied in dependency on the factors sex of the fly (male/female) and diet (full diet, i.e., a protein plus sugar diet /protein-deprived diet, i.e., a sugar only diet). The flies were kept in 66 cages, each containing approximately 6,000 flies, about evenly divided between males and females. The male and female flies in each cage form a cohort, for a total of 132 cohorts. Half of the cages were assigned to the full diet (protein plus sugar), and the other half to the protein-deprived diet (sugar only). More details about the experiment can be found in Müller et al. (1997a).

It was known from previous experiments that under full diet conditions, life expectancy for medfly females is longer than life expectancy for medfly males, in accordance with the situation in most species; this is called a female advantage in the life expectancy sex differential. A reversal of the life expectancy sex differential occurs if a female advantage in the life expectancy sex differential turns into a male advantage, that is, when males have longer life expectancy as compared to females. It was established in Müller et al. (1997a) that a reversal of the life expectancy sex differential occurs for medflies under protein deprivaton: Under full diet, female life expectancy exceeded male life expectancy by 1.30 ± 0.18 days (mean $\pm$ SD). In contrast, under protein deprivation, male life expectancy exceeded female life expectancy by 2.24 ± 0.27 days, and this reversal was highly significant.

This reversal phenomenon was tied to an early surge in mortality with a corresponding peak in the hazard rate (a hazard rate function is defined as $h(x) = \frac{f(x)}{1-F(x)}$, where f and F denote the density and the distribution function of the lifetimes respectively). Such an early surge with a peak was almost exclusively observed for protein-deprived female cohorts. This was interpreted as a vulnerable period in the early life cycle of female medflies, which was assumed to be caused by the activation of the reproductive system for individual females with resulting diversion of resources from repair and maintenance to the maturation of eggs. This leads to sharply increased mortality under protein deprivation. The peak in hazard rates was found to be narrowly confined in time, supporting this interpretation. There was no counterpart of such an early vulnerable period for male medflies.

The vulnerable periods and corresponding early peaks in hazard rates for female medflies are clearly visible when the samples of estimated hazard functions for each of the four groups [(a) females, protein-deprived; (b) females, full diet; (c) males, protein-deprived; (d) males, full diet] are plotted. These plots are shown in Figure 1, and are similar to those in Müller et al. (1997a). Refined dynamics can be discerned from Figure 2, where the samples of estimated first derivatives of the hazard functions are displayed. The early peaks in the hazard rates which are so prominent in the female protein-deprived group are resolved into a sine type wave, showing acceleration of hazard followed by deceleration, in accordance with the peak feature in the hazard rate itself. In fact, the derivatives reveal that these peaks are not quite symmetric: They rise faster than they decline.

Both curve and derivative estimates are valuable exploratory tools which allow one to gain insight into the relevant features of hazard rate functions. Methods for inference based on samples of curves will be discussed in the two subsequent sections. We emphasize here an important distinction of the curve data approach as compared to the common practice which ignores the cohort effects and combines the different cohorts into one large cohort. Key features in individual cohorts such as the peaks of mortality in Figure 1, may be

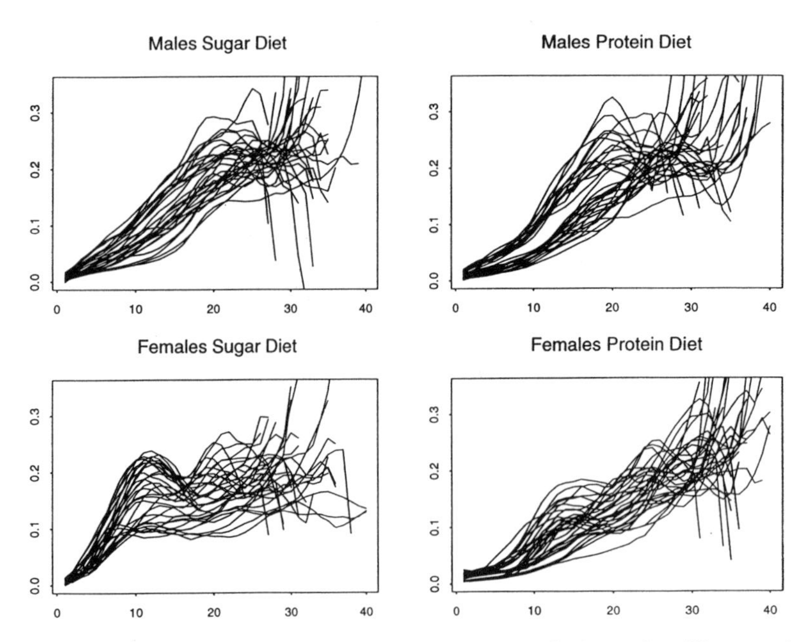

Figure 1. Samples of smoothed hazard functions for four groups of cohorts of medflies. Protein-deprived males, upper left; full diet males, upper right; protein-deprived females, lower left; full diet females, lower right. Each plot consists of 33 curves, each of which is constructed from a cohort of approximately 3,000 medflies of the same sex. Corresponds to Figure 1 of Müller et al. (1997a).

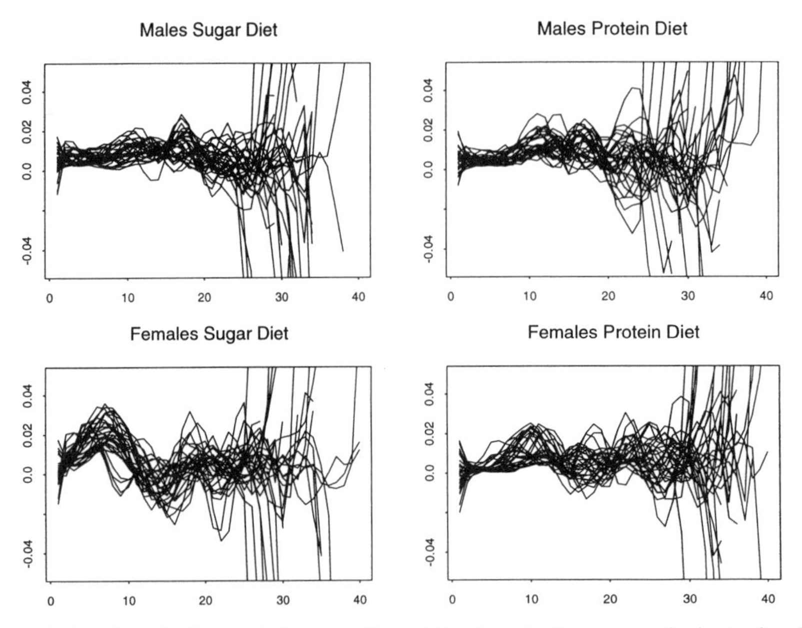

Figure 2. Samples of estimated derivatives of hazard functions for four groups of cohorts of medflies, arranged as in Figure 1.

masked or even lost in the resulting combined large single cohort due to time shifts or scale changes of the peaks across cohorts.

The remainder of this section is devoted to a review of the statistical methods used to generate Figures 1 and 2.

Given a scatterplot (X_i, Y_i), $i = 1,\ldots, n$, of data with x-coordinates X_i, and y-coordinates Y_i, we use local linear fitting to smooth the data. We can define a curve smoother as

$$S_0(x,b,(X_i,Y_i)_{i=1,\ldots,n},W_i) = \hat{a}_0 \tag{1}$$

where $\hat{a}_0$ minimizes the weighted least squares expression

$$\sum_{i=1}^{n} (Y_i - (a_0 + a_1(X_i - x)))^2 K\left(\frac{x-X_i}{b}\right) W_i \tag{2}$$

with respect to a_0. Here, the W_i are case weights associated with the observation (X_i, Y_i), and $K\left(\frac{x-X_i}{b}\right)$ is a kernel weight. The kernel weights are derived from a kernel function K and a bandwidth b which serves as the scaling factor of the kernel function. Common kernel functions are

$$K(x) = \begin{cases} 1-x^2 & |x| \le 1 \\ 0 & |x| > 1 \end{cases}, \text{ and} \qquad K(x) = e^{-x^2}. \tag{3}$$

The bandwidth has to be provided by the user or can be estimated from the data by procedures such as cross-validation. More details on this type of smoothing by fitting local lines by weighted least squares and related kernel estimation methods can be found in the monographs by Fan and Gijbels (1996) or Müller (1988).

For the estimation of derivatives, it is advantageous to fit local quadratic polynomials, so that one obtains

$$S_1(x,b,(X_i,Y_i)_{i=1,\ldots,n},W_i) = \hat{a}_1 \tag{4}$$

where $\hat{a}_1$ is the minimizer of the weighted least squares problem

$$\sum_{i=1}^{n} \left(Y_i - (a_0 + a_1(X_i - x) + a_2(X_i - x)^2)\right)^2 K\left(\frac{x-X_i}{b}\right) W_i \tag{5}$$

with respect to a_1.

We note that for estimating the curve itself, the locally fitted polynomial should be of odd order (the default being a linear polynomial), while for estimating the first derivative, the locally fitted polynomial should be of even order (the default being a quadratic polynomial). Different choices for these polynomials will lead to different estimates, and for most data, the default choices as above are sufficient. More important for the quality of the curve estimate is a good bandwidth choice. Various methods, including cross-validation, have been proposed for data-based bandwidth selection.

For the smooth estimation of hazard functions, we observe that the available data are of the form (t_j, d_j), where t_j is the jth day of age for the cohort and d_j is the number of flies which died on that day. We compute the central mortality rates $q_c(t_j) = \frac{2d_j}{n_j + n_{j+1}}$, where n_j are the flies surviving to the beginning of day j.

Applying the transformation $\psi(x) = -\log\left(\frac{2+\mathrm{x}}{2-\mathrm{x}}\right)$, we obtain the scatterplot $(t_j, \psi(q_c(t_j)))$. Smoothing this scatterplot with smoother S_0 (1), while employing case weights $W_j = n_j$, provides an estimate of the hazard rate function which enjoys good statistical properties (see Müller et al. 1997b).

Thus the hazard rate estimates are

$$\hat{h}(x) = S_0\left(x, b, (t_j, \psi(q_c(t_j)))_{j=1,2,\ldots}, n_j\right) \tag{6}$$

for the hazard function $h(x)$, and

$$\hat{h}^{(1)}(x) = S_1\left(x, b, (t_j, \psi(q_c(t_j)))_{j=1,2,\ldots}, n_j\right) \tag{7}$$

for the first derivative $\frac{d}{dx}h(x) = h^{(1)}(x)$.

COMPARISONS USING CHARACTERISTIC FEATURES

In this technique, we use the nonparametric curve estimates of the hazard rate functions displayed in Figures 1 and 2 to extract characteristic features which can then be compared across the four groups by classical statistical techniques. This approach was developed in Gasser et al. (1984) in the context of analyzing features of the human growth curve from samples of growth curves. This analysis revealed the existence of a midgrowth spurt which occurs around 6-8 years in children, well ahead of the pubertal growth spurt.

We illustrate the feature-based approach with two examples from the medfly cohort data. In the first example, we use as features the size and timing (location) of the first peak in the sample of hazard rate curves. This peak is of interest as it is the salient feature in the hazard rate curves obtained for the female protein-deprived group. But we see from Figure 1 that the other hazard curves also have peaks, which however occur later.

A peak is best quantified by two main characteristics: The timing of the maximum and the size or amplitude of the peak. For each estimated curve, we first obtain the corresponding two-vector consisting of (estimated peak timing, estimated amplitude). The scatterplots of amplitude versus timing (location) for the four groups (protein-deprived males, full diet males, protein-deprived females, and full diet females) are shown in Figure 3.

The obvious differences are mainly due to the protein-deprived females. This becomes even clearer in Figure 4, where amplitudes versus peak locations are plotted for all four groups together, using a different symbol for the protein-deprived females. This group clearly stands out in terms of early timing of the peak.

We can now enter these two-dimensional estimated *biological parameters* defined as estimated values of the characteristic features (peak timing, peak amplitude) into a multivariate analysis of variance (MANOVA). For the diet effect, we find an F statistic of $F = 0.30976$ at (2,127) degrees of freedom (d.f.) corresponding to a p-value of 0.74.

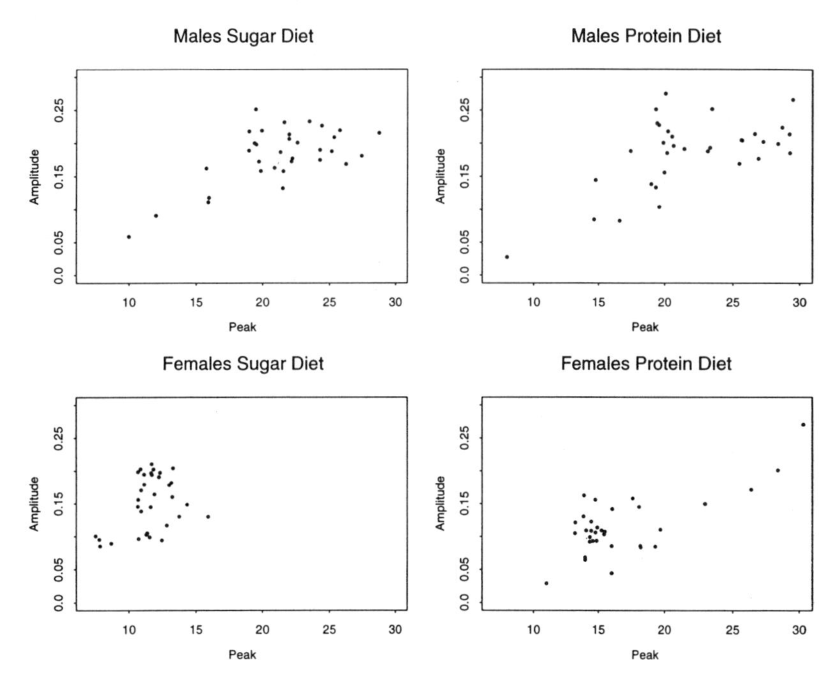

Figure 3. Scatterplots of amplitude of leftmost peak versus timing of leftmost peak, where peak coordinates are estimated from the smoothed curves in Figure 1.

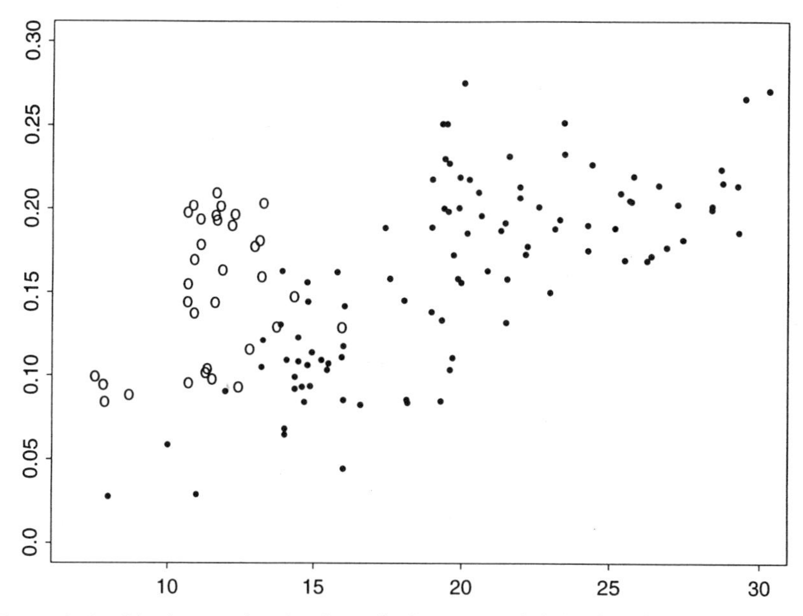

Figure 4. Combined scatterplot of peak amplitude versus peak timing for leftmost peak, using open circles for protein-deprived females and closed circles for the other three groups.

For the sex (male/female) effect, $F = 56.3$ at (2,127) d.f., $p < 0.0001$; and for the interaction of the two, $F = 20.1$ at (2,127) d.f., $p < 0.0001$. Tukey's studentized range test shows that the two male groups cannot be distinguished while the two female groups differ significantly among themselves and also from the males. We conclude that the diet effect on the first peak in the hazard rate is significantly different for male flies as compared to female flies, and that diet matters for females.

A similar comparison can be carried out by taking the estimated first derivative of the hazard function at day 5 and the level of the hazard function at day 5 as characteristic features. Day 5 was chosen because it is in the acceleration phase of an early mortality peak and therefore values for the derivative and level of the hazard function at day 5 are expected to characterize the early dynamics of the hazard function.

The scatterplots of estimated derivatives versus estimated levels are shown in Figures 5 and 6, again highlighting the protein-deprived female medflies in Figure 6. Clearly, the protein-deprived female medflies stand out, as they exhibit much higher rises in early mortality in terms of derivatives compared to any other group. This confirms the previous analysis based on peak characteristics.

COMPARISONS USING PRINCIPAL COMPONENTS

The Principal Components approach does not require that different curves in the sample share similar features such as peaks. Instead, each curve is regarded as the realization of a stochastic process. Under mild assumptions, a stochastic process $X(t)$ allows a Karhunen-Loève decomposition into eigenfunctions (see Karhunen, 1947, and Ash and Gardner, 1975, for the underlying theory).

$$X(t) = \mu(t) + \sum_{j=1}^{\infty} \varepsilon_j \rho_j(t) \tag{8}$$

Here $\mu(t)$ is a mean trajectory, the ε_i are uncorrelated random variables with zero expectation and finite variances, and the $\rho_j(\cdot)$ are the eigenfunctions of the process. This decomposition can also be interpreted as a representation of the covariance structure of the process $X(\cdot)$, given by $cov(X(t), X(s))$, for all s, t in the domain. The interpretation of the eigenfunctions depends on the specific application.

The functional principal components approach has been studied by Castro et al. (1986), with industrial applications; by Kirkpatrick and Heckman (1989), who emphasized biological applications; and more recently by Rice and Silverman (1991), who discussed the importance of smoothing methods in this context. With suitable estimates of $\hat{\mu}, \hat{\rho}_j$ of functions μ and ρ_j, and a cut-off point M, an observed process for the ith subject out of N subjects, $X(\cdot)$, can be approximated by

$$X_i(t) = \hat{\mu}(t) + \sum_{j=1}^{M} \hat{\varepsilon}_{ij} \hat{\rho}_j(t) \qquad i = 1, \ldots, N \tag{9}$$

with estimated coefficients $\hat{\varepsilon}_{ij}$. Here $\hat{\varepsilon}_{ij}$ is the jth principal component (PC) for the ith subject.

This is a generalization of classical principal component analysis in multivariate analysis (see, e.g., Johnson and Wichern, 1992), to the case of functional data. Details of estimating $\hat{\mu}(\cdot)$, the $\hat{\rho}_j(\cdot)$, and the $\hat{\varepsilon}_{ij}$ can be found in Capra and Müller (1997) and

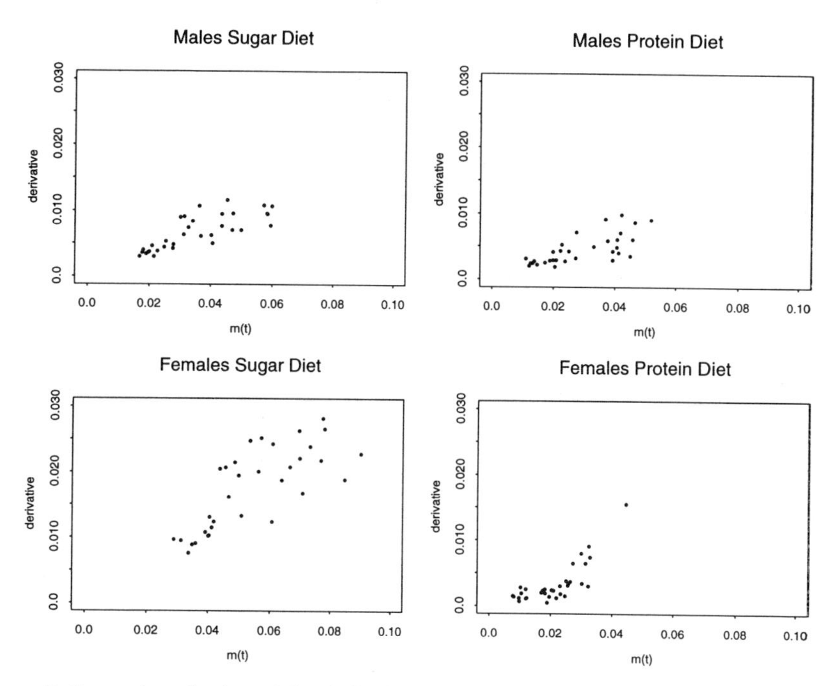

Figure 5. Scatterplots of estimated first derivatives of hazard function at day 5 versus estimated levels of hazard function at day 5, separated according to group.

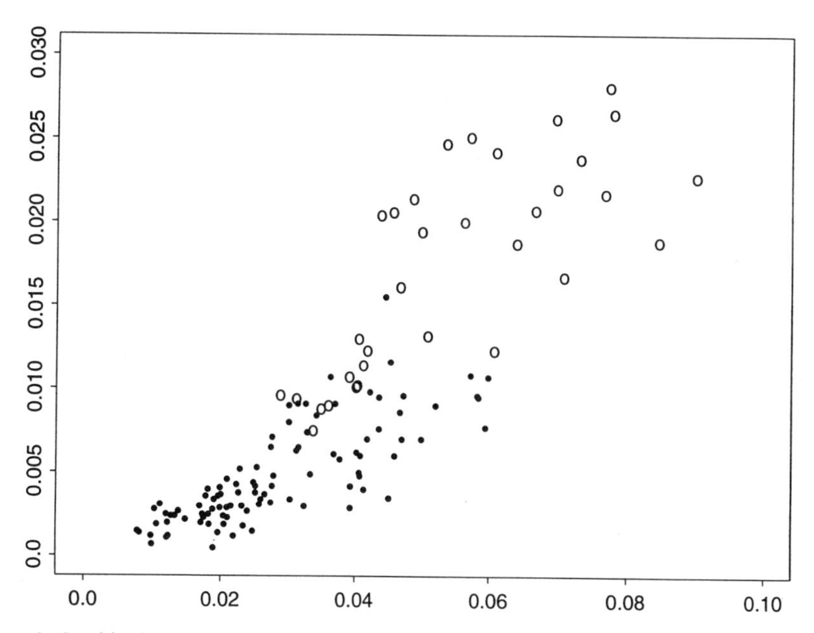

Figure 6. Combined scatterplot of estimated first derivatives versus estimated levels of hazard function at day 5. Open circles correspond to cohorts of protein-deprived females, closed circles to cohorts of the other three groups.

Ramsey and Silverman (1997). The latter authors explore also the issue of 'registration,' i.e., time shifts for the individual processes, before carrying out the functional version of Principal Components Analysis.

Applying this concept to the medfly data, we pool all groups together under the assumption of no group effect and then find that the first three PCs explain 48%, 18%, and 6% (together 72%) of the variance, so the choice $M = 3$ is natural. The corresponding first three estimated eigenfunctions are shown in Figure 7.

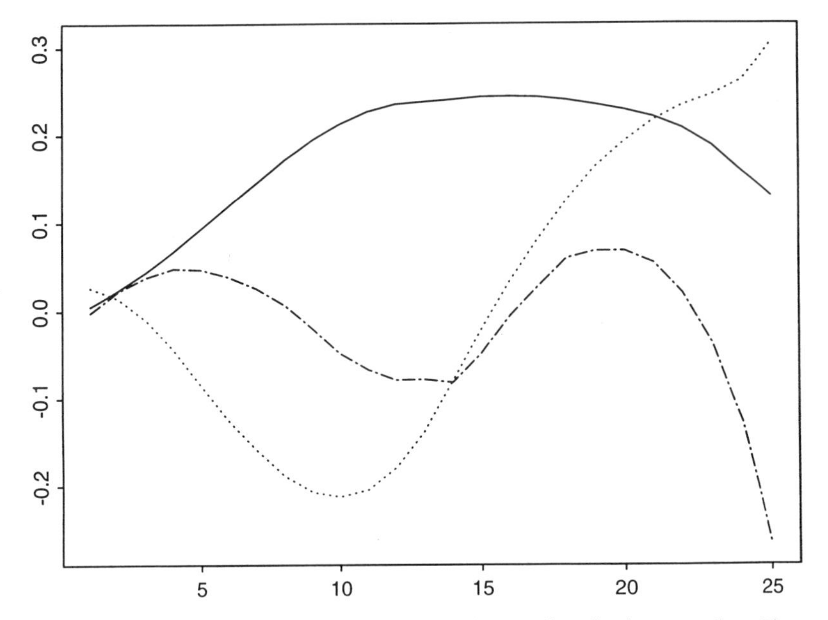

Figure 7. Estimated first three eigenfunctions obtained by pooling all cohorts together. First eigenfunction is solid, second eigenfunction dotted, and third is dash-dotted.

The first PC, corresponding to the coefficient $\hat{\varepsilon}_{i1}$, indicates how strongly an overall rise (first eigenfunction) is represented in a given hazard rate function. The second PC loads mainly on the early mortality peak (the sign of the eigenfunctions is arbitrary, so that a trough is as good as a peak), while the third PC loads on a somewhat later peak. The second and third eigenfunctions differ in the right tail. For each subject we obtain a vector of three principal components $(\hat{\varepsilon}_{i1}, \hat{\varepsilon}_{i2}, \hat{\varepsilon}_{i3})$, $i = 1,\ldots, N$. These PC vectors can then be entered into a MANOVA. The results are similar to those reported above for characteristic points. Plots for PC2 versus PC1 for the four groups are shown in Figure 8, and combining these and marking the protein-deprived female group separately leads to Figure 9.

The protein-deprived female group tends to have negative values for the second PC, corresponding to a more expressed early mortality peak, and relatively large positive values for the first PC, indicating a well-developed overall rise in the hazard function. It is also of interest to note that a more expressed early mortality peak is associated with a decline in hazard later on, as indicated by the shape of the second eigenfunction.

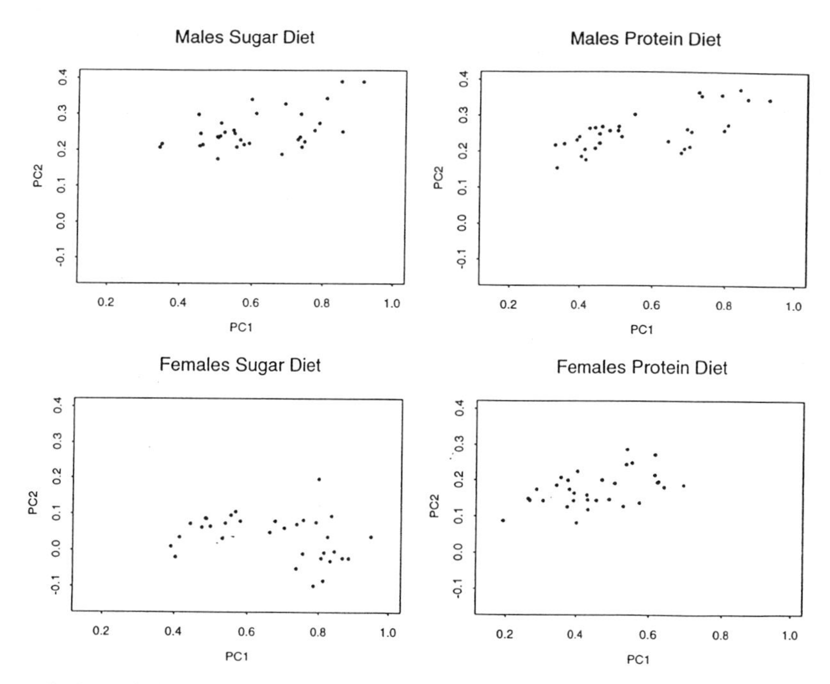

Figure 8. Second principal component versus first principal component, separated according to group.

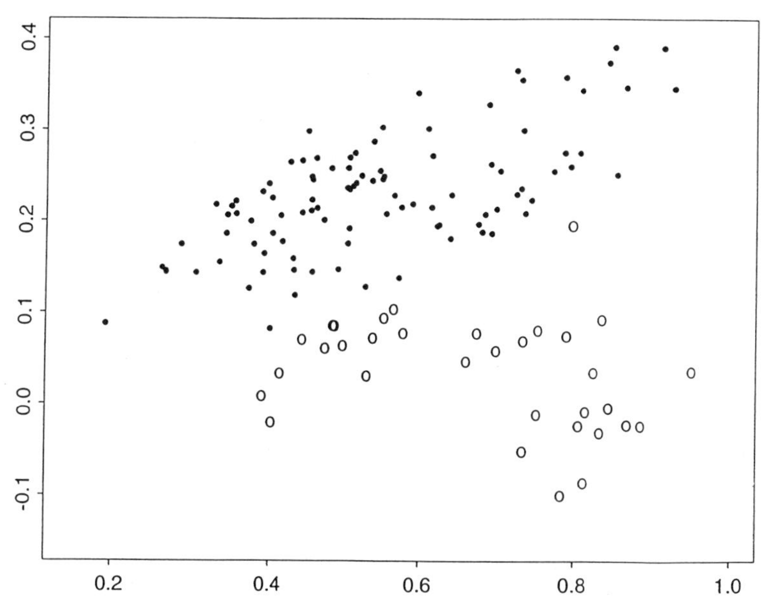

Figure 9. Second principal component versus first principal component, combining all groups. Open circles correspond to cohorts of protein-deprived females, closed circles to cohorts of the other three groups.

CONCLUSIONS

We have discussed in this chapter methods to analyze samples of curves, in this case hazard functions. Basic exploratory tools are plots of smoothed curves and of estimated derivatives. For purposes of statistical inference, we encourage the use of characteristic features of the curves, whenever such 'biological parameters' can be readily defined and make biological sense. Another, more universally applicable, method is based on the Karhunen-Loève decomposition of stochastic processes and the identification of corresponding eigenfunctions and principal components. The first few principal components together with the eigenfunctions allow a parsimonious description of the sample of curves, and the vectors of principal components can then be further analyzed with low-dimensional multivariate analysis techniques.

ACKNOWLEDGMENTS

We thank W.B. Capra for help with the figures, J.R. Carey for making the medfly data available, and a referee for detailed comments which led to improvements in the presentation. This research was supported in part by NSF grants DMS-93-12170, DMS-94-04906 and DMS-96-25984.

CORRESPONDING AUTHOR

Please address all correspondence to:
Hans-Georg Müller
Division of Statistics
University of California, Davis
One Shields Avenue
Davis, CA 95616

REFERENCES

Ash RB; Gardner MF. *Topics in Stochastic Processes*. Academic Press: New York. 1975.

Capra WB; Müller HG. An accelerated time model for response curves. *J Am Stat Assoc,* 1997, 92:72-83.

Castro PE; Lawton WH; Sylvestre EA. Principal modes of variation for processes with continuous sample curves. *Technometrics*, 1986, 28:329-337.

Davidian M; Giltinan DM. *Nonlinear Models for Repeated Measurements Data.* Chapman and Hall: London. 1995.

Fan J; Gijbels I. *Local Polynomial Modeling and its Applications.* Chapman Hall: London. 1996.

Gasser T; Kneip A. Searching for structure in curve samples. *J Am Stat Assoc*, 1995, 90:1179-1188.

Gasser T; Müller HG; Köhler W; Molinari L; Prader A. Nonparametric regression analysis of growth curves. *Ann Stat,* 1984, 12:210-229.

Guardabasso V; Munson PJ; Rodbard D. A versatile method for simultaneous analysis of families of curves. *The FASEB Journal,* 1988, 2:209-215.

Johnson RA; Wichern DW. *Applied Multivariate Statistical Analysis. 3rd Ed.* Prentice-Hall: Englewood Cliffs, New Jersey. 1992.

Karhunen K. Über lineare Methoden in der Wahrscheinlichkeitsrechnung. *Ann Acad Scient Fenn,* 1947, 37:1-79.

Kirkpatrick M; Heckman N. A quantitative genetic model for growth, shape and other infinite-dimensional characters. *J Math Bio,* 1989, 27:429-450.

Kirkpatrick M; Lofsvold D. Measuring selection and constraint in the evolution of growth. *Evolution*, 1992, 46:954-971.

Lindstrom MJ. Self-modeling with random shift and scale parameters and free-knot spline shape function. *Stat in Med,* 1955, 14:2009-2021.

Müller HG. *Nonparametric Regression Analysis of Longitudinal Data.* Springer: New York. 1988.

Müller HG; Remschmidt H. Nonparametric methods for the analysis of longitudinal medical data, with an application to the prognosis of anorectic patients from longitudinal weight measurements, in: *Proceedings of the 14th International Biometric Conference.* Société Adolphe Quetelet: Gembloux, Belgium. 1988. pp. 229-240.

Müller HG; Wang JL; Capra WB; Liedo P; Carey JR. Early mortality surge in protein-deprived females causes reversal of sex differential of life expectancy in Mediterranean fruit flies. *Proc Nat Acad Sci USA,* 1997a, 94:2762-2765.

Müller HG; Wang JL; Capra WB. From lifetables to hazard rates: The transformation approach. *Biometrika,* 1997b, (in press).

Ramsay JO; Silverman BW. *Functional Data Analysis.* Springer: New York. 1997.

Rice J; Silverman BW. Estimating the mean and covariance structure nonparametrically when the data are curves. *J Royal Stat Soc B,* 1991, B53:233-243.

Verotta D; Sheiner LB. Comparing responses when each response is a curve. *Am J Phys,* 1992, 263:206-214.

Part III

APPLICATIONS OF MODELING

DEVELOPMENT OF A COMPARTMENTAL MODEL DESCRIBING THE DYNAMICS OF VITAMIN A METABOLISM IN MEN

Doris von Reinersdorff,[1] Michael H. Green,[2] and Joanne Balmer Green[2]

[1]Hoffmann-La Roche
Basel, Switzerland and Nutley, New Jersey
[2]Nutrition Department
The Pennsylvania State University
University Park, PA

ABSTRACT

Model-based compartmental analysis was used with the Simulation, Analysis and Modeling (SAAM) computer programs to analyze data on plasma retinoid kinetics in adult male subjects for 7 d after a single oral dose of 105 μmol of [8,9,19-^{13}C]retinyl palmitate. We present here the data for one subject and discuss in detail the steps taken to develop a physiologically-based compartmental model that describes the dynamic behavior of plasma retinyl esters, [^{12}C]retinol, [8,9,19-^{13}C]retinol, and the sum of [^{12}C] and [^{13}C]retinol.

First an absorption model was developed to fit data on the plasma appearance and disappearance of retinyl esters; this was used as input in development of models for labeled and unlabeled retinol. The large oral load of labeled vitamin A perturbed the unlabeled tracee system, and thus parallel models for tracer and tracee were developed; and a time-variant fractional transfer coefficient was incorporated into the tracee model. Following the absorption model, four-compartment models were developed to describe the dynamics of both labeled and unlabeled retinol.

These models predict that, in spite of the large vitamin A load, the absorption efficiency was 34%; hepatic (presumably parenchymal cell) processing of the absorbed dose was essentially complete by 24 h; and, by 7 days, ~80% of the absorbed dose was in a compartment that presumably represents stored liver retinyl esters. The model also predicts that ~50 μmol of retinol passed through the plasma each day, compared to an estimated utilization rate of 4 μmol/day. This project provides unique and important information about whole-body vitamin A dynamics in humans, and presents approaches to specific modeling issues that may be encountered by others.

INTRODUCTION

At the Fifth Conference on Mathematical Modeling in Experimental Nutrition in 1994, M. H. Green discussed the application of mathematical modeling to research on vitamin A metabolism in the rat (Green and Green, 1996). As indicated there, much has been learned about whole-body vitamin A metabolism by applying model-based compartmental analysis to data on tracer and tracee kinetics in plasma and tissues after administration of [^{3}H]vitamin A to rats at various levels of vitamin A status.

This work by Green and colleagues has contributed to the current realization that whole-body vitamin A metabolism is more complex and fascinating than previously believed. For example, rather than being simply absorbed, processed through the liver for secretion into plasma or for storage, and transported to target tissues for utilization, kinetic studies in the rat indicate that there is extensive recycling of plasma retinol before it is irreversibly utilized. This recycling presumably contributes to the homeostatic control over plasma retinol concentrations. In addition, these experiments indicate that much of the input of retinol to plasma is from nonhepatic tissues and that there are significant pools of vitamin A in extrahepatic tissues.

In spite of keen interest in all aspects of vitamin A metabolism in humans, only a limited number of kinetic studies have been published to date (Sauberlich et al., 1974; Bausch and Rietz, 1977; Furr et al., 1987; v. Reinersdorff et al., 1996). This is no doubt largely due to limitations in the use of radioisotopes in humans. The first application of compartmental analysis to data on vitamin A dynamics in humans was recently published by Green and Green (1994a). Data collected in 1965 by Goodman and colleagues on plasma retinol kinetics after administration of [^{14}C]retinol to three men (personal communication to M.H.G.) were analyzed retrospectively by model-based compartmental analysis, and a "working hypothesis," three-compartment model was developed. The model has many similarities to models proposed for vitamin A-sufficient rats (Green and Green, 1994b). Also relevant is a study on the plasma dynamics of β-carotene, a nutritionally-important precursor of vitamin A, in an adult volunteer (Novotny et al., 1996; Novotny et al., 1995).

Here we discuss a project in which we are applying model-based compartmental analysis to recently-published data (v. Reinersdorff et al., 1996) on plasma retinoid kinetics in adult male subjects after a single oral dose of [8,9,19-^{13}C]retinyl palmitate. The data for one representative subject will be presented and are used to illustrate the steps taken to develop a physiologically-based compartmental model that describes the dynamic behavior of both the tracer and tracee after a perturbing dose of vitamin A. Because of several practical limitations in the experimental data that will likely arise in other stable isotope studies in humans, and because of the complexity of the processes being modeled, some of our discussion centers on our approaches to dealing with these constraints.

EXPERIMENTAL DATA

This project uses data on plasma vitamin A kinetics in men that were collected by v. Reinersdorff et al. (1996). To summarize, healthy, normal-weight, fasting, young adult male volunteers (n=11) ingested a gelatin capsule containing 105 μmol of [8,9,19-^{13}C]retinyl palmitate in coconut oil (1:7, w/w) along with 0.5 L of whole milk. Blood samples (n=25) were collected between 0.5 h and 7 days; 20 of the samples were collected during the first 24 h, with collections at 30 min intervals for the first 6 h. Low-fat meals were given 4 and 8 h after the dose, and only low vitamin A-containing foods were consumed during the 7-day kinetic study. After the 24-hour blood collection, samples were obtained after an overnight fast. Aliquots of plasma were extracted and analyzed for retinyl esters by reverse phase HPLC. Other aliquots were extracted, derivatized, and analyzed by GC/MS for [^{12}C]retinol, [8,9,19-^{13}C]retinol, and certain retinol metabolites.

Data on plasma retinyl esters, [^{12}C]retinol, [8,9,19-^{13}C]retinol, and total retinol (the sum of [^{12}C] and [^{13}C]) were plotted as concentrations versus times, and were analyzed by noncompartmental methods. Isotope kinetics were compatible with current understanding of the metabolism of dietary- and endogenous vitamin A in plasma. Retinyl esters peaked in plasma between 3.5 and 12 h after dosing. The large mass of vitamin A in the dose caused an early increase in plasma [^{12}C]retinol, and then a decrease as the levels of [8,9,19-^{13}C]retinol increased. Labeled retinol peaked at ~51% of total plasma retinol concentrations between 10 and 24 h after dosing. Subsequently, concentrations decreased, reaching 13% on day 7, indicating a slow final elimination of labeled retinol from plasma.

MODELING METHODOLOGY

In order to obtain more quantitative and predictive information from these unique data on plasma retinol kinetics, we used model-based compartmental analysis as described by Foster and Boston (1983), and Green and Green (1990). We first reviewed the plasma tracer and tracee response curves for all 11 men and choose one representative subject (# 5) as the focus of this project.

First, data for tracer and tracee were transformed to account for natural abundance in the dose as described by Cobelli et al. (1987). Transformed data on plasma concentrations of each retinoid were converted to plasma pool sizes, assuming that plasma volume is 4.35% of body weight (Diem, 1962). Based on current understanding of vitamin A digestion and absorption, chylomicron metabolism, and plasma retinol dynamics, we postulated a simple starting model to describe the tracer response (see next section). Because the administered tracer perturbed the endogenous vitamin A (tracee) system, we decided at the outset that we would need to develop parallel models for the tracer and tracee systems. This required that we estimate initial conditions in both systems. For the tracee, we used literature values of 126 mg vitamin A/kg liver in victims of accidental death in the U.S. (Underwood, 1984) and a liver weight of 2.5% of body weight (Arias et al., 1982) to estimate a liver vitamin A content of 770 μmol . We assumed that 95% of this was present in stellate cells (95% in retinyl esters and 5% in retinol) and 5% was in hepatocytes; and that the level of vitamin A in extrahepatic tissues was 10% that in liver. For initial conditions for the tracer, we used the known amount of the dose (105 μmol) and the measured abundance of the M+3 ion (271) in the two baseline plasma retinol samples.

Then we used CONSAM 31 (Berman et al., 1983), the conversational version of the Simulation, Analysis and Modeling computer program (SAAM) (Berman and Weiss, 1978) run on an IBM 80486 to compare observed data for plasma retinyl esters and the three retinol species to predictions of the starting model. We sequentially adjusted the model parameters and structure in physiologically-reasonable ways to achieve an adequate match between observed and simulated data. Once a good fit was obtained, we used weighted nonlinear regression analysis in CONSAM to calculate the model parameters (fractional transfer coefficients [L(I,J)s], or the fraction of compartment J's tracer or tracee transferred to compartment I per unit time). [NOTE: The 'L(I,J)' nomenclature used in the SAAM program is equivalent to the 'k_{ij}' designation often seen in the tracer kinetic field.] A fractional standard deviation of 0.05 was assigned to each data point as the weighting factor. Transfer rates [R(I,J)s] were calculated from the L(I,J)s and estimated or predicted pool sizes. Because the model-predicted fractional transfer coefficients depend on the estimated initial conditions as well as the dynamic responses, we consider the calculated transfer rates as the model predictions of most interest, especially for the tracee system.

OVERVIEW OF VITAMIN A METABOLISM

As background, it may be useful to overview some aspects of vitamin A metabolism that are relevant to modeling of the current data (Fig. 1). See reviews by Blomhoff et al. (1991), and Blaner and Olson (1994). We assumed that [^{13}C]retinyl palmitate from the oral dose was hydrolyzed in the small intestinal lumen, the resulting retinol was absorbed into enterocytes and most was reesterified, and retinyl esters plus some labeled retinol were packaged into absorptive lipoproteins (mainly chylomicrons). After secretion of chylomicrons into lymph and transport to plasma, a majority of the chylomicron triglycerides were presumably cleared by peripheral tissues as a result of the action of lipoprotein lipase. Chylomicron remnants (containing most of the newly absorbed vitamin A) were presumably cleared primarily into liver hepatocytes. There retinyl esters would have been hydrolyzed. Some of the resulting [^{13}C]retinol would have been processed for secretion into plasma in a complex with retinol-binding protein (RBP) for transport to tissues; some would have been esterified, and some transferred to perisinusoidal stellate cells for reesterification and storage. Based on current understanding of the metabolism of plasma retinol, we assumed that RBP-bound [^{13}C]retinol would recycle among plasma and tissues several times before irreversible utilization. We assumed that some intracellular retinol might cycle between retinol and retinyl ester pools and that, finally at some point after uptake, intracellular retinol would be oxidized to retinoic acid, which then would interact with one of several nuclear receptors to regulate gene transcription.

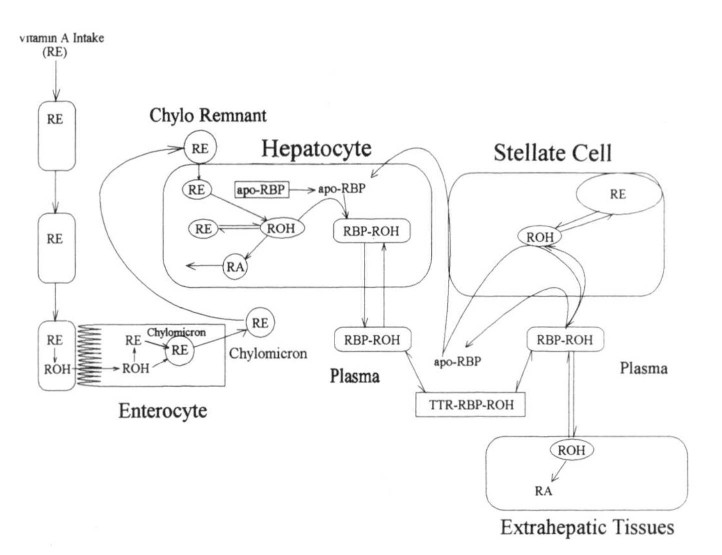

Figure 1. Schematic overview of whole-body vitamin A metabolism. RE, retinyl esters; ROH, retinol; chylo, chylomicron; RBP, retinol-binding protein; TTR, transthyretin; RA, retinoic acid.

MODEL-BASED COMPARTMENTAL ANALYSIS

Limitations in the Data

Before describing development of our model, it is important to comment on several experimental issues that distinguish this study from experiments on vitamin A kinetics in rats. First, as always seems to be the case in kinetic studies, modeling would have been helped by a longer study duration. For human studies, the length of experiments may always need to be balanced by factors of subject availability, economics, and limits in isotope detection.

One significant issue in this study was the large mass of labeled retinyl palmitate (105 μmol or 30 mg) which had to be administered to ensure detection of labeled retinoids during the one-week study. Because the administered dose of [^{13}C]retinyl palmitate was approximately 24 times the mass of the plasma retinol (tracee) pool, the isotope introduced a significant perturbation of the tracee system and was therefore not an ideal tracer. One approach to circumventing this problem in future studies of vitamin A dynamics in humans might be to use uniformly-labeled [^{13}C]vitamin A, as proposed by Swanson et al. (1996) for β-carotene. By increasing the number of ^{13}Cs per molecule and analyzing for $^{13}CO_2$, a lower amount of isotope could be administered.

Another complicating feature is that, in most vitamin A kinetic studies done in rats by Green and collaborators, the complexities and variability in absorption, and in absorptive lipoprotein metabolism, have been bypassed by administering the label as [^{3}H]retinol in its physiological plasma transport complex. In contrast, the most feasible route of dose administration in human studies is oral ingestion. Thus, sufficient data must be collected during the absorptive phase to identify this aspect of the model. Fortunately, in this study, blood was sampled frequently during the first 6 h, but modeling the absorption of the dose was one of the more challenging aspects of this project.

A final limitation we encountered was that tissue vitamin A levels, including liver stores of vitamin A, were not known. Liver vitamin A content has been implicated in rat studies to be a key determinant of plasma vitamin A kinetics, influencing the geometry of the plasma tracer response curves. Recently an isotope dilution method for estimating liver vitamin A content in rats was described (Adams and Green, 1994); it is hoped that this technique may be useful in human studies as well. Because our model development required estimates of vitamin A levels in liver and extrahepatic tissues, we assumed, based on case histories and clinical profiles (v. Reinersdorff et al., 1996), that subject 5 had normal liver vitamin A stores. Thus we used reference values from the literature to calculate estimates of tissue vitamin A.

Model Development

Shown in Table 1 and Figure 2 are the observed data for plasma retinyl esters, [^{13}C]retinol (tracer) and [^{12}C]retinol (tracee) in subject 5. During the first 2 h after dose administration, no changes were observed in baseline levels of plasma retinyl esters. Then there was a rapid rise followed by a gradual decline in retinyl ester levels. [^{13}C]Retinol also began to increase by 2 h, reaching a maximum at 10-11 h after dose administration, and then levels slowly fell. The endogenous [^{12}C]retinol began increasing almost immediately after the [^{13}C]retinyl palmitate dose. It peaked at 3.5 h, decreased to a nadir at 18 h, then showed a slow rise back to its baseline level by 4 days.

We next drew a simple compartmental model (Figure 3) that was the starting point for model development. The initial model shows three compartments in a cascade to simulate retinyl ester movement down the gastrointestinal tract, and digestion and absorption;

plasma compartments for chylomicron-bound retinyl esters and RBP-bound retinol; two compartments in liver (one for processing chylomicron retinyl esters and the other for liver vitamin A pools); and one extravascular/ extrahepatic retinol compartment.

Table 1. Observed plasma vitamin A pool sizes for subject 5[a]

Time (h)	RE	[^{13}C]Retinol	[^{12}C]Retinol
0.0	17.3	3.27	1449
0.5	15.3	2.99	1474
1.0	14.0	3.39	1603
1.5	15.3	2.91	1529
2.0	17.8	6.37	1616
2.5	49.4	17.84	1663
3.0	239.3	44.42	1592
3.5	2834	99.19	1718
4.0	3882	171.3	1671
4.5	3876	263,2	1677
5.0	4225	427.8	1589
5.5	3046	569.1	1492
6.0	1985	730.0	1492
7.0	1140	938.1	1371
8.0	871.7	974.9	1153
9.0	682.6	992.1	1051
10.0	628.6	1020	917.9
11.0	570.4	1025	856.6
12.0	491.6	954.3	818.4
18.0	207.9	959.6	771.0
24.0	111.9	935.8	817.8
48.0	39.8	617.6	1094
72.0	27.5	468.4	1346
96.0	33.1	368.3	1468
120.0	ND	338.3	1569
168.0	14.3	260.1	1453

[a]Data are estimated plasma pool sizes (nmol) for retinyl esters (RE), [^{13}C]retinol, and [^{12}C]retinol (i.e., observed values for plasma concentrations * estimated plasma volume) at each time.

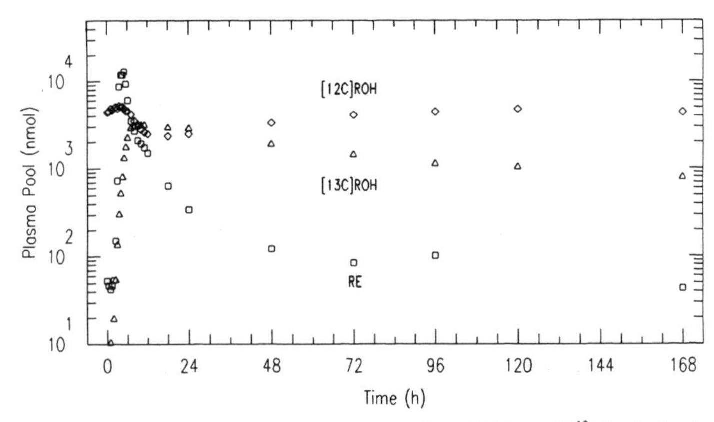

Figure 2. Plasma retinoid kinetics for 7 days after ingestion of 105 μmol [^{13}C]retinyl palmitate in subject 5. Shown are the estimated plasma pool sizes (nmol) for retinyl esters (□), [^{12}C]retinol (◇), and [^{13}C]retinol (Δ).

We began model development by considering the absorption and processing of the administered dose. Compatible with known physiology, the data indicated that there was a delay before appearance of retinyl esters in plasma, corresponding to the time for absorption of [^{13}C]retinyl palmitate, its processing and packaging into chylomicrons, and chylomicron secretion into plasma from lymph. There was a rapid rise in plasma retinyl esters at about 4 h to a maximum of 10% of the dose (Fig. 2). At a later stage in the modeling process, the model predicted that the total absorption of the dose was ~34%, higher than might have been expected for such a large load. The subsequent peak in plasma retinyl esters at 5 h represents a decrease in the rate of appearance of chylomicrons and the continued rapid clearance of chylomicrons. We hypothesized that the shoulder between 9 and 18 h corresponds to secretion of retinyl esters in hepatic very low density lipoproteins (VLDL) and that this was a consequence of the large size of the administered dose. Plasma retinyl esters finally returned to a low basal level by 7 days. We hypothesized that this basal level corresponds to a small amount of retinyl esters that normally circulate as components of longer-lived lipoproteins, as also seen in rats (Green et al., 1985).

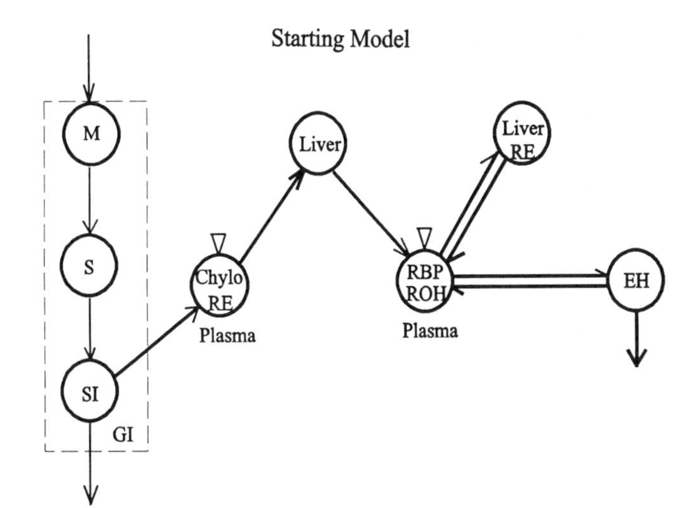

Figure 3. Starting model for compartmental analysis. M, mouth; S, stomach; SI, small intestine; GI, gastrointestinal tract; chylo, chylomicron; RE, retinyl esters; RBP, retinol-binding protein; ROH, retinol; EH, extra-hepatic tissues. The triangles indicate the site of sampling (plasma) and the analytes measured (retinyl esters and retinol).

To fit the retinyl ester data, we had to modify the starting model in several ways. First, we needed to add several gastrointestinal components (compartments 1,2 and 4, and delay component 3) to represent the delay in absorption (Fig. 4) and we needed two kinetically-distinct components (compartments 5 and 6) that presumably represent sites in the small intestine that are involved in vitamin A absorption. After the delay, the initial shape of the increase in plasma retinyl esters, followed by a rapid rise, indicated that we needed a delay following each of these small intestinal feeder compartments (presumably enterocyte pools). These are shown as delay elements 10 and 15 in Figure 4. Retinyl esters entered the chylomicron retinyl ester pool (compartment 20) from delay component 10; component 10 processed the majority (>98%) of the absorbed dose. From compartment 20, chylomicron-remnant retinyl esters were cleared into liver compartment 42. Because there was a rather rapid initial rise in labeled retinol in plasma, we assume that a small portion of the dose (~0.8%) was absorbed as chylomicron-bound retinol (compartment 30).

We fixed liver uptake of retinol [L(42,30)] to equal that for retinyl esters [L(42,20)], because we assumed that remnants are cleared as particles. A fraction of compartment 30 was transferred directly into the plasma RBP-retinol pool [L(62,30)], to obtain the observed initial rise in the plasma [^{13}C]retinol pool (compartment 62). We modeled the later shoulder and fall to baseline levels in plasma retinyl esters (Fig. 2) by adding a plasma VLDL-retinyl ester compartment (compartment 25, Fig. 4), with retinyl ester recycling to plasma from extrahepatic tissues in longer-lived lipoproteins (low- and high-density lipoproteins; compartment 26). Because such a small portion of the dose was processed through this pathway, we constrained its output [L(0,26)] for leaving the system with its unknown amount of tracer and tracee. As we developed the absorption model, we found it useful to simulate the various compartments that comprised the retinyl ester pool. This enabled us to obtain a model that provided a better fit to the shape of the data.

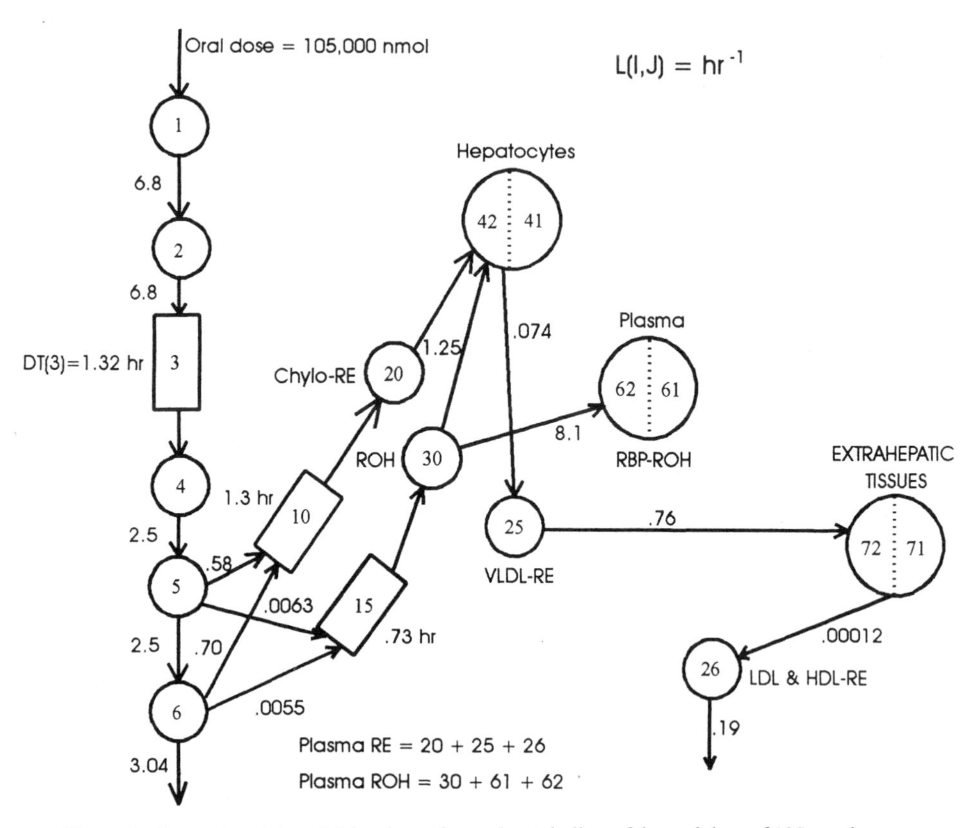

Figure 4. Compartmental model for absorption and metabolism of the oral dose of 105 μmol [^{13}C]retinyl palmitate. Circles represent compartments, rectangles are delay components, vectors represent fractional transfer coefficients. Chylo, chylomicron; RE, retinyl esters; ROH, retinol; RBP, retinol-binding protein; VLDL, very low density lipoproteins; LDL, low density lipoproteins; HDL, high density lipoproteins. Compartments/components 1-6 are assumed to be in GI tract, 42 and 41 are in liver, 72 and 71 are in extrahepatic tissues; others are in plasma.

The fit of the retinyl ester data to the absorption model is shown in Figure 5. Once we had obtained a satisfactory match between the observed and predicted data, we fixed the model parameters related to absorption and retinyl ester turnover, and turned our attention to the data for plasma retinol tracer and tracee (Fig. 6). The plasma response curve for [^{12}C]retinol increased for a short time, then fell to ~50% of its baseline level and then slowly rose back to baseline. This pattern indicated that we would need to develop a nonsteady state model in which certain rate constants for the tracee were time variant. Because of both the dynamic responses and the expected behavior of the tracee after input of the dose, it was clear that we would need to develop parallel models for the tracer and tracee that would be interdependent but could be separately manipulated. We are not aware of any published work that incorporates a similar approach, but, given current technologies, it is not hard to imagine the same scenario in certain future studies that use stable isotopes.

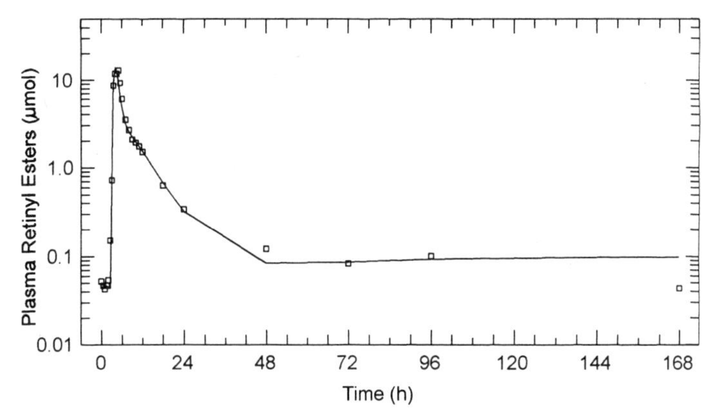

Figure 5. Observed data (symbols) versus model-predicted values (lines) for plasma retinyl esters for 7 days after ingestion of 105 μmol [^{13}C]retinyl palmitate in subject 5.

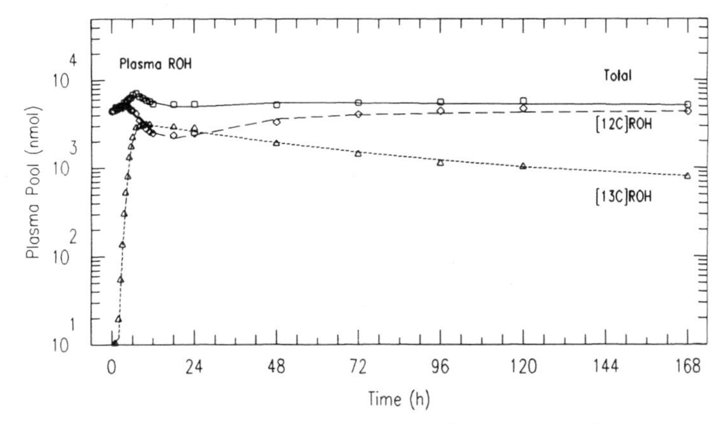

Figure 6. Observed and model-predicted data for plasma [^{13}C]retinol (Δ), [^{12}C]retinol (◇) and total [^{13}C+^{12}C]retinol (ρ) for 7 days after ingestion of 105μmol [^{13}C]retinyl palmitate in subject 5. The fits to the data are for the final model (Fig.11).

The perturbation in the tracee system resulting from the mass of the labeled retinyl palmitate in the ingested dose is plotted in Figure 7 as a fraction of the baseline tracee pool. We knew that compartment 41 (Fig. 4) would be most perturbed by the incoming vitamin A load. We decided to fit the data in Figure 7 to a mathematical function and then apply that function to the fractional transfer coefficient representing hepatocyte secretion of [^{12}C]retinol bound to RBP. Using previous empirical analyses of kinetic data as a guide, we chose to use an exponential function, assuming the components might have biological meaning. We knew that we would need at least a three-component exponential function to simulate the rise and fall, and then the later rise, in the data. In fact, we needed a four-component exponential ("perturbation function") to fit the data.

Figure 8 is a diagram that may explain in physiological terms what occurred in the liver tracee/tracer system to cause the observed perturbation of the tracee system in plasma.

We hypothesize that, as the large load of tracer entered hepatocytes, the incoming retinyl esters were hydrolyzed to retinol, increasing the size of the retinol pool and resulting in the initial displacement of more tracee retinol onto RBP. This increased the secretion of tracee into plasma, and caused the rise in plasma [^{12}C]retinol and total retinol. Then, the tracer began to compete with tracee and dominated sites for RBP binding. Thus, secretion of tracer increased, resulting in rising plasma tracer levels as plasma tracee levels fell. We assume that the excess retinol tracer was reesterified and stored in hepatocyte retinyl ester pools until more apoRBP became available. After ~18 h, the tracee level in plasma began to rise again, probably reflecting the ability of hepatocytes to process the incoming tracer and the lack of further input of appreciable amounts of vitamin A from the diet. Without new input, endogenous vitamin A would be mobilized from stores which were dominated by [^{12}C]retinol.

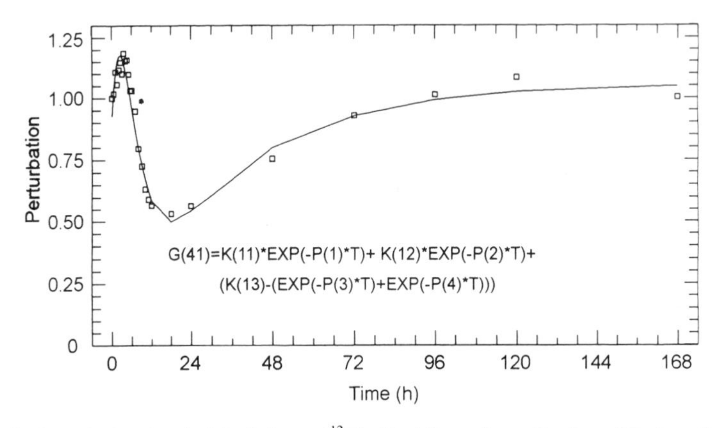

Figure 7. Perturbation (symbols) of plasma [^{12}C]retinol (tracee) as a fraction of the baseline tracee pool versus time after ingestion of 105 μmol [^{13}C]retinyl palmitate. The line represents the fit of the observed data to the four-component exponential equation, G(41). In the equation, G is a user-defined function; 41 is an arbitrary name; Ks are intercepts (exponential constants), Ps are slopes (exponential coefficients), and T is time (h). K(11)=59.879, K(12)=58.011, K(13)=1.0582, P(1)=0.2374, P(2)=0.2470, P(3)=0.1214, P(4)=0.02845.

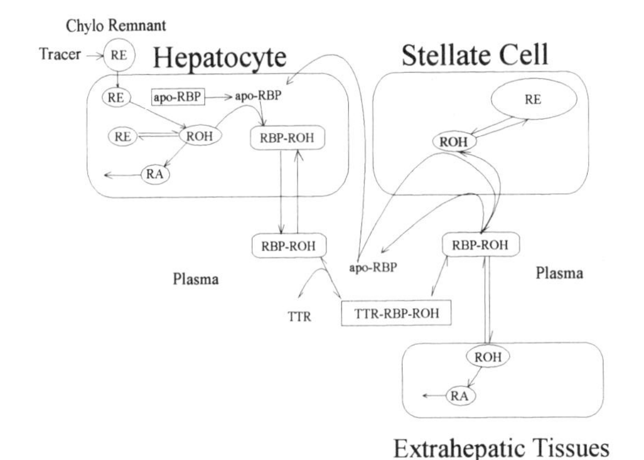

Figure 8. Diagram showing hypothesized effect of the perturbing dose of retinyl palmitate on liver vitamin A metabolism. Chylo, chylomicron; RE, retinyl esters; ROH, retinol; RBP, retinol-binding protein; RA, retinoic acid; TTR, transthyretin.

As model development continued, it was difficult to simultaneously fit tracer/tracee responses in the plasma retinol pool. In order to get a better appreciation of candidate input pools, we simulated rates of tracer/tracee movement into and out of the plasma retinol pools. This allowed us to uncouple the model simulation of the plasma tracer/tracee pool into its component inputs/outputs. By so doing, we were able to find a model that provided a better fit to the shape of all of the data for the tracer and the tracee. We were fortunate to be working with a program such as CONSAM which has the capacity to handle so many compartments/components (75) and such a large number of observed or simulated data points (5000).

After obtaining a satisfactory fit for all data for subject 5 by iteratively adjusting the model structure and parameter values, applying the perturbation function to key tracee L(I,J)s, and keeping in mind current knowledge/hypotheses about liver handling and storage of vitamin A, we arrived at the "Atomium" model shown in Figure 9. This model structure is nicknamed for the symbol of the 1958 World Expo in Brussels, Belgium. Although this model gave a very good fit to the data (and also for data from several other subjects), we felt that it was overly complicated. Furthermore, although the model predicted extensive recycling of plasma retinol relative to the rate of utilization (as found repeatedly in models developed to explain vitamin A kinetics in rats) (Green and Green, 1994b; Green et al., 1985; Green et al., 1987), it was not robust enough to maintain reasonable levels of vitamin A in the liver. Specifically, when the compartment representing liver stores of vitamin A was simulated over time, the model predicted that stores were depleting rapidly, with vitamin A being transferred to the compartment representing extrahepatic retinol pools. Because this prediction was unrealistic and because we realized that it would be difficult to correct the liver vitamin A depletion problem while maintaining a good fit to the data, we were dissatisfied with the "Atomium" model. Thus we decided to try again to find a simpler model that could explain the tracer and tracee data.

Final Model

In the last stages of model development, we modified our goal of obtaining a purely physiologically-based model. Instead, we reviewed the approach used to develop the model for Goodman's human data (Green and Green, 1994a) and for rat data (Green and Green, 1994b). In those cases, plasma tracer responses following an IV injection of labeled retinol were initially fit by empirical compartmental analysis, and then model-based compartmental analysis, to a three- or four-compartment model for vitamin A kinetics as viewed from the plasma space. Here, as in the development of the "Atomium" model, we used the absorption model (Fig. 4) to deliver new tracer input into the liver (compartment 42, Fig. 10).

The data indicated that, for both the tracer and the tracee, we needed liver feeder compartments (compartment 41 for tracee and 42 for tracer) to be one source of delivery of retinol into plasma (compartment 61 for tracee and 62 for tracer), and at least two compartments to exchange directly or indirectly with the plasma retinol pool. One of these would be turning over more slowly (presumably representing stores of retinyl esters; compartment 65 for tracee and 66 for tracer) and the other would contain faster turning-over extravascular retinol pools (compartment 63 for tracee and 64 for tracer). Thus, two parallel models with appropriate dependency relationships for the L(I,J)s were developed (Fig. 10). For the tracee's model, we applied the perturbation function [G(41); Fig. 7] to the L(I,J) describing input of retinol into plasma compartment 61 from liver tracee compartment 41. That is, L(61,41) equaled (adjustable parameter) * G(41). Realizing that the dynamic behavior of the tracer system would provide the most identifiable L(I,J)s related to input and output from the plasma RBP-bound retinol pool (compartment 61 for tracee and 62 for tracer), we made those L(I,J)s adjustable in the tracer model and used the same values for the tracee (Fig. 10). As we modeled, the value for L(62,66) tended toward zero. Because some of the later input to the plasma retinol pool for the tracee had to come from liver stores of vitamin A (compartment 65), we added a delay (component 40) between storage compartment 65 and compartment 41, the source of secretion of RBP-bound retinol. With so many dependent parameters in the model, we were unable to adequately fit either the tracer or tracee data. When we removed the dependencies between the parallel models, except for that between L(65,61) and L(66,62), the fit for tracer data (but not tracee) improved.

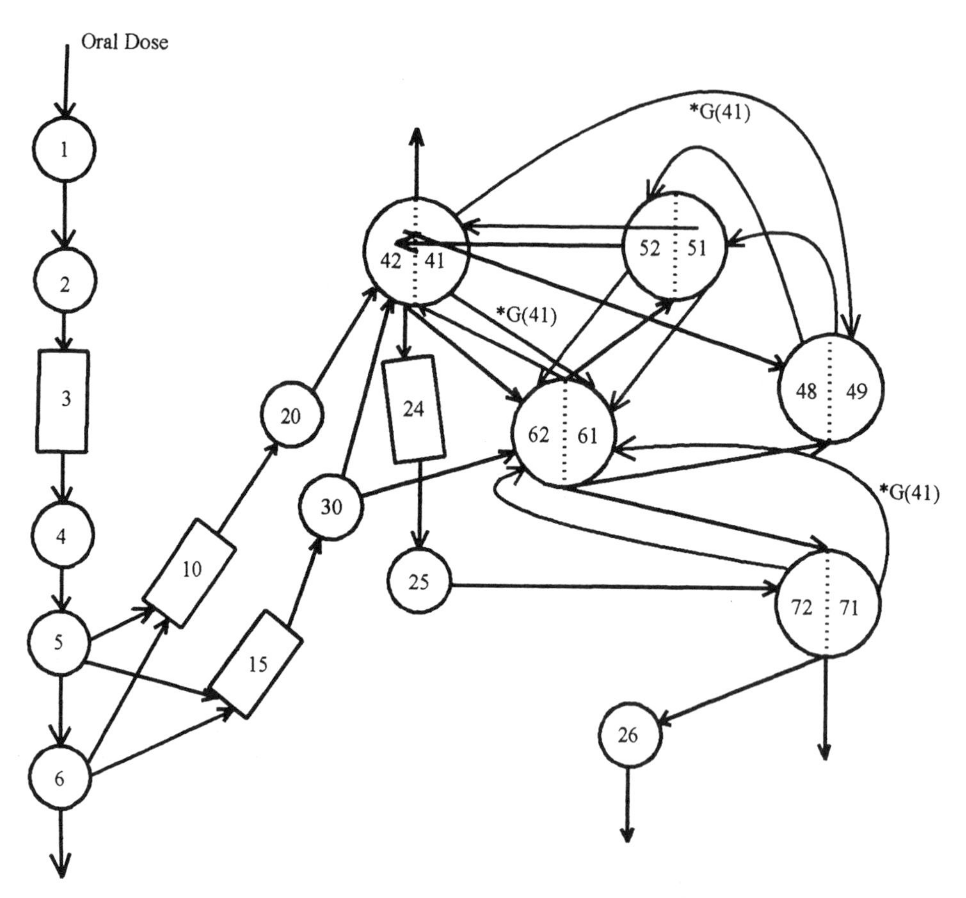

Figure 9. The "Atomium" model developed to fit data on plasma retinoid kinetics after ingestion of 105 μmol [^{13}C]retinyl palmitate in subject 5. Circles represent compartments, rectangles are delay components, vectors represent fractional transfer coefficients, and G(41) is the four-component exponential equation developed to fit the perturbation in the plasma tracee pool (Fig.7).

In view of our inability to adequately fit the tracee data over the 7-day period, we wondered whether our assignment of initial conditions for the tracee might be incorrect. Thus, in the next step of model development, we made the initial conditions in compartments 41 and 63 adjustable. This improved the fit for the tracee data, and the initial condition in compartment 63 dropped to zero. However, with this model, the mass of vitamin A in the large storage pool (compartment 65) dropped too fast and compartment 63 expanded too rapidly, reminiscent of the problem with the "Atomium" model. Working momentarily with only the tracee system, we combined compartments 65 and 63 into one storage compartment (component 70) and included a reasonable estimate for vitamin A depletion rate (1.05 μmol /day). These adjustments eliminated the excessive loss of vitamin A stores. In order to keep most of the storage of tracee in compartment 65, we included L(65,63) in the model to return retinol to stores and to inhibit compartment 63 from expanding so much. This adjustment implies that part of compartment 63 represents liver stellate cell retinol and fast turning-over retinyl esters in exchange with plasma retinol and slower turning-over liver retinyl esters. When we again included both the tracer and tracee systems, the fit for the tracer was not as good as before. Thus, we added L(66,64) and reincluded L(42,66) in the tracer's model, in analogy to the parameters in the tracee system. At first, we included a delay between compartments 66 and 42, but this was not needed.

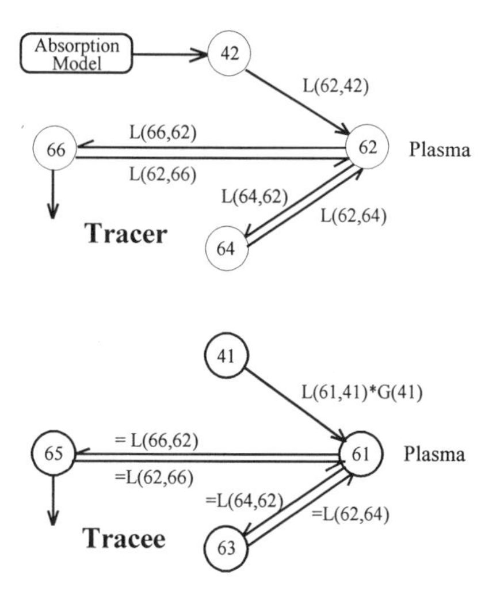

Figure 10. First post-"Atomium" compartmental model proposed for fitting kinetic data on plasma [^{12}C]retinol (tracee) and [^{13}C]retinol (tracer) after ingestion of 105 μmol [^{13}C]retinyl palmitate in subject 5. The top panel shows the model for the tracer system, the lower panel shows the parallel model developed for the tracee system. Circles represent compartments, the rectangle is a delay component, vectors represent fractional transfer coefficients [L(I,J)s or the fraction of compartment J's tracer or tracee transferred to compartment I each hour], G(41) is the four-component exponential equation developed to fit the data on perturbation of the plasma tracee pool (Fig.7).

These changes brought us to the final simple model presented in Figure 11. As shown by the model simulations in Figure 6, this model provided a satisfactory fit to the plasma retinol data. After iteration, we found that most of the tracee system's L(I,J)s were well identified [i.e., their fractional standard deviations (FSDs) were less than 0.8], except that the initial condition in compartment 41 was not well identified, nor were the irreversible loss parameters [L(0,65) and L(0,26)]. A longer study duration would have helped in the accurate identification of these parameters. For the tracer system, all parameters were well identified (FSDs <0.2) except for a few related to the early portion of the absorption model. Because an absorption model will be part of most human retinol and carotenoid models (assuming doses will be given orally), more extensive sensitivity analyses should be done to determine the minimal data sets needed to identify this aspect of the model. Not doing so may compromise the overall model usefulness, that is, its reliability.

Some predictions of the final model for the time-variant changes in extravascular tracer and tracee pools are shown in Figures 12 and 13, respectively. Although the dose represented about 1 month's worth of vitamin A intake, the model predicts that the hepatocyte processing of chylomicron-derived vitamin A (36 μmol) through compartment 42 was almost complete by 24 h (Fig. 12). That is, the constitutive synthesis of RBP and the recycling of apoRBP must have been rapid enough to handle such a large load of dietary vitamin A. In spite of the load, the perturbation in the plasma retinol pool, while evident, was not tremendous. The model predicts that the large mass was first transferred primarily to compartment 64 and then to the presumed retinyl ester storage compartment (compartment 66). By 7 days, the model predicts that 27% of the oral dose (~80% of the absorbed dose) was in compartment 66; this percentage appeared to still be increasing at the end of the study.

We also simulated the size of the model-predicted extravascular pools of tracee during the 7-day study (Fig. 13). The model predicts that the mass in compartment 41 decreased during the first 12 hours after the large tracer load and then stabilized. The mass in compartment 63 apparently increased over time to ~40 μmol by 7 days. This may reflect the need to mobilize endogenous stores of vitamin A to maintain plasma retinol levels and meet tissue needs for vitamin A. Compartment 65, the large storage compartment for retinyl esters, was predicted to decrease as compartment 63 increased, and the sum of the two slowly decreased over time, reflecting the slow rate of irreversible utilization (estimated to be 1.05 μmol /day) of vitamin A.

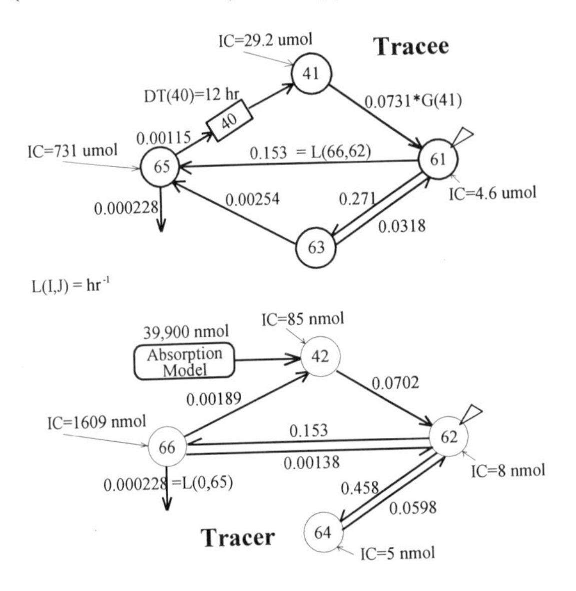

Figure 11. Final post-"Atomium" compartmental model developed to fit kinetic data on plasma [^{12}C]retinol (tracee) and [^{13}C]retinol (tracer) after ingestion of 105 μmol [^{13}C]retinyl palmitate in subject 5. The top panel shows the model for the tracee system, the lower panel shows the parallel model developed for the tracer system. Circles represent compartments, the rectangle is a delay component, vectors represent fractional transfer coefficients [L(I,J)s or the fraction of compartment J's tracer or tracee transferred to compartment I each hour], G(41) is the four-component exponential equation developed to fit the data on perturbation of the plasma tracee pool (Fig.7). IC, initial conditions.

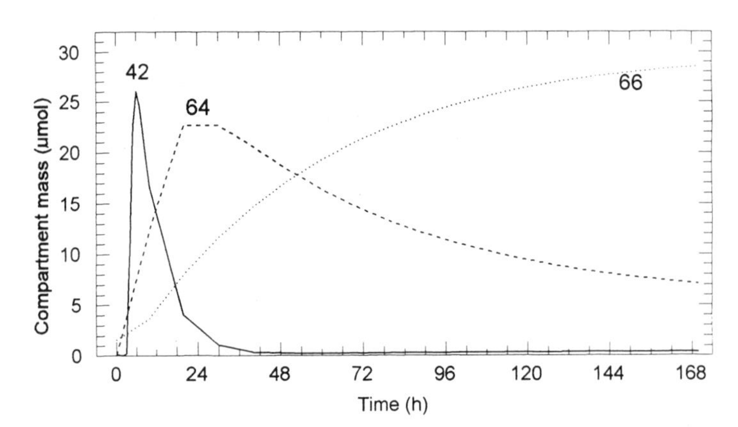

Figure 12. Simulations of the final model of changes in extravascular tracer pools. The model is shown in Fig. 11.

In Figure 14 are plotted the time-variant transfer rates [R(I,J)s] for tracee retinol input into the plasma retinol pool (compartment 41, Fig. 11). The model predicts plasma retinol transit times of 2.4 and 1.6 h for the tracee and tracer, respectively. Similar values have been calculated for rats (Green and Green, 1994b; Green et al., 1987) and humans (Green and Green, 1994a) in other studies. The model also indicates that ~50 μmol of retinol passed through the plasma each day, in comparison to the estimated utilization rate of 4 μmol /day. Thus, as shown in rats (Green et al., 1985), the extensive recycling of plasma vitamin A is much higher than the disposal rate.

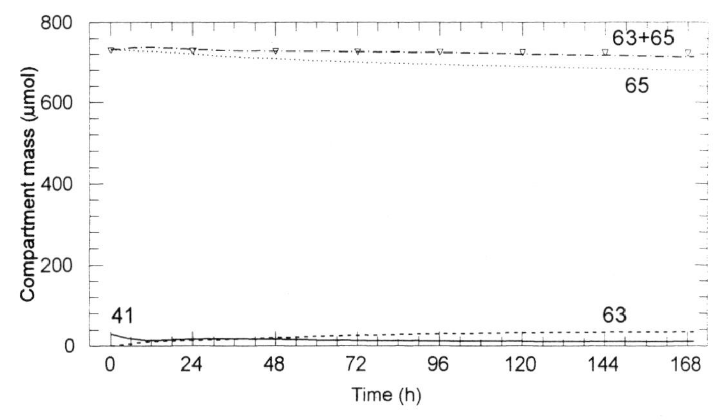

Figure 13. Simulations of the final model of changes in extravascular tracee pools. The model is shown in Fig. 11.

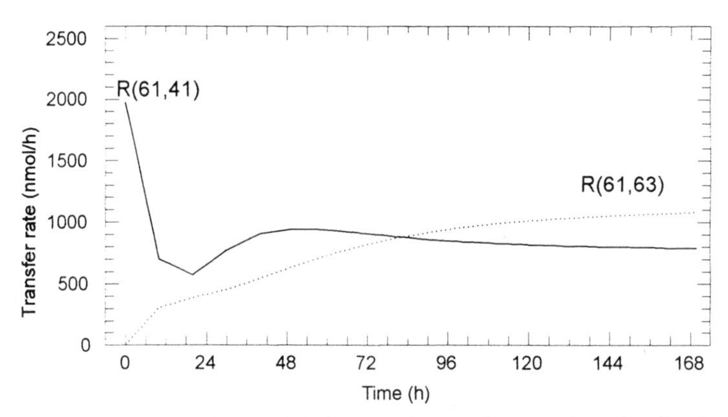

Figure 14. Simulations of the final model of the transfer rates for tracee input into the plasma retinol compartment. The model is shown in Fig. 11.

FUTURE DIRECTIONS

As model development progressed for the data from subject 5, we also worked on model identification for data from the other 10 subjects, including development of the perturbation functions, absorption models, and "Atomium" models. Our goal is to find one common model that fits data for all subjects so that we can apply CONSAM's multiple studies feature (Lyne et al., 1992) to obtain population estimates of the model L(I,J)s, and thence transfer rates and other interesting parameters (v. Reinersdorff et al., in preparation). At this point, we are confident that data for at least a majority of the other subjects will be compatible with the same, or a similar, simple model as is shown in Figure 11. Thus, ultimately, the results of this project will provide some unique and important information on whole-body vitamin A dynamics in humans. In addition, our modeling work on these data has provided an opportunity to work on several novel problems that may prove relevant to other modelers.

ACKNOWLEDGMENTS

This work was supported by special funds provided by Hoffmann-La Roche, Nutley, NJ. Present address for D. v. Reinersdorff is Jago Pharma AG, Eptingerstr. 51, Muttenz, CH-4132, Switzerland.

CORRESPONDING AUTHOR

Please address all correspondence to:
M.H. Green
Pennsylvania State University
Nutrition Department
S-126 Henderson South
University Park, PA 16802.

REFERENCES

Adams WR; Green MH. Prediction of liver vitamin A in rats by an oral isotope dilution technique. *J Nutr,* 1994, 124:1265-1270.

Arias IM; Popper H; Schachter D; Shafritz DA, Eds. *The Liver: Biology and Pathobiology*. Raven Press: New York. 1982.

Bausch J; Rietz P. Method for the assessment of vitamin A liver stores. *Acta Vitamin Enzymol,* 1977, 31:99-112.

Berman M; Weiss MF. *SAAM Manual*. DHEW Pub.# 78-180. US Government Printing Office, Washington, DC. 1978.

Berman M; Beltz WF; Greif PC; Chabay R; Boston RC. *CONSAM User's Guide*. PHS Pub.# 1983-421-132:3279, US Government Printing Office, Washington, DC. 1983.

Blaner WS; Olson JA. Retinol and retinoic acid metabolism, in: *The Retinoids: Biology, Chemistry, and Medicine,* 2nd Ed. Sporn MB; Roberts AB; Goodman DS; Eds. Raven Press: New York. 1994. pp. 229-255.

Blomhoff R; Green MH; Green JB; Berg T; Norum KR. Vitamin A metabolism: New perspectives on absorption, transport, and storage. *Physiol Rev,* 1991, 71:951-990.

Cobelli C; Toffolo G; Bier DM; Nosadini R. Models to interpret kinetic data in stable isotope tracer studies. *Am J Physiol,* 1987, 253 (Endocrinol Metab.16):E551-E564.

Diem K; Ed. *Documenta Geigy: Scientific Tables*. 6th Ed. Geigy Pharmaceuticals: Ardsley, NY. 1962.

Foster DM; Boston RC. The use of computers in compartmental analysis: The SAAM and CONSAM programs, in: *Compartmental Distribution of Radiotracers*. Robertson JS; Ed. CRC Press: Boca Raton. 1983. pp. 73-142.

Furr HC; Clifford AJ; Bergen HR; Jones AD; Olson JA. The in vivo kinetic pattern of ingested tetradeuterated vitamin A in the serum of an adult human. *Fed Proc,* 1987, 46:1335, Abstract 5946.

Green MH; Uhl L; Green JB. A multicompartmental model of vitamin A kinetics in rats with marginal liver vitamin A stores. *J Lipid Res,* 1985, 26:806-818.

Green MH; Green JB; Lewis KC. Variation in retinol utilization rate with vitamin A status in the rat. *J Nutr,* 1987, 117:694-703.

Green MH; Green JB. The application of compartmental analysis to research in nutrition. *Ann Rev Nutr,* 1990, 10:41-61.

Green MH; Green JB. Dynamics and control of plasma retinol, in: *Vitamin A in Health and Disease*. Blomhoff R; Ed. Marcel Dekker: New York. 1994. pp. 119-133.

Green MH; Green JB. Vitamin A intake and status influence retinol balance, utilization and dynamics in rats. *J Nutr,* 1994, 124:2477-2485.

Green MH; Green JB. Quantitative and conceptual contributions of mathematical modeling to current views on vitamin A metabolism, biochemistry, and nutrition. *Adv Food Nutr Res,* 1996, 40:3-24.

Lyne A; Boston R; Pettigrew K; Zech L. EMSA: A SAAM service for the estimation of population parameters based on models fit to identically replicated experiments. *Comput Methods Prog Biomed,* 1992, 38:117-151.

Novotny JA; Dueker SR; Zech LA; Clifford AJ. Compartmental analysis of the dynamics of β-carotene metabolism in an adult volunteer. *J Lipid Res,* 1995, 36:1825-1838.

Novotny JA; Zech LA; Furr HC; Dueker SR; Clifford AJ. Mathematical modeling in nutrition: Constructing a physiologic compartmental model of the dynamics of β-carotene metabolism. *Adv Food Nutr Res,* 1996, 40:25-54.

Sauberlich HE; Hodges RE; Wallace DL; Kolder H; Canham JE; Hood J; Raica Jr N; Lowry LK. Vitamin A metabolism and requirements in the human studied with the use of labeled retinol. *Vitam Horm,* 1974, 32:251-275.

Swanson JE; Wang Y-Y; Goodman KJ; Parker RS. Experimental approaches to the study of β-carotene metabolism: potential of a ^{13}C tracer approach to modeling β-carotene kinetics in humans. *Adv Food Nutr Res,* 1996, 40:55-76.

Underwood BA. Vitamin A in animal and human nutrition, in: *The Retinoids*, Vol. 1. Sporn MB; Roberts AB; Goodman DS; Eds. Academic Press: New York. 1984. pp. 281-392.

v. Reinersdorff D;Bush E; Liberato DJ. Plasma kinetics of vitamin A in humans after a single oral dose of [8,9,19-^{13}C]retinyl palmitate. *J Lipid Res,* 1996, 37:1875-1885.

COMPARTMENTAL MODELS OF VITAMIN A AND β-CAROTENE METABOLISM IN WOMEN

Betty Jane Burri and Jin-Young K. Park

Western Human Nutrition Research Center
U.S. Department of Agriculture
Agricultural Research Service
Presidio of San Francisco, CA 94129

INTRODUCTION

Vitamin A has several essential functions, including roles in dark adaptation, cell growth and maintenance, and immunological functions (Underwood, 1978; Bauernfeind, 1986; National Research Council, 1989; Sommer and West, 1996; Gerster, 1997). Vitamin A deficiency is the leading cause of preventable blindness in the world (Sommer, 1982). Furthermore, milder vitamin A inadequacy has been implicated as a leading cause of death in poor children of Southern Asia and Africa (Sommer et al., 1983; Sommer and West, 1987). A majority of people in the world derive most of their vitamin A from carotenoids found in many fruits and vegetables (Narasinga, 1991; Solomons and Bulux, 1993; Solomons and Bulux, 1994; FAO/INMU/South and East Asia Nutrition Research-cum-Action Network, 1995; Seshadri, 1996).

Beta-carotene (βC) is the most commonly consumed and easily metabolized carotenoid with vitamin A activity. Conversion of βC to vitamin A appears to be poor: the effectiveness of carotenoid-rich foods in improving vitamin A status has been investigated in several epidemiological studies (De Pee et al., 1995; De Pee and West, 1996), with disappointing results. Furthermore, eating large amounts of preformed vitamin A leads to a variety of well-defined toxicity symptoms, ranging from genetic malformations to joint pains and skeletal deformations, to death in extreme cases (National Research Council, 1989). Eating large amounts of βC, that should theoretically supply the same toxic amounts of vitamin A, do not cause these symptoms. Still, βC is the major source of vitamin A for most of the people in the world.

Little is known about factors affecting the conversion of βC to vitamin A in humans. The reasons for this lack of knowledge are largely methodological. There are few good animal models to study. Serum vitamin A concentrations are well-regulated and do not change with dietary intakes under most conditions. Until recently, it was not possible to trace newly absorbed vitamin A non-invasively in humans. We have begun to study

Mathematical Modeling in Experimental Nutrition
Edited by Clifford and Müller, Plenum Press, New York, 1998

vitamin A and βC metabolism in humans using kinetic models based on new stable isotope research data. Our objectives are to develop compartmental models for vitamin A and βC metabolism in men and women. We hypothesized that pre-formed vitamin A (from meat and milk) might be metabolized differently than vitamin A derived from βC. We further hypothesized that dietary intake levels of βC may be one of the important factors regulating vitamin A metabolism.

ISOTOPE STUDIES OF VITAMIN A AND β-CAROTENE

Recently, isotope dilution techniques in combination with mathematical equations have been used to estimate the dynamics of vitamin A metabolism as well as tissue reserves in a few surgical patients (Bausch and Rietz, 1977; Furr et al., 1989; Reinersdorff et al., 1996; Haskell et al., 1997). Also, a kinetic model of vitamin A metabolism has been based on a series of investigations in rodents using a radioactive tracer of retinol (Lewis et al., 1981; Lewis et al., 1990; Duncan et al., 1993; Green et al., 1993; Adams and Green, 1994; Green and Green, 1996). Finally, a compartmental analysis has been reported in a few men using radioisotopes (Green and Green, 1994) or stable isotopes (Novotny et al., 1995) of vitamin A or βC. These analyses have been conducted in people living in uncontrolled conditions. Their diet and health histories, food consumption, exercise, and activity patterns are largely unknown. Thus, investigation of factors influencing vitamin A and βC metabolism were difficult.

The present chapter discusses our results from two metabolic research unit studies. They were performed between 1992 and 1994 at the Western Human Nutrition Research Center in San Francisco, CA. A total of fourteen healthy adult women participated in these studies and lived under controlled conditions. Serum levels of deuterated stable-isotopes of retinol (retinol-d_4) and βC (βC-d_8) were measured using a high-precision isotope-ratio mass-spectrometry technique (Dueker et al., 1994). We modeled this data using a computer kinetic program (Simulation, Analysis, and Modeling kinetic computer program, RFKA, University of Washington, Seattle). As far as we know, these studies are representative of the limited number of controlled studies performed in humans to investigate the role of βC intake on the dynamics and status of vitamin A, and are the first controlled studies of this metabolism in women.

Vitamin A (Retinyl Acetate) Isotope Methods

Subject selection criteria, diet, body composition, and metabolic status of the participants have been described previously for our retinyl-d_4 acetate study (Dixon et al., 1994). In brief, nine premenopausal women (18 to 42 years) completed the study. Subjects lived in the metabolic unit of the Western Human Nutrition Research Center for 100 days. Their diet contained 1100 RE per day of pre-formed retinol, most given in the highly accessible form of a vitamin pill supplement. (The recommended daily allowance (RDA) for women is 800 RE per day). Subjects were given a low-carotene diet throughout the study (60 mg βC per day), supplemented with 1,500 μg βC per day during the four days of baseline, and 15,000 μg βC per day during the last 28 days of the study. βC repletion was done by supplementing the low βC diet with dry βC beadlets (Hoffmann-La Roche Inc., Nutley, NJ). Twelve days before the end of the repletion period, mixed carotenoid supplements (Carotenoid Complex, lot 38654, Neo-life Company of America, Fremont, CA) were given daily until the end of the study period.

At the beginning of the depletion period, a bolus dose of deuterated retinyl acetate (retinyl-d_4 acetate, 20,000 μg per subject) in a capsule containing corn oil was orally ingested before breakfast. Blood was collected either by catheter on the day the stable isotope-labeled vitamin A was given, or by single sticks after the first day of feeding. Fifteen mL of blood was collected each time. Blood was collected just prior to feeding (0 d), then 0.2, 0.67, 0.83, 1, 2, 3, 4, 7, 22, 29, 36, 43, and 57 days after the feeding. The same amount of retinyl-d_4 acetate was given at the beginning of the repletion period of βC. Then, blood was collected at 0, 0.2, 0.67, 0.79, 1, 2, 3, 4, 7, 19, 20 and 27 days after the feeding of retinyl-d_4 acetate. Diet and the resulting serum concentrations of vitamin A and β-carotene in the Vitamin A-d_4 study are summarized below.

Table 1. Diet and serum levels of vitamin A and β-carotene in the Vitamin A-d_4 study.

	Depletion of βC	Repletion of βC
Number of days	68	28
Days of study	d5 to d72	d73 to d100
Serum vitamin A (μM)	1.58[a]	1.57[b]
Serum β-carotene (μM)	0.3[a,c]	2.3[b,c]
Dietary vitamin A (μg/d)	1100	1100
Dietary β-carotene (μg/d)	60	15,000

[a]Measured on study day 72 [b]Measured on study day 100
[c]p value (paired t-test) < 0.0001

β-CAROTENE ISOTOPE METHODS

In the β-carotene-d_8 study, five healthy adult women ate self-selected diets from a limited menu. Diets were weighed and recorded, and the dietary intake of nutrients calculated by the NESSY system and by weighed food record collections (Fong and Kretsch, 1990). These diets ranged from 1,700 to 17,000 μg per day of vitamin A and from 600 to 17,000 μg per day of βC. Subjects lived on the metabolic unit for 25 days. A bolus dose of deuterated β-carotene-d_8 (40,000 μg βC-d_8 per subject) in a capsule containing corn oil was orally ingested before breakfast. Blood was collected at 0, 0.13,0.21, 0.29, 0.46, 0.63, 1, 2, 3, 4, 7, 14, and 21 days after the feeding of isotope. Subjects were screened to exclude habitual use of alcohol and drugs, hepatitis, syphilis, pregnancy, and chronic diseases prior to entering the study.

Table 2. Diet and serum levels of vitamin A and β-carotene in the β-carotene-d_8 study

Subjects	% Body Fat	μM, Serum vitamin A	μM, Serum β-carotene	μg/d, Dietary vitamin A	μg/d, Dietary β-carotene
1	23	1.49	0.89	2,139	1,167
2	35	2.05	1.60	3,648	2,665
3	20	1.43	0.76	1,661	1,226
4	20	2.21	5.56	17,260	16,562
5	44	1.22	0.25	1,566	593
Means	28	1.68	1.81	5,255	4,443
SD	11	0.43	2.15	6,763	6,818

Subject #4 had the highest intake and serum concentrations of both vitamin A and βC. She was attempting to maintain a recent weight-loss, and consumed a lot of carrots and celery when she felt hungry. Her skin temporarily turned yellow during the study. (This skin discoloration, carotenodermia, is a classic symptom of βC overload. It is the only known consequence of overloading on carotenoids.) Subject #5 had the lowest intake and serum concentrations of retinol and βC. She was overweight compared to the other women volunteers and was using a contraceptive drug.

Stable isotopes for both studies were synthesized by ARC (Netherlands), shipped to Cambridge Isotope (Boston, MA), then to San Francisco. The preparations were purified at the University of California, Davis by HPLC prior to use, and checked for purity by NMR, HPLC, and GC-MS. Blood serum was analyzed for the concentrations of stable isotope-labeled retinol-d_4, retinol, βC-d_8, and βC. Concentrations of retinol-d_4 and total retinol in serial blood serum samples were determined by gas chromatography-mass spectrometry (GC-MS) (Dueker et al., 1993). Concentrations of βC-d_8 and βC were determined by high performance liquid chromatography (HPLC)-UV detection (Dixon et al., 1994; Dueker et al., 1994).

RESULTS

Model-based Compartmental Analysis

We used the compartmental modeling methods of Green and Novotny (Green and Green, 1994; Novotny et al., 1995) to develop initial kinetic models of retinol and βC metabolism in women. We used SAAM II for kinetic analyses and simulation. Initial parameters were estimated and the SAAM program adjusted the parameters to obtain a least-squares fit to the observed data.

Assumptions that we used in the developing the compartmental model of vitamin A kinetics depicted in Figure 2 included the following.

1. The experiment was performed in a steady-state condition as a constant dietary intake of vitamin A was provided throughout the study.
2. The depletion/repletion of βC and the introduction of retinol-d_4 constituted a transient perturbation, and the system quickly reached a steady-state after any perturbation. As a result, fractional transfer coefficients were constant over the course of the experiment in each individual.
3. The contents of each compartment were homogeneously distributed.
4. The volume of distribution was 4.5 percent of body weight.
5. The fractional transfer coefficients before and after the delay term were the same.
6. All errors were in dependent variables, such as the serum concentrations of isotopes.
7. Fractional standard deviations (FSD) of the dependent variables were specified as 0.05 on the basis of the analytical variability of the individual measurements.

The same assumptions listed in (1) through (6) for developing the vitamin A kinetics were also applied in developing the working model for βC and vitamin A kinetics, depicted in Figure 6. Two additional assumptions for this working model are that the standard deviation of the dependent variables was 0.05; and that the fractional transfer coefficient between compartment 8 (βC in tissues) and 6 (serum βC in chylomicrons) was constrained as 34.75 day^{-1} (Novotny et al., 1995).

Model of pre-formed vitamin A metabolism

Total serum vitamin A (retinol) concentrations did not change during these studies. This was expected, because retinol concentrations are highly regulated and are not influenced by differences in dietary intakes except during extreme deficiency or toxicity. Fig.1 shows typical serum concentration curves for the stable isotope of pre-formed vitamin A. A bolus of pre-formed vitamin A was given to each subject twice, on study days 4 and 72. The first bolus dose was metabolized at the beginning of carotenoid depletion, when βC concentrations were decreasing rapidly. The second dose was metabolized during βC supplementation, when concentrations were increasing rapidly. The serum concentrations of βC 10 days after the changes in carotenoid intake were much higher during the second (carotenoid repletion) phase of the experiment than the first (carotenoid depletion) phase. The large differences in serum carotenoid concentrations due to the significant changes in dietary carotenoid intake appear to have some effects on the kinetic model. These effects are on the equilibration time of the retinol-d_4 and on its amount in serum at the time of equilibration (Fig.1).

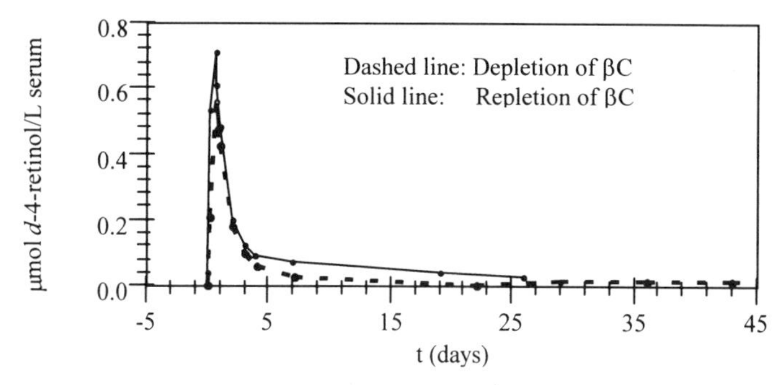

Figure 1. Metabolism of pre-formed-isotope-labeled retinol.

The compartmental model of vitamin A metabolism when it was given as pre-formed retinol (in a bolus of retinyl-d_4 acetate) is shown in Figure 2. A four-compartment model was chosen as the simplest 'best-fit' for the observed data for the women who participated in this study (Fig. 2). The model contains the gastrointestinal tract (GI tract) compartment (compartment 1), a serum compartment (compartment 3), a tissue compartment (compartment 4), and a delay (delay 2). We have tried two tissue compartments in the model, one with faster exchange and another with slower exchange of retinol, and an irreversible output of retinol to the outside from the serum compartment or from one of the tissue compartments. These variations did not improve the fit of the data. Our results suggest, however, that our model may have applied only to the faster-exchange compartments of the tissues in healthy women with adequate intakes of vitamin A. Results may have been different in women with more marginal vitamin A intakes.

The simple four-compartment model accurately describes the kinetics (the CV of each kinetic parameter is < 0.2) of the first bolus of vitamin A given to a representative subject (Fig.3). This same four-compartment model accurately describes the kinetics (CV < 0.2) of eight of the nine women when they were given the first dose of retinol isotope, and it describes the kinetics of all nine women when they were given the second bolus dose. The serum concentration of retinol-d_4 of all subjects, except subject #9 for the first dose, were equilibrated within 30 days.

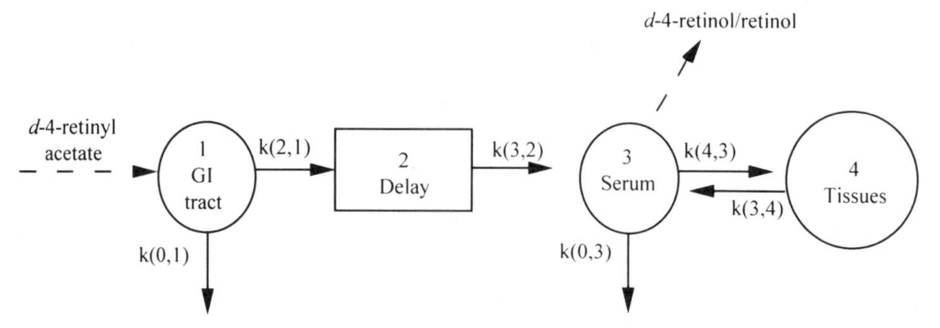

Figure 2. Compartmental model of retinol after oral administration of retinyl-d_4 acetate. Enclosed spaces represent compartments that are kinetically homogenous and distinct from other compartments. A square compartment designated as delay consists of 10 delay terms between when retinol leaves the GI tract compartment and when it enters the serum compartment. Numbers within the spaces correspond to compartment numbers in the model. Arrows represent fractional transfer (solid) of retinol, input of retinyl-d_4 acetate (dashed), and collection of serum sample (dashed). The $k_{i,j}$ represents fractional transfer coefficients of mass transferred from compartment *j* to compartment *i*.

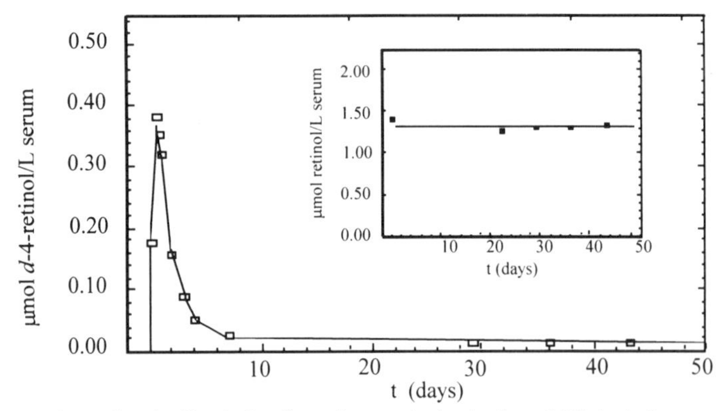

(3A) After the first bolus dose given at the beginning of βC depletion.

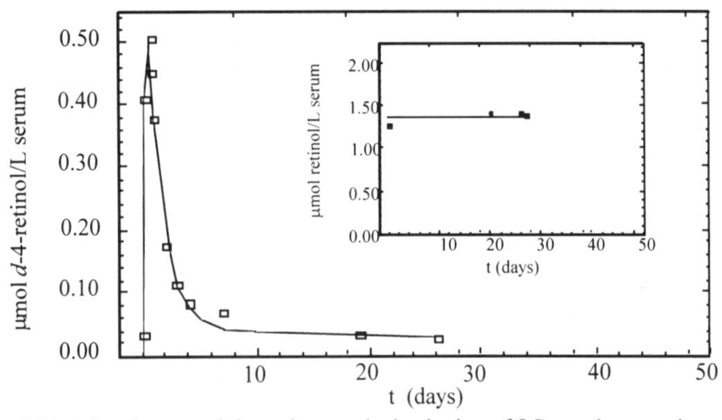

(3B) After the second dose given at the beginning of βC supplementation

Figure 3. Computer-simulated serum concentration curves of retinol-d_4. Retinol-d_4 (a solid line) and total retinol (a solid line in insert) in serum from a representative subject (A) after the first dose of retinyl-d_4 acetate with decreasing intake of βC, and (B) after the second dose with increasing intake of βC. Lines are the model prediction and squares are experimental observation.

The first dose was probably influenced by the subjects' diets prior to the beginning of the study, while the second bolus dose was given after they had eaten 1,100 RE of vitamin A for more then 10 weeks. Thus, all of the women studied had very similar kinetics after they had eaten a study diet with an adequate vitamin A for 10 weeks.

Fractional transfer coefficients of retinol between compartments, turnover rate, steady-state masses, and residence times of retinol in each compartment were estimated from the model-based compartmental analysis. The model predicts a mean bioavailability of retinol-d_4 of 7 % (± 2 % SD) and an irreversible turnover rate of 4 μmol/d (± 0.1 SD). The rapid increase of βC during repletion did not influence the turnover rate of retinol in any compartment. However, it did result in changes ($p<0.05$) of the mean bioavailability of retinyl-d_4 acetate, fractional transfer coefficients, steady-state masses and residence times of retinol in several compartments including 'Tissues' and 'GI tract.' The changes were different depending upon the initial status of mass in 'Tissues.' Both an equation-based analysis (Furr et al., 1989) and area-under-serum-concentration-curve analysis showed the same conclusions.

Most of the dynamics that form the basis of the kinetic model occurred during the first 10 days after the oral administration of the isotope when serum concentrations of βC were changing rapidly. Serum concentrations of βC approached a plateau within a week of depleting or repleting humans, and about 50% of the plateau was reached within a day. Therefore, we could assume that serum concentrations of vitamin A and βC reached steady states quickly. We also were able to assume that the affect of conversion of βC into retinol on the kinetics of vitamin A metabolism during its repletion period would be negligible. This assumption is based on data from our study on βC-d_8 metabolism. Less than 5% of the ingested dose was absorbed, and less than 30% of the absorbed dose was converted into retinol. Assuming similar kinetics during the vitamin A study, only 1.5% of 15,000 μg βC (0.8 μmol of retinol) could have been formed from βC, while 3.86 μmol of preformed retinol per day was being fed directly. When we ran the SAAM program adjusted for the additional 0.8 μmol of retinol coming from βC, changes of kinetic parameters were insignificant. Therefore, we were able to use steady-state assumptions for retinol. This greatly simplified the calculations used for the model. We are now conducting studies where both βC and retinol isotopes are administered to the same people, which will allow us to assess the consequences of making these assumptions for the model.

The woman (subject 9) who could not be described adequately with this model the first time she was given the stable isotope appeared to have entered the study with very low vitamin A stores. Her initial vitamin A status, estimated by either the kinetic model or by equation (Furr et al., 1989) was less than 10 μg per gram liver. Furthermore, her dietary intake questionnaires (NCI Health Habits and History Questionnaire, v. 2.1; collected after the study began) suggested that she ate a nutritionally poor diet that was low in vitamin A and carotenoids and high in fat and sugar. Her vitamin E and C serum concentrations increased while she was on the study, and her blood pressure, cholesterol, and triglyceride concentrations decreased. In addition, anecdotal evidence suggests that she may have been a recent smoker (contrary to the selection criteria of subjects); but no physiological evidence was ever found of smoking behavior during the study. The kinetics of retinol-d_4 metabolism for subject 9 during the first part of the study is shown below, along with the kinetics of a typical subject (subject 6).

Subject 9's vitamin A metabolism was normalized by the time the second dose was given. Although an affect of smoking cannot be entirely ruled out, it is most likely that subject 9 had different kinetics initially because she had marginal vitamin A status. After eating a nutritionally adequate diet relatively rich in vitamin A, her status improved and her metabolism became similar to the other subjects in the study. This suggests that the metabolism of pre-formed vitamin A is affected by low vitamin A status; and that the kinetics of vitamin A metabolism in people with poor vitamin A status may be different and more complex than metabolism during vitamin A adequacy.

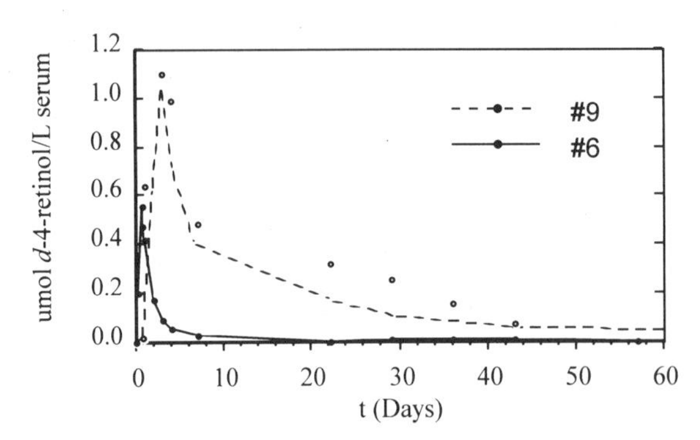

Figure 4. Metabolism of pre-formed retinol during the first study period (βC depletion) in subject #9, compared to another subject. Lines are the model predictions and dots are experimental observations. Subject #6 had the typical metabolic pattern for pre-formed retinol.

Metabolism of Vitamin A Formed from β-Carotene.

Figures 5A through 5E show the concentration curves of retinol-d_4 derived from a bolus dose of β-carotene-d_8 (βC-d_8). There are several noteworthy features of these concentration curves. First, the metabolism of vitamin A generally appears more complex when it is derived from its precursor βC. We observed multiple peaks of retinol-d_4 when βC-d_8 was the parent compound. Second, these subjects show obvious individual differences in the appearance and serum concentration curves of retinol-d_4. Third, it is interesting that some of the retinol-d_4 derived from βC-d_8 appears in the serum sooner then would retinol-d_4 derived from a bolus dose of retinyl-d_4 acetate.

Clearly, the metabolism of vitamin A formed from βC (Fig.5) differs from the simple metabolism of vitamin A given as a bolus dose of retinyl acetate. Thus, it is not surprising that the model of retinol from βC is more complex than the one from pre-formed retinol (Fig. 2). The kinetics of retinol from βC suggests that a portion of βC is converted to vitamin A in the intestine, and that this vitamin A appears in the blood stream rapidly. Other portions of βC are converted to vitamin A and metabolized much more slowly, appearing as broad peaks over a period of several days. The simplest working model we have derived for this metabolism is shown in Figure 6. This model has a delay term before the 'enterocytes' compartment and three compartments representing βC metabolism after the serum compartments. This is a reconstructed model similar to that described by Novotny et al. (Novotny et al., 1995). Figure 7 shows that the model describes the first (sharp) and second (major broad) peaks of βC and a rising peak of retinol, but not the multiple peaks of retinol.

We do not know the reasons for the high variability of vitamin A metabolism when it is derived from its precursor. However, several intriguing possibilities are apparent. First, subject 4 had an extremely high dietary intake of βC (approximately 17,000 μg per day), mostly in the form of carrots. Her skin turned orange (carotenodermia), the main symptom associated with very high carotenoid consumption. She also had the simplest and one of the most rapid concentration curves for vitamin A formation. This may indicate that dietary βC is rapidly metabolized and excreted under conditions of extremely high intake.

This would not be surprising. A feedback reaction of that sort, where high intakes stimulate high rates of metabolism and excretion, is fairly common for many water-soluble nutrients. Second, the concentration curve for subject 5, who was on a contraceptive, was also unusually rapid and relatively simple. Contraceptives have been reported to influence both βC and vitamin A status and metabolism (Anonymous, 1979; Vahlquist et al., 1979; Boots et al., 1983).

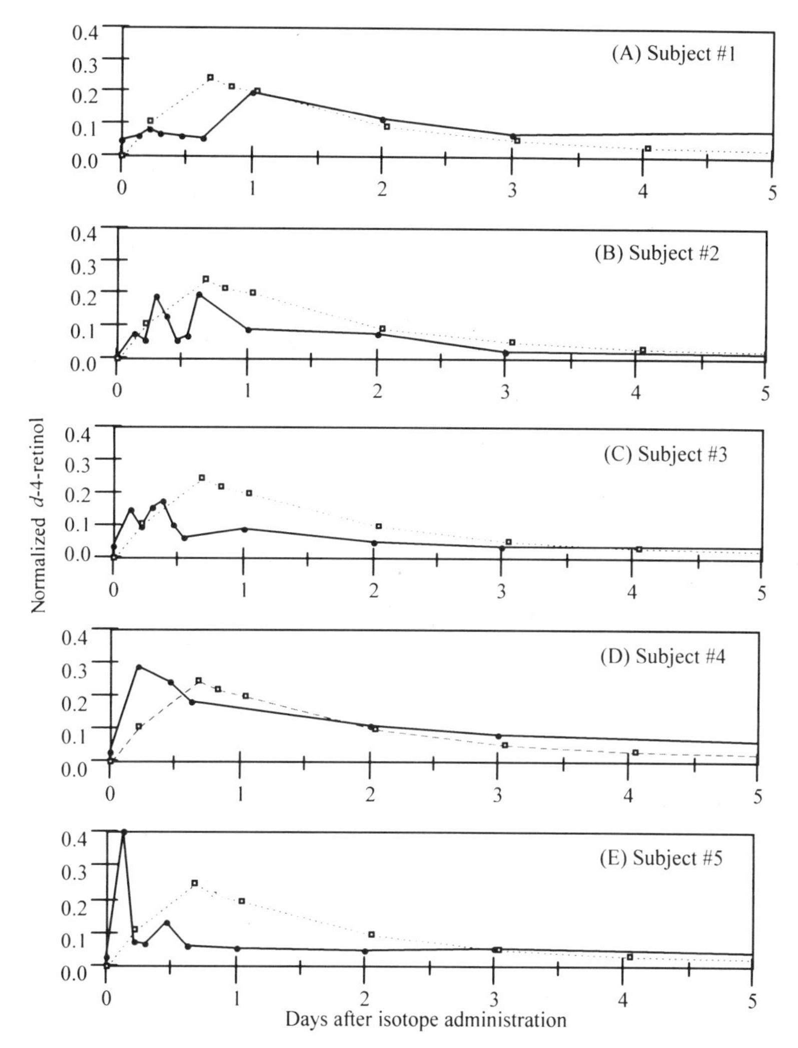

Figure 5. Kinetics of retinol-d_4 metabolism in serum after oral administration of βC-d_8 (solid lines) were compared to retinyl-d_4 acetate given as the pure compound (from the first experiment, dashed lines). Concentrations (μmol/L serum) of retinol-d_4 were normalized to the total concentration observed in the experiment for that individual.

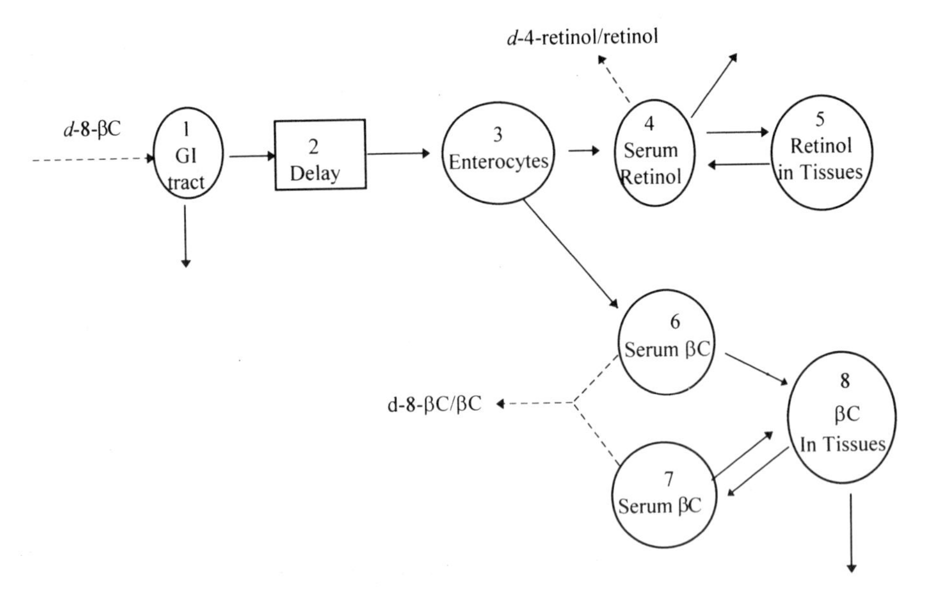

Figure 6. A working compartmental model of βC and retinol after administration of βC-d_8. Enclosed spaces represent compartments that are kinetically homogenous and distinct from other compartments. A square compartment consists of 10 delay terms between when retinol leaves GI tract compartment and when it enters the enterocytes compartment. Numbers within the spaces correspond to compartment numbers in the model. Solid arrows represent fractional transfer of retinol or βC. Dashed arrows represent either an input of βC-d_8 or collections of serum sample of βC-d_8, retinol-d_4, βC, and retinol.

However, our data shows that any conclusions about the causes of the diversity in vitamin A metabolism should be made cautiously. For example, the serum concentration curves for subjects 1 and 3 differ substantially, even though these subjects were very similar. They had similar ages, heights, weights, body compositions, and vitamin A and carotenoid intakes (Table 2). Both had unremarkable health histories, and remained healthy throughout the study. Neither smoked, nor were they on any medications. Both had lived in the San Francisco Bay Area for several years. Neither had current parasite infestations. Factors that may influence carotenoid metabolism have been grouped under the mnemonic term 'SLAMANGHI' (De Pee and West, 1996). SLAMANGHI stands for **s**pecies (type) of carotenoid, molecular **l**inkage, **a**mount of carotene in meal, **m**atrix in which the carotenoid is incorporated, **a**bsorption modifiers, **n**utrient status of host, **g**enetic factors, **h**ost-related factors (health history, age, sex, menstrual cycle status), and **i**nteractions. For these women, most of these factors were controlled, or were quite similar. Pre-formed vitamin A appears to be metabolized similarly in individuals under these conditions. However, it seems to be different when vitamin A is formed from βC. We cannot completely rule out the possibility that differences in the subjects' health history, intestinal flora, nutrient matrix, or menstrual cycle had influenced their metabolism. However, the differences observed between subjects 1 and 3 (Table 2) seem too small to be physiologically important. This suggests that some of the differences in βC metabolism between these two individuals were hereditary.

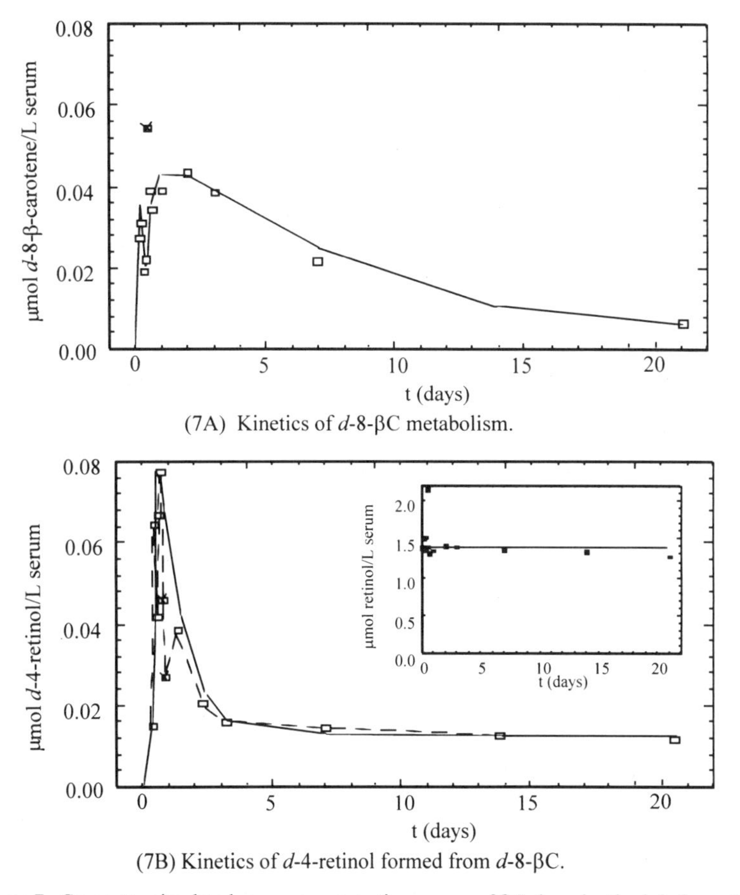

(7A) Kinetics of *d*-8-βC metabolism.

(7B) Kinetics of *d*-4-retinol formed from *d*-8-βC.

Figure 7. Computer-simulated serum concentration curves of βC-d_8 and retinol-d_4 formed from βC-d_8. Simulated curves based on the model depicted in Fig. 6 of (7A) βC-d_8 (a solid line) and (7B) retinol-d_4 (a solid line) and total retinol (a solid line in insert) in serum from a representative subject (#3). Squares represent the observed data points. Squares with a cross inside were not included in the fitting process. A dashed line connecting the two adjacent squares represent the observed data.

SUMMARY

We have developed compartmental models of vitamin A and β-carotene (βC) metabolism in women living under controlled conditions on diets with known concentrations of vitamins and carotenoids. Fourteen healthy adult women were given either retinyl-d_4 acetate, or βC-d_8 before breakfast. Natural and stable-isotopes of retinol and βC were collected in serum for up to 95 days or 20 days, respectively. Stable isotopes were separated from other components and measured by GC-MS or HPLC-UV. Pre-formed retinyl-d_4 acetate metabolism in all women tested can be accurately described by a simple four-compartment model. However, the model did not fit one women initially, when she had marginal vitamin A status. We tested the hypothesis that dietary changes of βC intake have important roles on the kinetics of vitamin A metabolism. Dietary changes of βC intake did not influence the turnover rate of retinol in any compartment. However, it did result in changes in steady-state masses and residence times of retinol in several compartments.

A working compartmental model for βC metabolism was developed. The kinetics of retinol-d_4 formed from βC is more complicated than the pre-formed retinol-d_4. Results suggest that βC-d_8 readily converts into retinol-d_4 with high inter-individual variability.

ACKNOWLEDGMENTS

The authors are grateful to Drs. Adrianne Bendich and Hemmige Bhagavan of Hoffmann-La Roche Ltd., Nutley, NJ for the β-carotene capsules and technical assistance. We thank Dr. Ariana Carughi of GNLD, San Jose, CA for the mixed carotenoid supplements. We also thank Drs. David Foster and Hugh Barrett of the Resource Facility for Kinetic Analysis, Center for Bioengineering, University of Washington, Seattle, WA for help in setting up the initial model for the compartmental analysis and providing valuable comments for this manuscript.

CORRESPONDING AUTHOR

Please address all correspondence to:
Betty Jane Burri
Western Human Nutrition Research Center
USDA/ARS/PWA
PO Box 29997
Presidio of San Francisco, CA 94129

TEXT FOOTNOTES

Portions of this work were presented at Experimental Biology, April 6-9, 1997, in New Orleans, Louisiana (Park JYK; Burri BJ; et al.; Plasma bioavailability of vitamin A-d_4 stable isotope in women: Depletion/repletion study of β-carotene. *FASEB J.* # 50345), and at The Third International Congress on Vegetarian Nutrition, March 24-26, 1997, in Loma Linda, California (Burri BJ; Park JK; et al.; Influence of carotenoid depletion and repletion on serum carotenoid and vitamin A concentrations, # P19).

REFERENCES

Adams WR; Green MH. Prediction of liver vitamin A in rats by an oral isotope dilution technique. *J Nutr,* 1994, 124:1265-70.

Anonymous. The effect oral contraceptives on blood vitamin A levels and the role of sex hormones. *Nutr Rev,* 1979, 37:346-48.

Bauernfeind J. *Vitamin A Deficiency and Its Control.* Academic Press: Orlando. 1986.

Bausch J; Rietz P. Method for the assessment of vitamin A liver stores. *Acta Vitaminol Enzymol,* 1977, 31:99-112.

Boots LR; Cornwell PE; et al. Vitamin fluctuations in the blood of female baboons in relation to normal menstrual cycles, treatments with Lo-Ovral or Depo-Provera and a selected vitamin supplement. *Am J Clin Nutr,* 1983, 37:518-31.

De Pee S; West CE. Dietary carotenoids and their role in combating vitamin A deficiency: A review of the literature. *Eur J Clin Nutr,* 1996, S38-53.

De Pee S; West CE; et al. Lack of improvement in vitamin A status with increased consumption of dark-green leafy vegetables [see comments]. *Lancet,* 1995, 346:75-81.

Dixon ZR; Burri BJ; et al. Effects of a carotene-deficient diet on measures of oxidative susceptibility and superoxide dismutase activity in adult women. *Free Radic Biol Med,* 1994, 17:537-44.

Dueker SR; Jones AD; et al. Stable isotope methods for the study of beta-carotene-d8 metabolism in humans utilizing tandem mass spectrometry and high-performance liquid chromatography. *Anal Chem,* 1994, 66:4177-85.
Dueker SR; Lunetta JM; et al. Solid-phase extraction protocol for isolating retinol-d4 and retinol from plasma for parallel processing for epidemiological studies. *Clin Chem*, 1993, 2318-22.
Duncan TE; Green JB; et al. Liver vitamin A levels in rats are predicted by a modified isotope dilution technique. *J Nutr,* 1993, 123:933-9.
FAO/INMU/South and East Asia Nutrition Research-cum-Action Network. Empowering vitamin A foods: A food-based process for the Asia and Pacific Region. *Inst. Nutr Maihidol U,* 1995.
Fong AK; Kretsch MJ. Nutrition Evaluation Scale System reduces time and labor in recording quantitative dietary intake. *J Am Diet Assoc,* 1990, 90:664-70.
Furr HC; Amedee MO; et al. Vitamin A concentrations in liver determined by isotope dilution assay with tetradeuterated vitamin A and by biopsy in generally healthy adult humans. *Am J Clin Nutr,* 1989, 49:713-6.
Gerster H. Vitamin A – Functions, dietary requirements and safety in humans. *Internat J Vit Nutr Res,* 1997, 67:71-90.
Green M; Green J. Dynamics and control of plasma retinol, in: *Vitamin A in Health and Disease.* Marcel Dekker: New York. 1994.
Green MH; Green JB. Quantitative and conceptual contributions of mathematical modeling to current views on vitamin A metabolism, biochemistry, and nutrition. *Adv Food Nutr Res,* 1996, 40:3-24.
Green MH; Green JB; et al. Vitamin A metabolism in rat liver: A kinetic model. *Am J Physiol,* 1993, 264:G509-21.
Haskell MJ; Handelman GJ; et al. Assessment of vitamin A status by the deuterated-retinol-dilution technique and comparison with hepatic vitamin A concentration in Bangladeshi surgical patients. *Am J Clin Nutr,* 1997, 66:67-74.
Lewis KC; Green MH; et al. Retinol metabolism in rats with low vitamin A status: A compartmental model. *J Lipid Res,* 1990, 31:1535-48.
Lewis KC; Green MH; et al. Vitamin A turnover in rats as influenced by vitamin A status. *J Nutr,* 1981, 111:1135-44.
Narasinga R. *Use of β-carotein-rich foods for combating vitamin A deficiency.* Bull Nutrition Foundation: India. 1991.
National Research Council. *Recommended Dietary Allowances.* 10th Ed. National Academy Press: Washington DC. 1989.
Novotny JA; Dueker S; et al. Compartmental analysis of the dynamics of beta-carotene metabolism in an adult volunteer. *J Lipid Res,* 1995, 36:1825-38.
Seshadri S. *Use Of Carotene-Rich Foods to Combat Vitamin A Deficiency in India: A Multicenter Study.* Media Workshop Nutrition Foundation, India. 1996,
Solomons NW; Bulux J. Plant sources of provitamin A and human nutriture. *Nutr Rev,* 1993, 51:199-204.
Solomons NW; Bulux J. Plant sources of vitamin A and human nutrition revisited: Recent evidence from developing countries. *Nutr Rev,* 1994, 62-4.
Sommer A. *Nutritional Blindness, Xeropthalmia, and Keratomalacia.* Oxford Press: New York. 1982.
Sommer A; Tarwotjo I; et al. Increased mortality in children with mild vitamin A deficiency. *Lancet,* 1983, 2:585-8.
Sommer A; West KJ. Impact of vitamin A on childhood mortality. *Indian J Pediatr,* 1987, 54:461-3.
Sommer A; West KJ. *Vitamin A Deficiency, Health, Survival, and Vision.* Oxford Press: New York. 1996.
Underwood B. Hypovitaminosis A and its control. *Bull WHO,* 1978, 56:525-541.
Vahlquist A; Johnsson A; et al. Vitamin A transporting plasma proteins and female sex hormones. *Am J Clin Nutr,* 1979, 32:1433-38.
v. Reinersdorff DV; Bush E; et al. Plasma kinetics of vitamin A in humans after a single oral dose of [8,9,19-13C]retinyl palmitate. *J Lipid Res,* 1996, 37:1875-85.

THE DYNAMICS OF FOLIC ACID METABOLISM IN AN ADULT GIVEN A SMALL TRACER DOSE OF ^{14}C-FOLIC ACID

Andrew J. Clifford,[1] Ali Arjomand,[1] Stephen R. Dueker,[1]
Philip D. Schneider,[2] Bruce A. Buchholz,[3] and John S. Vogel[3]

[1]Department of Nutrition
University of California
Davis, CA 95616
[2]Cancer Center
University of California Davis Medical Center
Sacramento, CA 95817
[3]Center for Accelerator Mass Spectrometry
Lawrence Livermore National Laboratory
Livermore, CA 94551

ABSTRACT

Folate is an essential nutrient that is involved in many metabolic pathways, including amino acid interconversions and nucleotide (DNA) synthesis. In genetically susceptible individuals and populations, dysfunction of folate metabolism is associated with severe illness. Despite the importance of folate, major gaps exist in our quantitative understanding of folate metabolism in humans. The gaps exist because folate metabolism is complex, a suitable animal model that mimics human folate metabolism has not been identified, and suitable experimental protocols for *in vivo* studies in humans are not developed.

In general, previous studies of folate metabolism have used large doses of high specific activity tritium and ^{14}C-labeled folates in clinical patients. While stable isotopes such as deuterium and ^{13}C-labeled folate are viewed as ethical alternatives to radiolabeled folates for studying metabolism, the lack of sensitive mass spectrometry methods to quantify them has impeded advancement of the field using this approach.

In this chapter, we describe a new approach that uses a major analytical breakthrough, Accelerator Mass Spectrometry (AMS). Because AMS can detect attomole concentrations of ^{14}C, small radioactive dosages (nCi) can be safely administered to humans and traced over long periods of time. The needed dosages are sufficiently small that the total radiation exposure is only a fraction of the natural annual background radiation of Americans, and the generated laboratory waste may legally be classified non-radioactive in many cases.

The availability of AMS has permitted the longest (202 d) and most detailed study to date of folate metabolism in a healthy adult human volunteer. Here we demonstrate the

Mathematical Modeling in Experimental Nutrition
Edited by Clifford and Müller, Plenum Press, New York, 1998

feasibility of our approach and illustrate its potential by determining empirical kinetic values of folate metabolism. Our data indicate that the mean sojourn time for folate is in the range of 93 to 120 d. It took ≥ 350 d for the absorbed portion of small bolus dose of ^{14}C-folic acid to be eliminated completely from the body.

INTRODUCTION

Folate is an essential nutrient that serves as a single-carbon donor/acceptor in several physiologic pathways, including amino acid interconversions and nucleotide (DNA) synthesis. The connection between an adequate intake of dietary folate, a functional folate metabolism, and optimal health is well established. Furthermore, severe health problems (neural tube defects, heart disease, and cancer) can be associated with modest declines in folate nutritional status when mutations in folate enzymes are present. These health problems are a product of complex genetic traits that "load the gun" and environmental factors that "pull the trigger".

A significant portion (20-40%) of the US population is characterized as marginally deficient in folate nutriture as indicated by low blood folate concentrations. This marginal status can be attributed to factors such as an inadequate dietary folate intake, an inefficient/dysfunctional folate metabolism, conditions of increased folate need such as pregnancy, and the use of antifolates and alcohol. Despite the importance of folate nutritional status and an optimum folate metabolism, there is a paucity of quantitative information and few reliable protocols for assessing either the dynamics or the nutritional status of folate of humans.

The slow progress towards understanding the dynamics of folate metabolism can be attributed to three major factors. First, the complexity of folate metabolism (see reviews by Steinberg, 1984; Shane, 1989; Shane, 1990; Stokstad, 1990; Shane, 1995; Wagner, 1995; Ma et al., 1997). Second, the lack of a suitable animal model that mimics human folate metabolism. Third, the lack of suitable experimental protocols for *in vivo* studies in humans. Useful protocols require both an appropriately-labeled folate that can safely be administered to humans, and physico/chemical analytical methods which are more sensitive and precise than current bioassays. These methods would be used to measure trace amounts of the labeled folate in key tissues for long periods of time after the administration of physiologic doses of the labeled folate.

In general, previous studies of folate metabolism have used large doses of high specific activity tritium and ^{14}C-labeled folates. In a short term (3 d) study on an adult man suffering from chronic lymphatic leukemia and given a pharmacologic dose (10 μmol) of ^{14}C-folic acid, Butterworth et al. (1969) demonstrated that folic acid monoglutamate was almost completely (90%) absorbed. Furthermore, a large portion of the dose (44% of the administered dose or 49% of the absorbed dose) was lost in urine in 3 days. In a long-term (129 d) study on an adult woman in remission with Hodgkin's disease who was given a physiologic dose (0.725 μmol) of ^{14}C-folic acid, Krumdieck et al. (1978) confirmed that folic acid monoglutamate was also completely (92.2%) absorbed, though only a small portion of the dose (3.6% of the administered dose or 3.9% of the absorbed dose) was lost in urine in 6 d. They also noted the nature of the loss of ^{14}C in urine over time was biphasic with both short-lived and long-lived phases. They determined the $t_{1/2}$ of the short-lived and long-lived phases to be 31.5 h and 100 d, respectively. These two ^{14}C-folate studies in clinical patients, along with data from an earlier folate depletion/repletion study in a healthy man by Herbert (1962), still serve as the basis for much of our present knowledge of folate metabolism and turnover in humans.

Stable isotopes such as deuterium and ^{13}C-labeled nutrients are viewed as ethical alternatives to radiolabeled nutrients for studying the absorption and metabolic fate of nutrients, and stable isotope tracers have largely replaced radiolabeled ones for human studies. Still, a scarcity of stable isotope-labeled folates, the lack of sensitive mass spectrometry methods to quantify them, and the possibility of significant isotope effects from deuterium has impeded advancement using this approach. The preparation and use of $^{2}H_2$-folic acid was first described by Rosenberg et al. (1973). Even though stable isotope-labeled folates appeared promising, their applications to date have yielded little new insight into folate metabolism (Stites et al., 1997). The lack of sensitive mass spectrometry methods to measure the stable isotopes in biological specimens has proven to be a major obstacle.

In this chapter, we describe a new approach that uses a major analytical breakthrough of a decade ago: Accelerator Mass Spectrometry (AMS). An excellent review of the application of AMS to nutrition research is presented elsewhere in this volume by Vogel and Turteltaub (1998). This particular type of isotope ratio mass spectrometry can detect attomole quantities (1×10^{-18} mol) of long-lived radioisotopes in milligram-sized samples with high precision. Because AMS can detect attomole concentrations (1 part in 10^{18} parts) of ^{14}C, small radioactive dosages (nCi) can now be administered to humans and traced over long time periods. AMS obviates the need for high specific activity dosage regimens.

In the present study, the ^{14}C of a 100 nCi dose of ^{14}C-folic acid was traced in the plasma, erythrocytes, urine, and feces of a healthy adult for 202 days. The needed dosages are sufficiently small that the radiation exposure is only a fraction of the natural annual background radiation of Americans, and the generated laboratory waste may legally be classified non-radioactive in many cases (10 Consolidated Federal Register 20.2005).

The availability of AMS allowed us to conduct the longest and most detailed study to date of folate metabolism in a healthy adult human volunteer. We have described our approach and illustrate its potential for obtaining the sort of data needed to determine the dynamics of folate metabolism in an adult man. A better understanding of folate metabolism is prerequisite for more reliable estimates of the dietary folate requirements for optimal health, and for understanding the factors that influence the requirements and the consequences of altering them.

EXPERIMENTAL METHODS

Under ideal conditions, investigations of nutrient metabolism require the administration of a physiologic amount (a tracer) of a strategically labeled nutrient, adequate sampling density of appropriate tissues, mass balance of the administered tracer nutrient, and analytical methods that reliably measure the tracer in sampled tissues for long periods after dosing. Elements of our approach for studying the dynamics of nutrient metabolism *in vivo* in humans are illustrated using the nutrient folic acid as an example. The study was approved by the Human Subjects Review Committees of the University of California, Davis, CA, and the Lawrence Livermore National Laboratory, Livermore, CA.

Tracer Synthesis

Pteroyl-[^{14}C(U)]-glutamic acid (^{14}C-folic acid) was synthesized according to the method of Plante et al. (1980) with some modifications. Pteroic acid (0.789 mmol) was dissolved in 10 mL trifluoroacetic acid (TFA). Once in solution, 10 mL trifluoroacetic anhydride was added, and the solution refluxed for 1.5 h. The solvent was evaporated, the residue was suspended in water, and the white precipitate (N^{10}-trifluoroacetylpteroic acid) was collected and dried. Two hundred-fifty μCi [^{14}C(U)]-glutamic acid (180-220 Ci/mol manufacturer specification) was diluted with 130 μmol non-labeled glutamic acid, dried, suspended in ethanol with 1% acetyl chloride and incubated for 2 h at 60°C and dried to yield [^{14}C(U)]-glutamate diethyl ester-HCl. This intermediate was coupled to N^{10}-trifluoroacetylpteroic acid by a mixed anhydride synthesis using isobutylchloroformate in dimethylformamide. The free acid form was restored by alkaline saponification and crystallized overnight at 4°C in water. The yellow solid that contained pteroyl-[^{14}C(U)]-glutamic acid was collected and dried.

The dried yellow solid was dissolved in TFA and pteroyl-[^{14}C(U)]-glutamic acid was isolated by reverse-phase HPLC using a semi-prep C18 column and an isocratic mobile phase of 850 mL 8.8 mmol TFA/L H_2O with 150 mL CH_3CN pumped at 2 mL/min. The radioactive peak corresponding to pteroyl-[^{14}C(U)]-glutamic acid was collected, dried, and stored at 4°C. Chemical purity of the pteroyl-[^{14}C(U)]-L-glutamic was confirmed by coelution with authentic standards on two different chromatographic systems. The radioactivity of a known mass of pteroyl-[^{14}C(U)]-L-glutamic from the HPLC was used to calculate specific activity. Using this approach, the specific activity of the administered pteroyl-[^{14}C(U)]-L-glutamic acid was 1.25 Ci/mol.

Subject, Diet, and Dose

The subject was a well-educated, consenting, healthy 57-year-old male weighing 91 kg, whose typical diet consisted mostly of 'convenience foods'. The subject consumed that diet throughout the 202-day study. Diet folate intake was estimated using food folate values from the USDA Food Composition Handbook # 8 and Perloff and Butrum (1977). A complete two-day collection of the diet was made, composited, and analyzed for dry matter (lyophilize), fat (ether extract), and protein (Kjeldahl N x 6.25). The subject ingested an 80 nmol (100 nCi) oral dose of pteroyl-[^{14}C(U)]-glutamic in 125 mL water just before a light breakfast consisting of a doughnut and coffee (~10 g fat and ~300 calories). The radiation effective dose to the volunteer was only 1.1 mrem.

Specimen Collection and Handling

Blood drawing was conducted at the University of California Davis, Cancer Center, Sacramento. Separation and washing of erythrocytes, and initial processing of urine and feces specimens were performed in a special clean room at the University of California Davis that had not been previously exposed to radiocarbon work. After the initial processing, the specimens were transferred to the Center for Accelerator Mass Spectrometry (CAMS), at the Lawrence Livermore National Laboratory (LLNL) for final processing and ^{14}C analysis.

Blood One blood sample drawn just before dosing served a background control. Fifty-eight serial blood specimens were drawn in tubes with EDTA over a 202-day period after the initial ingestion of the ^{14}C-folic acid. The intensity of sampling was high immediately after dosing in order to monitor rapidly changing blood concentrations during the initial absorptive and disposition phase. Twenty specimens were drawn during the first 12 hours after dosing, and four specimens during the period 12 to 24 hours after dosing. Packed cell volume (hematocrit) was measured in selected specimens drawn during the first several days to ensure the hematocrit values did not change from the blood loss. A 0.5 mL aliquot of whole blood was removed and stored at –20°C. The remainder was centrifuged to separate the plasma. The 0.5 mL-aliquots of whole blood and plasma were stored at –20°C. The leukocyte layer (buffy coat atop sedimented erythrocytes) was also removed. The erythrocytes were washed four times in buffered saline (150 mmol NaCl + 10 mmol K_2HPO_4 + 0.05 mmol EDTA/L H_2O, pH 7.4). They were then resuspended to ~35% hematocrit in buffered saline and stored at –20°C.

Urine One complete 24-hour collection was made just before dosing. Complete 6-hour collections (4) were made during the first 24-hour period after dosing. Complete 24-

hour collections were made on days 2 through 43, 64 through 75, and 174 through 177 after dosing. Immediately after collection, the mass of each collection was measured, duplicate aliquots (~40 mL) from each collection were stored at –20°C, and the remainder was discarded.

Feces One collection was made just before dosing. Complete collections (27) were made on days 1 through 43 since dosing. Complete collections (7) were also made on days 65 through 75. Finally, complete collections (3) were made on days 174 through 177 after dosing. Each stool was collected in a plastic vessel containing a one liter volume of 0.5 mol KOH/L H_2O and the mixture was stirred with a magnetic stirrer for 6 h at room temperature (23°C). The material was dispersed with a spatula periodically during the first two hours of stirring. Two aliquots (10 mL and 40 mL) of the suspension were placed in plastic screw cap tubes and stored at –20°C. An additional 1 mL aliquot was placed in microcentrifuge tubes for transport to LLNL. The remainder of the suspension was discarded.

Laboratory Analysis

Total Carbon and Folate A 75 μL aliquot of each thawed specimen was lyophilized and total carbon was measured with a Carlo Erba carbon/nitrogen analyzer (Pella 1990) at the University of California, Davis, Division of Agricultural and Natural Resources Laboratory. Concentrations of total folate were measured at the University of California, Davis, Clinical Nutrition Research Unit using the *Lactobacillus casei* assay (Tamura, 1990).

Tracer Analysis Triplicate aliquots of each whole blood (20 μL), plasma (20 μL), lysed erythrocyte (20 μL), urine (100 μL) and feces supernatant extract (50 μL) were placed in individual quartz tubes. The aliquots were dried under vacuum, combusted, reduced to graphite (Vogel, 1992), and measured for ^{14}C using AMS as described by Vogel and Turteltaub (1992) and Creek et al. (1994). AMS measures the abundance of ^{14}C relative to total carbon. Naturally occurring ^{14}C (1.2 parts/trillion or 1.2 parts in 10^{12} parts) was subtracted from the measured isotope ratio. Excess isotope concentration was converted to folate equivalents (parent compound and all metabolites) using the specific activity of the dosed folic acid, its molecular weight, and the carbon content of the tissue. Our results are expressed as pmol or nmol ^{14}C-folate/g tissue-carbon inferred from the ^{14}C/total carbon measurements. The nmol ^{14}C-folate (including catabolites) per g urine-carbon times grams total carbon/d in urine was used to determine the nmol ^{14}C-folate loss in urine; similar calculations were made for feces.

Only ^{14}C atoms are counted by AMS and, because it is a combustive process, no molecular information is obtained. Thus, radiocarbon levels in the specimens represent the ^{14}C in both the folate and its catabolites, ρ-aminobenzoylglutamate and its acetamido derivative (McPartlin et al., 1992). Because the ^{14}C-folic acid was prepared as pteroyl-^{14}C(U)-glutamic acid, the radiolabel would not be separated or recycled into other biomolecules during metabolism. Radiocarbon levels are thus assumed to represent the amount of ^{14}C-folate and ^{14}C-folate-catabolites, ^{14}C-ρ-aminobenzoylglutamate and its labeled acetamido derivative, present in the specimens.

Empirical Modeling

The plasma ^{14}C-folate, urine ^{14}C-folate, and feces ^{14}C-folate concentration-time data were described using an empirical multiexponential description of the data by means of a weighted, nonlinear least squares regression using the PC-SAS NLIN procedure

(*SAS/STAT® User's Guide*, v. 6, 4th Edition, SAS Institute Inc., Cary, NC). Each observation was weighted by the reciprocal of its predicted value.

The area under the concentration-time curve (AUC) was calculated as AUC = $\int_0^\infty y(t)dt$, and the area under the moment curve (AUMC) was calculated as AUMC = $\int_0^\infty ty(t)dt$ where y(t) describes the tissue concentration as a function of time. The AUMC/AUC is a commonly used means to estimate the mean sojourn time (MST) which may be related to half-life values using compartmental models.

RESULTS

Subject and Diet

Metabolism of the physiologic 80 nmol (100 nCi) oral dose of ^{14}C-folic acid (1/6 the RDA) in water was determined from the ^{14}C-folate concentration profiles in serial plasma, erythrocyte, urine, and feces over an 202-d study period in a healthy adult male using AMS. The study subject kept a daily log of his diet and activities. He was also a creature of habit. This trait enabled him to adhere to a standard diet on a very regular schedule throughout the 202-d study period. His folate intake was 275 μg/d (624 nmol/d). His diet had 20% fat (ether extract) and 17.6% protein (Kjeldahl-N times 6.25).

These regular habits also enabled the subject to adhere to a schedule for collecting blood, urine, and feces samples. The estimated daily intake of dry matter, fat, protein, and carbohydrate were 636 g, 127 g, 112 g, and 387 g, respectively. The mean intake of dietary folate was 275 μg/d (624 nmol/d). The mean daily intake of carbon was 315 g/day (127g fat*0.76 + 112g protein*0.5 + 387g carbohydrate*0.42).

The total body carbon mass of the study subject was estimated to be 20.7 kg (23% of body mass). Mean (± SEM) carbon loss in feces was 9.8 ± 0.7 g per 24-h day. Mean daily urine carbon loss was 9.6 ± 0.2 g. Mean concentration of carbon (C) in all plasma samples taken during the study was 4.42 ± 0.02 g/100 mL. Carbon concentration of all washed erythrocyte preparations was 6.86 ± 0.23 g/100 mL. Mean hemoglobin concentration in all erythrocyte preparation was 11.7 ± 0.4 g/100 mL. Mean carbon concentration in all urine collections was 0.68 ± 0.03 g/100 mL urine. Mean carbon concentration in all homogenized feces preparations (suspensions) was 1.23 ± 0.06 g/100 mL. There were no significant changes or trends by time since dosing in the mass of C excreted in urine or feces (C mass = C concentration × volume/mass). There were also no significant changes in the hematocrit or hemoglobin values due to the volume of blood drawn. Finally, there were no significant trends in plasma or erythrocyte total folate concentration by time since dosing.

Sensitivity and Precision for Tracer Analysis

The AMS detection limits for plasma, urine, and feces were 0.4, 0.04, and 0.12 fmol ^{14}C-folate/mL, based on double the uncertainty in the background. These respective concentrations correspond to absolute detection limits of 0.008, 0.004, and 0.006 fmol ^{14}C folate in neat plasma, urine, and feces preparations.

Time Course Plots

The time since dose plots of the ^{14}C-folate (including catabolites) are summarized in Figure 1. Peak labeling in plasma occurred at one to two hours after dosing. There was a 4-day delay before the ^{14}C-folate appeared in erythrocytes, and peak labeling of this tissue occurred at ~25 d after dosing. Cumulative urine ^{14}C-folate loss (including catabolites) through the first 43 d of the study was 6.6 nmol. Cumulative feces ^{14}C-folate loss (including catabolites) through the first 43 d of the study was 11 nmol. Data were cumulative through the first 43 d only are presented because collections were not made from d 44 through 64 or from d 76 through 173.

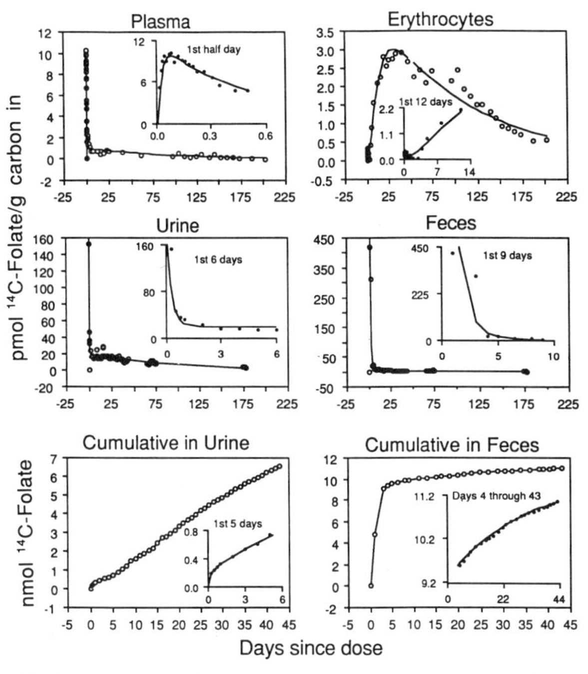

Figure 1. Tissue ^{14}C-folate concentrations by time since ingesting an 80 nmol dose of ^{14}C-folic acid (top four panels). Peak labeling of plasma was at 1–2 h after dosing. Peak labeling of erythrocytes was at ~25 d, after a 3-d delay. Cumulative loss of ^{14}C-folate (+ catabolites) in urine during the first 43 d after dosing are in the bottom left panel. By day 43, 6.5 nmol ^{14}C-folate (+ catabolites) was lost in urine, with 0.53 nmol lost in the first 3 d after dosing. Cumulative loss of ^{14}C-folate (+ catabolites) in feces during the first 43 d are in the bottom right panel. By day 43, 11 nmol was lost by this route. Nine of the 11 nmol was lost in the first 3 d after dosing and was assumed to represent the unabsorbed portion of the 80 nmol dose. Only 2 nmol of absorbed ^{14}C-folate was lost in feces between days 4 and 43, indicating that absorbed folate is finally eliminated mainly in urine despite underlying extensive biliary recycling..

^{14}C-Folate Concentration Kinetic Profiles

Plasma The plasma ^{14}C-folate concentration-time data (top left panel) were described by the three-term exponential equation below.

$$y(t) = -13.89e^{-48.43t} + 10.70e^{-2.078t} + 1.03e^{-0.010t} \quad (1)$$

The AUC, AUMC, and MST for plasma ^{14}C-folate were 105 pmol*day/g plasma-carbon, 9679 pmol*day^2/g plasma-carbon and 93 days, respectively.

Erythrocyte The erythrocyte ^{14}C-folate concentration-time data (top right panel) were described by the following three-term exponential equation.

$$y(t) = -6.64e^{-0.100t} + 2.43e^{-0.315t} + 4.44e^{-0.010t} \quad (2)$$

The AUC, AUMC, and MST for erythrocyte ^{14}C-folate were calculated to be 409 pmol*day/g erythrocyte-carbon, 48613 pmol*day^2/g erythrocyte-carbon and 120 days, respectively. There was a four day delay after administering the dose before ^{14}C-folate labeled erythrocytes appeared in the circulation.

Urine and Feces A three-term equation was not feasible for urine and feces, as the peak occurred at the first post-dose collection. Therefore, the urine ^{14}C-folate concentration-time data (Fig. 1, middle left panel) were described with the two-term exponential equation below. Because urine and feces may not represent pools, MST may not be a meaningful concept and is not reported for these specimens.

$$y(t) = 179.76e^{-3.630t} + 20.33e^{-0.011t} \quad (3)$$

The AUC and AUMC were calculated to be 1856 pmol*day/g urine-carbon and 160622 pmol*day^2/g urine-carbon, respectively.

The feces ^{14}C-folate concentration-time data (middle right panel) were also described with the two-term exponential equation below.

$$y(t) = 1571e^{-0.966t} + 6.91e^{-0.016t} \quad (4)$$

The AUC and AUMC were calculated to be 2058 pmol*day/g feces-carbon and 28690 pmol*day^2/g feces-carbon, respectively.

Cumulative Loss and Mass Balance of ^{14}C-Folate

Urine The cumulative loss of ^{14}C-folate (including catabolites) in urine and feces are plotted in the bottom two panels of Figure 1. Even though the cumulative urine ^{14}C-folate (including catabolites) loss by time was exponential, this loss through the first 43 d after dosing could be approximated by the simple linear regression equation below.

$$y = 0.101 + 0.156t, R^2 = 0.997 \quad (5)$$

By day 44, 6.6 of the 80 nmol dose of ^{14}C-folate (including catabolites) was lost in urine.

Feces Change in the cumulative fecal ^{14}C-folate (including catabolites) loss by time since dosing was described by the two-term exponential equation below; AUC, AUMC, and MST are not appropriate for cumulative data.

$$y(t) = 11.62 - 10.23e^{-1.562t} - 2.31e^{-0.032t} \quad (6)$$

By day 44, 11 of the 80 nmol dose of ^{14}C-folate (including catabolites) was lost in feces; 9 of the 11 nmol were lost in first 3 d and represented unabsorbed ^{14}C-folate (including catabolites).

DISCUSSION

Study Subject and Radiation Dose

The volunteer received an radiation effective dose of only 1.1 mrem. This exposure compares to the 150-300 mrem/y natural background radiation exposure in the US. The low levels of ^{14}C used in the present study requires careful selection of prospective study subjects, fastidious laboratory practices, and a dedicated laboratory to avoid contamination of the specimens with exogenous ^{14}C. Preferred study volunteers would not have worked where ^{14}C-radioisotopes had been used. Generated laboratory waste may legally be classified non-radioactive in many cases (10 Consolidated Federal Register 20.2005), and this can have significant cost savings.

The subject's trait of regularity enabled him to adhere to a schedule throughout the study, stay in a steady state, and avoid significant changes or trends by time since dosing in nutrient intake. Because the volume of blood drawn altered neither the hematocrit nor hemoglobin values, and there were no trends in plasma or erythrocyte total folate concentrations by time since dosing, the blood loss appeared to have no significant effect on either the general health or folate metabolism of the study subject.

AMS Characteristics

AMS is a special isotope ratio mass spectrometer method that measures the abundance of ^{14}C relative to total carbon. The mass of ^{14}C was converted to folate equivalents using a ^{14}C-folate specific activity of 1.25 Ci/mol. Because tissue folate was not isolated prior to AMS, and because the official *L. casei* assay for folate suffers from difficulties with precision, the results of the present study are expressed as pmol or nmol ^{14}C-folate/g tissue-carbon rather than pmol or nmol ^{14}C-folate/g tissue-folate. A future study will be needed to compare both ways of presenting our results. The nmol ^{14}C-folate (including catabolites) per g urine-carbon times g total carbon/d in urine gave nmol ^{14}C-folate loss in urine daily: similar calculations were made for feces. In summary, the abundance of ^{14}C relative to total carbon must be multiplied by the total carbon in urine (and feces) to quantify urine and stool loss of ^{14}C-folate (including its catabolites).

Kinetic Profiles and Mass Balance of ^{14}C-Folate

Feces The ^{14}C-folate (including catabolites) in the first two stools (collected over the first 3 d after dosing) represent the unabsorbed portion of the 80 nmol dose. During this period, 9 of the 80 nmol ^{14}C-folate was recovered in feces. Mass balance calculations for the ^{14}C-folate over this period indicated that 9 (11%) of the 80 nmol administered bolus of ^{14}C-folic acid was not absorbed. The remaining 71 nmol (89%) represents the absorbed portion. The 89% value compares well with the 90% and 92% values already reported by Butterworth et al. (1969) and Krumdieck et al. (1978). The ~90% value is also consistent with the widely held view that fully oxidized folic acid is almost completely absorbed.

By day 44 (through day 43) of the present study, 11 nmol (the unabsorbed 9 + only 2 more) of ^{14}C-folate (including catabolites) was recovered in feces. Since only two more nmol ^{14}C-folate including catabolites (2.8% of the absorbed dose) was recovered in feces over the 41 days, it can be concluded that feces was a minor route of elimination of absorbed folate under the conditions of the present experiment.

Urine Of the 71 nmol ^{14}C-folate that was absorbed, 1.0 nmol (1.5% of the absorbed dose or 1.3% of the administered dose) was recovered in urine in the first 6 d since dosing. This 1.5% value compares to the 3.9% value reported by Krumdieck et al. (1978). At the same time, the 1.3% value is much smaller than the 44% value reported by Butterworth et al. (1969). The difference is probably related to the differences in the mass of administered ^{14}C-folate between the two studies. Butterworth et al. (1969) administered a 10 μmol dose of folic acid, whereas the present study administered only 80 nmol. It is widely accepted that the portion lost in urine is directly proportional to the mass of the administered dose. Also, the 44% value may be high because the study subject suffered from chronic lymphatic leukemia. By day 44 of the present study, the cumulative loss of ^{14}C-folate (including catabolites) in urine was 6.6 nmol (9.3% of the absorbed dose). This is ~3 times the amount lost in feces over the same period.

The enterohepatic circulation of folate plays an important role in human folate homeostasis (Steinberg, 1984). Large amounts of methylated (conjugated) folates are excreted into bile and then reabsorbed into the systemic circulation from the intestine. The usual enterohepatic circulation involves excretion of conjugated folate, hydrolysis of the conjugates in the intestine, and reabsorption of the parent folate. It is interesting that absorbed folate, like other compounds such as fenofibrate (Weil et al., 1988), is finally eliminated mainly in the urine despite undergoing extensive biliary cycling.

The cumulative loss of ^{14}C -folate (including catabolites) in urine and feces through the first 43 d was 6.6 and 11 nmol, respectively. This translates to an average daily loss of ~0.15 nmol/d in urine and ~0.05 nmol/d in feces, for a total loss of ~0.2 nmol/d. Seventy-one of the 80 nmol dose was absorbed. By extrapolating beyond the first 43 days following dosing, and using the mass balance of ^{14}C -folate (including catabolites) as a criteria, ≥ 355 d (71/0.2) would be required for the 71 nmol absorbed dose to be eliminated from the body (more about this later under erythrocytes).

Krumdieck et al. (1978) reported $t_{1/2}$s of 31.5 h and 100 d for ^{14}C -folate (including catabolites) in the urine of a 36-year-old woman in remission from Hodgkin's disease who was given 40 μCi ^{14}C -folic acid. The corresponding $t_{1/2}$s of 4.6 h (ln(2)/3.63) and 63 d (ln(2)/0.011) for the man in the present study are based on the exponents of Equation (3). The 4.6 h value in the present study is much shorter than the 31.5 h value from the earlier study. This difference may be related to differences in the density of blood sampling between the two studies. A special effort was made to incorporate high density sampling immediately after dosing in the present study, thus the 4.6 h value is probably more reliable because it is based on a greater number of data points.

The $t_{1/2}$ of 63 d for ^{14}C -folate (including catabolites) in urine in the present study is reasonably similar to the 100 d value already reported by Krumdieck et al. (1978). The apparent difference between the studies may be related to health status and gender of the study subjects. Administration of diphenylhydantoin, a drug known to interact with folate metabolism (Lewis et al., 1995; Carl et al., 1997), to the woman may also be a factor in the earlier study.

The $t_{1/2}$s for ^{14}C -folate (including catabolites) in feces in the present study are estimated to be 17 h (ln(2)/0.966) and 43 d (ln(2)/0.016) based on the exponents in Equation (4). The 43 d value for feces may be compared to the 63 d value for urine.

Plasma The MST of ^{14}C-folate in plasma in the present study was 93 days. Exponents in Equation (1) can be used to calculate the $t_{1/2}$s of the ^{14}C-folate in plasma: 8 h (ln(2)/2.078) and 69 d (ln(2)/0.010). These plasma $t_{1/2}$s are similar to the urine $t_{1/2}$s and may suggest that the folate metabolism of these tissues may be related to one another even though the major labeled elements in urine are ^{14}C-ρ-aminobenzoylglutamate and its acetamido derivative, while those in plasma would be ^{14}C-5-methyltetrahydrofolate. Similar $t_{1/2}$s for plasma and urine might be expected because plasma concentration determines glomerular filtration rate, about one half of filtered folate is reabsorbed, and folate excretion is independent of urine volume (Chanarin and Bennett, 1962).

Erythrocyte As new erythrocytes are needed, erythropoietin spurs the erythroid colony-forming units to produce a proerythroblast that makes hemoglobin, incorporates folate, undergoes four rounds of cell division over 4 days to make 16 cells that eject their nuclei, shrink and enter the blood stream. The 4-d delay before ^{14}C-folate appears in circulating erythrocytes in the present study is consistent with the role of folate during maturation (rather than development) of erythroid progenitor cells (Bills et al., 1992).

The MST of ^{14}C-folate in erythrocytes in the present study was ~120 d, even though Equation (2) was a suboptimal fit. That value coincides well with the life-span of erythrocytes and is consistent with the concept that folate acquired initially as erythroid progenitor cells are maturing is polyglutamated and unable to escape for lack of γ-glutamyl hydrolase. The 120-day value is also consistent with the 17-week-period needed to drop erythrocyte folate concentration of a healthy adult man to below normal (i.e., to make anemic) by feeding a diet very low in folate (Herbert, 1962).

The final exponent in Equation (2) can be used to calculate the approximate $t_{1/2}$ of the ^{14}C-folate in erythrocytes. Using this approach, the long-lived $t_{1/2}$ of the ^{14}C-folate in erythrocytes was 69 d (ln(2)/0.010).

Herbert (1968) determined that 137 d was needed to eliminate 90% of the body folate by extrapolating the rate of erythrocyte folate depletion, and by assuming that the behavior of erythrocyte folate mimicked that of body tissues. As the 90% of body folate would be lost in ~3.5 $t_{1/2}$s, the 137 d period (to eliminate 90% of the body folate) would ≈3.5 $t_{1/2}$s, or one $t_{1/2}$ would be ~39 d long. The ~39 d *versus* ~69 d $t_{1/2}$s of erythrocyte folate in the previous *versus* present study is likely due to differences in the folate nutritional status of the study subjects between the studies

Mass Balance *vs.* Kinetic Profile Data

As indicated above, by extrapolating the mass balance of ^{14}C-folate (including catabolites) beyond the first 43 days following dosing, the whole body turnover time of the absorbed dose might be 355 d (71/0.2) or longer. At the same time, the MST for ^{14}C-folate (including catabolites) based on kinetic profiles in plasma and erythrocytes were 93 d and 120 d, respectively. The difference between MSTs of 93 to 120 d and the ~355 d may be reconciled by appreciating that there are several points (paths) where irreversible losses of administered folic acid (including catabolites) occur during metabolism, and they are not necessarily accounted for when relying on kinetic profiles alone. Because a compartmental physiologic model can account for multiple exit paths of a system, such a model is currently under construction.

Despite the fact that the present study is based on only one adult subject, that shortcoming is offset by the high density of sampling that allowed a more detailed examination of the nature of the kinetic and mass balance patterns of a physiologic bolus of ^{14}C-folic acid. Also, some features of human folate metabolism can be suggested by the present study. These include additional estimates of the MST from kinetic data and an estimate of the long time needed for complete elimination of the absorbed portion of an oral bolus dose using mass balance data.

CONCLUSION

In conclusion, AMS detection allowed us to conduct the longest (202 d) and most detailed physiologic study to date of folate metabolism in a healthy adult human volunteer. Plasma and erythrocyte kinetic profile data, indicate the MST for folate is in the range of 93 to 119 d. Folate mass balance data, indicate that ≥ 350 d are needed for a physiologic oral bolus dose of ^{14}C-folate to clear the body. Pteroyl-[$^{14}C(U)$]-L-glutamic acid with a modest specific activity (1.25 Ci/mol) was suitable for dosing and could be traced for long periods after physiologic doses. The small (80 nmol) oral dose of ^{14}C-folic acid also served as a true tracer because its mass was ≤ 0.15% of the total body folate store of 60 μmol (~25 mg) and did not disturb the dynamics of folate metabolism (Hoppner and Lampi, 1980). For human nutrition, the bioavailability of folate is readily determinable and mass balance data permits quantitative estimates of whole body turnover. AMS and mathematical modeling offer the opportunity to make better estimates of diet folate need, and provide a more scientific basis for adjusting the needs for individuals and populations at increased risk of severe health problems associated with dysfunctional folate metabolism.

ACKNOWLEDGMENTS

Work was performed under the auspices of US Department of Energy at LLNL under contract W-7405-Eng-48. Supported by the University of California CLC Program, LLNL UC-DRD Program, NIH; DK 45939 and DK-53801, and USDA W-143 Regional Research. Total carbon was measured at the UCD-DANR Laboratory. Total folate was measured at the UCD-CNRU laboratory. The authors thank Janet Peerson for conducting the PC-SAS NLIN procedures and our reviewer.

CORRESPONDENCE

Please address all correspondence to
Andrew J. Clifford
Department of Nutrition
University of California
One Shields Avenue
Davis, CA 95616-8669
Ph 530-752-3376, fax 530-752-8966
ajclifford@ucdavis.edu

REFERENCES

Bills ND; Koury MJ; Clifford AJ; Dessypris EN. Ineffective hematopoiesis in folate deficient mice. Blood, 1972, 79:2273-2280.

Butterworth CE; Baugh CM; Krumdieck CL. A study of folate absorption and metabolism in man utilizing carbon-14-labeled polyglutamates synthesized by the solid phase method. J Clin Invest, 1969, 48:1131-1142.

Carl GF; Hudson FZ; McGuire BS. Phenytoin induced depletion of folate in rats originates in liver and involves a mechanism that does not discriminate folate form. J Nutr, 1997, 127:2231-2238.

Chanarin I; Bennett M. The disposal of small doses of intravenously injected folic acid. Brit J Haematol, 1962, 8:28-35.

Creek MR; Frantz CE; Fultz E; Haack K; Redwine K; Shen N; Turteltaub KW; Vogel JS. C-14 AMS quantification of biomolecular interactions using microbore and plate separations. Nuc Inst & Meth, 1994, B92:454-458.

Herbert V. Experimental nutritional folate deficiency in man. Trans Assoc Am Phys, 1962, 75:307-320.

Herbert V. Nutritional requirements for vitamin B12 and folic acid. Am J Clin Nutr, 1968, 21:743-752.

Hoppner K; Lampi B. Folate levels in human liver from autopsies in Canada. Am J Clin Nutr, 1980, 33:862-864.

Krumdieck CL; Fukushima K; Fukushima T; Shiota T; Butterworth CE. Long-term study of the excretion of folate and pterins in a human subject after ingestion of 14C-folic acid, with observations on the effect of diphenylhydantoin administration. Am J Clin Nutr, 1978, 31:88-93.

Lewis DP; Van Dyke DC; Willhite LA; Stumbo PJ; Berg MJ. Phentoin-folic acid interaction. Ann Pharmacotherapy; 1995, 29:726-735.

Ma J; Stampfer MJ; Giovannucci E; Artigas C; Hunter DJ; Fuchs C; Willett WC; Selhub J; Hennekens CH; Rozen, R. Methylenetetrahydrofolate reductase polymorphism, dietary interactions, and risk of colorectal cancer. Cancer Res, 1997, 57:1098-1102.

McPartlin J; Courtney G; McNulty H; Weir D; Scott J. The quantitative analysis of endogenous folate catabolites in human urine. Anal Biochem, 1992, 206:256-261.

Pella E. Elemental organic analysis. Am Lab, 1990; 22:116-125.

Perloff BP; Butrum RR. Folacin in selected foods. J Am Diet Assoc, 1977, 70:161-172.

Plante LT; Williamson KL; Pastore EJ. Preparation of folic acid specifically labeled with carbon-13 in the benzoyl carbonyl. Meth Enzym, 1980, 66: 533-535.

Rosenberg IH; Hachey DL; Beer DE; Klein PD. Proceedings of the First International Conference on Stable Isotopes in Chemistry, Biology and Medicine. Klein PD; Peterson SV; Eds. USAEC Conference Series # 730525. USAEC, Office of Information Services, Technical Information Center: Springfield, VA. 1973, pp. 421-427.

Shane B. Folylpolyglutamate synthesis and role in the regulation of one-carbon metabolism. Vit Horm. 1989, 45:263-333.

Shane B. Folate metabolism, in: Folic Acid Metabolism in Health and Disease. Piciano MF; Stokstad ELR; Gregory JF; Eds. Wiley-Liss, Inc.: New York. 1990, pp. 65-78.

Shane, B. Folate chemistry and metabolism, in: Folate in Health and Disease. Bailey LB; Ed. Marcel Dekker, Inc.: New York. 1995, pp. 1-22.

Steinberg SE. Mechanisms of folate homeostatsis. Am J Physiol, 1984, 246: G319-G324.

Stites TE; Bailey LB; Scott KC; Toth JP; Fisher WP; Gregory JF III. Kinetic modeling of folate metabolism through use of chronic administration of deuterium-labeled folic acid in men. Am J Clin Nutr, 1997, 65:53-60.

Stokstad ELR. Historical perspective on key advances in the biochemistry and physiology of folates, in: Folic Acid Metabolism in Health and Disease. Piciano MF; Stokstad ELR; Gregory JF; Eds. Wiley-Liss, Inc.: New York. 1990, pp. 1-21.

Tamura T. Microbiological assay of folates, in: Folic Acid Metabolism in Health and Disease. Piciano MF; Stokstad ELR; Gregory JF; Eds. Wiley-Liss, Inc.: New York. 1990, pp. 121-137.

Vogel JS. Rapid production of graphite without contamination for biological AMS. Radiocarbon, 1992, 34:344-350.

Vogel JS; Turteltaub KW. Biomolecular tracing through accelerator mass spectrometry. Trends Anal Chem, 1992, 11:142-149.

Vogel JS; Turteltaub KW. Accelerator mass spectrometry as a bioanalytical tool for nutrition research. Adv Exptl Med Biol, 1998; this volume.

Wagner C. Biochemical role of folate in cellular metabolism, in: Folate in Health and Disease. Bailey LB; Ed. Marcel Dekker, Inc.: New York. 1995, pp. 23-42.

Weil A; Caldwell J; Strolin-Benedetti M. The metabolism and disposition of fenofibrate in rat. Drug Metab Dispos, 1988; 16:302-309.

HUMAN ZINC METABOLISM:
ADVANCES IN THE MODELING OF STABLE ISOTOPE DATA

Leland V. Miller, Nancy F. Krebs, and K. Michael Hambidge

Center for Human Nutrition
University of Colorado Health Sciences Center
Denver, CO 80262

ABSTRACT

Compartmental modeling is a useful tool for investigating metabolic systems and processes. We and others have applied it to the study of zinc metabolism in humans. Because existing models could not be accurately fitted to our data, we have developed a new model of human zinc metabolism based on stable isotope tracer data from studies of five healthy adults. Multiple isotope tracers were administered orally and intravenously and the resulting enrichment measurement in plasma, erythrocytes, urine, and feces. These tracer kinetic data, along with other measured and calculated tracee and steady-state data, were used to develop the model. A single model structure composed of fourteen compartments was found to be suitable for all subjects. Model development and fitting of data and model for each subject were accomplished using the SAAM/CONSAM computer programs. The model development and fitting processes are described and exemplified using data from one of the subjects. While identifiability could not be demonstrated *a priori* due to the model's complexity, parameter statistics for the fitted models did show most parameters to be adequately identified *a posteriori*.

INTRODUCTION

Although zinc is well-established as a micronutrient of major practical importance in human nutrition, our understanding of how to prevent or to detect and treat human zinc deficiency continues to be hampered by our limited knowledge of zinc metabolism under various dietary circumstances and host conditions. To better understand human zinc metabolism, we and other investigators have applied various mathematical models to tracee and isotope tracer data, in order to derive information on zinc balance, absorption, etc. There is well-understood value in the development of a whole-body model that integrates and accounts for all the various pieces of currently available information, as well as

Mathematical Modeling in Experimental Nutrition
Edited by Clifford and Müller, Plenum Press, New York, 1998

provides additional information regarding aspects of the metabolic system which cannot be directly observed or measured (Novotny et al., 1996).

Given our current knowledge and assumptions regarding biological organisms and their representation, the most appropriate mathematical conceptualization is that of the compartmental model. Several compartmental models of human zinc metabolism have been published (Fairweather-Tait et al., 1993; Foster et al., 1979; Jackson et al., 1984; Lowe et al., 1993; Lowe et al., 1997), most notably that of Wastney (Wastney et al., 1986). Development of Wastney's model was based on data from radioisotope studies of normal and dysfunctional populations over both short- and long-term study periods. This well-founded model provides particularly detailed information on the distribution and storage of zinc in the body, unattainable with the use of stable isotopes in living subjects. Nonetheless, its representation of the system-intestinal interchange and excretory pathways is not adequate to serve our research interests. Furthermore, our stable isotope tracer data are not accurately fitted by this model.

For these reasons and others, e.g. our interest in investigating the characteristics of a minimally complex model adequate to fit/describe our tracer and tracee data, we elected to develop our own compartmental model of human zinc metabolism. It was perceived that the resulting model would make numerous contributions to our various zinc research interests and goals. It would, for example, (1) perform an explanatory and evaluative role in the examination of steady-state parameter calculation methods, e.g. that for estimating the mass of the rapidly exchanging zinc; (2) constitute a preliminary phase in the ongoing development of a representative normal population model having general predictive as well as comparative value when other populations are studied; (3) contribute to the validation of improved indices of zinc nutritional status; and (4) form the basis for more detailed modeling of the interchange between the system and the small intestine using data from extensive multi-site sampling in current tracer studies of zinc homeostasis and transport in the small intestine.

Our development of a compartmental model of zinc metabolism was based on data from studies of adults in which multiple isotope tracers were administered orally and intravenously, and the resulting simultaneous enrichments measured in plasma, erythrocytes, urine, and feces. These enrichment data were fitted to a consistent compartmental model using the SAAM ("Simulation, Analysis, and Modeling") and CONSAM ("Conversational SAAM") computer programs (Berman and Weiss, 1978; Berman et al., 1983; CONSAM User's Manual, 1990). Additional information, including calculations of fundamental tracee steady-state parameters derived from algebraic and graphical calculations and measurements of tracee intake and excretion, were accounted for in the fitting process. After extensive manual fitting, we implemented more refined data-weighting schemes and other statistical constraints to facilitate computer fitting.

Throughout this exposition, *manual fitting* refers to the efforts of a human modeler using SAAM/CONSAM to effect an optimum fit of model and data by the manual adjustment of parameters and variables and interpretation of results. *Computer fitting* in this instance refers to the use of SAAM's iterative parameter adjustment algorithms to produce the best fit to the data. During the development and fitting processes model and parameter identifiability issues were examined. We are also using the Extended Multiple Studies Analysis (EMSA) facility of SAAM to produce an aggregate of the individual subject models and a single set of population statistics.

MATERIALS AND METHODS

Subjects

Five healthy adults (four females and one male) have been studied thus far. All subjects gave informed consent. None of the women were pregnant or lactating. None were taking mineral supplements. Data from one representative female subject (Subject 1) and the process of its analysis and modeling will be described and discussed here. Subject 1 was 33 years of age, with a body weight of 47 kg and a body mass index of 19 kg/m^2.

Diet

Diet records were obtained and a constant daily diet typical of normal eating habits was planned for ingestion during the study. Subject 1 was calculated to have a typical dietary zinc intake of 7.0 mg/d.

Tracer Administration

On the first day of the study, 0.446 mg of ^{70}Zn-enriched zinc was administered intravenously and 1.00 mg of ^{67}Zn-enriched zinc was administered orally while the subject was in the post-absorptive (fasting) state prior to breakfast. On the second day, 3.07 mg of ^{68}Zn-enriched zinc in water was given orally with the three main meals of the day. This tracer was sipped at regular intervals during the meals, with the quantity of isotope given with each meal being in proportion to the natural zinc content of the meal. Details of the preparation and administration of the isotopically enriched material have been described elsewhere (Miller et al., 1994). Due to inferior ^{68}Zn isotope ratio measurement sensitivity and blood sampling frequency limitations on the second day, this protocol was subsequently modified to give the ^{67}Zn tracer with meals on the first day and to eliminate use of the second oral tracer. Although this change resulted in improved data on tracer absorption with meals, the earlier studies, exemplified here by Subject 1, provided data on both post-absorptive and prandial (with meals) absorption processes.

Sample Collection, Preparation, and Analysis

Baseline blood, urine, and fecal samples were collected and the hematocrit measured prior to tracer administration. Frequent blood sampling was begun immediately after administration of the IV tracer. The initial sampling rate was every two minutes, with the interval increasing to eight hours by the end of the first 24 hours, and finally to daily sampling after two days until the completion of the study. Blood samples were separated into their plasma and erythrocyte components immediately after collection. Complete quantitative collection of urine and feces began at the time of tracer administration and continued until the end of the study. While the duration of the study of Subject 1 was nine days, the study period was lengthened in later studies. After samples were prepared for analysis, zinc concentrations and total zinc content of samples were determined using atomic absorption spectrometry, and isotope enrichment was measured by fast atom bombardment-secondary ion mass spectrometry. Details of the sample collection, storage, preparation and analysis procedures have been described elsewhere (Krebs et al., 1995; Miller et al., 1994).

Data Calculations and Analyses

All our isotope enrichment values and calculations are based on the following definition:

$$^{x}\text{Zn enrichment} = \frac{\text{total mass of Zn from } ^{x}\text{Zn enriched souce}}{\text{total mass of Zn from all natural and enriched sources}} \quad (1)$$

Total dietary zinc and total zinc excreted in urine and feces were calculated as per-day rates. The mass of circulating plasma zinc was estimated by two methods and the average of the two results was used. First, the y-intercept (enrichment at t = 0) of the sum of exponentials function best fitting the initial IV tracer enrichment data from the plasma was determined and the plasma zinc mass calculated as

$$\text{Plasma Zn} = \frac{\text{Dose}}{\text{Enrichment}_{t=0}} - \text{Dose} \quad (2)$$

In the second method, the mean measured plasma zinc concentration was multiplied by the estimated plasma volume. This estimated value was the mean of the results of five different calculations relating plasma volume to gross physical characteristics (height, weight, body surface area) for normal adult women (Geigy Scientific Tables). The mass of circulating erythrocytic zinc was estimated in the same manner, i.e., the mean measured erythrocyte zinc concentration was multiplied by the estimated erythrocyte volume. Again, this volume estimation was the mean of the results of different calculations of erythrocyte and whole blood volumes, where the whole blood volume values were then multiplied by the measured hematocrit to derive additional erythrocyte volume estimates. When erythrocyte zinc concentration measurement data were not available, a normal population value from the literature (Geigy Scientific Tables) was used. Whole body zinc mass was estimated as 20 mg per kg of whole body mass.

Several important tracee steady-state parameters were also calculated using previously described algebraic, graphical, and other methods. Fractional absorption (FAZ) of oral tracers was calculated using four mostly independent methods:

- Measurement of the excretion of oral tracer in the feces (Krebs et al., 1995; Van Dokkum et al., 1996).
- Measurement of the ratio of IV and oral tracer enrichment in urine (Friel et al., 1992; Krebs et al., 1995; Van Dokkum et al., 1996).
- Measurement of the ratio of IV and oral tracer enrichment in plasma (similar to urine method).
- Deconvolution of the appearance of oral tracer in the plasma.

The deconvolution method utilizes the convolution integral

$$r(t) = \int_0^t u(\tau)\, w(t-\tau)\, d\tau$$

where $w(t)$ is the observed plasma response to the IV tracer bolus input, $u(t)$ is the rate of absorbed oral tracer input into the plasma and $r(t)$ is the observed plasma response to the oral tracer input. Because $w(t)$ and $r(t)$ are both measured, $u(t)$ can be obtained by deconvolution of this integral (Carson et al., 1983; Jacquez, 1985; Shipley and Clark, 1972). Fractional absorption is then derived from $u(t)$. We performed the deconvolution using SAAM as described in the SAAM Users Manual (Berman and Weiss, 1978; Berman, 1978).

Fecal excretion of endogenous zinc (EFZ) was determined by measuring excretion of IV tracer in the feces, taking into account the average IV tracer enrichment in the plasma or urine during the same time period (Krebs et al., 1995). The size of the "pool" of zinc that exchanges rapidly (i.e., within several days) with the plasma (EZP) was estimated using the y-intercept ($t = 0$) of a regression of the IV tracer enrichment data (from plasma or urine) between three and ten days after tracer administration (Miller et al., 1994).

The development of the model structure and the initial manual fitting of the data and model were accomplished using CONSAM. The computer fitting of data and models was done primarily in batch-mode SAAM. Occasionally other software packages were also used to perform sum of exponentials analysis (exponential curve stripping and nonlinear regression), other linear and nonlinear regression analyses, and the graphical presentation of data: RSTRIP (MicroMath Scientific Software, Salt Lake City, UT) and GraphPad Prism (GraphPad Software, San Diego, CA).

DEVELOPMENT AND MANUAL FITTING OF THE MODEL

Development of Model Structure

While the model structure development and data-model fitting processes are in reality intertwined efforts, we will describe these aspects separately for the sake of clearer exposition. Preliminary model development began with examination of the IV tracer enrichment data from the plasma. Because plasma is the major route of zinc transport in the body (Cousins, 1989), the disappearance of tracer from the plasma after administration of an IV bolus provides the clearest picture of zinc distribution within the body attainable with practical stable isotope techniques. Given our general knowledge of human zinc physiology, e.g., that plasma is the major route of zinc transport, a mammillary-type structure (each compartment exchanges only with a central compartment; Jacquez, 1985) with plasma zinc as the central compartment is the most appropriate starting point.

The next step is to determine how many peripheral compartments are required to fit the tracer data. Sum of exponentials analyses of the data demonstrated that four or five exponential decay terms were required to adequately mimic the data. While five terms appeared to provide superior fit to more than offset the cost of two additional degrees of freedom, such analyses were not taken to be conclusive, given the potential pitfalls accompanying analyses involving large number of exponential terms (Shipley and Clark, 1972). Four compartments in addition to the plasma were, indeed, required to adequately fit the data from Subject 1, as well as that from the other subjects.

The original mammillary structure was then modified to that shown by the solid lines in Figure 1. This was required to make the modeling results consistent with knowledge of the systemic transport and distribution of zinc. The most rapid uptake of zinc circulating in the plasma occurs in the liver (Cousins, 1985; Cousins, 1989), and we know roughly the amount of hepatic zinc (Geigy Scientific Tables; Jackson, 1989). With the pure mammillary structure, the most rapidly exchanging compartment was much smaller than expected for a liver compartment, and too large to be an additional plasma compartment since the plasma zinc mass already calculated was within the expected range for total plasma zinc. Furthermore, *in vitro* investigations have shown that the kinetics of zinc uptake by hepatic cells is two-phase (Cousins, 1985). Therefore, we investigated the rearrangement of two compartments into a catenary structure (each compartment exchanges only with the ones immediately adjacent; Jacquez, 1985), to represent primarily fast and slower liver uptake and storage of zinc. This did in fact result in more realistic sizes for these "liver" compartments. This imposition on the model development and fitting process of an expectation based on independent knowledge of the system being modeled is the first example of numerous informal and formal constraints placed on the model structure and parameter space.

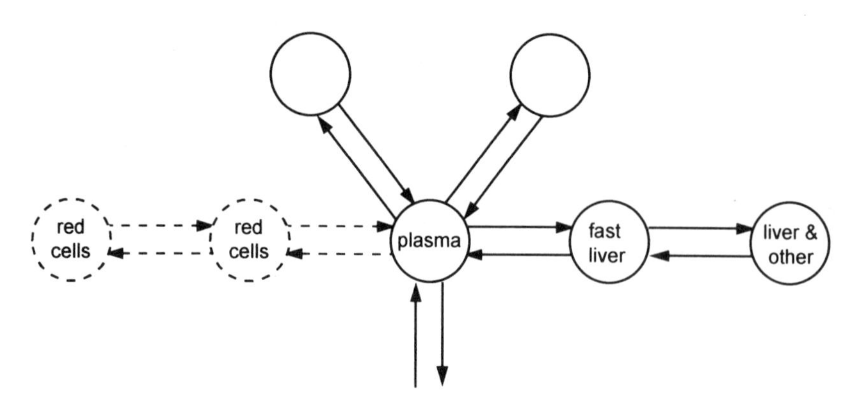

Figure 1. Modeling of IV tracer in plasma and erythrocytes.

None of the existing compartments corresponded to erythrocytic zinc so the modeling of the IV tracer enrichment in erythrocytes involved the addition of (two) more compartments. Because the mass of these two compartments is relatively small, their addition had a minimal effect on the existing modeling of the plasma data. It was evident from the modeling of the erythrocyte data that the two-compartment catenary structure shown in Fig. 1 (dashed lines) was minimally required to produce an adequate fit. This is consistent with *in vitro* investigations of zinc uptake showing that erythrocytes also exhibit two-phase kinetics (Van Wouwe et al., 1990). The calculated total erythrocyte zinc, described above, also provides an additional constraint on the possible fitting solutions.

The two unlabeled compartments, required for fitting the plasma data, are yet to be given physiological designations. The larger and more slowly exchanging of the two probably represents primarily bone and muscle. The other represents unknown tissue which exchanges more rapidly with the plasma.

To the extent that this model of systemic transport and distribution of zinc is appropriate, it should also adequately describe the behavior of absorbed oral tracer that has reached the systemic circulation.

The modeling of urinary and fecal excretion of the IV tracer required additional compartments having obvious correspondence to physiological structures, shown in Fig. 2. Urinary and fecal tracer data are calculated as cumulative excretion of tracer modeled by compartments with no outputs. While the urine excretion modeling is straight-forward, that of the fecal excretion is more involved because the small intestine also plays a role in absorption of dietary zinc. The model of fecal excretion shown in Fig. 2 is the minimal structure that is compatible with the additional compartments and transfer pathways necessary to describe the absorption process. Note that the colon is represented by a (nonmixing) delay element. Also, endogenous zinc losses from the body to the intestine are via the fast "liver" compartment. It was necessary to model it this way to best fit the initial appearance of IV tracer in feces. The same effect could be accomplished by manipulations of the small intestine parameters, but not without compromising the modeling of the absorption and excretion of oral tracer. The liver is a route of zinc transfer to the small intestine, but it is one of several and is relatively minor. Therefore, the fast "liver" compartment has been renamed "liver & other" to acknowledge that it consists of more than just hepatic zinc, e.g. it is likely that it includes some pancreatic zinc.

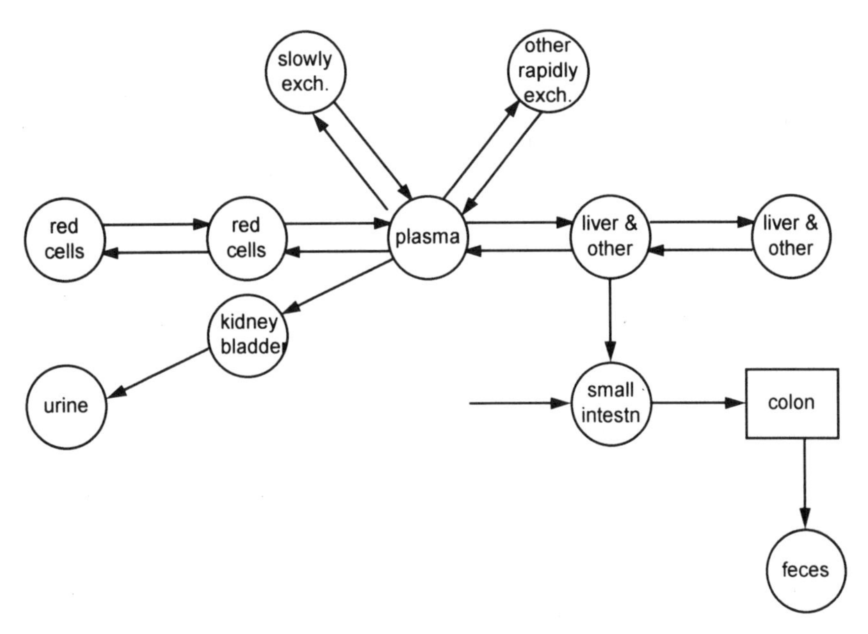

Figure 2. Modeling of IV tracer in plasma, erythrocytes, urine, and feces.

With the inclusion of oral tracer data into the development and fitting process, the quantity of kinetic information is doubled (and more so when the second oral tracer is also considered). This new information provides the basis for completing the modeling of the GI tract and subsequently validates the model structure that has already been developed.

Figure 3 shows the final model structure. The addition of the stomach compartment is necessary to adjust for the delay in initial appearance of oral tracer in plasma. And the existence of two small intestine compartments provides the needed versatility to fit the temporal characteristics of the appearance of oral tracer(s) in the plasma, while at the same time fitting the characteristics of the appearance of the IV and oral tracers in the feces. Again, this structure is supported by the knowledge that zinc absorption occurs at multiple sites in the small intestine (Cousins, 1985; Lönnerdal, 1989). Furthermore, we understand that any endogenous zinc secreted into the intestine may be reabsorbed at a different rate, and at different sites, than dietary (exogenous) zinc. This structure gives us at least limited ability to set different absorption rates and sites for endogenous verses exogenous zinc.

One aspect of zinc absorption that this model does not account for is the transport of absorbed zinc via the portal circulation. Our experiments with adding an absorption route to the liver to reflect zinc from the portal circulation taken up by hepatocytes during "first pass" through the liver demonstrated no improvement in fit of the data to justify the increased complexity. This issue will be revisited in our more detailed investigations of the small intestine.

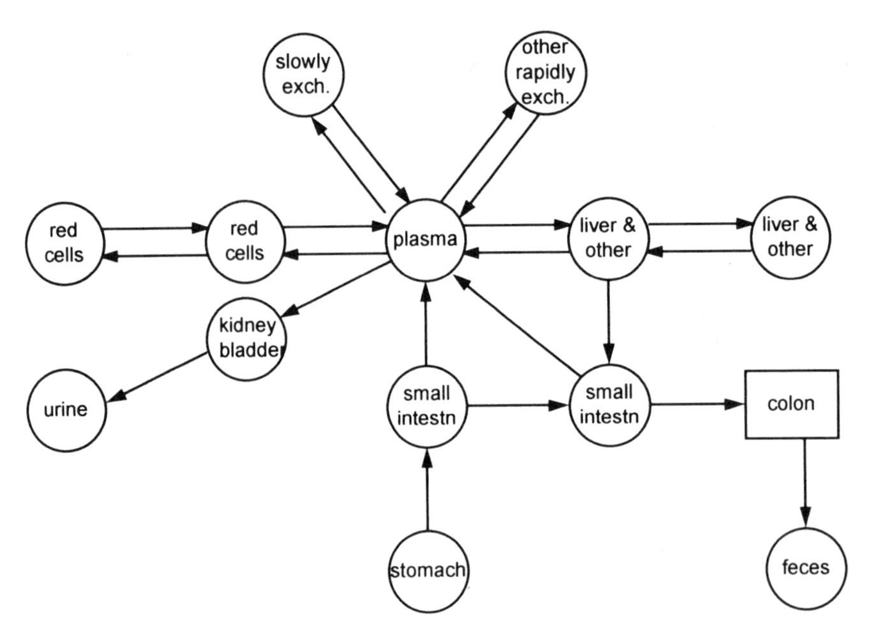

Figure 3. Complete model structure, based on IV and oral tracer data measured in plasma, erythrocyte, urine, and feces.

A Priori Identifiability

With the development and application of a model structure, identifiability becomes a pertinent issue. An important concern is whether, in an effort to develop a model that is capable of accurately fitting our data, we have created a model that is more complex than is justified by the quality and quantity of the available data. A negative consequence of this would be the existence of unidentifiable parameters, i.e., those for which values can not be adequately determined. This would lead to an unidentifiable model. Is it possible, before we model any actual data, to evaluate the proposed model structure and experimental design to determine whether there are any inherent structure-related limitations that would prevent us from meaningfully estimating all the parameters of the model, even with ideal data? This is the issue of *a priori* identifiability, which may also be referred to as *structural* or *theoretical* identifiability (Carson et al., 1983; Jacquez, 1985; Walter, 1987).

There are several easily evaluated yet necessary conditions for *a priori* identifiability. The structure must have the characteristics of input and output connectability (Carson et al., 1983; Cobelli et al., 1979), also called input/output reachability (Jacquez, 1985). That is, administered tracer (inputs) must reach all compartments and that all compartments must have an effect on at least one of the tracer measurement sites (outputs). The model structure and experimental design we have described here do meet these conditions. Another necessary condition is that the number of unknown parameters, $\mathbf{n_u}$, must be less than or equal to the number of relationships, $\mathbf{n_e}$, among those unknown parameters. The determination of $\mathbf{n_e}$ is described by Cobelli (Carson et al., 1983; Cobelli and DiStefano, 1980). Our model and experiment also meet this condition. Though algorithms exist for proving the *a priori* identifiability of simpler models (Carson et al., 1983; Cobelli and DiStefano, 1980; Godfrey and DiStefano, 1987), such a formal proof for a model of this complexity has not been feasible. This situation may be changing, though, as Cobelli has recently discussed a more powerful version of his GLOBI program, perhaps capable of evaluating the *a priori* identifiability of our model (Cobelli, 1997).

Although general *a priori* identifiability of this model has not been demonstrated, and may not be demonstrable practically speaking, the model may still have significant value. Some parameters may be identifiable, and the unidentifiable parameters may not affect the characteristics of the model that are of real interest (Berman, 1982), and/or the whole model may prove to be identifiable, but only for a parameter space restricted by additional information on the system under study. Although several conceptions of limited or contingent identifiability have been described, their formal application to this model is again precluded by its complexity (DiStefano, 1983; Godfrey and DiStefano, 1987; Jacquez and Perry, 1990). With the consideration of real information and data, the discussion of identifiability falls into the realm of *a posteriori* identifiability which will be discussed below.

Manual Fitting and Data and Model–Parameter Estimation

After development of an appropriate model structure, the final fitting of data and model may be viewed as a parameter estimation process. It is by means of the adjustment of the model's parameters (fractional transfer coefficients) and other adjustable variables that the data and model are made to be consistent. The fractional transfer coefficients (FTCs) quantify the flow of material between compartments in terms of fraction of source compartment content per unit of time (days in this case). It is these parameters, as well as certain variables and functions which are dependent on them, which are of fundamental interest to metabolism researchers.

Because the manual fitting process entails a great deal of interaction with the software, it was accomplished exclusively in CONSAM, the conversational interface to SAAM. The fitting of data and a model for Subject 1 involved fitting data from the three tracers to a single common model. In SAAM/CONSAM, this can be accomplished by either of two different methods: solution interrupts or simultaneous analysis of parallel models (Foster et al., 1989). We found it most effective to use the latter method, constructing three separate models, one for each tracer, and establishing all the parameter relationships necessary to make the models essentially identical. The models were then all solved and fitted simultaneously. The three models were identical except for six FTCs describing movement between the small intestine and other compartments. These parameters had to be free to vary separately for the two oral tracers because the kinetics of zinc absorption and related intestinal processes differ depending on whether the zinc is ingested by itself or with other nutrient material. This is most evident from the difference in fractional absorption measurements listed in Table 1. The requirement that data from three separate tracers administered via different routes or under different conditions fit a single model is a rigorous one. If simultaneous fitting of all data is successful, it lends significant credibility to the resulting model and parameter estimations.

While the kinetic data (shown fitted to the model in Fig. 4) contain the most information and are the focus of the modeling process, the consideration of all available additional data and information about the system being modeled is critically important to the success of complex model development. Table 1 shows the measured and estimated tracee steady state data also used in the modeling of Subject 1. These data were used as criteria to evaluate and guide the manual fitting process and most were then used as formal statistical constraints in the computer fitting process, as described below.

All data values are well within normal ranges except the dietary intake, which is considered to be marginally low, and the urinary zinc output, which is also low. Generally, there is very good agreement among results where different methods are used to calculate a single parameter. The exception here is the disparity in the plasma zinc values. We have used these two methods for determining plasma zinc in a dozen studies thus far and observed that they produce similar results (within 10%) in three out of four cases. The difference seen here is, as yet, unexplained and, because we have no basis for attributing greater accuracy to either method, we use the mean of the values, 1.7 mg, as the plasma zinc mass. In SAAM/CONSAM, this value is assigned to the fixed steady-state variable representing the mass of the plasma compartment in the model.

Table 1. Measured and estimated tracee steady state data. Estimated values, i.e. those derived in part from population data or a calculation method known to have limited accuracy, are indicated by an asterisk (*).

Dietary Zn intake	7.0 mg/d
Fecal Zn excretion	7.2 mg/d
Urinary Zn excretion	.062 mg/d
Plasma Zn mass	
from plasma kinetic data	2.1 mg
meas. concentration × est. volume	1.2 mg*
Erythrocyte Zn mass	17 mg*
Whole body Zn mass	930 mg*
Fractional absorption of fasting tracer	
fecal excretion of oral tracer	.77
oral/IV tracer ratio in plasma	.74
oral/IV tracer ratio in urine	.72
deconvolution of oral tracer	.71
Fractional absorption of prandial tracer	
fecal excretion of oral tracer	.31
oral/IV tracer ratio in plasma	.31
Fecal excretion of endogenous Zn	
based on plasma enrichment	2.6 mg/d
based on urine enrichment	2.7 mg/d
Mass of rapidly exchanging Zn	
indicated by plasma enrichment	180 mg*
indicated by urine enrichment	180 mg*

The results of the manual fitting effort are shown in Figs. 4 and 5. Figure 4 shows the kinetic data for the three tracers and the quality of fit to the model, represented by the superimposed curves. The kinetic data and model agree very well except for data from the oral tracer given with meals as measured in the plasma, erythrocytes, and urine. These aberrant data appear to indicate a lower level of absorption of the tracer than the model shows. The inaccuracy of the fit to these data was the cost paid to make the model consistent with all the other data, including the steady state information in Table 1. It is likely that the deviation of these data is related to the inferior analytical accuracy and precision of our ^{68}Zn isotope ratios measurements, particularly in plasma and erythrocytes where the concentration of zinc is low. Because of this analytical limitation, these data were, in effect, given less weight than the other data in the manual fitting process.

Figure 5 shows the final values for the parameters (FTCs and colon delay time), steady state variables (compartment masses and input and excretion flow rates) and parameter-dependent functions (fractional absorption, endogenous fecal zinc, and rapidly exchanging zinc). Compartment masses are based on the pre-determined plasma zinc mass. Note the six FTCs describing the upper GI tract that have two values, as explained above. Values in brackets are for the modeling of the oral tracer administered during fasting. Also note the addition of the (dashed) integumental losses compartment. Though not used here, this feature of our general model provides for integumental and other zinc losses in subjects where data indicate that these are significant. The compartments that are included as part of the rapidly exchanging zinc (EZP) are marked with an asterisk (*). Initially, the criterion for inclusion was based on the FTCs between a compartment and the plasma. Subsequently, the criterion was defined to be any compartment with IV tracer enrichment equal to that in the plasma ± 25% by 72 hours after tracer administration. The data generation feature of SAAM was used to confirm that the originally selected compartments did meet the criterion.

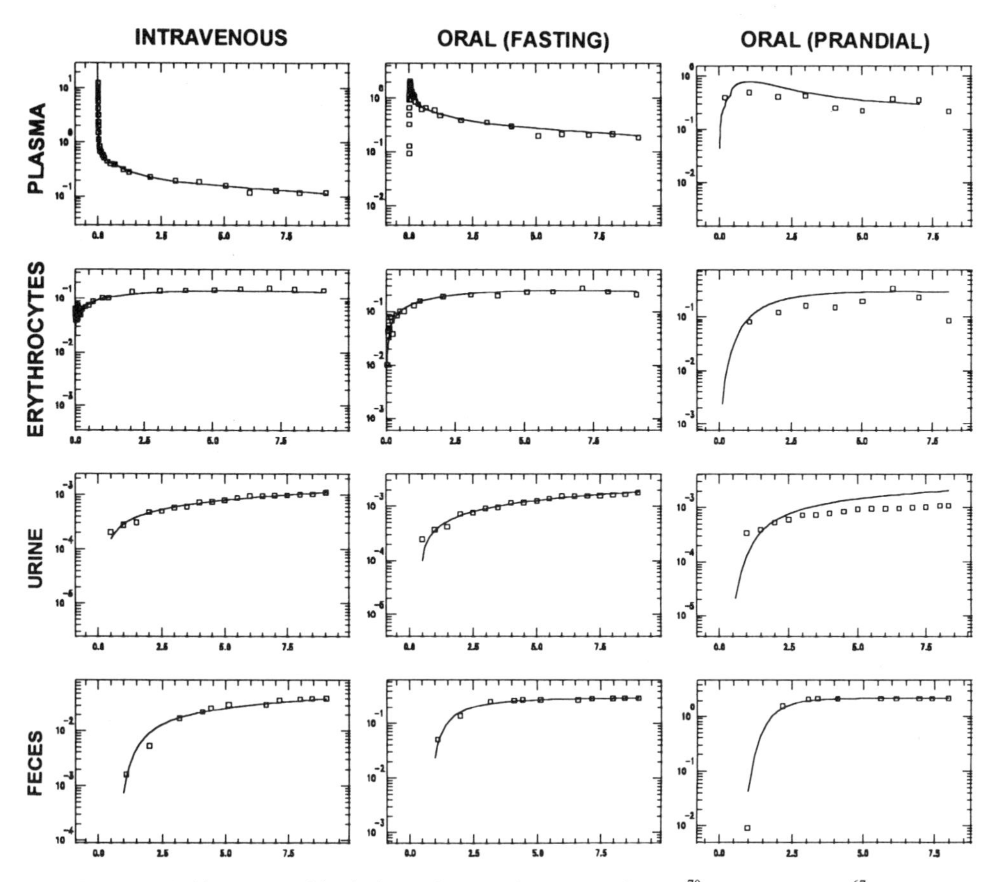

Figure 4. Subject 1 tracer kinetic data. The tracer isotopes used were: ^{70}Zn (intravenous), ^{67}Zn (oral, fasting) and ^{68}Zn (oral, prandial). The plasma and erythrocyte data are in units of % enrichment, and the urine and fecal data are shown as cumulative tracer mass (mg). Time is in days.

All steady-state results of this modeling agree well with the previously measured and estimated steady state data (see Table 1), except for whole body zinc and the rapidly exchanging zinc. The whole body zinc had not been estimated and had not been taken into account during the manual fitting work. The mass of the rapidly exchanging zinc determined by the estimation method is 29% higher than that of the model. This deviation is consistent with hypothetical data based on the Wastney model which predict that the estimation method will overestimate the actual mass of rapidly exchanging zinc by 27% on average (Miller, unpublished data).

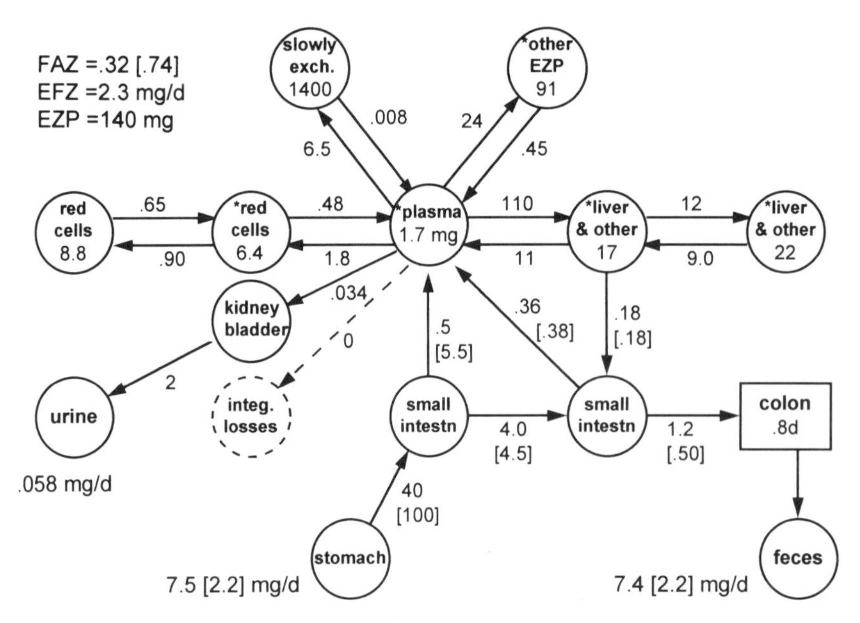

Figure 5. Results of manual fitting of model and data. Fractional transfer coefficients (FTCs) associated with the upper GI tract have two values, one for each oral tracer. The values shown in brackets are for the fasting tracer. All other FTC values apply to all tracers. Various steady-state data for the model are also shown, including intake and excretion rates, fractional absorption (FAZ), endogenous fecal zinc (EFZ), and mass of rapidly exchanging zinc (EZP). All compartment masses are in milligrams. Compartments included in the EZP are indicated with an asterisk.

While such detailed manual fitting is not necessary given the powerful computer fitting capability of the SAAM/CONSAM programs, we thought it important to attain the manual modeling experience and familiarity with the data in the initial studies. Furthermore, the results of this effort provide the initial parameter estimates and other starting point data for the subsequent computer fitting of the data and model.

Computer Fitting of Data and Model

While the results of the manual-fitting effort look very successful in terms of the agreement between the data and model, the process is by its nature quite subjective and we cannot, in the end, claim to have met some objective/quantitative criteria of optimal/best fit, particularly for a model of this complexity. Nor can we provide useful statistical information on our results. These tasks are, of course, what SAAM/CONSAM were designed to accomplish. To perform the computer fitting, we relied primarily on the batch mode operation of SAAM, although the work could also have been accomplished from CONSAM.

While SAAM/CONSAM's iterative fitting of the data and model are exhaustive and objective, the programs nonetheless require additional information beyond the fundamental kinetic and steady state data with which to guide the process. There are three basic types of information used by SAAM/CONSAM to guide fitting: initial parameter values, limits on the permissible range of values of adjustable parameters and variables, and statistical information about the kinetic and other data used as the basis for data weighting.

The initial parameter values provide the basis for the initial solution of the model from which the fitting process begins. The initial values must, themselves, produce a reasonably good fit to the data. The parameter values from our manual fitting more than met this requirement. Determining appropriate limits for parameters and variables is more problematic unless there exists adequate information on possible values. Where this information was not available, we arbitrarily set the parameter limits at ± 50% of the initial values. We also took into account the range of parameter values experienced in the manual fitting of all subjects.

Data weighting is a more involved issue, and is very important to obtaining appropriate results, particularly if data are of different types or have greatly varying magnitude or precision. Kinetic data must be explicitly assigned weights. Typically these are in the form of standard deviations (SDs) or fractional standard deviations (FSDs). When SDs or FSDs are used, SAAM assigns a statistical weight to each datum which is proportional to the inverse of the square of the SD (standard weighted least squares). Because reliable precision data is often unavailable and the SAAM/CONSAM documentation says little about data weighting, the modeler must exercise care in assigning precision or weight values to data. If, as is the case with much analytical data, variance is known or predicted to be proportional to magnitude, the use of FSDs produces the appropriate weighting scheme. This is the method used in most examples and tutorials in the SAAM/CONSAM documentation. Knowing the actual data precision is, obviously, the preferred situation, but, short of that, reasonable estimates are probably adequate. More critical are considerations of relative precision, and weights, within the data.

For instance, although actual precision data was not readily available, we took into account the relative precisions of our enrichment measurements for the different isotopes. Also we found several situations where the use of SDs was preferable. If data forming a plateau are very noisy (random) or the data form a sharp peak, the use of FSDs may cause the fitting of the plateau data to be biased towards the low values or the fitting of the peak to be dampened. If the data being fitted are cumulative, as is the case with our urine and fecal data, the precision of the data may improve as the magnitude increases, making the use of FSDs inappropriate. If data vary greatly in magnitude due to unit differences, for example, care must be taken to insure that the statistical weights assigned to the data are appropriate, particularly in relative terms.

In addition to the kinetic data, one can introduce other data with statistical quantification. These data are also assigned weights and fitted. In this way, additional information may be used as statistical constraints during the fitting process. For this model, we used the data on fractional absorption, endogenous fecal zinc, rapidly exchanging zinc, erythrocyte zinc, and whole body zinc as statistical constraints. Dietary intake data was also informally taken into consideration in evaluating the fitting results.

In evaluating the results of a computer fitting run we examined five kinds of information:

1. Absolute and relative parameter values. Were they reasonable, based on our knowledge? Were any values at the permissible range limits? This is usually not allowed to happen, even though on occasion letting a particular parameter value run to its limit had some utility in terms of the overall result. When a parameter value reaches the limit of its permissible range, some statistical information on the estimation of that parameter cannot be determined, and, as a consequence, there is a gap in the statistical data used in subsequent calculations.
2. Parameter FSDs. The program provides statistical information on the certainty of the parameter estimations. As mentioned in the following discussion of identi-

fiability, parameter values having an FSD of ≤ 0.5 were judged to be adequately estimated.

3. Statistical constraint values. Are they close to the previously measured and calculated data?
4. Sums of squares of residuals from the fitting of data associated with compartments. How have the sums of squares changed? Have any gotten worse?
5. Appearance of data-model plots. Does visual inspection of fit show something not apparent in the numbers?

Figure 6 shows the final outcome of the computer fitting process. It is the product of multiple SAAM analyses. After each analysis, the results were evaluated and improved information was provided to SAAM for the subsequent analysis, the goal being an optimal fit of the model parameters to all available data. The steady-state data are similar to those of the manual fitting except the whole body zinc is closer to the estimated value. Again, the parameter values and other data shown in brackets are for the oral tracer administered during fasting. The values in parentheses are the SDs of the parameters. Precision data has not been calculated for the compartment masses, or the steady state values, because the necessary covariance information is not available. The fit of this model to the kinetic data remains essentially unchanged from that shown in Fig. 4.

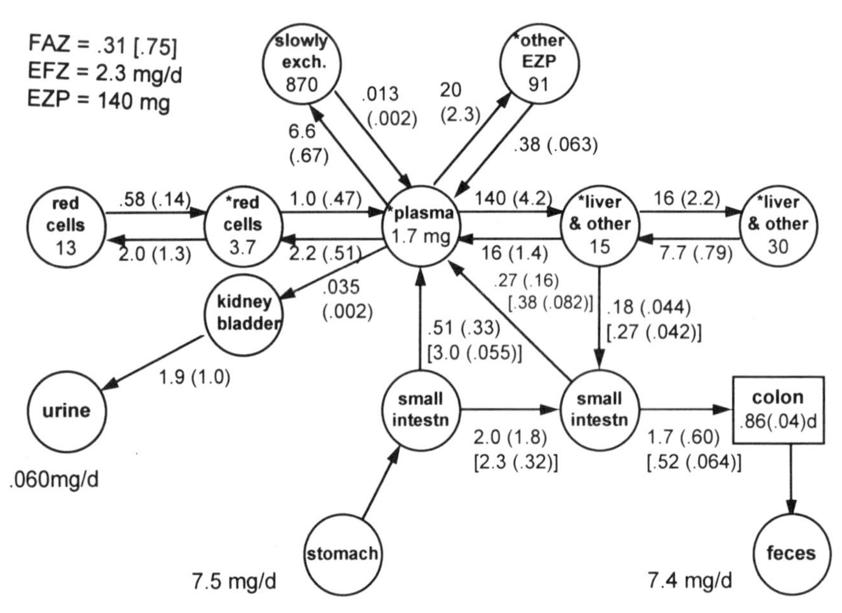

Figure 6. Results of computer fitting of model and data. See caption to Fig. 5 for explanation. Standard deviations of parameter values are shown in parentheses.

A Posteriori Identifiability

Although the *a priori* identifiability of our model was not demonstrated, we may nonetheless evaluate the identifiability of the resulting parameters and model after their application to actual data. In other words, given the model structure, the experimental design, and the available real data being fitted, are the resulting parameter estimations known with enough certainty that they provide meaningful information about the system under study? This is variously called *a posteriori*, numerical, or practical identifiability (Carson et al., 1983). SAAM/CONSAM provide information from the fitting process, e.g. the covariance matrix and parameter SDs/FSDs, which can be used to assess how well parameters have been identified/estimated (Berman, 1982; Jacquez and Perry, 1990; Novotny et al., 1996). A common and easily applied criterion of acceptability is a parameter FSD of ≤ 0.5. In examining Fig. 6, it is evident that a majority of the parameter estimations have SDs that are less than half of the parameter value. The exceptions to this positive finding include several of the FTCs describing movement of prandially administered oral tracer within and out of the small intestine. Given how well the corresponding FTC values for the fasting tracer are identified, the uncertainty of these determinations is probably related to the infrequent sampling and inferior analytical quality of the ^{68}Zn kinetic data. The high SDs associated with the FTCs between the erythrocyte compartments and between the bladder and urine are probably due to characteristics of the erythrocyte and urine data, which makes their fitting less sensitive to the adjustment of those parameters. The apparent limitation on our ability to adequately estimate these two parameters is of no consequence to the overall utility of the model or relevant knowledge gained from these results.

SUMMARY DISCUSSION

We have developed a model of zinc metabolism in the human and fit this model to data from studies of five healthy adult subjects. Although there were several minor variations in the models that initially evolved from fitting the data from the different subjects, a single optimal model structure was used for the computer fitting of all data. A common model structure for all studies was necessary in preparation for any subsequent population model development. Data from one of the subjects has been described in detail as a representative example of the modeling process experienced with all the subjects studied. The processes of model structure development, manual data-model fitting and parameter estimation, and computer data-model fitting and parameter estimation have been carefully performed, culminating in the model shown in Figure 6. The model is based on data from the administration of three stable isotope tracers measured in only circulatory and excretory material. These data were augmented with independently measured and calculated data on tracee zinc intake, absorption, circulation, storage, and excretion, as well as information on zinc metabolism from the published literature. Although the use of stable isotopes and practical sampling protocols imposed significant limitations on the data that could be collected, the employment of multiple tracers (administered orally and intravenously) and the inclusion of the additional tracee metabolic data resulted in a model that is relatively detailed and complex. And, although the identifiability of the model has not been formally demonstrated, the model's complexity generally appears to be warranted, and its utility established, by the parameter statistics provided by SAAM.

We have yet to fully examine the details of the models and their implications for our understanding of zinc metabolism, and for our current and future research efforts. It is significant that the structure of our model is very similar to that of Wastney's (Wastney et al., 1986). The differences that do exist between the models are in the details. Wastney's model provides a more detailed look at the distribution of zinc in the body, attained with regional counting and other radioisotope techniques. In contrast, our model exhibits more detail in the modeling of the upper GI tract and urinary and fecal excretory pathways. This is due to our experimental design and sampling protocols, which focused on tracer absorption and excretion. Further comparisons and reconcilement of any significant differences will await another of our goals, the completion of development of a normal population model. To that end, we have submitted the individual modeling results from the five studies completed thus far to Extended Multiple Studies Analysis (EMSA). EMSA is a recently developed facility of SAAM which derives a single set of population parameters and statistics from a collection of individual studies (Boston et al., 1994; Lyne et al., 1992). The results of this analysis are being reported elsewhere.

ACKNOWLEDGMENTS

This research was supported by grants from the National Institutes of Health, General Clinical Research Centers, RR00069 and RR00051; from the National Institutes of Diabetes, Digestive and Kidney Diseases, KO8-DK02240 and Clinical Nutrition Research Units, P30-DK48520.

CORRESPONDING AUTHOR

Please address all correspondence to:
Leland V. Miller
Campus Box C225
Center for Human Nutrition
University of Colorado Health Sciences Center
4200 E. Ninth Avenue
Denver, CO 80262
Leland.Miller@uchsc.edu

REFERENCES

Berman M; Weiss MF. *User's Manual for SAAM.* US Department of Health, Education, and Welfare, Publication No. (NIH) 78-180, Washington, DC. 1978.

Berman M. A deconvolution scheme. *Math Biosci*, 1978, 40:319-323.

Berman M. Kinetic analysis and modeling: theory and applications to lipoproteins, in: *Lipoprotein Kinetics and Modeling.* Berman M; Grundy SM; Howard BV; Eds. Academic Press: New York. 1982. pp. 3-36.

Berman M; Beltz WF; Greif PC; Chabay R; Boston RC. *CONSAM User's Guide.* US Department of Health and Human Services, Washington, DC. 1983.

Boston R; Lyne A; McNabb T; Pettigrew K; Greif P; Ramberg C; Zech L. Kinetic models to describe populations: A strategy for summarizing the results of multiple studies, in: *Kinetic Models of Trace Element and Mineral Metabolism During Development.* Siva Subramanian KN; Wastney ME; Eds. CRC Press: Boca Raton, FL. 1995. pp. 359-372.

Carson ER; Cobelli C; Finkelstein L. *The Mathematical Modeling of Metabolic and Endocrine Systems.* John Wiley & Sons: New York. 1983.

Cobelli C; Lepschy A; Jacur GR. Identifiability of compartmental systems and related structural properties. *Math Biosci*, 1979, 44:1-18.

Cobelli C; DiStefano JJ III. Parameter and structural identifiability concepts and ambiguities: A critical review and analysis. *Am J Physiol*, 1980, 239 (*Regulatory Integrative Comp Physiol* 8):R7-R24.

Cobelli C. Compartmental models: Theory and practices using the SAAM II software system. In this volume.

CONSAM User's Manual. Resource Facility for Kinetic Analysis: Seattle. 1990.

Cousins RJ. Absorption, transport, and hepatic metabolism of copper and zinc: Special reference to metallothionein and ceruloplasmin. *Physiol Rev*, 1985, 65:238-309.

Cousins RJ. Systemic transport of zinc, in: *Zinc in Human Biology.* Mills CF; Ed. Springer-Verlag: London. 1989. pp. 79-93.

DiStefano JJ III. Complete parameter bounds and quasi-identifiability conditions for a class of unidentifiable linear systems. *Math Biosci*, 1983, 65:51-68.

Fairweather-Tait SJ; Jackson ML; Fox TE; Wharf SG; Eagles J; Croghan PC. The measurement of exchangeable pools of zinc using the stable isotope ^{70}Zn. *Br J Nutr*, 1993, 70:221-234.

Foster DM; Aamodt RL; Henkin RI; Berman M. Zinc metabolism in humans: A kinetic model. *Am J Physiol,* 1979, 237(5):R340-349.
Foster DM; Boston RC; Jacquez JA; Zech LA. *The SAAM Tutorials: An Introduction to using Conversational SAAM Version 30.* Resource Facility for Kinetic Analysis: Seattle. 1989.
Friel JK; Naake VL; Miller LV; Fennessesy PV; and Hambidge KM. The analysis of stable isotopes in urine to determine the fractional absorption of zinc. *Am J Clin Nutr,* 1992, 55:473-477.
Geigy Scientific Tables, 8th Ed. Lentner C; Ed. Ciba-Geigy Corporation: West Caldwell, New Jersey. 1984.
Godfrey KR; DiStefano JJ III. Identifiability of model parameters, in: *Identifiability of Parametric Models.* Walter E; Ed. Pergamon Press: Oxford. 1987.
Jacquez JA. *Compartmental Analysis in Biology and Medicine.* University of Michigan Press: Ann Arbor. 1985.
Jacquez JA; Perry T. Parameter estimation: local identifiability of parameters. *Am J Physiol,* 1990, 258 (*Endocrinol Metab* 21):E727-E736.
Jackson ML; Jones DA; Edwards RHT; Swainbank IG; Coleman ML. Zinc homeostasis in man: Studies using a new stable isotope-dilution technique. *Br J Nutr,* 1984, 51:199-208.
Jackson ML. Physiology of zinc: general aspects, in: *Zinc in Human Biology.* Mills CF; Ed. Springer-Verlag: London. 1989. pp. 1-14.
Krebs NF; Miller LV; Naake VL; Lei S; Westcott JE; Fennessey PV; Hambidge KM. The use of stable isotope techniques to assess zinc metabolism. *J Nutr Biochem,* 1995, 6:292-301.
Lönnerdal B. Intestinal absorption of zinc, in: *Zinc in Human Biology.* Mills CF; Ed. Springer-Verlag: London. 1989. pp. 33-55.
Lowe NM; Green A; Rhodes JM; Lombard M; Jalan R; Jackson ML. Studies of human zinc kinetics using the stable isotope ^{70}Zn. *Clin Sci,* 1993, 84:113-117.
Lowe NM; Shames DM; Woodhouse LR; Matel JS; Roehl R; Saccomani MP; Toffolo G; Cobelli C; King JC. A compartmental model of zinc metabolism in healthy women using oral and intravenous stable isotope tracers. *Am J Clin Nutr,* 1997, 65:1810-1819.
Lyne A; Boston R; Pettigrew K; Zech L. EMSA: A SAAM service for the estimation of population parameters based on model fits to identically replicated experiments. *Comput Methods Prog Biomed,* 1992, 38:117-151.
Miller LV; Hambidge KM; Naake VL; Hong Z; Westcott JL;. Fennessey PV. Size of the zinc pools that exchange rapidly with plasma zinc in humans: Alternative techniques for measuring and relation to dietary zinc intake. *J Nutr,* 1994, 124:268-276.
Novotny JA; Zech LA; Furr HC; Dueker SR; Clifford AJ. Mathematical modeling in nutrition: Constructing a physiologic compartmental model of the dynamics of β-carotene metabolism, in: *Advances in Food and Nutrition Research, Vol. 40, Mathematical Modeling in Experimental Nutrition: Vitamins, Proteins, Methods.* Coburn SP; Townsend DW. Eds. Academic Press: San Diego. 1996. pp. 25-54.
Shipley RA; Clark RE. *Tracer Methods for In Vivo Kinetics.* Academic Press: New York. 1972.
Van Dokkum W; Fairweather-Tait SJ; Hurrell R; Sandström B. Study techniques, in: *Stable Isotopes in Human Nutrition: Inorganic Nutrient Metabolism.* Mellon F; Sandström B; Eds. Academic Press: London. 1996. pp. 23-42.
Van Wouwe JP; Veldhuizen M; De Goeij JJM; Van den Hamer CJA. In vitro exchangeable erythrocytic zinc. *Biol Trace Element Res,* 1990, 25:57-69.
Walter E; Ed. *Identifiability of Parametric Models.* Pergamon Press: Oxford. 1987.
Wastney ME; Aamodt RL; Rumble WF; Henkin RI. Kinetic analysis of zinc metabolism and its regulation in normal humans. *Am J Physiol,* 1986, 251 *(Regulatory Integrative Comp Physiol* 20):R398-R408.

KEY FEATURES OF COPPER VERSUS MOLYBDENUM METABOLISM MODELS IN HUMANS

Judith R. Turnlund,[1] Katherine H. Thompson,[2] and Karen C. Scott[3]

[1]USDA/ARS
Western Human Nutrition Research Center
PO Box 29997
Presidio of San Francisco, CA 94129
[2]Chemistry Department
The University of British Columbia
Vancouver, British Columbia, Canada V6T 1Z1
[3]Department of Small Animal Clinical Sciences
University of Florida
Gainesville, FL 32610-0126

INTRODUCTION

Copper and molybdenum are essential nutrients for humans. Both trace elements are required in the diet in small amounts, are toxic in excess, and body stores are low. But there are major differences in their metabolism. Metabolic studies were conducted in the human nutrition suite of the Western Human Nutrition Research Center. Young men were confined to the unit for 2 to 90 days in studies of copper metabolism and for 120 days for studies of molybdenum metabolism. Stable isotopes were used as tracers to follow the metabolic fate of these elements. ^{65}Cu was administered orally and intravenously in the copper studies. ^{100}Mo was administered orally and ^{97}Mo was administered intravenously in the molybdenum studies.

Kinetic models were developed to gain a better understanding of the metabolism of these essential trace elements. The long-term goals of the model development were to estimate pool sizes and rates of transfer between compartments which could not be sampled directly, to identify points of metabolic regulation, and then to elucidate the mechanisms of regulation. Compartmental models were developed using CONSAM. A five-compartment model with two delay components and two routes of excretion was developed for copper and a model with seven compartments and two routes of excretion was developed for molybdenum. Key features of these models and differences between the metabolism of copper and molybdenum are discussed below.

Mathematical Modeling in Experimental Nutrition
Edited by Clifford and Müller, Plenum Press, New York, 1998

DIETARY RECOMMENDATIONS AND BODY STORES

United States dietary recommendations were first established for copper and molybdenum in 1980 (National Research Council, 1980). These recommendations were in the form of "Estimated Safe and Adequate Daily Dietary Intakes of Selected Vitamins and Minerals." Because there was less information on which to base allowances than there was for other nutrients with Recommended Dietary Allowances (RDAs), the figures were not included in the main RDA tables and were provided as ranges of recommended intakes. The range for adults was 2–3 mg/d for copper and 0.15–0.5 mg/d per day for molybdenum. The ranges were changed for both in 1989 (National Research Council 1989). The lower end of the range was revised downward for copper (1.5–3 mg/d) and both the upper and lower ends were revised downward for molybdenum (0.075–0.250 mg/d).

There are no storage sites for large amounts of either mineral in the body. This is very different from many other essential minerals. For example, RDA for calcium for men is 800 mg/d. The body contains about 1500 g of calcium, which is 1900 times the amount in the diet. The recommended allowance for iron recommendation is 10 mg/d, with total body content, 4 g, about 400 times greater than dietary intake. The dietary recommendation is 1.5 mg for copper and the body contains only 100 mg, 66 times the amount in the diet. The recommendation for molybdenum is 0.075 mg and the body contains only 2.5 mg, 33 times the amount in the diet. If these minerals were completely absorbed from the diet, their body stores would be equivalent to five years of dietary calcium and one year of dietary iron, but only one month of dietary molybdenum and two months of dietary copper.

COPPER NUTRITION

Background

Copper compounds were prescribed as early as 400 BC to treat pulmonary and other diseases (Mason, 1979). Their use continued into the nineteenth century, but declined when treatments were not successful and as other more effective ways of treating diseases became available. In the late nineteenth century, copper was identified as a constituent of blood and its toxicity was observed. In the early twentieth century, it was found to be essential in animals when anemia not reversed by iron supplements responded to copper. Genetic defects were also associated with copper metabolism early in the twentieth century. The association between anemia in humans and copper was explored in the 1930s, with inconclusive results. The first copper balance studies in infants and adults were reported at about the same time. Dietary copper deficiency in humans was first documented in 1964 (Cartwright and Wintrobe, 1964).

Enzymes and Physiological Roles

Numerous copper-containing enzymes and copper-binding proteins are found in humans (Turnlund, 1998). These include the amine oxidases: monoamine oxidase, diamine oxidase, lysyl oxidase, and peptidylglycine-α-amidating monooxygenase (PAM). Two ferroxidases, ceruloplasmin (ferroxidase I) and ferroxidase II, contain copper. Other copper-containing enzymes are cytochrome c oxidase, dopamine β-monooxygenase, superoxide dismutase, and monophenol monoxygenase. Copper binding proteins include metallothionein, albumin, transcuprein, blood clotting factor V, amino acids and peptides.

Copper plays a number of critical physiological roles (Turnlund, 1998). Through lysyl oxidase, it is essential for cross-linking of collagen and elastin. Collagen is a component of connective tissue throughout the body, including bone and the cardiovascular system. Copper plays a role in iron metabolism, central nervous system integrity, melanin pigment formation, cardiac function and cholesterol metabolism. Roles for copper in thermal regulation, glucose metabolism, blood clotting, antioxidant status, and immune function have also been suggested.

Indices of Copper Status

Severe copper deficiency can be detected easily. Serum copper and ceruloplasmin concentrations decline well below the normal range and are rapidly reversed by copper supplementation. Erythrocyte superoxide dismutase activity also declines, but normal levels have not been established in humans. The normal range for serum copper is approximately 10 to 25 μmol/L and the range is 180 to 400 mg/L for ceruloplasmin, but normal ranges vary among laboratories (Turnlund, 1998). These indices may not be sensitive to marginal copper status. They are not reliable in the presence of conditions which elevate ceruloplasmin, an acute phase reactant. Other suggested indices include cytochrome c oxidase in erythrocytes, platelets, or white cells, leukocyte copper, and changes in lysyl oxidase activity (Turnlund, 1998). Stable isotopes and compartmental models of copper kinetics may be useful in evaluating copper status by estimating total body copper, the exchangeable copper pool size, and copper turnover.

Copper Deficiency and Toxicity

Frank copper deficiency is relatively rare in humans. It has been observed in individuals receiving enteral or parenteral diets without copper and in infants recovering from malnutrition. Serum copper and ceruloplasmin fall to very low levels and anemia, leukopenia, and neutropenia are observed. Osteoporosis is observed when bones are still growing (Turnlund, 1998).

Copper toxicity is more common in animals than in humans. Sheep are especially vulnerable to copper poisoning. Acute copper poisoning in humans has been observed after consumption of several grams of copper from sources such as contaminated drinking water and consuming acidic foods that were stored in copper containers. Chronic poisoning has been observed in dialysis patients when copper tubing was used and in vineyard workers exposed to copper-containing pesticides. Current WHO recommendations suggest intakes below 12 mg/d for men (Anonymous, 1996). Little is known about the subtle effects of moderately high copper intake. A number of conditions appear to alter copper metabolism and increase plasma copper and ceruloplasmin levels markedly. These appear to be a redistribution of endogenous copper and not an effect of toxicity. These conditions include pregnancy, inflammatory conditions, infectious diseases, cardiovascular diseases, and many other diseases (Turnlund, 1998). Two well-known genetic defects are associated with altered copper metabolism (Turnlund, 1998). Menkes' disease, an X-linked disorder, is associated with abnormal distribution of copper and is usually fatal by 3 years of age. Serum, brain, and liver copper levels are low, but copper accumulates in kidney, muscle, spleen, and intestinal mucosa. Wilson's disease is an autosomal recessive disease in which copper accumulates in liver, brain, and the cornea of the eye. Urinary copper excretion is high and ceruloplasmin levels are low. Early diagnosis and treatment can prevent serious consequences of the disease.

MOLYBDENUM NUTRITION

Background

Very little was known about the role of molybdenum in nutrition until recently and much remains to be learned about its metabolism. Molybdenum was found to be present in plants and animals in 1932. In the 1950s it was discovered that two enzymes require molybdenum for activity and a dietary molybdenum requirement was established for chicks. In 1967, a sulfite oxidase deficiency was observed in humans and, in 1971, it was discovered that sulfite oxidase required molybdenum for activity. Because sulfite oxidase is required by humans, the essentiality of molybdenum in humans was established. A dietary deficiency of molybdenum in a human was observed in 1981. The first work with molybdenum in humans was conducted in cancer patients using the radioisotope ^{99}Mo in the 1970s (Rosoff and Spencer, 1973), but very little information was available on metabolism in humans until recently (Turnlund et al., 1995c; Cantone et al., 1993; Cantone et al., 1995).

Enzymes and Physiological Roles.

Three enzymes present in humans are associated with molybdenum: xanthine oxidase, aldehyde oxidase, and sulfite oxidase (Rajagopalan, 1988). All of these catalyze chemical reactions in humans, but only sulfite oxidase is required by humans (Abumrad, 1984). The enzymes are associated with what little is known about the physiological roles of molybdenum. Xanthine oxidase catalyzes the transformation of xanthine and hypoxanthine to uric acid. Lack of xanthine oxidase activity results in accumulation of xanthine and hypoxanthine. Aldehyde oxidase catalyzes the peroxidation of purines and pyrimidines. Sulfite oxidase is required to transform sulfite to sulfate. Lack of this enzyme results in inability to detoxify sulfite.

Molybdenum Status, Deficiency, and Toxicity

Little is known about methods of assessing molybdenum status. Blood levels are so low that they are very difficult to measure, but they appear to reflect recent dietary intake rather than status. A number of possibilities have been suggest for assessing molybdenum status, based primarily on individuals with metabolic defects in molybdenum cofactor.

The first and only clear case known of dietary Mo deficiency was observed in an individual on long-term parenteral nutrition (Abumrad et al., 1981). He had Chron's disease and a history of multiple bowel resections. Biochemical abnormalities, reversed with molybdenum supplementation, included high plasma methionine, low serum uric acid, high urinary sulfate, and elevated xanthine and hypoxanthine excretion. In a molybdenum depletion study, we used load tests in an attempt to stress molybdoenzymes (Chiang, 1991; Chiang et al., 1989) and identify sensitive indices of molybdenum status. After 102 days of depletion, loading doses of AMP resulted in increased xanthine and hypoxanthine excretion, compared to before depletion, and loading doses of cysteine resulted in increased sulfate excretion.

Molybdenum toxicity is much more likely than deficiency. Toxicity has been observed in animals grazing on high molybdenum soil (Mills and Davis, 1987). Humans consuming 10–15 mg of molybdenum per day have reported gout-like symptoms (National Research Council, 1989).

A metabolic defect, molybdenum cofactor deficiency, results in the biochemical symptoms similar to those observed in the dietary deficiency described above and is related to enzyme inactivities, as well as neurological symptoms and mental retardation (Johnson et al., 1980; Johnson, 1988). The cofactor, molybdopterin, is absent in this syndrome.

STABLE ISOTOPES AND THE METABOLIC FATE OF DIETARY MINERALS

Stable isotopes of copper and molybdenum have been used to follow the metabolic fate of these elements. There are a number of important considerations to keep in mind when selecting a stable isotope to use as a tracer. Ideally, the chosen isotope is relatively rare in natural abundance. Analytical methods must also be considered and are covered in other papers in this volume. One or more stable isotopes of multiple elements can be administered simultaneously. Details on the use of multiple isotopes of zinc, copper, and molybdenum for our modeling studies have been published (Turnlund, 1995).

Studies with Copper Isotopes

Copper has only two stable isotopes, ^{63}Cu and ^{65}Cu, with natural abundances of 30.8% and 69.2%, respectively. The number of isotopes limits the way in which the isotopes can be used. Only one of the two can be enriched, so simultaneous administration via oral and intravenous routes is not possible, nor can intrinsic and extrinsic food tracers be administered at the same time. In our studies, we fed and infused ^{65}Cu at different times or fed the isotope to some subjects and infused it in others. Enrichment has been measured in blood plasma and stool samples. Urinary copper is too low to measure isotopic enrichment by thermal ionization mass spectrometry (TIMS).

Studies with Molybdenum Isotopes

Molybdenum has seven isotopes with abundances ranging from 9 to 24%. The isotopes and their abundances are: ^{92}Mo – 14.8%, ^{94}Mo – 9.2%, ^{95}Mo – 15.9%, ^{96}Mo – 16.7%, ^{97}Mo – 9.6%, ^{98}Mo – 24.1%, and ^{100}Mo – 9.6%. The large number of these isotopes make it possible to use more than one isotope simultaneously. Different isotopes can be fed and infused. A number of isotopically enriched diets or foods can be fed simultaneously or in close proximity to one another. We administered ^{100}Mo orally and ^{97}Mo intravenously to investigate molybdenum absorption and kinetics (Turnlund et al., 1995; Thompson et al., 1996). In another study, we fed foods enriched with one of the isotopes ^{96}Mo, ^{97}Mo, and ^{100}Mo simultaneously to compare intrinsic and extrinsic tracers in foods, then used ^{94}Mo as an isotopic diluent (Turnlund et al., 1996). We measured the tracers in urine and stool samples and will measure plasma enrichment using ICP/MS, once we have methods in place to do so.

Isotope Analysis

To date we have used magnetic sector, thermal ionization mass spectrometry (TIMS) for isotope analysis (Turnlund and Keyes, 1990). This is an excellent analytical method, with very high precision, but relatively slow and requiring extensive sample purification. Another approach used is inductively coupled plasma/mass spectrometry (ICP/MS) (Crews et al., 1994), which we will use in future work not requiring the highest precision of ratio measurement.

Validity of Extrinsic Tracers

The approach we have usually used for stable isotope work employs stable isotopes of copper and molybdenum added as extrinsic labels. This raises the question of the validity of extrinsic labels as tracers of food sources of elements. Studies of copper absorption have validated the approach, showing that absorption is similar, if not identical from foods intrinsically labeled with copper and extrinsic tracers (Johnson et al., 1988).

We conducted studies with two foods intrinsically labeled with isotopes of molybdenum to test the validity of extrinsic labels (Turnlund et al., 1996). The study demonstrated that while absorption of molybdenum in kale was identical to that of the extrinsic label, the molybdenum in soy was less well absorbed than either the extrinsic label or the molybdenum in kale. However, the urinary excretion patterns of absorbed isotopes was identical for all three. This suggests that extrinsic tracers of molybdenum do not always reflect absorption, but extrinsic tracers appear to be valid for studies of utilization following absorption.

Copper Homeostasis

Balance studies have demonstrated that there are homeostatic mechanisms regulating the retention of copper and molybdenum in the body. The regulation is achieved in different ways for the two minerals. Copper absorption is regulated such that as intake increases, the fraction of dietary copper absorbed declines (Turnlund et al., 1989; Turnlund et al., 1995a). The regulation through absorption is incomplete, however, and initially after dietary copper is increased, individuals retain a considerable amount of copper. However, excretion into the gastrointestinal tract then increases, so that individuals go into negative balance as excess copper absorbed is excreted into the gastrointestinal tract (Turnlund et al., 1995b). The regulation via absorption and excretion into the gastrointestinal tract protects against copper deficiency and toxicity. Recent research suggests that the change in the rate of copper excretion is key to maintaining constant body stores and that copper excretion is very low when dietary intake is low (Turnlund et al., 1995b). This excretory regulation has also been observed in rats (Levenson and Janghorbani, 1994). Urinary copper is consistently low and does not increase to eliminate excess copper (Turnlund et al., 1997; Turnlund et al., 1990). Except in depletion with very low copper diets, indices of copper status remain relatively constant in individuals over a broad range of intakes (Turnlund et al., 1990; Turnlund et al., 1997).

Molybdenum Homeostasis

In contrast, molybdenum retention is not regulated through absorption (Turnlund et al., 1995; Turnlund et al., 1995c). About 90% of dietary molybdenum is absorbed over a broad range of intakes. However, as the amount absorbed increases, urinary excretion increases. Thus, molybdenum retention is regulated via urinary excretion. Also in contrast to copper, fecal molybdenum is very low compared to urinary molybdenum and little absorbed molybdenum is excreted into the gastrointestinal tract.

MODELING MINERAL METABOLISM

We conducted studies in the Human Nutrition Suite of the Western Human Nutrition Research Center in order to develop models of copper, zinc, and molybdenum metabolism (Scott and Turnlund, 1994a; Scott and Turnlund, 1994b; Thompson and Turnlund, 1996;

Thompson et al., 1996). The CONSAM software package, Version 30.1 for IBM-compatible personal computers, was used for model development (Berman et al.,1983). The initial copper model was developed using average data, then models were fit to individuals in order to refine and further develop the models. For molybdenum, the initial model was developed with individual data, then fractional transfer coefficients were hand-fit and integrated to converge toward a sample mean.

Copper Model

We developed a model of copper metabolism in humans from data acquired in a study of young men (Scott and Turnlund, 1994a). A 90-d study was conducted during which three levels of dietary copper were fed. ^{65}Cu was administered orally four times and intravenously three times during the study. Models of copper metabolism have been developed in laboratory animals, sheep, and dairy cows (Buckley, 1991; Dunn, 1995; Weber et al., 1980), and these were used as a starting point in the development of our human model.

A model, shown in Fig.1, was developed from plasma enrichment data that contained five compartments, two delay components, and two excretion pathways. These compartments represented two plasma compartments, two liver compartments, an other-tissue compartment, and fecal and urinary excretion routes. Dietary copper level influenced the flow from the second liver compartment to the second plasma compartment and from the second plasma compartment to an other-tissues compartment.

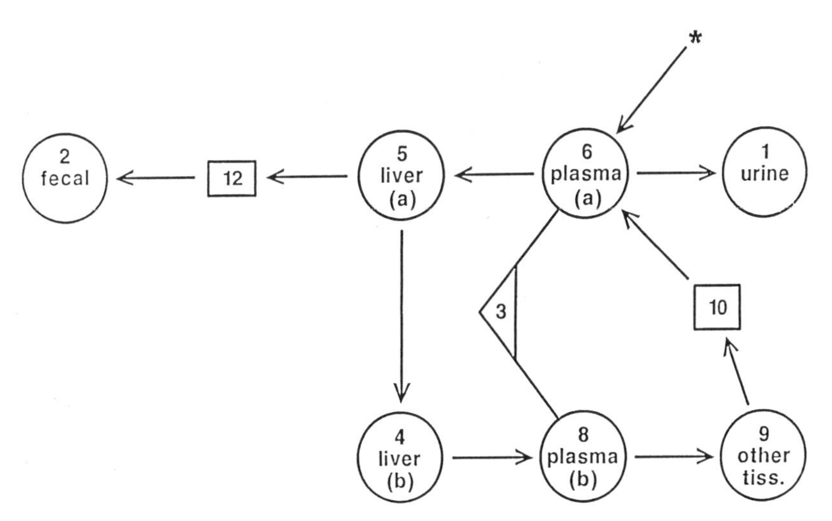

Figure 1. Compartmental model of copper metabolism in adult men (Scott and Turnlund, 1994a). Circles represent compartments, rectangles are delay elements, triangle is the sum of the indicated compartments, * is the site of ^{65}Cu input. Plasma samples were taken from compartment 3. Compartment 8 is hypothesized to be ceruloplasmin copper and compartment 6 nonceruloplasmin copper. Delay 10 is 4-7 hours, delay 12 is 72 hours.

It appeared that metabolism of oral and intravenous copper was different. The flow from plasma to the first liver compartment differed between routes of administration. The second liver compartment was identified as the major storage site, with more copper in the delay components than in compartments other than the liver.

The model predicted 4.5% of total body copper was in the plasma compared to literature estimates of 2 to 6%. Total body stores were estimated in the range of 1.18 to 1.54 μmol of copper, compared to literature estimates of 0.77 to 1.85 μmol. Small changes in the rate constants for movement of copper from plasma compartment 6 (see Fig.1) to liver compartment 5, from liver compartment 4 to plasma compartment 8, and from plasma compartment 8 to other-tissue compartment 9, resulted in marked changes in the shape of the curves, suggesting possible regulatory sites.

Our copper model was based on plasma enrichment only (Scott and Turnlund, 1994a). It is a simple model and not all physiological pathways were included due to limitations of the data. We have conducted further studies to expand this model and plan to include one or more gastrointestinal compartments, based on fecal data. Studies were done in which oral copper was administered to some subjects and intravenous copper to others. Thus, enrichment from one route can be followed for a longer period of time without interference from the other. We also plan to refine the model based on data from more frequent sampling of plasma to determine if a decline in plasma enrichment followed by an increase (Scott and Turnlund, 1994a) can be confirmed. Ceruloplasmin enrichment data will be added as well.

Molybdenum Model

We used the data from two 120-d studies conducted in young men to model molybdenum metabolism (Thompson et al., 1996; Thompson and Turnlund., 1996). One study was with a low molybdenum diet for 108 d followed by 18 d of a high molybdenum intake, and the other included five 24-d periods of increasing levels of dietary molybdenum. Two stable isotopes, ^{100}Mo, administered orally, and ^{97}Mo, administered intravenously, were used to trace molybdenum metabolism. A model containing seven compartments and two excretory pathways (Fig.2) simulated the data for both oral and intravenous administrations from both studies. The compartments in this model were an oral, three equivalent gastrointestinal compartments, fast- and slow-turnover tissue compartments, and a central plasma compartment. The model predicted total body molybdenum stores of 2.2 to 2.8 mg (Thompson and Turnlund, 1996; Thompson et al., 1996). This is less than one published estimate of 10 mg (Schroeder et al., 1970) made prior to the use of more accurate methods of molybdenum measurement, which have resulted in downward revision of tissue molybdenum contents (Mills and Davis, 1987). Rapid clearance of molybdenum from plasma circulation (10 to 23 min half-life) was predicted.

Unlike the copper model, increased dietary levels of molybdenum did not result in decreased absorption; instead absorption remained high, but urinary excretion increased proportionately more than fecal excretion. There was an increased turnover and decreased residence time in fast turnover tissues (compartment 8) as dietary molybdenum increased. Fractional transport coefficients that were most sensitive to increased dietary molybdenum were those representing flux from fast-turnover tissue compartment 8 to plasma compartment 6, and presumed biliary secretion from tissue compartment 7 to GI compartment 2. Residence time in the slow-turnover tissue compartment increased by a factor of 8 as dietary molybdenum increased, whereas residence time decreased for fast-turnover tissue with increased dietary molybdenum.

No plasma data were available for the development of this model because plasma molybdenum levels are so low we could not measure isotopic ratios by TIMS. We have acquired an ICP/MS and will measure isotopic enrichment of molybdenum by ICP/MS in future work. This should permit further refinement of the model.

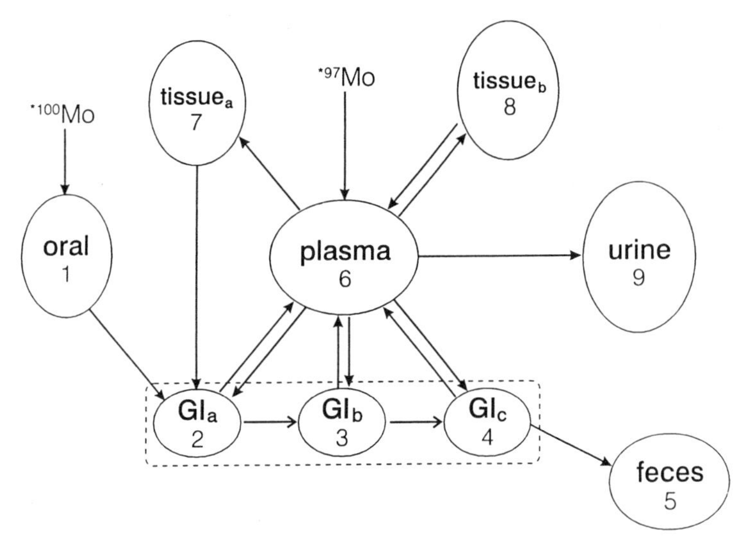

Figure 2. Compartmental model of molybdenum metabolism in adult men (Thompson and Turnlund, 1996). Enclosed spaces represent compartments that are kinetically homogenous and distinct from other compartments. Numbers within these spaces correspond to compartment numbers in the mathematical model. Arrows represent transfer and input pathways. GI_a, GI_b, and GI_c are equivalent gastrointestinal compartments. $Tissue_a$ represents a slow-turnover tissue compartment (partly hepatic) and $tissue_b$ represents a fast-turnover tissue compartment.

Compartment-based model analyses of both copper and molybdenum isotopic decay curves have demonstrated the importance of recirculation pathways in tracing the uptake and excretion of both metals. They have highlighted the key differences in homeostatic mechanisms, with molybdenum showing little variability in uptake, but a changing ratio of urinary to fecal excretion, and copper showing declining absorption with increasing intake, but no change in urinary excretion. Molybdenum is very rapidly cleared from plasma, whereas copper circulates from plasma to liver and back to plasma, possibly indicating a greater potential for conservation of copper at low levels of dietary intake (Thompson et al., 1996; Scott and Turnlund., 1994a).

FUTURE PLANS

The modeling of copper and molybdenum is just beginning. We expect that with time, further development of the models will result in clearer definition of the points of regulation in tissues we cannot sample. Once the points of regulation are established, we will attempt to elucidate the mechanisms of regulation.

ACKNOWLEDGMENTS

The authors wish to thank William Keyes for mass spectrometric analysis.

CORRESPONDING AUTHOR

Please send all correspondence to:
Judith R. Turnlund, PhD
USDA/ARS
Western Human Nutrition Research Center
PO Box 29997
Presidio of San Francisco CA 941229
ph: 415-556-5662
fax: 415-556-1432
Jturnlun@whnrc.usda.gov

REFERENCES

Anonymous. *Trace Elements in Human Nutrition and Health.* World Health Organization: Geneva. 1996. pp. 1-343.

Abumrad NN. Molybdenum: Is it an essential trace metal? *Bul NY Acad Med,* 1984, 60:163-170.

Abumrad NN; Schneider AJ; Steel D; Rogers LS. Amino acid intolerance during prolonged total parenteral nutrition reversed by molybdate therapy. *Am J Clin Nutr,* 1981, 34:2551-2559.

Berman M; Beltz WF; Greif PC; Chabay R; Boston, RC. *CONSAM User's Guide.* National Institutes of Health: Bethesda, MD. 1983.

Buckley WT. A kinetic model of copper metabolism in lactating diary cows. *Can J Anim Sci,* 1991, 71:155-166.

Cantone MC; de Bartolo D; Gambarini G; Giussani A; Ottolenghi A; Pirola L. Proton activation analysis of stable isotopes for a molybdenum biokinetics study in humans. *Med Phys,* 1995,22:1293-1298.

Cantone MC; de Bartolo D; Molho N; Priola L; Gambarine G; Hansen C; Roth P; Werner E. Response to a single oral test of molybdenum stable isotopes for absorption studies in humans. *Physiol Meas,* 1993, 14:217-225.

Cartwright GE; Wintrobe MM. The question of copper deficiency in man. *Am J Clin Nutr,* 1964, 14:94-110.

Chiang. G. *Studies of Biochemical Markers Indicating Molybdenum Status in Human Subjects Fed Diets Varying in Molybdenum Content.* Doctoral thesis, University of California, Los Angeles.1991.

Chiang G; Swendseid ME; Turnlund JR. Studies of biochemical markers indicating molybdenum (Mo) status in humans. *FASEB J,* 1989, 3:1073 (abs.).

Crews HM; Ducros V; Eagles J; Mellon FA; Kastenmayer P; Luten JB; McGaw BA. Mass spectrometric methods for studying nutrient mineral and trace element absorption and metabolism in humans using stable isotopes. *Analyst,* 1994, 119:2491-2514.

Dunn MA. Historical overview of copper kinetics, in: *Kinetic Models of Trace Element and Mineral Metabolism During Development.* Subramanian KNS and Wastney ME; Eds. CRC Press: Boca Raton. 1995. pp. 171-185.

Johnson JL. Molybdenum, in: *Methods in Enzymology, Vol. 158.* Riordan JF and Vallee BL; Eds. Academic Press: San Diego. 1988. pp. 371-382.

Johnson JL; Waud WR; Rajagopalan KV; Duran M; Beemer FA; Wadman SK. Inborn errors of molybdenum metabolism: Combined deficiencies of sulfite oxidase and xanthine dehydrogenase in a patient lacking the molybdenum cofactor. *Proc Natl Acad Sci,* 1980, 77:3715-3719.

Johnson PE; Stuart MA; Hunt JR; Mullen L; Starks TL. ^{65}Cu absorption by women fed intrinsically and extrinsically labeled goose meat, goose liver, peanut butter, and sunflower butter. *J Nutr,* 1988, 118:1522-1528.

Levenson CW; Janghorbani M. Long-term measurement of organ copper turnover in rats by continuous feeding of a stable isotope. *Anal Biochem,* 1994, 221:243-249.

Mason KE. A conspectus of research on copper metabolism and requirements of man. *J Nutr,* 1979, 109:1979-2066.

Mills CF; Davis GK. Molybdenum, in: *Trace Elements in Human and Animal Nutrition*, Mertz W; Ed. Academic Press: San Diego. 1987. pp. 429-463.

National Research Council. *Recommended Dietary Allowances; 9.* National Academy of Sciences: Washington DC. 1980.

National Research Council. *Recommended Dietary Allowances; 10.* National Academy Press: Washington, DC.1989.

Rajagopalan KV. Molybdenum: An essential trace element in human nutrition. *Ann Rev Nutr,* 1988, 8:401-427.
Rosoff B; Spencer H. The distribution and excretion of molybdenum-99 in mice. *Health Physics,* 1973, 25:173-175.
Schroeder HA; Balassa JJ; Tipton IH. Essential trace metals in man: Molybdenum. *J Chron Dis,* 1970, 23:481-499.
Scott KC; Turnlund JR. Compartmental model of copper metabolism in adult men. *J Nutr Biochem,* 1994a, 5:342-350.
Scott KC; Turnlund JR. Compartmental model of zinc metabolism in adult men fed three levels of dietary copper. *Am J Physiol,* 1994b, 267:E165-E173.
Thompson KH; Scott KC; Turnlund JR. Molybdenum metabolism in young men fed a range of intakes of molybdenum: Changes in kinetic parameters. *J Appl Physiol,* 1996, 81:1404-1409.
Thompson KH; Turnlund JR. Kinetic model of molybdenum metabolism developed from dual stable isotope excretion in men consuming a low molybdenum diet. *J Nutr,* 1996, 126:963-972.
Turnlund JR. Stable isotopes of copper, molybdenum, and zinc used simultaneously for kinetic studies of their metabolism, in: *Kinetic Models of Trace Element and Mineral Metabolism During Development.* Subramanian KNS; Wastney ME; Eds. CRC Press: Boca Raton. 1995. pp. 133-143.
Turnlund JR. Copper, in: *Modern Nutrition in Health and Disease.* Shils ME; Olson JA; Shike M; . ɔss AC; Eds. Williams & Wilkins: Baltimore. 1998.
Turnlund JR; Keen CL;. Smith RG. Copper status and urinary and salivary copper in young men at three levels of dietary copper. *Am J Clin Nutr,* 1990, 51:658-664.
Turnlund JR; Keyes WR. Automated analysis of stable isotopes of zinc, copper, iron, calcium, and magnesium by thermal ionization mass spectrometry using double isotope dilution for tracer studies in humans. *J Micronutrient Analysis,* 1990, 7:117-145.
Turnlund JR; Keyes WR; Anderson HL; Acord LL. Copper absorption and retention in young men at three levels of dietary copper by use of the stable isotope ^{65}Cu. *Am J Clin Nutr,* 1989, 49:870-878.
Turnlund JR; Keyes WR; Peiffer GL. Copper absorption and retention from diets low and adequate in copper. *Am J Clin Nutr,* 1995a, 61:908 (abs.).
Turnlund JR; Keyes WR; Peiffer GL. Copper excretion into the gastrointestinal tract at three levels of dietary copper studied with the stable isotope ^{65}Cu. *FASEB J,* 1995b, 9:A725 (abs.).
Turnlund JR; Keyes WR; Peiffer GL. Molybdenum absorption, excretion, and retention studied with stable isotopes in young men at five intakes of dietary molybdenum. *Am J Clin Nutr,* 1995c, 62:790-796.
Turnlund JR; Keyes WR; Peiffer GL; Chiang G. Molybdenum absorption, excretion, and retention studied with stable isotopes in young men during depletion and repletion. *Am J Clin Nutr,* 1995, 61:1102-1109.
Turnlund JR; Scott KC; Peiffer GL; Jang AM., Keen CL; Sakanashi TM. Copper status of young men consuming a low copper diet. *Am J Clin Nutr,* 1997, 65:72-78.
Turnlund JR; Weaver CM; Kim SK; Keyes WR; Peiffer GL. Absorption and utilization of molybdenum from soy, kale and an extrinsic label. *FASEB J,* 1996,A818 (abs.).
Weber KM; Boston RC; Leaver DD. A kinetic model of copper metabolism in sheep. *Aust J Agric Res,* 1980, 31:773-790

INSIGHTS INTO BONE METABOLISM FROM CALCIUM KINETIC STUDIES IN CHILDREN

Steven A. Abrams, MD

USDA/ARS Children's Nutrition Research Center
Department of Pediatrics
Baylor College of Medicine and Texas Children's Hospital
Houston Texas

ABSTRACT

Changes in the mineralization rate of the skeleton during childhood are related to normal growth and pubertal development. These may be affected by genetic factors, including race and gender, and by the presence of abnormalities of growth or hormonal abnormalities such as occur in children with chronic illnesses. We have used multi-compartmental studies to examine calcium kinetics in healthy children ranging in age from premature infants of 1-2 kg body weight through adolescence. These studies are performed using orally and intravenously administered stable isotopes of calcium. Sample collection requires multiple blood samples to be obtained during the initial time period after isotope dosing, which is feasible in older children using an indwelling catheter. We have found that the peak rate for both bone calcium deposition and removal occurs in girls during the year before menarche. Peak kinetic rates decrease in an exponential fashion post-menarche. On a body-weight basis, the greatest rates of bone calcium deposition and removal are in infants, especially premature infants.

INTRODUCTION

Although measurements of net calcium retention provide information regarding calcium homeostasis, they provide no information about the dynamics (kinetics) of bone formation and resorption. Whereas these values may be assessed indirectly using biochemical measurements such as serum osteocalcin or urinary crosslinks, the use of calcium isotopes and multi-compartmental models allows for a direct measurement of the rate of bone calcium kinetics.

For many years such tracer studies have been performed, primarily in adults using radioactive calcium tracers (Heaney and Whedon, 1958; Neer et al., 1967). A few early studies in children were also performed using radioactive tracers. These include a study of calcium kinetics in adolescent boys performed in the 1950s by Bronner and Harris (1956). The development of stable isotope techniques during the last 30 years has obviated the need to expose subjects to radiation to perform these studies and led to an increase in their application in children as well as pregnant and lactating women (Heaney and Skillman, 1971; Moore and Machlan, 1972; Moore et al., 1985). These studies have provided a new understanding of the processes of mineral metabolism and bone development in children.

Although neutron activation analysis may be used to analyze stable isotopes (Heaney and Skillman, 1971), these measurements are more readily made using mass spectrometry. In 1972, Moore and Machlan reported a mass spectrometric technique to measure calcium stable isotopes from blood and urine samples using a specially constructed magnetic sector thermal ionization mass spectrometer (1972). Measurement precision for isotope ratios of <0.1% was achieved, which is comparable to the precision obtained currently with commercially available magnetic sector thermal ionization mass spectrometers. This method continues to be the reference method for the analysis of mineral stable isotopes. Other mass spectrometric techniques, including fast-atom-bombardment mass spectrometry and inductively coupled-plasma mass spectrometry, have also been applied to calcium isotope ratio analysis. However, the precision and accuracy of these techniques is generally less than magnetic sector thermal ionization mass spectrometers (Hachey et al., 1987).

During the last 10 years, our research group has utilized calcium stable isotopes to assess calcium kinetics in premature infants, and in children 3-16 years of age. More than 250 children and adolescents have participated in these studies without any complications related to the studies. Our primary interest has been to assess the developmental changes in calcium kinetics, especially as related to growth and pubertal development (Abrams et al., 1992a; Abrams et al., 1994a; Yergey et al., 1995). Other studies have considered the relationship of chronic illness and medication usage to calcium kinetics (Abrams et al., 1993; Mauras et al.,1994). In this chapter, we summarize key findings related to calcium kinetics during growth and development, and consider areas in which future investigations can utilize this powerful, noninvasive technique.

METHODS

Study Subjects

Calcium kinetic studies described in this report were conducted in groups of 13 premature infants (Abrams et al., 1993) and 62 girls 5-16 years of age (Abrams et al., 1996). Comparison data are provided based on the literature for premenopausal women (Table 1). Premature infants were healthy and receiving full oral feeds of human milk or infant formula at the time of the studies. Older girls were healthy and denied substance abuse. None had a history compatible with an eating disorder, and none was receiving any form of steroid hormones or mineral supplements. Written informed consent was obtained from a parent or legal guardian for each study; written assent was obtained from girls 7 yr and older. The Institutional Review Board of Baylor College of Medicine approved all study protocols.

Study Protocol

Older Children. The approach we used to conduct calcium absorption and kinetic studies in children was similar to that described previously for radioisotope calcium kinetic studies in adults (Neer et al., 1967) and is described below briefly. All research studies were conducted within an inpatient metabolic research unit at the Children's Nutrition Research Center in Houston, TX. Early in the morning of the first day of the study, a heparin lock catheter was inserted and 10 mL of blood was withdrawn from the catheter and used for routine laboratory tests and hormone levels. Subjects were instructed to empty their bladders and then were given a breakfast meal. Toward the end of the meal, the subjects were given an isotope of calcium which had been premixed (and allowed to equilibrate in the refrigerator for 12-24 h) with 120 mL of milk. Either ^{44}Ca (0.5 mg/kg) or ^{46}Ca (0.5 μg/kg) is usually used for oral administration. After breakfast, a different calcium isotope (0.08 mg/kg of ^{42}Ca) was administered intravenously over 5 minutes via the heparin lock catheter. After infusion, the heparin lock was infused with 3 mL saline and the hub of the catheter was replaced. Subsequently, 2 mL blood was removed for mass spectrometric analysis at 6, 12, 20, 30, 45, 60, 120, 180, 240, and 480 min after completion of the infusion.

Beginning with breakfast, a complete 24-h urine collection was performed in 8-hour aliquots. A complete record including weighed food intakes was maintained during this 24-hour period. At the end of the 24-hour period, the subjects were discharged and asked to collect three random urine samples each day for 5 additional days and to provide a 48-h dietary record. Nutrient intakes were calculated using a standard nutrient database system. For studies in which endogenous fecal calcium excretion was directly determined, a larger dose of ^{42}Ca (0.25-0.3 mg/kg) was infused intravenously and a complete in-patient 7-day urine and stool collection was performed.

Premature Infants. Premature infants were studied during their hospitalization in the nurseries of Texas Children's Hospital. Approximately 18 hours before the beginning of the study, one-half volume of a single feeding was mixed with ^{44}Ca (0.8 mg/kg) and refrigerated until the study began. The next morning, the infants were given ^{46}Ca (0.01 mg/kg) intravenously over 3-5 minutes using a butterfly infusion set. The premixed feeding with the added ^{44}Ca was then fed. The remainder of the feeding was given with the same syringe and feeding tube or bottle.

At 6 and 10 minutes after completion of the isotope infusion, blood samples were obtained by venipuncture and the serum analyzed for isotope ratio measurements. The infants were then placed on a metabolic bed for the continuous collection of urine in 6-h pools for 72 hours. At that point, the infants were removed from the metabolic bed and three random urine samples were collected daily for the next 3 days. Feces were collected in plastic bags that were attached to the buttocks for a total of 120 hours after isotope administration. Each stool was analyzed separately. Milk intake was determined from samples obtained from tared feeding syringes or bottles.

Analytical Methods

Calcium was isolated from serum and urine samples by precipitation using ammonium oxalate (Yergey et al., 1995). After precipitation, calcium samples were baked in a muffle furnace and resuspended in dilute nitric acid. Five microliters of suspension were loaded onto a multi-sample turret and placed in the mass spectrometer for analysis. Samples were analyzed for isotopic enrichment with a Finnigan MAT 261 (Bremen, Germany) magnetic sector thermal ionization mass spectrometer. Each sample was analyzed for the ratio of $^{42}Ca/^{43}Ca$ and $^{46}Ca/^{43}Ca$ with correction for fractionation to the reference $^{44}Ca/^{43}Ca$ ratio (older children). Comparison of the ion intensity of an administered isotope vs. a non-administered, naturally occurring isotope (^{43}Ca) allows for increased measurement precision. Accuracy of this technique for natural abundance samples compared to standard data is 0.1%. Precision, including sequential measurement of the same sample (on different filaments) over a period of time, is 0.15% or better.

Selection of Isotopes

There are multiple considerations to balance in choosing which of the five minor stable isotopes of calcium to administer in a dual-tracer study. The first of these is the analytical capacity and accuracy of the mass spectrometer. For example, some mass spectrometers are unable to measure ^{46}Ca accurately. For other mass spectrometers, certain combinations of administered isotopes may be more rapidly and/or more readily measured than others. A second consideration is ensuring that the dose of isotope given intravenously is not so large as to perturb calcium homeostasis. Finally, consideration must be given to the cost and availability of isotopes. Although recently there was significant concern regarding this issue (Abrams et al., 1992b), currently these isotopes are readily available, although their cost remains substantially greater than would be desired for maximum research utilization.

We will consider these issues for a typical, widely utilized type of mass spectrometer used for mineral isotope ratio analyses, the multi-cup magnetic sector thermal ionization mass spectrometer such as the Finnigan MAT 261 (or 262, Finnigan, Inc., Bremen, Germany). In general, these mass spectrometers are constructed so as to have a central, non-movable Farraday cup and 3-4 Farraday cups on either side of this cup. For calcium, a typical mass range which can be measured simultaneously would be 4 atomic mass units (i.e. ^{42}Ca-^{46}Ca).

The isotope mass distributions for calcium are approximately as follows: ^{40}Ca – 96.9%; ^{42}Ca – 0.646%; ^{43}Ca – 0.13%; ^{44}Ca – 2.08%; ^{46}Ca – 0.0032%; ^{48}Ca – 0.18% (Moore and Machlan, 1972; Moore et al., 1985). Because of the large abundance of ^{40}Ca, it is not usually administered in tracer studies. Therefore some combination of ^{42}Ca, ^{43}Ca, ^{44}Ca and ^{46}Ca is usually administered. It should be noted that masses outside this range (i.e. ^{48}Ca) can be measured using special "jumping" programs, but this process is more time-consuming and less accurate than taking simultaneous measurements of the ^{42}Ca–^{46}Ca range.

In a dual-tracer study, it is optimal to administer two isotopes to the study subject and measure the ratio of two different unadministered isotopes. The ratio of the unadministered isotopes is utilized for fractionation correction. Using this approach, increased measurement precision and an accuracy of <0.2% can routinely be obtained for all of the calcium isotopes, including ratios involving ^{46}Ca.

We usually use ^{46}Ca as the first of the two isotopes in our studies. This is done to prevent having to use a very small ratio, such as the $^{46}Ca/^{43}Ca$ ratio for fractionation correction. The choice of the second isotope to administer is generally related to the cost and availability of these isotopes. ^{42}Ca or ^{43}Ca is usually chosen as the second isotope in order to minimize the total calcium dose administered to the subjects, which might need to be relatively high if ^{44}Ca was given intravenously.

Once the two isotopes to be administered are chosen, it is not critical which is given orally and which is administered intravenously. When ^{42}Ca is given intravenously, the total dose for a 7-day kinetic study is approximately 0.1 mg/kg in pubertal children. As the ^{42}Ca is generally at least 80% enriched ^{42}Ca, the total calcium given remains small relative to the initial distribution volume for calcium. The exact doses of isotopes utilized in any given study will depend on the age and pubertal status of the subject. In general, for kinetic studies in children and adolescents, we administer a dose of ^{42}Ca of 0.08 to 0.1 mg/kg intravenously and for ^{46}Ca we use a dose of 0.4 - 0.5 μg/kg orally.

Modeling

The compartmental model used for calcium kinetics is similar to that described by Neer et al. (1967). Our model is based on three sequential pools prior to calcium deposition in the "deep" bone calcium pool. Bone calcium deposition (V_O+) is the rate of flow of calcium to the final pool. The mass of the exchangeable calcium pool (EP) is determined as the sum of the masses of the three pools determined from the model.

Endogenous fecal excretion of calcium was estimated in our studies of older children as 1.5 mg/kg/day (Abrams et al., 1996). Errors in kinetic rates related to this estimate are likely to be minimal, as even an improbably large error in this estimate (i.e., 0.5 mg/kg/day) would have at most a 2-3% net effect on the calculated kinetic parameters in the older adolescents in this study. In premature infants, endogenous fecal calcium excretion was directly measured using the fecal recovery of the intravenously administered tracer (Abrams et al., 1994b).

Compartmental modeling of the data was performed with the aid of the SAAM (Simulation, Analysis, and Modeling) program (Berman and Weiss, 1978). Details of this program and its application to calcium metabolism have been described (Neer et al., 1967).

RESULTS AND DISCUSSION

Summary data for studies in 13 premature infants and 62 older children are shown in Table 1. Comparison data are included for adults. Data for V_O+ and EP from the 43 postmenarcheal girls are shown in Figs.1 and 2. These data show a gradual decline in these values throughout the initial time period post-menarche. These data and limited longitudinal data suggest that the peak time for bone calcium deposition is shortly prior to menarche. Although a large decline in V_O+ occurs during this period, the rate of V_O+ at 60 months post-menarche remains well above the value for perimenopausal women (approximately 350 mg/day) (Heaney et al., 1978). The decline we found in V_O+ post-menarche is similar to data from Wastney et al. encompassing both adolescents and young adult women (1996).

Table 1. Developmental changes in calcium kinetics

Study Group	n	Age (yr)	Wt (kg)	*Vo+* (g/d)	*Vo–* (g/d)	EP (g)	*Vo+*/EP (d^{-1})
Preterm Infants[a]	13	–	1.3 ± 0.03	0.21 ± 0.009	0.12 ± 0.008	0.65 ± 0.008	0.32
Prepubertal Girls (Tanner 1)	25	8.3	25.4 ± 1.2	1.42 ± 0.06	1.30 ± 0.06	2.13 ± 0.11	0.67
Pubertal Girls (Tanner 2/3/4)	29	12.7 ± 0.4	43.7 ± 1.6	1.92 ± 0.11	1.76 ± 0/11	3.37 ± 0.17	0.57
Post-pubertal Girls (Tanner 5)	8	15.5 ± 0.6	52.4 ± 3.1	1.01 ± 0.13	0.96 ± 0.11	2.45 ± 0.29	0.41
Adults	>200	~40	–	0.34[b]	0.36[b]	–	–

[a]Abrams et al., 1994. [b]Heaney, 1978

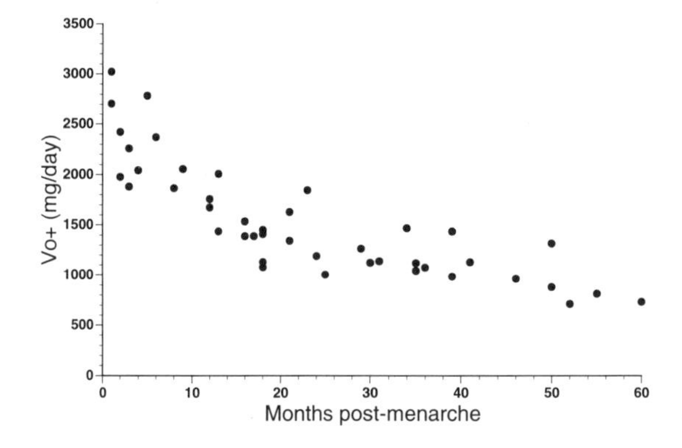

Figure 1. Relationship between the number of months post-menarche and the bone calcium deposition rate, V_O+ The linear relationship between these is $V_O+ = -28.2 * (\text{months}) + 2176$, $r = -0.82$, $p < 0.01$).

Although this study did not relate calcium kinetic findings to the hormonal status of the study subjects, it is interesting to consider our data in light of recent findings regarding the importance of estrogen in skeletal metabolism. For example, it has recently been shown that in the absence of estrogen, a healthy male failed to have epiphyseal fusion (Smith et al., 1994). The decrease in V_O+ with advancing puberty seen in this study may be an effect of estrogen on mineralizing new bone and subsequent closure of epiphyseal plates. This view is supported by the recent report of Blumsohn et al. (1994) in which a negative correlation was found between estradiol levels and biochemical markers of bone turnover.

The differences in kinetic findings between adults and premature infants may be related to the relatively greater rate at which absorbed calcium is used by premature infants for mineralization of new bone (Abrams et al., 1994b). In older individuals, new bone formation accounts for only a small proportion of bone turnover and probably cannot be greatly increased. Therefore, an increase in absorbed calcium seems to be handled by a decrease in bone resorption (Bronner, 1994). This is similar to findings from an animal model in which rats, raised on a Ca-sufficient diet, maintained constant levels of plasma calcium as their absorbed calcium increased by inhibiting bone calcium resorption, while bone calcium deposition remained virtually unchanged (Sammon et al., 1970). In contrast, immature rats increased their bone calcium deposition rate in response to an increase in absorbed calcium (Hurwitz et al., 1969).

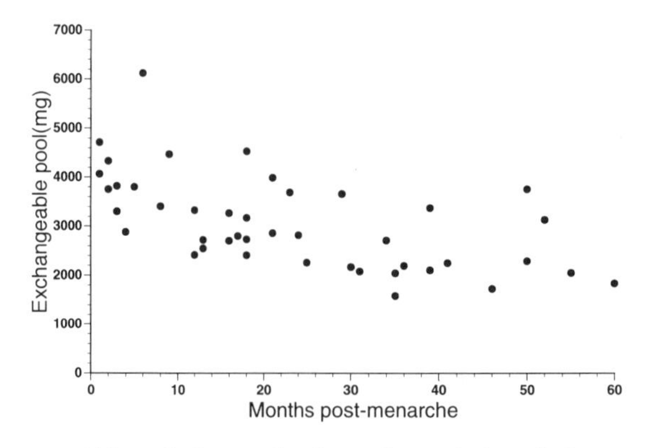

Figure 2. Relationship between the number of months post-menarche and the mass of the exchangeable pool, EP. The linear relationship between these is $EP = -33.7 * (\text{months}) + 3834$, $r = -0.59$, $p < 0.01$)

Although the mechanisms for variations in bone accretion and resorption with age and dietary intake are unknown, hormonal factors may be involved. Bronner has suggested that parathyroid hormone (PTH) may act to alter the regulatory relationship between V_O+ and V_O-. For example, the relative importance of V_O+ compared to V_O- in response to alterations in dietary calcium intake is increased in parathyroidectomized animals compared with euparathyroid animals (Bronner et al., 1982). The slowing of bone resorption in the face of increased absorbed calcium appears to be partly mediated by the release of calcitonin which occurs in response to a transient increase in plasma calcium (Talmage et al., 1975).

We also found a decrease in the EP mass post-menarche (Fig.2). Even more striking is the decrease in EP mass on a body-weight basis between the third trimester fetus (as represented by the data in premature infants, Table 1) and in older children. These results are similar to our earlier reports in older infants and to data from early studies in infants (Abrams et al., 1992a; Bauer et al., 1957). Based on estimates of total body calcium at different stages of development, the EP in premature and older infants represents approximately 5% of total body calcium compared to 0.4-0.5% in adults (Bronner, 1962; Bronner, 1982).

As the EP is a mathematically derived entity, its exact anatomical representation is uncertain. It probably consists primarily of calcium in metabolically active bone which is in equilibrium with plasma and extracellular fluid calcium (Abrams et al., 1994a). Dickerson has shown that mineralization of long bones during childhood proceeds most rapidly in the epiphyseal areas (Dickerson, 1992). These data are consistent with reports by Bauer et al. (1961), demonstrating in rat tibias that approximately 5% of the calcium in the ends and 1% in the shafts were part of the EP. It is therefore likely that the EP is principally located at these and other trabecular bone sites. Autoradiographic studies using ^{45}Ca in animals have demonstrated more rapid uptake of ^{45}Ca in recently formed osteons than in older bone tissue, further supporting the concept of a metabolically active EP (Ampino, 1952; Arnold, 1956).

Our data demonstrate the dynamic changes in calcium metabolism that occur during childhood which may be explored using kinetic techniques and mathematical modeling. Future studies will need to focus on the physiological basis for these changes and the consequences of acute and chronic illnesses on these changes and on ultimate bone mineralization. For example, conditions such as Turner's Syndrome and anorexia nervosa represent frequently occurring conditions of hormonal deficiency in children and adolescents. The widespread use of medications such as growth hormone and cortico-steroids may also be evaluated for their effects on mineral kinetics in children. Finally, multimineral kinetic studies, including calcium, magnesium, and zinc, are now feasible in children, and may be applied to the evaluation of the relationship among these minerals in health and disease.

ACKNOWLEDGEMENTS

This work is a publication of the U.S. Department of Agriculture (USDA)/Agricultural Research Service (ARS) Children's Nutrition Research Center, Department of Pediatrics, Baylor College of Medicine and Texas Children's Hospital, Houston, TX. This project has been funded in part with federal funds from the USDA/ARS under Cooperative Agreement number 58-6250-1-003 and the NIH, NCRR General Clinical Research for Children Grant number M01RR00188. Contents of this publication do not necessarily reflect the views or policies of the USDA, nor does mention of trade names, commercial products, or organizations imply endorsement by the U.S. Government.

CORRESPONDING AUTHOR

All correspondence and reprint requests should be directed to:
Steven A. Abrams, M.D.
Children's Nutrition Research Center
1100 Bates Street
Houston, Texas 77030-2600
Phone: (713) 798-7124
FAX (713) 798-7119
e-mail: sabrams@bcm.tcm.edu

REFERENCES

Abrams SA; Esteban NV; Vieira NE; Sidbury JB; Specker BL; Yergey AL. Developmental changes in calcium kinetics in children assessed using stable isotopes. *J Bone Min Res,* 1992a, 7:287-93.

Abrams SA; Klein PD; Young VR; Bier DM. Letter of concern regarding a possible shortage of separated isotopes. *J Nutr,* 1992b, 122:2053.

Abrams SA; Silber TJ; Esteban NV; Vieira NE; Stuff JE; Meyers R; Majd M; Yergey AL. Mineral balance and bone turnover in adolescents with anorexia nervosa. *J Pediatr,* 1993, 123:326-331.

Abrams SA; Schanler RJ; Yergey AL; Vieira NE; Bronner F. Compartmental analysis of calcium metabolism in very low birth weight infants. *Pediatr Res,* 1994, 36:424-8.

Abrams SA; O'Brien KO; Stuff JE. Changes in calcium kinetics associated with menarche. *J Clin Endocrinol Metab,* 1996, 81:2017-2020.

Amprino R. Autoradiographic analysis of the distribution of labelled Ca and P in bones. *Experimentia,* 1952, 8:20-2.

Arnold JS; Jee WSS; Johnson K. Observations and quantitative radioautographic studies of calcium[45] deposited in vivo in forming haversian systems and old bone of rabbit. *Am J Anat,* 1956, 99:291-308.

Bauer GCH; Carlsson A; Lindquist B. Bone salt metabolism in humans studied by means of radiocalcium. *Acta Medica Scandinavica,* 1957, 158:143-50.

Bauer GCH; Carlsson A; Lindquist B. Metabolism and homeostatic function of bone, in: *Mineral Metabolism: An Advanced Treatise, Vol. I, Principles, Processes, and Systems, Part B.* Comar Cl; Bronner F; Eds. Academic Press: New York, 1961. pp. 609-676.

Berman M; Weiss M. *SAAM Manual,* USDHEW Publication No. (NIH) 78-180, Washington DC, 1978.

Blumsohn A; Hannon RA; Wrate R, et al. Biochemical markers of bone turnover in girls during puberty. *Clin Endocrinol,* 1994, 40:663-70.

Bronner F; Harris RS. Absorption and metabolism of calcium in human beings, studied with calcium[45]. *Ann N Y Acad Sci,* 1956, 64:314-25.

Bronner F. Dynamics and function of calcium, in: *Mineral Metabolism: An Advanced Treatise, Vol. II, The Elements, Part A.* Comar CL; Bronner F; Eds. Academic Press: New York. 1964. pp. 341-344.

Bronner F. Calcium homeostasis, in: *Disorders of Mineral Metabolism, Vol. II: Calcium Physiology.* Bronner F; Coburn JW; Eds. Academic Press: New York. 1982. pp. 45-102.

Dickerson JWT. Changes in the composition of the human femur during growth. *Biochem J,* 1962, 32:56-63.

Hachey DL; Wong WW; Boutton TW; Klein PD. Isotope ratio measurements in nutrition and biomedical research. *Mass Spectrometry Rev,* 1987, 6:289-32.

Heaney RP; Whedon GD. Radiocalcium studies of bone formation rate in human metabolic bone disease. *J Clin Endo Met,* 1958, 35:1246-67.

Heaney R; Skillman TG. Calcium metabolism in normal human pregnancy. *J Clin Endocr,* 1971, 33:661-70.

Heaney RP; Recker RR; Saville PD. Menopausal changes in bone remodeling. *J Lab Clin Med,* 1978, 92:964-70.

Hurwitz S; Stacey RE; Bronner F. Role of vitamin D in plasma calcium regulation. *Am J Physiol,* 1969, 216:254-62.

Mauras N; Haymond MW; Darmaun D; Vieira NE; Abrams SA; Yergey AL. Protein and calcium kinetics in prepubertal boys: Positive effects of testosterone. *J Clin Invest,* 1994, 93:1014-1019.

Moore LJ; Machlan LA. High accuracy determination of calcium in blood serum by isotope dilution mass spectrometry. *Anal Chem,* 1972, 44:2291-6.

Moore LJ; Machlan LA; Lim MO; Yergey AL; Hansen JW. Dynamics of calcium metabolism in infancy and childhood. I. Methodology and quantification in the infant. *Pediatr Res,* 1985, 19:329-34.

Neer R; Berman M; Fisher F; Rosenberg LE. Multicompartmental analysis of calcium kinetics in normal adult males. *J Clin Invest,* 1967, 46:1364-78.
O'Brien KO; Abrams SA. Effects of development on techniques for calcium stable isotope studies in children. *Biol Mass Spectromet,* 1994, 23:357-61.
Sammon PJ; Stacey RE; Bronner F. Role of parathyroid hormone in calcium homeostasis and metabolism. *Am J Physiol,* 1970, 218:479-485.
Smith EP; Boyd J; Frank GR; et al. Estrogen resistance caused by a mutation in the estrogen-receptor gene in a man. *N Engl J Med,* 1994, 331:1056-61.
Talmage RV; Doppelt SH; Cooper CW. Relationships of blood concentrations of calcium, phosphate, gastrin, and calcitonin to the onset of feeding in the rat. *Proc Soc Exp Biol Med,* 1975, 149:855-859.
Wastney ME; Ng J; Smith D; Martin BR; Peacock M; Weaver CM. Differences in calcium kinetics between adolescent girls and young women. *Am J Physiol,* 1996, 271:R208-216.
Yergey AL; Vieira NE; Abrams SA; Marini J; Goans RE. Use of stable isotopic tracers in studies of whole body calcium metabolism. *Connective Tiss Res,* 1995, 31:291-293.

MODELING OF ENERGY EXPENDITURE AND RESTING METABOLIC RATE DURING WEIGHT LOSS IN HUMANS

Janet A. Novotny and William V. Rumpler

U.S. Department of Agriculture
Agricultural Research Service
Beltsville Human Nutrition Research Center
Diet and Human Performance Laboratory
Beltsville, MD 20705

INTRODUCTION

With obesity prevalent and increasing in the United States, there is substantial interest in human energy balance. According to results from the Third National Health and Nutrition Examination Survey (NHANES III), the proportion of overweight Americans has increased from 25% to 33% between 1980 and 1991 (Kuczmarski et al., 1994). Overweight and obesity are known to increase an individual's risk for diabetes (NIDDK, 1995), heart disease, hypertension, and gallbladder disease (Sjostrom, 1992a), and cancer mortality (Sjostrom, 1992b).

The economic costs associated with obesity are enormous. Colditz (1992) calculated the costs associated with obesity-related diseases for 1986. The overall cost of illnesses associated with obesity was $39.3 billion, 5.5% of total illness costs in 1986. For specific diseases, the costs were $11.3 billion for Non-insulin Dependent Diabetes Mellitus, $22.2 billion for cardiovascular disease, $2.4 billion for gall bladder disease, $1.5 billion for hypertension, and $1.9 billion for breast and colon cancer.

An additional cost related to obesity is that which individuals spend toward weight loss programs and aids. At any given time, 33-40% of adult American women and 20-24% of adult American men are attempting weight loss, often by enrolling in expensive weight reduction programs (National Institutes of Health Technology Assessment Conference Statement, 1992). Americans spend $33 billion annually on weight-reduction products and services, including diet foods, products, and programs (Colditz, 1992).

The prevalence, economic costs, and health implications of obesity continue to fuel scientific research in this area. Mathematical modeling is an important tool for use to elucidate the physiology related to weight gain and loss. The purpose of this chapter is to describe the essential features of energy modeling in humans and to illustrate a model which we have developed to investigate the influence of organ mass decreases on resting energy expenditure during weight loss.

Mathematical Modeling in Experimental Nutrition
Edited by Clifford and Müller, Plenum Press, New York, 1998

Mathematical modeling is an important tool for the investigation of complex systems such as human energy balance. As systems become increasingly complex, a system of bookkeeping is required to develop a unified theory about the system functioning, and mathematics provides that system of bookkeeping. A mathematical model serves as a tool which captures important aspects of a real world system within the framework of a mathematical apparatus such that the model provides a researcher with a means of exploring the system mirrored in the model. Because a mathematical model is a hypothesis about a system's functioning and produces simulation results which can be compared to observation, a model acts as a bridge between theory and experiment.

BACKGROUND: RESTING ENERGY EXPENDITURE DURING WEIGHT LOSS

Many studies have shown a decline in resting energy expenditure (REE) during weight reduction (Hill et al., 1987; Ravussin et al., 1985; Valtuena et al., 1995). Because resting energy expenditure accounts for 60% of total energy expenditure, a change in REE has significant potential to influence energy balance and success of weight reduction. For this reason, the mechanism behind the reduction in REE observed during weight loss has generated significant interest among scientists and health professionals.

The most important factor in determining resting energy expenditure is lean mass. The majority of resting energy expenditure (~60%) arises from organs, a component of lean mass which account for only about 5-6% of total body weight. Many researchers have observed a reduction in the ratio of resting energy expenditure per unit lean mass during weight reduction (Bessard et al., 1983; Elliot et al., 1989; Fricker et al., 1991; Hill et al., 1987; Valtuena et al., 1995; Wadden et al., 1990). Such a change may be caused by a decrease in metabolic activity per unit tissue or by a redistribution of the influence of high and low activity tissues on resting energy expenditure. Because fat-free mass is comprised of many tissue types, changes in high-activity tissues would strongly influence REE. For example, the average kidney metabolic activity is 440 kcal/kg/d, while the metabolic activity of skeletal muscle is only 13 kcal/kg/d. Therefore, a reduction in kidney mass would have a larger impact on REE than the same mass reduction in skeletal muscle.

Animal studies have shown that the masses of different lean tissue types have differing responses to varied levels of energy intake (Ferrell et al., 1986; Koong and Nienaber, 1985; Koong et al., 1983). Particularly sensitive to food intake level were the liver, GI tract, and kidneys. Based on these findings, we propose that the masses of different lean tissue types in adults respond differentially to energy deficit and subsequently lead to depression of resting energy expenditure during weight reduction.

In developing our model of energy expenditure, our goal was to investigate the changes in lean tissue mass which could account for the depression in resting energy expenditure during weight loss.

METHODOLOGY – MODEL DEVELOPMENT

When starting a modeling endeavor, a helpful initial step is to draw a diagram including all relevant components of the system. Once the important pools are determined, their interactions should be represented on the diagram by arrows. This facilitates inclusion of the important mathematical equations. A schematic of a simple energy balance model is shown in Figure 1. Energy in the form of food enters the available energy pool, and energy exits the energy pool as it is consumed by energy-expending processes. Energy in excess of that expended enters the fat mass and lean mass pools. When energy intake does not meet energy demand, the remaining energy required is provided by the fat and lean pools. The arrows from the energy pool to the fat pool and lean pool are dashed because in this model we are considering only the case of energy deficit, therefore energy is not stored.

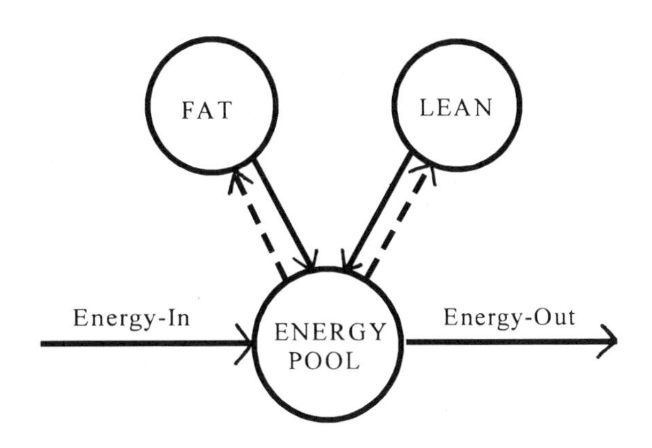

Figure 1. Schematic of the model system. Energy as food enters the *Energy Pool* via the arrow labeled *Energy-In*. *Energy-Out* represents total energy expenditure. When energy intake is less than energy expenditure, the remaining energy requirement is provided by the *Fat* and *Lean* pools. In the case of excess energy intake, the excess moves from the *Energy Pool* to the *Fat* and *Lean* pools. The arrows representing transfer of energy from the *Energy Pool* to the *Fat* and *Lean* pools are dashed because this model deals only with energy deficit.

Once the schematic is drawn, the proper form of each mathematical equation must be considered. Differential equations are derived and the resulting equations may be of many forms, such as linear, exponential, or of Michaelis-Menton formalism, as well as many other alternatives. There are several approaches to determine the mathematical relationships which describe dynamic components of a model, including direct observation, mechanistic derivations, probability-based derivations, and dimensional analysis. Dimensional or unit analysis can provide assistance in suggesting equation forms and should always be used to check derived equations. For example, given the mass of a tissue and the tissue's energy use per mass per unit time, dimensional analysis suggests that the total energy used by the tissue per unit time is described as follows:

$$\text{E Use (kcal/day)} = \text{Tissue Mass (kg)} \times \text{Rate of E Use (kcal/kg/day)} \tag{1}$$

Direct observation refers to development of equations by fitting existing data to an equation form. For example, many equations for resting energy expenditure are based on direct observation. Figure 2 shows the relationship between REE and body weight for subjects who have participated in studies at the Beltsville Human Nutrition Research Center. By direct observation and linear regression, one can see that REE is related to weight by the following equation:

$$\text{REE} = 362.19 + 14.92 \times \text{Weight.} \tag{2}$$

Mechanistic-based equations are derived directly from theoretical arguments. Again using the example of resting energy expenditure, one might assume that the important tissues determining basal metabolic rate are lean tissue and fat tissue. Therefore, resting energy expenditure should be the sum of the basal activities of those tissue types. Based on this argument, resting energy expenditure could be expressed as

$$REE = (Fat\ Mass) \times (Fat\ EE\ per\ Mass) + (Lean\ Mass) \times (Lean\ EE\ per\ Mass) \quad (3)$$

where EE represents energy expenditure. We tested an equation of this form for our energy modeling, but this equation did not provide us with accurate predictions of REE, as described in more detail below.

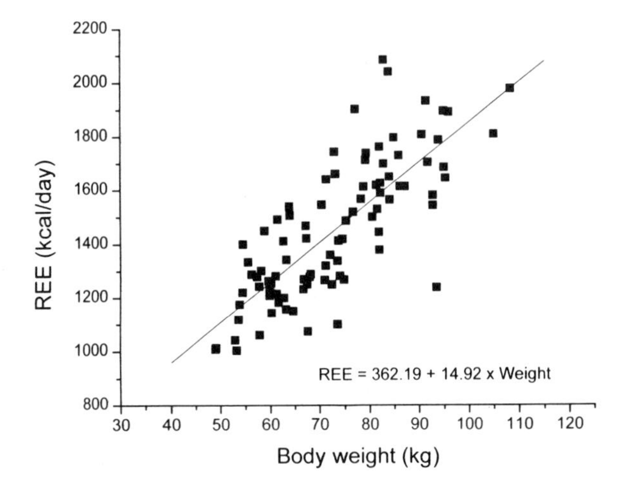

Figure 2. Resting energy expenditure as a function of body weight. The squares are observed data points, and the line represents a linear fit to that data. Regression for generation of mathematical relationships is an example of derivation of model equations by direct observation.

Probability relationships are developed by listing all possible outcomes and assigning a probability to each. Simulations are then performed using a random number generator associated with the probabilities. Probability relationships are particularly useful in population modeling. In developing our model of energy balance in humans, we used direct observation and mechanistic arguments to develop our equations, and dimensional analysis to check the derived relationships.

In the energy model shown in Figure 1, *Energy-In* is simply energy associated with food intake. It can be constant or varying during the weight loss period. We have described *Energy-Out* as total energy expenditure (TEE). Total energy expenditure is most often predicted as resting energy expenditure times an activity factor (National Research Council, 1989). A more physiologic means of estimating total energy expenditure is to predict the energy associated with various components of expenditure. In our model, TEE was expressed as the sum of resting energy expenditure (REE), dietary-induced thermogenesis (DIT), and activity:

$$TEE = REE + DIT + Activity \quad (4)$$

where TEE, REE, DIT, and Activity were expressed in MJ/day.

Resting energy expenditure is most commonly predicted by regression equations based on body weight or lean mass. To develop a model in which the equations were physiologically based, we first attempted to describe resting energy expenditure as fat mass times the metabolic activity of fat mass plus lean mass times the metabolic activity of lean mass (based on Eq. 3) as follows:

$$REE = (4.5\ kcal/kg/d) \times Fat\ Mass + (30\ kcal/kg/d) \times Lean\ Mass \quad (5)$$

where Fat Mass and Lean Mass were expressed in kg. The value of 4.5 kcal/kg/d for fat mass was reported by Elia (1992). We calculated the value of 30 kcal/kg/d to be the weighted average of lean mass activity from metabolic activities of organs and skeletal muscle tissue reported by Elia (1992).

Because fat mass and lean mass are readily measured in human subjects, this equation could be tested against experimental observation. This equation and several others were compared to REE measurements (n=93) for subjects participating in studies at the Beltsville Human Nutrition Research Center. Figure 3 shows the percent discrepancy for the equations tested. While this equation predicted resting energy expenditure fairly well, we found that several of the published regression equations for resting energy expenditure predicted more accurately. The poor prediction by this equation may have been caused by errors associated with measurement of lean and fat mass, errors associated with tissue activities, the varied composition of lean mass, and/or diversity in tissue activities for different subjects. For this reason, we continued model development using a regression equation.

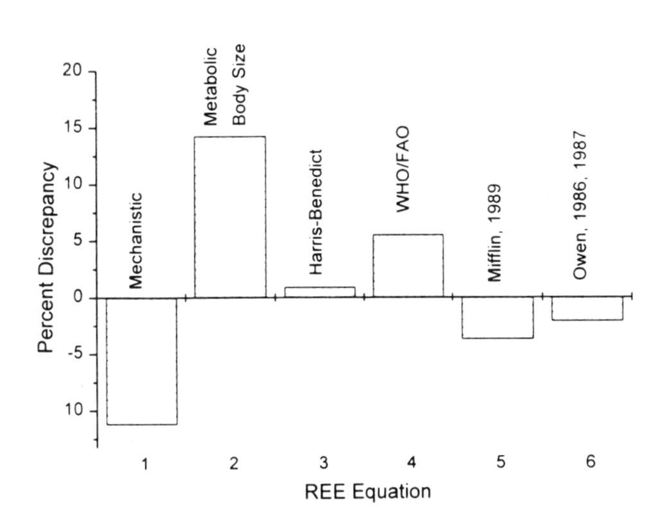

Figure 3. Percent discrepancy for several equations for prediction of individual resting energy expenditure. EQ. 1: Our mechanistic equation, 4.5 x Fat Mass + 30 x Lean Mass; EQ. 2: Harris Benedict equation (Harris and Benedict, 1919); EQ. 3: WHO/FAO regression equations (NRC, 1989); EQ. 4: Metabolic body size, 70 x Weight $(kg)^{3/4}$ (Guthrie, 1986); EQ. 5: Mifflin regression equations (Mifflin et al., 1990); EQ. 6: Owen regression equations (Owen et al., 1986; Owen et al., 1987).

Several regression equations were tested for accurate prediction of REE at weight maintenance for adult male subjects who had participated in studies at the Beltsville Human Nutrition Research Center. An equation developed by Cunningham (1980) was found to most accurately predict maintenance resting energy expenditure in our subjects, and this equation was used in the model to predict maintenance REE:

$$\text{REE} = (0.120 + 0.00516 \times \text{Lean Mass}) \tag{6}$$

where REE was expressed in MJ/d, and Lean Mass was expressed in kg.

To test our hypothesis that decreases in organ masses can account for the observed depression in resting energy expenditure during weight reduction, it was necessary to modify the REE equation for energy deficit. It was hypothesized that the drop in REE during the first week of energy deficit was the result of a decrease in size of one or more high-activity tissues. It has been shown that decreased level of food intake results in decreased visceral organ size in growing animals (Ferrell et al., 1986; Koong et al., 1983; Koong and Nienaber, 1985). Because liver, kidney, and GI tissues were all found to be very sensitive to level of food intake in animals, the model was modified to include a decrease in the masses of these tissues during energy deficit. To produce a decrease in REE of similar magnitude to that seen in human studies at our facility, we adjusted the REE equation to reflect a 7.7% decrease in total liver, kidney, and GI tract activity, based on the average organ size and metabolic activity of a 70 kg reference man (Elia, 1992). Therefore, for a 70 kg man, the liver would decrease from 1.8 kg to 1.66 kg, the GI would decrease from 1.4 kg to 1.29 kg, and the kidneys would decrease from 0.31 kg to 0.29 kg. These represent very small changes in visceral organ mass. The metabolic activities of these organs have been reported as follows: the liver consumes 200 kcal/kg/day, the GI tract consumes 240 kcal/kg/day, and the kidneys consume 440 kcal/kg/day (Elia, 1992). With this information incorporated, the adjusted REE equation was as follows:

$$
\begin{aligned}
\text{REE} = \; & (0.120 + 0.00516 \times \text{Lean Mass}) \\
& - 0.077 \times G(t) \times (1.8 \text{ kg} \times 200 \text{ kcal/kg/d}) \\
& - 0.077 \times G(t) \times (1.4 \text{ kg} \times 240 \text{ kcal/kg/d}) \\
& - 0.077 \times G(t) \times (0.31 \text{ kg} \times 440 \text{ kcal/kg/d})
\end{aligned} \quad (7)
$$

where G(t) was a function equal to (Days/Days+2) to allow the changes in visceral organ mass to occur gradually rather than instantaneously, with approximately 80% of the mass response occurring during the first week of energy deficit. The first term in the equation represents the maintenance resting energy expenditure. The second term in the equation subtracts energy related to decrease in liver mass, the third term subtracts energy related to decrease in GI tissue mass, and the fourth term subtracts energy related to decrease in kidney mass.

Dietary-induced thermogenesis (DIT), the second component of our total energy expenditure equation, represents the energy expended for food digestion. DIT ranges from about 10% to 15% of food intake (Miles et al., 1993; Romon et al., 1993). In our model, DIT was calculated as

$$\text{DIT} = 0.12 \times \text{Energy Value of Food Intake} \quad (8)$$

where the Energy Value of Food Intake was measured in MJ.

Energy expended by activity, the third component of our total energy expenditure equation, was scaled to body weight by the following equation:

$$\text{Activity (MJ/d)} = \text{Body Weight (kg)} \times \text{Activity Factor (MJ/d/kg)} \quad (9)$$

By scaling activity-related energy use to body weight, activity-related energy expenditure decreased with decreasing body weight during energy deficit, as expected for energy expended for activities of locomotion. The activity factor was calculated prior to weight loss as

$$\text{Activity Factor} = (\text{TEE} - \text{REE} - \text{DIT}) / \text{Body weight} \quad (10)$$

This equation requires a prior estimate of TEE, while REE and DIT can be calculated as described above. An alternative would be to estimate energy associated with activity by activity records prior to modeling, and calculate the activity factor as follows:

$$\text{Activity Factor} = \text{Activity EE/Body weight} \quad (11)$$

The activity factor was assumed to remain constant throughout the weight loss period.

Given total energy expenditure and energy intake, the energy deficit was calculated as the difference. During energy deficit, the energy required by an individual beyond that supplied by food intake may be provided by either fat mass or lean mass. Because the composition of weight loss varies among individuals and can range from 20% from fat to 100% from fat, it must be determined for each individual.

For this modeling study, we calculated composition of weight loss from measured weight and lean mass loss in individual subjects. We attempted to derive a means for prediction of the composition of weight loss from non-invasive measurements conducted on subjects before weight loss. We performed correlation analysis for composition of weight loss against several non-invasive measurements for 200 subjects who had participated in studies at BHNRC or elsewhere. The resulting correlations were not strong enough to supply a means of prediction of composition of weight loss. Therefore, we used the measured value of the composition of weight loss for individual subjects as input, and tested other aspects of the model for validation.

The composition of weight loss was entered as the fraction of total mass loss associated with lean tissue (F_L). Given F_L and the energy deficit, the equations for loss of energy from the fat compartment and the lean compartment could be calculated as follows:

$$\Delta \text{ Lean Energy Content (MJ/day)} = F_L \text{ x Energy Deficit (MJ/d)}$$
$$\Delta \text{ Fat Energy Content (MJ/day)} = (1\text{-}F_L) \text{ x Energy Deficit (MJ/d)} \quad (12)$$

These changes in lean energy content and fat energy content can be used to calculate changes in mass. The energy content of fat mass is 39 kJ/g and the energy content of lean mass is 5 kJ/g. Thus, the mass change associated with these energy changes was calculated as follows:

$$\Delta \text{ Lean Mass (g/day)} = \Delta \text{ Lean Energy Content (kJ/d) x 5 (kJ/g)}$$
$$\Delta \text{ Fat Mass (g/day)} = \Delta \text{ Fat Energy Content (kJ/d) x 39 (kJ/g)} \quad (13)$$

Therefore, the change in total body mass was calculated as follows:

$$\Delta \text{ Total body mass (kg/day)} = \Delta \text{ Lean Mass (kg/day)} + \Delta \text{ Fat Mass (kg/day)} \quad (14)$$

Total body mass, lean mass, and fat mass were updated at each time point with the value of each at the previous time point and the change of masses of each during the previous time interval.

Once the equations have been derived, one may assemble the equations in a modeling software package or one may construct a computer program *de novo* for the simulations. Commercial or prepackaged modeling software is useful when features of the modeling package are in accord with the requirements of a particular modeling project. In other cases, a project may have specific requirements with respect to input/output or equation forms such that it is preferable to assemble the equations into new computer code. For the energy modeling, we assembled our equations into self-written computer code. This choice offered us a great deal of flexibility in the execution of the model. The equations described above were assembled into a FORTRAN program. Output of computer simulations were written to ASCII files, which were subsequently imported to graphics or statistics programs. Model simulations were performed on a Dell Pentium 5166.

HYPOTHESIS TESTING – RESULTS AND DISCUSSION

Once again, the hypothesis to be tested by this model was that a small and relatively rapid decrease in organ mass could account for the depression in resting energy expenditure which has been observed in many subjects during weight reduction. A model is a hypothesis in mathematical form. Therefore, comparing the results of simulations (the predictions of a model) to experimental data is a means of testing the hypothesis or group of hypotheses and assumptions used to generate the model. To test our model, we compared output of weight, lean mass, and resting energy expenditure to measured values of subjects who had participated in a weight-reduction study at the Beltsville Human Nutrition Research Center.

The subjects were 8 moderately overweight adult males. The subjects were on average 39 years of age, 181 cm in height, 97 kg in body weight, and 28% body fat at the initiation of the study period. The subjects were fed at weight maintenance for 14d prior to the weight reduction period, followed by 28d at 50% maintenance. All meals consumed were provided to subjects by the Beltsville Human Nutrition Research Center Human Study Facility. Total energy expenditure at weight maintenance was determined by doubly-labeled water. Resting energy expenditure was determined during the first day of energy deficit, the seventh day of energy deficit, and the twenty-eighth day of energy deficit. REE was calculated from sleep energy expenditure. Sleep energy expenditure was determined for each subject by indirect calorimetry in a room-sized calorimeter. Resting energy expenditure was calculated as 5% above sleep energy expenditure based on results from previous studies at the Beltsville Human Nutrition Research Center. Body composition on Day 1 of energy deficit and Day 28 of energy deficit was determined by under-water weighing. All measurements and procedures were approved by a Human Subject Institutional Review Board.

Input values of body weight, fat mass, lean mass, energy intake, measured ratio of lean loss to total weight loss, and an activity factor were entered into the computer program for each simulation. Simulations were performed for each subject reflecting the energy deficit encountered on our weight loss study. In all cases, the results of model simulations were in good agreement with experimentally measured quantities.

Our specific interest in this modeling study was to compare resting energy expenditure calculated by the model to that measured for individual subjects. The mean REE for Day 1, Day 7, and Day 28 of energy deficit are shown in Table 1. The model and measured values for each subject were compared using a paired t-test. There were no significant differences between model and measured REE for any given day. Thus, comparison of the model results with experimental data has shown that the assumptions and hypotheses used to build the model are in accord with our observations.

Table 1. Comparison of model and measured REE.

Time of Measurement	Measured REE (MJ/d)	Model REE (MJ/d)
Day 1 E-deficit	8.05	8.14
Day 7 E-deficit	7.78	7.85
Day 28 E-deficit	7.62	7.72

Agreement between model prediction and experimental observation ideally leads to further testing of the model with additional experiments and elaboration of the model with increasing detail. Modeling not only provides important clues about the functioning of complex systems, but it also provides an excellent means for scientists to organize hypotheses into a unified theory to learn about a system in greater detail than would be possible otherwise.

Mathematical models now reach almost every field of biologic research. In many cases, mathematical models are no longer a luxury, but are an indispensable tool for the design and execution of experimental research and for the interpretation of their results.

CORRESPONDING AUTHOR

Please direct all correspondence to:
Janet A. Novotny
USDA, DHPL, Bldg. 308
Beltsville, MD 20705
ph: 301-504-8263
fax: 301-504-9098

REFERENCES

Bessard T; Schutz Y; Jequier E. Energy expenditure and postprandial thermogenesis in obese women before and after weight loss. *Am J Clin Nutr*, 1983, 38:680-693.

Colditz GA. Economic costs of obesity. *Am J Clin Nutr,* 1992, 55:503S-507S.

Cunningham JJ. A reanalysis of the factors influencing basal metabolic rate in normal adults. *Am J Clin Nutr,* 1980, 33:2372-2374.

Elia M. Organ and tissue contribution to metabolic rate, in: *Energy Metabolism: Tissue Determinants and Cellular Corollaries.* Kinney JM: Tucker HN; Eds. Raven Press: NY. 1992. pp. 61-77.

Elliot DL; Goldberg L; Kuehl KS; Bennett WM. Sustained depression of the resting metabolic rate after massive weight loss. *Am J Clin Nutr,* 1989, 49:93-96.

Ferrell CL; Koong LJ; Nienaber JA. Effects of previous nutrition on body composition and maintenance energy costs of growing lambs. *Br J Nutr*, 1986, 56:595-605.

Fricker J; Rozen R; Melchior JC: Apfelbaum M. Energy-metabolism adaptation in obese adults on a very low-calorie diet. *Am J Clin Nutr*, 1991, 53:826-830.

Garrow JS. *Energy Balance and Obesity in Man.* Elsevier/North Holland Biomedical Press: Amsterdam. 1978. pp. 113-144.

Guthrie HA. *Introductory Nutrition, Sixth Edition.* Times Mirror/Mosby College Publishing: St. Louis. 1986. p. 142.

Harris JA; Benedict FG. *A Biometric Study of Basal Metabolism in Man.* The Carnegie Institute: Washington, DC. 1919. pp. 1-266.

Hill OJ; Sparling PB; Shields TW; Heller PA. Effects of exercise and food restriction on body composition and metabolic rate in obese women. *Am J Clin Nutr*, 1987, 46:622-630.

Kuczmarski RJ; Johnson CL; Flegal KM; Campbell SM. Increasing prevalence of overweight among US adults. *JAMA,* 1994, 272:205-211.

Koong LJ; Nienaber JA. Changes of fasting heat production and organ size of pigs during prolonged weight maintenance, in: *Energy Metabolism of Farm Animals.* Moe PW; Tyrrell HF; Reynolds PJ; Eds. EEAP Publication No. 32. Beltsville, MD. 1985. pp. 46-48.

Koong LJ; Nienaber JA; Mersmann JH. Effects of plane of nutrition on organ size and fasting heat production in genetically obese and lean pigs. *J Nutr*, 1983, 113:1626-1631.

Mifflin MD: St. Jeor ST; Hill LA; Scott BJ; Daugherty SA; Koh Y. A new predictive equation for resting energy expenditure in healthy individuals. *Am J Clin Nutr*, 1990, 51:241-247.

Miles CW; Wong NP; Rumpler WV; Conway J. Effect of circadian variation in energy expenditure, within-subject variation and weight reduction on thermic effect of food. *Eur J Clin Nutr,* 1993, 47:274-284.

National Institutes of Diabetes and Digestive and Kidney Diseases. *Diabetes in America, 2nd Edition.* NIH publication number 95-1468. 1995.

National Institutes of Health. *Methods of Voluntary Weight Loss and Control.* NIH Technology Assessment Conference Statement: Bethesda, MD. March 30-April 1, 1992.

National Research Council. *Recommend Dietary Allowances, 10th Ed.* Peter FM; Ed. National Academy

Press: Washington, DC. 1989.
Owen OE; Kavle EC; Owen RS; Polansky M; Caprio S; Mozzoli MA; Kendrick ZV; Bushman MC; Boden GH. A reappraisal of the caloric requirement for women. *Am J Clin Nutr,* 1986, 44:1-19.
Owen OE; Holup JL; D'Alessio DA; Craig ES; Polansky M; Smalley KJ; Kavle EC; Bushman MC; Owen LR; Mozzoli MA; Kendrick ZV; Boden GH. A reappraisal of the caloric requirement for men. *Am J Clin Nutr,*. 1987, 46:875-895.
Ravussin E; Burnand B; Schutz Y; Jequier E. Energy expenditure before and during energy restriction in obese patients. *Am J Clin Nutr,*. 1985, 41:753-759.
Romon M; Edme JL; Boulenguez C; Lescroart JL; Frimat P. Circadian variation of diet-induced thermogenesis. *Am J Clin Nutr,* 1993, 57:476-480.
Sjostrom LV. Morbidity of severely obese subjects. *Am J Clin Nutr,*. 1992a, 55:508S-515S.
Sjostrom LV. Mortality of severely obese subjects. *Am J Clin Nutr,*. 1992b, 55:516S-523S.
Wadden TA; Foster GD; Letizia KA; Mullen JL. Long-term effects of dieting on resting metabolic rate in obese outpatients. *JAMA,* 1990, 264:707-711.
Valtuena S; Blanch S; Barenys M; Sola R; Salas-Salvad J. Changes in body composition and resting energy expenditure after rapid weight loss: Is there an energy metabolism adaptation in obese patients? *Int J Obesity,* 1995, 19:19-125.

DEVELOPMENT AND APPLICATION OF A COMPARTMENTAL MODEL OF 3-METHYLHISTIDINE METABOLISM IN HUMANS AND DOMESTIC ANIMALS

John A. Rathmacher[1,2] and Steven L. Nissen[1,2]

[1]Department of Animal Science
Iowa State University, Ames, Iowa 50011
[2]Metabolic Technologies Inc.
Ames, Iowa 50010

ABSTRACT

Measurement of urinary 3-methylhistidine (3MH) excretion is the primary *in vivo* method to measure skeletal muscle (myofibrillar) protein breakdown. This method requires quantitative collection of urine and is based on the assumption that no metabolism of 3MH occurs once it is released from actin and myosin. This is true in most species, but in sheep and swine a proportion is retained in muscle as a dipeptide, balenine. In neither of these species does urine 3MH yield any data on the metabolism of 3MH. We have conducted studies that propose that 3MH metabolism in humans, cattle, dogs, swine, and sheep can be defined from a single bolus infusion of a stable isotope 3-[methyl-$^{2}H_3$]-methylhistidine. Following the bolus dose of the stable isotope tracer, serial blood samples and/or urine was collected over three to five days.

A minimum of three exponentials were required to describe the plasma decay curve adequately. The kinetic linear-time-invariant models of 3MH metabolism in the whole animal were constructed by using the SAAM/CONSAM modeling program. Three different configurations of a three-compartment model are described: (A) A simple three-compartment model for humans, cattle, and dogs, in which plasma kinetics (3-[methyl-$^{2}H_3$]-MH/3MH) are described by compartment 1 and with one urinary exit from compartment 1. (B) A plasma-urinary kinetic three-compartment model with two exits was used for sheep with a urinary exit out of compartment 1 and a balenine exit out of a tissue compartment 3. (C) A plasma three-compartment model was used in swine with an exit out of a tissue compartment 3.

The kinetic parameters reflect the differences in known physiology of humans, cattle, and dogs as compared to sheep and swine that do not quantitatively excrete 3MH into the urine. Steady-state model calculations define masses and fluxes of 3MH between three compartments and, importantly, the *de novo* production of 3MH. The *de novo* production of 3MH for humans, cattle, dogs, sheep, and swine are 3.1, 6.0, 12.1, 10.3, and 7.2 $\mu mol \times kg^{-1} \times d^{-1}$, respectively.

The *de novo* production of 3MH as calculated by the compartmental model was not different when compared to 3MH production as calculated via traditional urinary collection. Additionally, data suggest that steady-state compartment masses and mass transfer rates may be related to fat free mass and muscle mass in humans and swine, respectively. In conclusion, models of 3MH metabolism have been developed in numerous species, and these models can be used for the assessment of muscle proteolysis and 3MH kinetics without the collection of urine. This methodology is less evasive and will be useful in testing further experimental designs that alter myofibrillar protein breakdown.

INTRODUCTION

The objective of this chapter is to discuss muscle proteolysis by taking an in-depth look at a novel approach to the traditional urinary 3-methylhistidine (3MH) method for estimating muscle protein turnover. We will describe the model developmental process and validation of an isotope kinetic model of 3MH metabolism. The model can be used to calculate the *de novo* production rate of 3MH, which can be used to estimate skeletal muscle proteolysis in domestic animals and humans. Applications where the model is useful for the study of alterations in muscle proteolysis will be discussed. Finally, we would like to discuss future research in the context of the model. The overall hypothesis of the described studies is that isotopic decay of a stable isotope tracer of 3MH can be described by a linear compartmental model. This model can then be used to accurately and easily estimate myofibrillar proteolysis in skeletal muscle in anabolic and catabolic situations.

Muscle Protein Turnover

In most young adult mammals, skeletal muscle constitutes about 45% of body weight, whereas with ruminants it only comprises 30 to 35% of body weight (Young, 1970). Skeletal muscle contains about 50% of the body protein and is, therefore, one of the most important tissues in protein metabolism (Simon, 1989). Schoenheimer and Rittenberg (1940) established 50 years ago that the accumulation of muscle protein depends on both the rate of muscle protein synthesis and the rate of muscle degradation. These processes occur simultaneously and occur in situations with protein gain. Despite this premise, most of the attention given by scientists during the last 50 years has focused on the relationship between the rate of growth and the rate of muscle protein synthesis. Less attention has been given to the mechanisms controlling muscle protein degradation as compared to protein synthesis, but the degradation process is potentially as important as protein synthesis in the control of muscle protein mass. This is especially true in cases of wasting and certain growth promoters.

The lack of attention placed on muscle protein degradation is primarily because methods for quantitating muscle protein degradation directly *in vivo* are limited. However, the mechanisms of protein synthesis and degradation are distinct (Reeds, 1989). Methods that have been used for studying protein metabolism *in vivo* include the use of forelimb and hind limb balances of amino acids in combination with tracer amino acids, where the net balance is the difference between the rate of tissue disposal of arterial amino acids and rate of tissue release of amino acids into the vein (Fryburg et al., 1990). However, most studies have involved indirect measurement of proteolysis, where protein synthesis can be directly measured while proteolysis is estimated by the difference from protein accretion.

Two methods commonly used to measure the fractional synthesis rate (FSR) of muscle protein are the *constant infusion* and *flooding dose* approaches. The constant infusion method determines the rate at which a constantly-infused amino acid tracer is incorporated into muscle. The specific activities of the tissue protein fractions are used in equations that are reviewed by Garlick and Clungston (1981). A second isotopic method used to measure the FSR of muscle protein synthesis is the large flooding dose (Garlick et al., 1994; Garlick et al., 1989). This method was developed to overcome precursor pool problems. A bolus of labeled amino acid (phenylalanine) is given along with a large (10X the plasma pool size) non-physiological dose of amino acid (phenylalanine). The fall in free amino acid enrichment is relatively small and linear over a 10-minute period after the injection. Muscle protein enrichment and precursor pool enrichment are used in the equations previously described.

The fractional breakdown rate of muscle protein can be estimated if the fractional accretion rate of muscle protein is known (fractional breakdown rate = fractional synthesis rate – fractional accretion rate; FBR = FSR – FAR) (Millward, 1975). This approach to the study of protein degradation is somewhat unsatisfactory. The main problem with this method arises from the time scale of measurements. Synthesis is measured over a period of minutes or hours, while growth is integrated over the day and measured over a period of days or months, thus estimation of protein synthesis will vary over the course of the day and before and after a meal. These changes in protein synthesis could in turn grossly over- or underestimate the degradation rate.

A more direct approach to measure muscle degradation is the *tracee release* method (Zhang et al., 1996). This approach involves infusing a labeled amino acid to an isotopic equilibrium and then observing the isotopic decay in arterial blood and muscle intracellular pool. The FBR is calculated at the rate at which tracee dilutes the intracellular enrichment. This method can be combined with a tracer incorporation method to measure both the FSR and FBR in the same study.

Finally, a direct method to measure myofibrillar protein is the quantitation of urinary 3MH (Figure 1). Urinary 3MH can be used as an index of myofibrillar protein degradation or if the precursor pool (3MH bound to muscle protein) can be estimated, the FBR can be calculated.

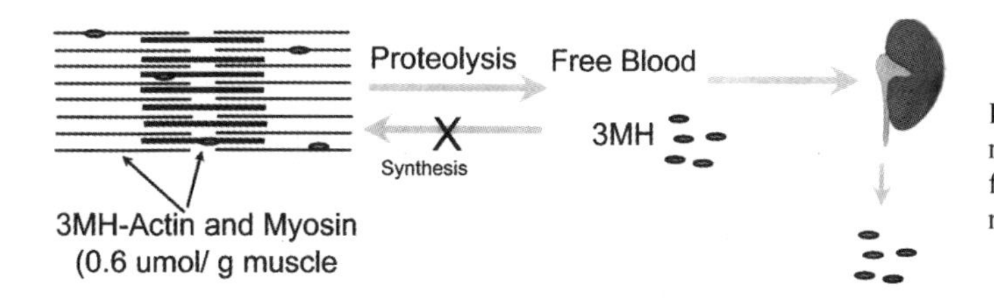

Figure 1. Metabolism of 3-methylhistidine(3MH) following the degradation of muscle protein.

3-METHYLHISTIDINE

The primary sequence in the myofibrillar protein actin and in fast-twitch, white myosin, contains the unique amino acid 3-methylhistidine (Nτ-methylhistidine) (3MH) (Johnson et al., 1967). 3MH is formed by a postranslational modification of one histidine residue in each protein. During degradation of these muscle proteins, free 3MH is released, but because 3MH does not have a specific *t*RNA, it is not reutilized for protein synthesis (Young et al., 1972). Instead of being used for protein synthesis, 3MH is quantitatively excreted in the urine of man, rat, cattle, and rabbit (Young et al., 1973; Young et al., 1972;

Harris and Milne, 1981; Harris et al., 1977) and is therefore thought to be a marker of skeletal muscle protein breakdown.

A FBR is calculated when an accurate estimate of the precursor pool is obtained. This method depends on quantitative urine collection and accurate measurement of urinary 3MH and on the assumption that no metabolism of 3MH occurs *in vivo*. In most species, no metabolism occurs, but in sheep and swine (Harris and Milne, 1980), a proportion of 3MH is thought to be retained in muscle as a dipeptide (balenine, b-alaninyl-3-methylhistidine) (Harris and Milne, 1987). In the rat, 3MH is transported to liver and acetylated. The N-acetyl-3-methylhistidine is the major form excreted in the rat (Young et al., 1972), whereas in the adult human, N-acetyl-3-methylhistidine accounts for less than 5% of the daily 3MH excreted (Long et al., 1975). In sheep and swine, therefore, urine 3MH cannot be used to estimate muscle protein breakdown, because 3MH production from muscle is not equal to urinary 3MH production. Although there is a substantial body of literature on the metabolism of 3MH, few reports have actually measured daily variability of endogenous 3MH excretion. Lukaski et al (1981) reported an intra-individual coefficient of variation of 4.5% (range 2.2 to 7.0%).

It has been debated as to whether urinary 3MH is primarily a product of skeletal muscle protein turnover or whether other tissues might contribute a significant amount to the daily production. Haverberg et al. (1975) showed that the mixed proteins in all of the organs sampled contained detectable levels of bound 3MH. However, when examining each organ as a whole, skeletal muscle contained the majority (98%) of the total amount. Nishizawa et al. (1977) concluded that the skin and intestine contributed up to 10% of the total body pool of 3MH. A study of humans with short-bowel syndrome indicated that skeletal muscle was the major source of urinary 3MH (Long et al., 1988). In human patients with varying degrees of infection (Sjölin et al., 1989), it was concluded that urinary 3MH was a valid marker of myofibrillar protein breakdown, because it was correlated with the release of 3MH from the leg. Furthermore, it was later shown with additional patients (Sjölin et al., 1990), that there was a significant linear relationship between the leg effluxes of tyrosine and phenylalanine and that of the leg efflux of 3MH and the resulting urinary excretion of 3MH. Therefore, urinary 3MH excretion is associated with net skeletal muscle protein breakdown. Based on previous studies, therefore, it is reasonable to assume that changes in 3MH production are largely reflective of muscle metabolism.

To date, no attempts at integrating tracer and tracee data into a comprehensive whole animal kinetic model of 3MH metabolism in any species have been reported. The initial studies showing the inadequacy of 3MH as an index of muscle protein breakdown required the intravenous administration of a dose of labeled 3MH, but the decay curve of [^{14}C]3MH in plasma was not characterized. A compartmental model for swine or sheep must include a compartment for 3MH metabolism other than excretion into a urine compartment. Swine excrete less than 2% of 3MH from muscle metabolism into the urine with the majority being retained in muscle as the dipeptide balenine. Therefore, swine not only have a large pool of free 3MH in muscle but also a large metabolic "sink" of 3MH in the form of balenine. Likewise, sheep excrete approximately 15% of 3MH in the urine with the remainder being retained in muscle as the dipeptide balenine. Hence a compartmental model describing the metabolism of 3MH in these two species must incorporate these metabolic differences as compared to humans, cattle, rats, and rabbits.

DEVELOPMENT OF THE ISOTOPE MODEL OF 3MH KINETICS

Urinary 3MH had been used in cattle and humans (Figure 1) as an index of muscle protein breakdown but was shown to be invalid for use in swine and lambs. 3MH is produced in these species but is not quantitatively excreted in the urine. In validating urinary 3MH as an index of muscle proteolysis, researchers have injected [^{14}C]-3MH intravenously and recovered the tracer in urine, but have never described its decay in plasma. 3MH is a histidine residue with one methyl group attached to the tau-nitrogen on the imidazole ring. To understand the metabolism of 3MH, we have used a dueterated molecule of 3MH (Figure 2), in which the three hydrogens of the methyl group have been replaced with three deuterium atoms, therefore the tracer is 3 mass units heavier than the naturally occurring 3MH and can be detected by gas chromatography-mass spectrometry (Rathmacher et al., 1992).

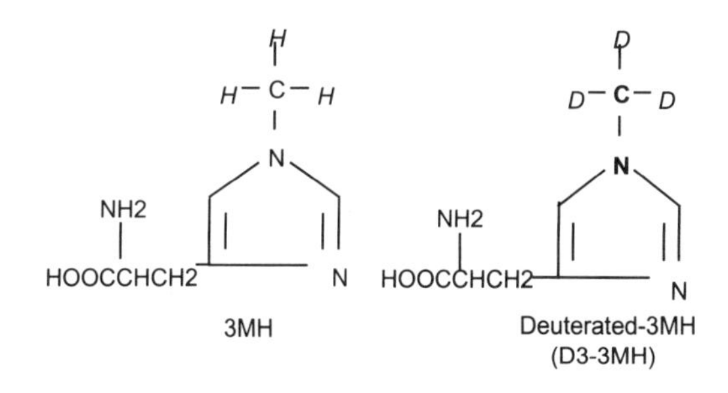

Figure 2. The structure of the 3-methylhistidine (3MH) molecule and the deuterated tracer of 3MH.

In constructing the three-compartment model, we kept in mind the known physiology of 3MH. It has been established that there are pools of 3MH in plasma, in other extra-cellular fluid pools, within muscle, and in other tissues. The primary fate of 3MH in humans, cattle, and dogs is into urine (model exit from compartment 1), but in sheep and swine there is a balenine pool in muscle that accumulates over time (model exit from compartment 3).

Table 1. Steps taken in the development of the 3-methylhistidine kinetic model.

Experimental Problem: 3MH is a valid muscle protein turnover model in cattle and humans but invalid in sheep and swine.
Step 1: Empirical modeling/first attempts
Step 2: Compartmental model -plasma (sheep)
Step 3: Compartmental model -plasma and urine (sheep)
Step 4: Compartmental model -plasma, urine, and muscle (sheep)
Step 5: Compartmental model -plasma (swine)
Step 6: Indirect validation -humans, cattle, and dogs

The 3MH kinetic model was developed from the need to measure muscle proteolysis directly in growing lambs. However, the problem was that urinary 3MH was a valid muscle protein turnover method in cattle and humans but was invalid in sheep and swine. Model development proceeded in the strategy outlined in Table 1. Our basic experimental design is illustrated in Figure 3, and involves the following: (1) intravenous bolus dose of tracer; (2) sampling (blood, urine, and muscle tissue); (3) 3MH isolation by ion exchange chromatography; (4) t-butyldimethylsilyl derivatization; (5) analysis by gas chromatography-mass spectrometry; and (6) compartmental modeling using the SAAM/CONSAM program (Berman and Weiss, 1978; Boston et al., 1981).

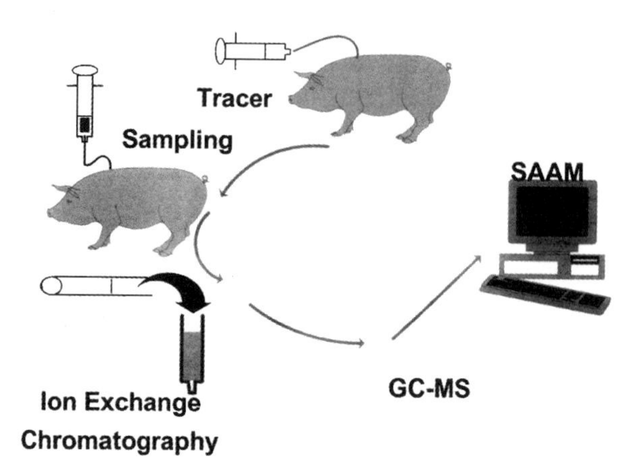

Figure 3. Experimental design of model development: (1) intravenous bolus dose of tracer; (2) sampling (blood, urine, and muscle tissue); (3) 3MH isolation by ion exchange chromatography; (4) t-butyldimethylsilyl derivatization; (5) analysis by gas chromatography-mass spectrometry; and (6) compartmental modeling using the SAAM/CONSAM program.

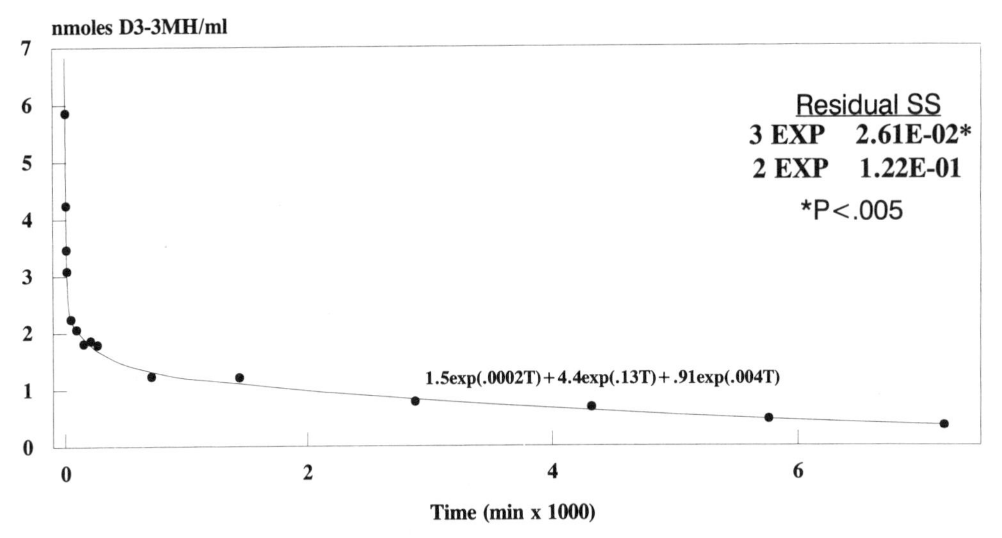

Figure 4. The fitting the time course of the deuterated tracer of 3MH in plasma of lamb. Observed values (●); values calculated by the exponential equation (—).

Our first two experiments (Rathmacher, unpublished data) were unsuccessful in growing lambs. We tried a constant infusion flux model, with different priming doses and infusion lengths, but could not reach a steady state after 12 h. The unsuccessful attempt will become obvious as illustrated in Figure 4. The second attempt was a bolus injection followed by blood samples over 5 d, but could not conclude if it was a two- or three-compartment model. The kinetics of 3MH in four lambs were studied (Rathmacher et al, 1993). The stable isotope disappearance in plasma and appearance in both urine and muscle were measured. The kinetic data were initially fitted to three exponential terms (Figure 4), indicating the possibility of a three-compartment model. A three-exponential equation was significantly better than two, and there was no advantage of adding a fourth exponential.

A series of linear time-invariant compartmental models were constructed based on the response of tracer in plasma, urine excreta, and muscle. Model code 10 of SAAM was used and the models were based on a set of linear differential equations having constant coefficients. A three-compartment model was the best model to describe the metabolism of 3MH (Figure 5), however, two- and four-compartment models as well as other configurations were also evaluated. A three-compartment model provided the best fit and was significantly different from a two-compartment model ($P < 0.05$); there was no advantage to adding a fourth compartment.

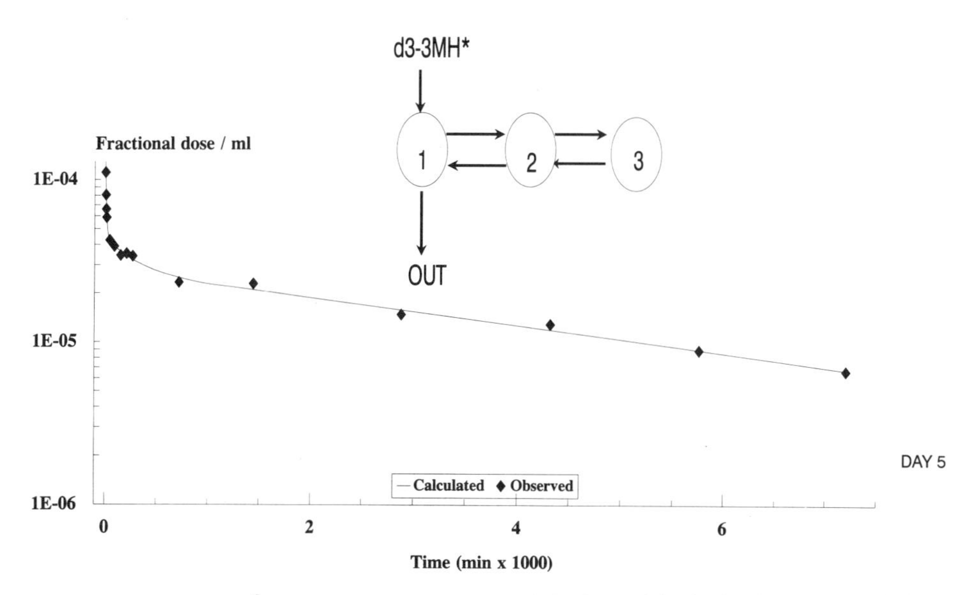

Figure 5. Plasma 3-[$^2H_3$-methyl]-histidine (d_3-3MH) following an injection in a lamb. A graphical representation of the kinetic model is presented. Observed values (◆); values calculated by the compartmental model (—).

The next question to answer was whether the accumulation of tracer in urine described that exit from compartment 1. The model was inadequate when the accumulative urine data were entered as the only exit out of compartment 1 (Figure 5). There was a significant decrease in sum of squares for the plasma fit. The model was then adjusted to include a urinary exit as well as a second exit out of pool 3 (Figure 6), which could be thought to be an exit into a balenine sink (Harris and Milne, 1987). The pool turns over very slowly or not at all during the time frame of this study. This model was adequate for describing both plasma and accumulation of urine. The final phase in the 3MH kinetic model development was to use the enrichment of tracer in muscle biopsies of the *longissimus dorsi*. These data could be best described as a sum of compartments 2 and 3 (Figure 7). This model was adequate for fitting the response of tracer in plasma, urine excreta and muscle.

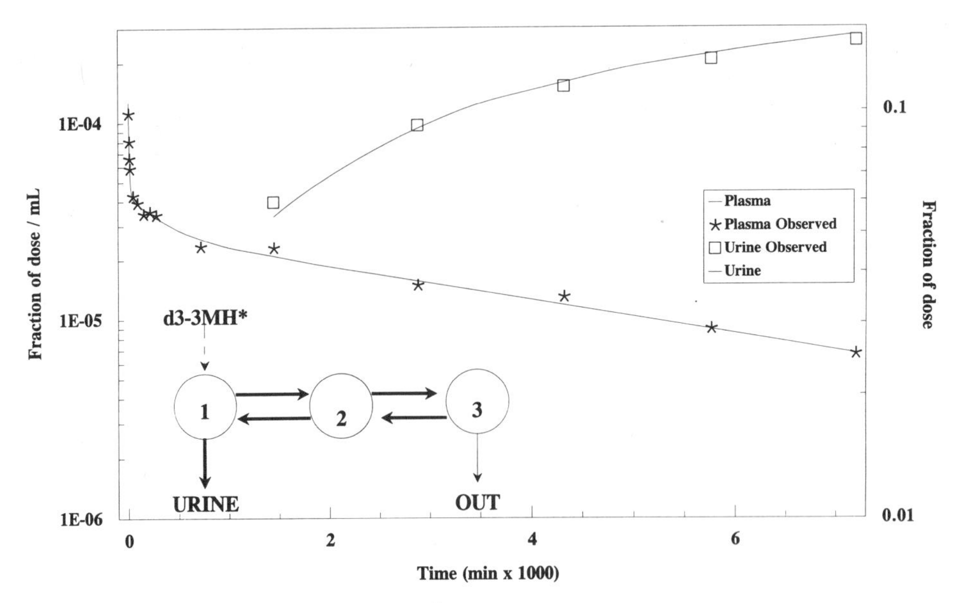

Figure 6. A simultaneous fit of Plasma 3-[$^2H_3$-methyl]-histidine (d_3-3MH) following an injection and accumulation of d_3-3MH in urine in a lamb. A graphical representation of the kinetic model is presented. Plasma (★) and urine (□) observed values; plasma (—) and urine (—) values calculated by the compartmental model.

The next step in model development (Table 1) was the development of a model for swine. The question was: Will the model we developed in sheep work in swine? The metabolism of 3MH is similar (balenine production), but there are differences in 3MH physiology as described by Harris and Milne (1981). In our first attempt, we tried to fit the data using the sheep plasma model (Figure 5), but were unable to come up with are reasonable fit. This was expected based on known 3MH physiology of the swine, as very little 3MH is excreted in the urine (<1-2%), but is conjugated with β-alanine in the muscle forming balenine. This dipeptide accumulates in the muscle and has a very slow turnover. The model (Rathmacher et al., 1996) was adjusted as depicted in Figure 8. The tracer was injected into compartment 1, with an exit out of a muscle compartment-3 into balenine sink.

Next, studies were designed utilizing the decay of a tracer of 3MH over a 3 to 5 day period. From these data, a compartmental model in lambs was developed and a steady-state production rate of 3MH was estimated. However, it is difficult to validate most isotopic models because there generally are no non-isotopic methods to validate them. The approach taken to validate the 3MH model in sheep and swine was comparative, using species in which urinary 3MH is a valid index of muscle proteolysis. Studies in cattle (Figure 9, Rathmacher et al.), humans (Figure 10, Rathmacher et al., 1995), 1992), and dogs (Rathmacher, 1993) were used to compare the estimates of 3MH as estimated from the model to urinary 3MH production (Figure 11).

The models for these species are identical. The curves illustrated were generated using the initial three-compartment plasma model for sheep (Figure 5) with the exit out of compartment 1. The curves for human and cattle are similar in shape to the curves generated in sheep and swine. We defined this exit as a urinary exit (Figure 9), thereby validating our model of 3MH metabolism in species that quantitatively excrete 3MH into urine. We feel this model adequately describes the kinetics of distribution, metabolism, and *de novo* production of 3MH in humans and cattle. Sampling and tracer administration was into compartment 1, (based on size was similar to the extracellular water space). Compartments 2 and 3 are likely intracellular pools of free 3MH (based on lamb data and theoretical calculations in humans), with *de novo* production into compartment 3. Succeeding studies have used this methodology in experimental designs where other methods have shown the fractional breakdown rate to be affected. Such effectors included: β-adrenergic agonists, trenbolone acetate and dietary protein.

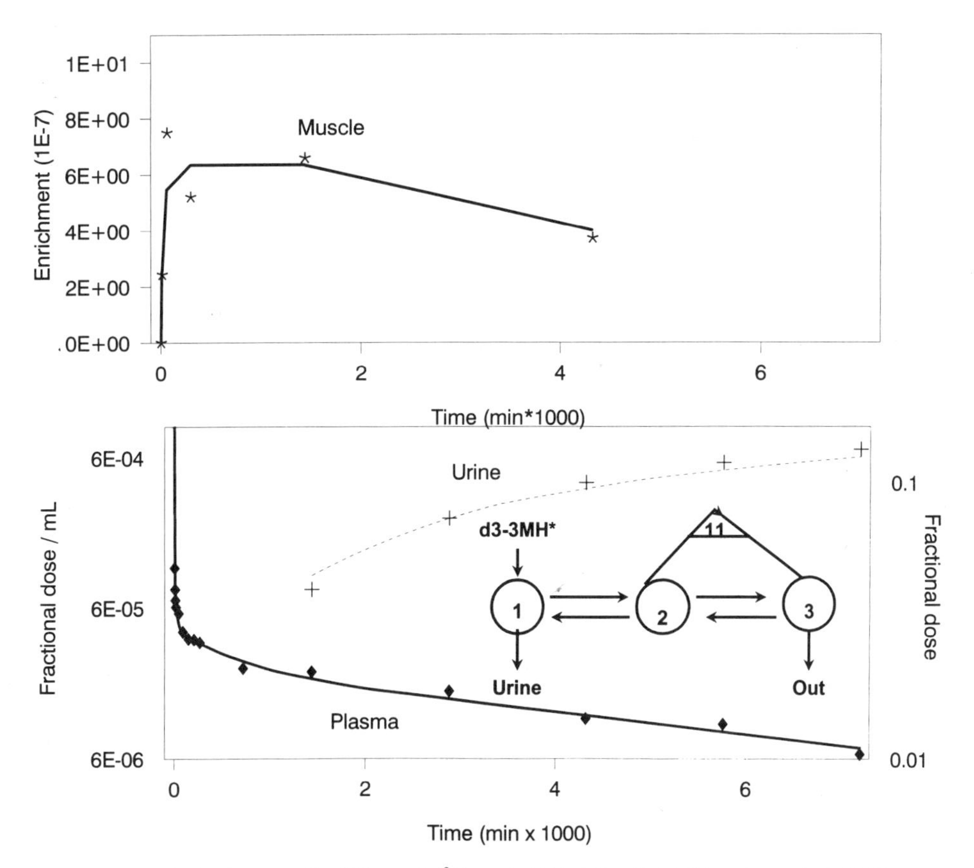

Figure 7. A simultaneous fit of Plasma 3-[$^2H_3$-methyl]-histidine (d_3-3MH) and muscle d_3-3MH enrichment following an injection and accumulation of d_3-3MH in urine in a lamb. A graphical representation of the kinetic model is presented. Plasma (♦), muscle (★), and urine (+) observed values; plasma (—), muscle, and urine (- - -) values calculated by the compartmental model.

Modeling Accomplishments

The modeling process presented in this chapter has demonstrated that the decay of a tracer of 3MH can be described by a three-compartment model even for sheep and swine, two species for whom urinary 3MH is an invalid index of muscle proteolysis. In humans, cattle, and dogs, the model has estimated 3MH production that was qualitatively close to the measured urinary 3MH production.

These combined models, and the information on *de novo* production rates they provide, have proven useful for calculating the fractional breakdown rate in experimental

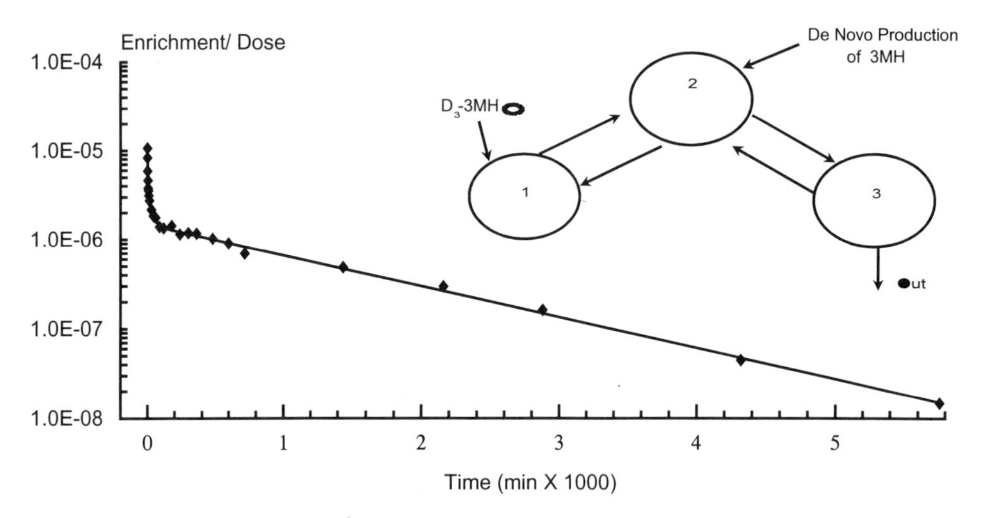

Figure 8. Disappearance of 3-[$^2H_3$-methyl]-histidine (d_3-3MH):3-methylhistidine in plasma described by a three-compartment model in a pig. A graphical representation of the kinetic model is presented Observed values (◆); values calculated by the compartmental model (—).

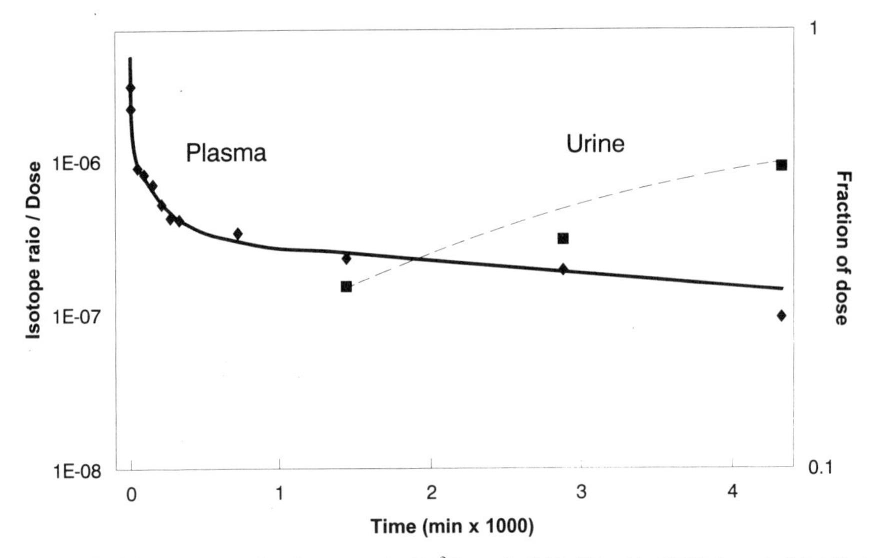

Figure 9. Disappearance of an isotope ratio (3-[$^2H_3$-methyl]-histidine (d_3-3MH):3-methylhistidine) in plasma described by a three-compartment model in a steer followed by the accumulation of d_3-3MH in urine. A graphical representation of the kinetic model is presented. Plasma (◆) and urine (■) observed values; plasma (—) and urine (- - -) values calculated by a three-compartment model.

situations where muscle mass is altered. They will continue to prove helpful in the validation of certain assumptions about 3MH metabolism.

The latter portion of the chapter will detail the development of a minimal one-compartment model based on the terminal slope of the complete model. We will also relate how the model parameters, and steady-state compartment masses and fluxes—as determined by the model—have been related to determinations of muscle mass.

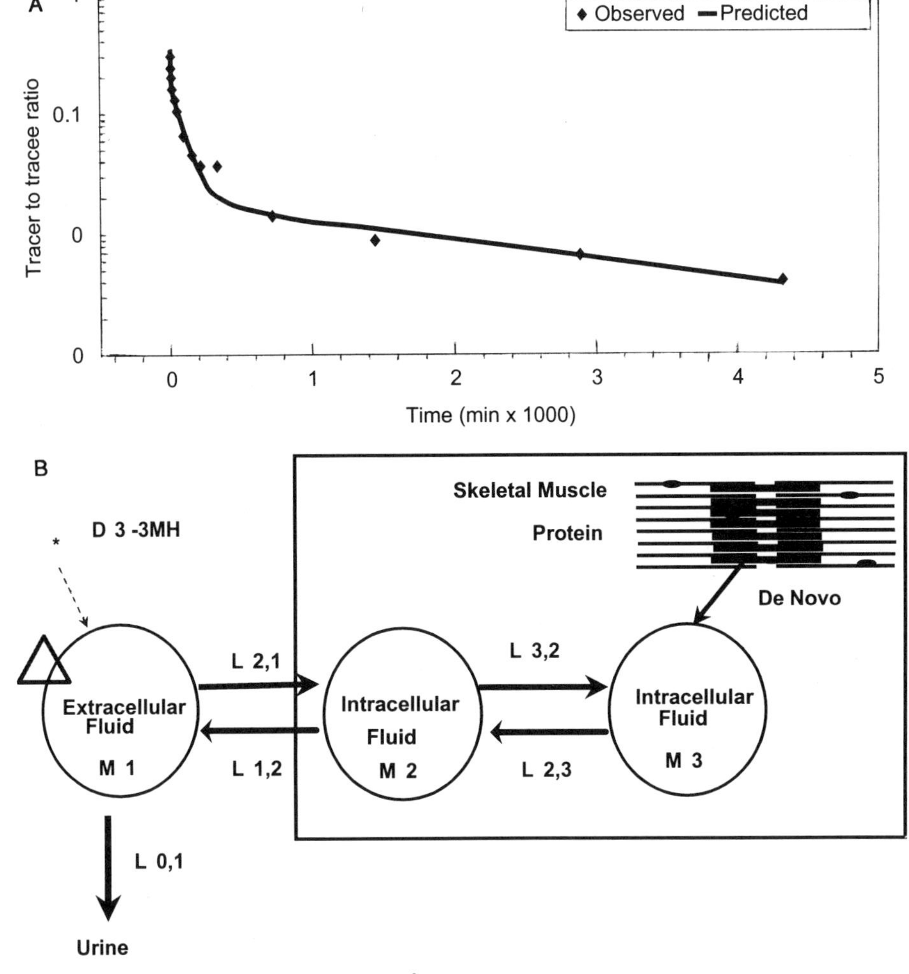

Figure 10. (10A) Disappearance of an tracer (3-[$^2H_3$-methyl]-histidine (D_3-3MH) in plasma described by a three-compartment model in a human subject. (10B) A graphical representation of the kinetic model is presented. M_1, M_2, and M_3 represent the mass of 3MH in compartments 1, 2, and 3, respectively. $L_{2,1}$, $L_{1,2}$, $L_{0,1}$, $L_{2,3}$, and $L_{2,3}$ are fractional transfer rate coefficients of 3MH within the system. The tracer, D_3-3MH was injected into compartment 1. Sampling was performed from compartment 1 (∇). *De novo* production of 3MH was into compartment 3.

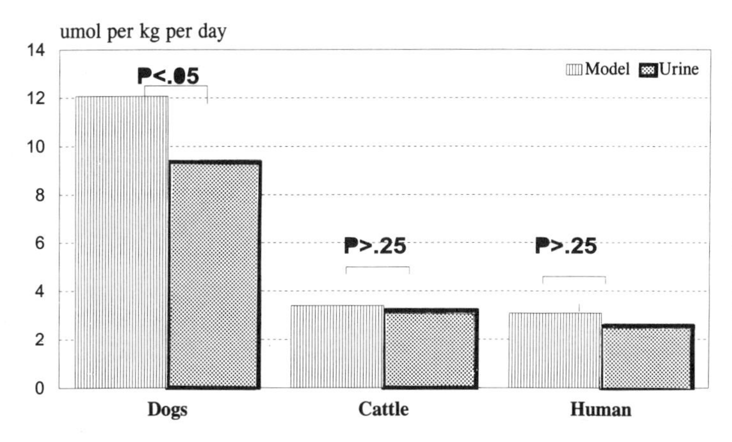

Figure 11. Daily 3MH production expressed as $\mu mol.kg^{-1}.d^{-1}$ for dogs, cattle, and humans as calculated from urinary excretion (solid bars) and by a three-compartment model of 3MH production (striped bars). *There was no mean difference between urinary and model 3MH production for cattle and humans, p >0.25.

Modeling Assumptions

There are three assumptions that must be accepted when using linear compartmental models. First, the volume or mass of the compartment is assumed to to constant. Secondly, the compartments are well-stirred. That is, when you sample a compartment, a representative sample of the entire compartment is taken. Finally, the rate constants remain constant.

The 3MH model in general also has specific assumptions:

- 3-Methylhistidine is not reutilized to a significant extent; there is no tRNA for 3MH.
- The precursor pool does not change; the myofibrillar protein-bound 3MH does not change with the experimental conditions. There is some indication that the pool increases shortly after birth. Our data would indicate that neither pharmacological manipulation nor dietary manipulation change the concentration in skeletal muscle (Rathmacher, unpublished data). However, there may be some difference between muscles (i.e., *longissimus dorsi* vs. *semitendinosus*) (Rathmacher, unpublished data).
- 3MH is quantitatively excreted in the urine, or, if metabolic products of free 3MH do occur, they are accounted for in the model. This true for humans, cattle, and dogs, but not for sheep and swine. In sheep and swine, a large proportion is retained in muscle as balenine and the model was adjusted to explain this process.
- Renal absorption does not change or is similar between treatments. This may be one explanation why sheep and swine do not quantitatively excrete 3MH in the urine. These species may selectively reabsorb 3MH.
- There is no 3MH in the diet or, if it is present, 3MH calculations must be corrected for dietary 3MH.
- The primary source of 3MH is from skeletal muscle myofibrillar protein. This assumption has caused the most controversy between researchers. On an organ basis, skeletal muscle contains more than 90% of the protein-bound 3MH. However, researchers who dispute the validity of the 3MH model maintain that the myofibrillar proteins in the other tissues probably turn over faster than skeletal myofibrillar protein. In response to this, V.R. Young (1978) and C.I. Harris (1981) criticized the experimental designs of these researchers and have presented data to show otherwise. (See the 3MH Introduction to this chapter for discussion.) Our data using the 3MH kinetic model in portal vein-cannulated swine (Van den Hemel et al., 1997), suggest that 3MH production from the gastro-intestinal tract is not increased in protein-free-fed swine. The FBR of the whole body was 2.16 and 2.56 for control and protein-free-fed swine, respectively, and the percentage from the gastro-intestinal tract was less than 6% for both treatments.

Species Comparison of 3MH Kinetics

Figures 12 and 13 and Tables 2 and 3 are a summary of the efforts to model 3MH metabolism using a three-compartment model in humans, cattle, and dogs which quantitatively excrete 3MH in urine, as compared to sheep and swine which do not. Figure 13 is a comparison of model structures between the species. The diversity of models between humans, cattle, and dogs, and sheep and swine reflects differences in known species physiology. In each species, the tracer is injected into compartment 1 which, based on the size (volume and mass), is similar to plasma and extracellular water space. Compartment 1 was the sampling compartment and the compartment from which the steady-state calculations were initiated. All models for all species but one can be resolved by sampling only plasma. The exception, sheep, required the sampling of both plasma and urine (Figure 13). However, the sheep model can be resolved from plasma kinetic of 3MH if the rate of exit from compartment 1 is fixed.

From the steady-state calculations the *de novo* production of 3MH was obtained into compartment 3 for humans, cattle, and dogs. The *de novo* production of 3MH for sheep and swine could be placed as a entry into compartment 2 and identical rates could be calculated. The compartment identities of compartment 2 and 3 are the intracellular pools of 3MH. The metabolic form of 3MH in these compartments may not be identical nor is the identity of compartment 2 or 3 for one species the same identity for another species (i.e, cattle vs. sheep). The models also depict differences in the route by which 3MH exits the system. In humans, cattle, and dogs, 3MH is quantitatively excreted in the urine, as illustrated by the exit from compartment 1. This urinary exit has been confirmed by comparison of urinary excretion of 3MH and model calculated values (Figure 11). Sheep excrete only 15% of total daily 3MH produced in the urine, swine excrete even less, 1.5 %/d. Therefore, an accurate accounting of 3MH production in sheep and swine requires an exit out of the system from compartment 3. This exit accounts for appreciable loss of 3MH into a balenine 'sink' which turns over slowly or not at all during the time frame of the study.

Representative plasma decay curves following a single dosing of 3-[methyl-$^2H_3$]-methylhistidine tracer is illustrated in Figure 12. In general, each species exhibited a similar exponential decay, characterized by rapid decay over the first 2-3 hours, followed by a slower decay through 12 hours, and a steady-state decay over the remainder of the study. The decays of tracer are representative of the models used. Humans, cattle, and dogs exhibit very similar decays while sheep and swine are very different.

Table 2 lists the model parameters and the fractional transfer rates ($L_{i,j}$ from compartment j to i). The fractional standard deviation of the parameters range from 5 to 50% and, in general, $L_{2,1}$, $L_{1,2}$, and $L_{0,1}$ or $L_{0,3}$ are solved with a higher precision than $L_{3,2}$ and $L_{2,3}$. Table 3 compares the compartment masses and mass transfer rates between compartments for each species. Also listed is the *de novo* production rate calculated by the model for each respective species. An important feature of these models is the description of 3MH metabolism within the body. The significance of mass transfer rates and compartment sizes is not fully understood. However, the model parameters and mass transfer rates may explain the failure of sheep and swine to quantitatively excrete 3MH in the urine (Rathmacher and Nissen, 1992). Three mechanisms may explain this phenomenon: (1) 3MH transport between the compartments limits the excretion of 3MH; (2) 3MH may be avidly reabsorbed by the kidney; and (3) the enzymatic conversion of 3MH to balenine is enhanced. An examination of the data from Tables 1 and 2 reveals that the low rate of 3MH excretion in sheep and swine is not due to impaired transfer of 3MH out of and between compartments. Cattle appear to have slower exchange of 3MH between tissues despite of near-quantitative urinary excretion. The most likely reason for 3MH sequestering in sheep and swine is that the kidneys in these species are very efficient in conserving 3MH, which in turn increases compartment size and plasma concentration, and through mass action could increase the synthesis of balenine.

A Minimal Model

There may be times, due to time limitations or experimental design constraints, when it may be impractical to take the number of samples necessary to resolve the three-compartment model. Another objective of this research was to be able to use this model in more more practical situations, such as when only four or five blood samples are taken (vs. the usual 14-25). For these instances, we have minimized the standard model from three compartments to only one. This minimal model has been evaluated in two studies to date.

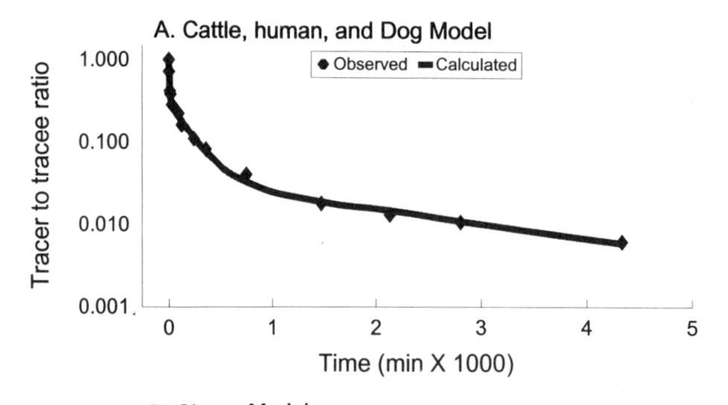

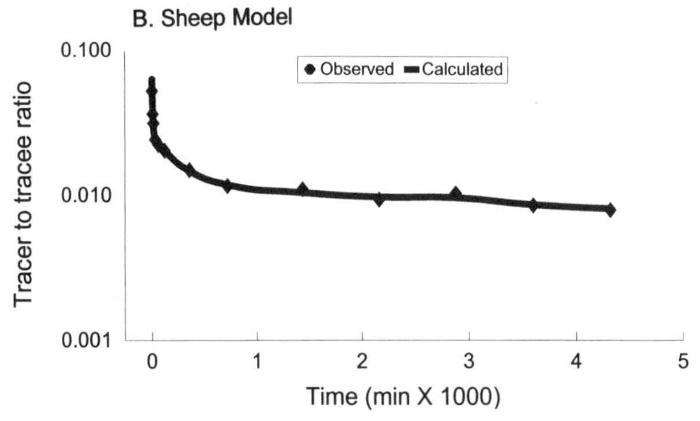

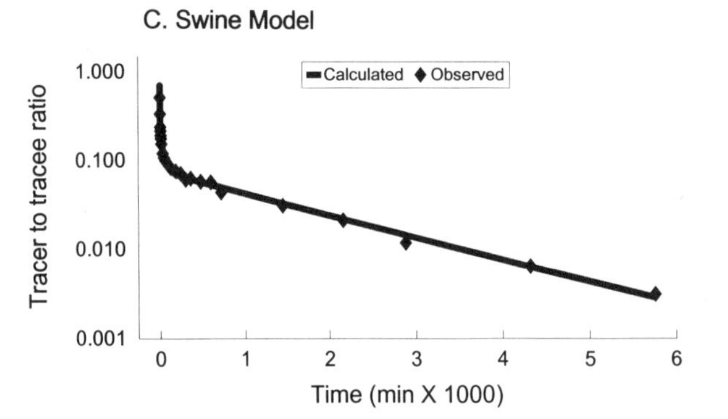

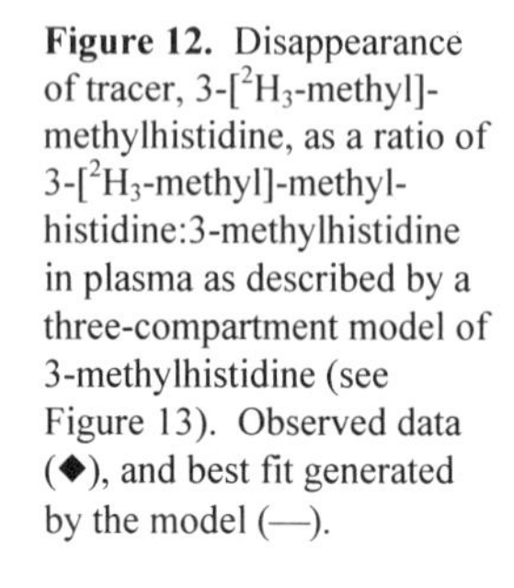

Figure 12. Disappearance of tracer, 3-[$^2H_3$-methyl]-methylhistidine, as a ratio of 3-[$^2H_3$-methyl]-methylhistidine:3-methylhistidine in plasma as described by a three-compartment model of 3-methylhistidine (see Figure 13). Observed data (◆), and best fit generated by the model (—).

Table 2. Comparison of 3-methyhistidine (3MH) kinetic parameters.

	Species				
Parameter	Cattle	Humans	Dogs	Swine	Sheep
Animals, n	39	4	5	20	40
Urinary 3MH loss[1]	100	100	100	1	17
Fractional transfer rate (min)					
$L_{2,1}$[b]	0.18	0.08	0.11	0.23	0.21
$L_{1,2}$	0.06	0.06	0.06	0.09	0.08
$L_{3,2}$	0.003	0.009	0.006	0.014	0.007
$L_{2,3}$	0.002	0.002	0.008	0.006	0.005
$L_{0,3}$	NA[c]	NA	NA	0.0009	0.0004
$L_{0,1}$	0.006	0.004	0.02	NA	0.0003

[a]Percent of total
[b]Fractional transfer rate ($L_{i,j}$) from compartment j to i.
[c]Not applicable.

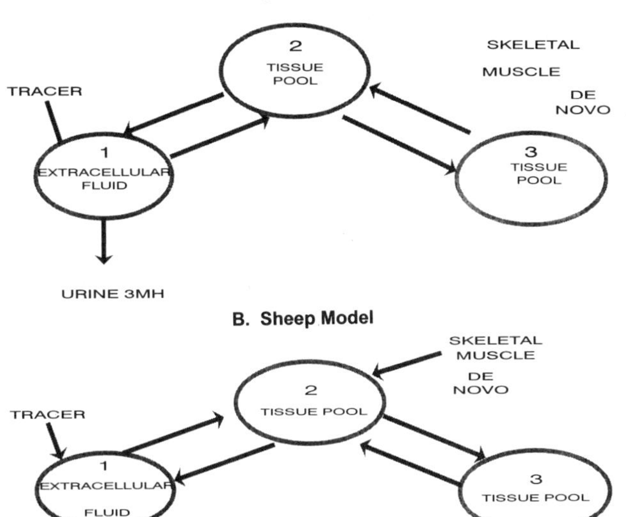

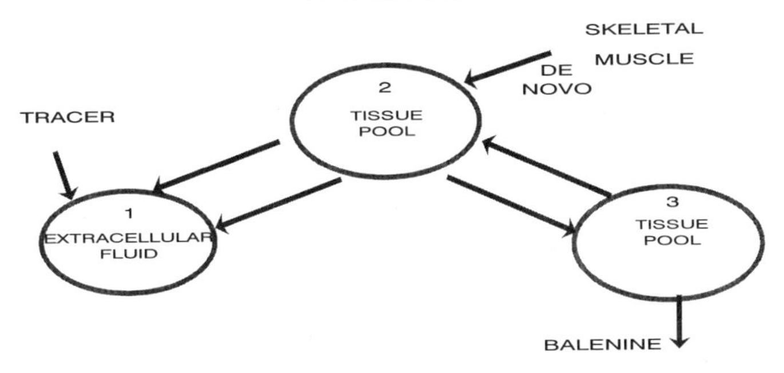

Figure 13. Schematic of the three-compartment models used to analyze the kinetics of distribution, metabolism, and *de novo* production of 3-methylhistidine (3MH). The tracer, 3-[$^2H_3$-methyl]-methylhistidine (D_3-3MH), was injected into compartment 1. Sampling was performed from compartment 1. *De novo* production of 3MH and the exit from the system are dependent on the physiology of the species.

Table 3. Steady-state compartment masses and mass transfer rates for a three-compartment model of 3-methylhistidine (3MH) metabolism.

	Species				
Parameter	Cattle	Humans	Dogs	Swine	Sheep
Animals, n	39	4	5	20	40
Plasma 3MH, μM	8.6	2.9	21.8	10.4	36.9
M_1, nmol·kg^{-1} [a]	807	603	3227	1110	5308
M_2, nmol·kg^{-1}	2291	912	7973	2857	12483
M_3, nmol·kg^{-1}	8079	7938	9261	6151	17017
R_{21}, nmol·kg^{-1}·min^{-1} [b]	101	51	319	247	944
R_{12}, nmol·kg^{-1}·min^{-1}	105	53	329	247	946
R_{32}, nmol·kg^{-1}·min^{-1}	5.8	7.9	56	37	81
R_{23}, nmol·kg^{-1}·min^{-1}	4.1	2.2	9	NA	1.4
R_{01}, nmol·kg^{-1}·min^{-1}	NA	NA	NA	5.0	5.8
R_{03}, nmol·kg^{-1}·min^{-1}	6.0	3.1	12	7.2	10.3
3MH production [c]	100	100	100	1	17

[a] M_i = compartment mass i. [b] R_{ij} = mass transfer from compartment j to i.
[c] 3MH production was obtained from the model; μmol·kg^{-1}·d^{-1}.

For this minimal model, the three pools of the three-compartment model have been combined to form one homogeneous pool of 3MH, characterized by a *de novo* production into the body pool and one exit from the model. The kinetic data between 720 and 4320 min was used for the analysis. This time frame corresponds to the apparent linear steady-state portion of the decay curve. The limitation of the one-compartment model is that we lose the ability to describe the entire metabolism of 3MH. Only the rate of proteolysis and a total pool size can be estimated. Similar results were obtained in dogs when the complete (three-compartment) model is compared to the minimal model (Rathmacher, 1993). However, in cattle the minimal model estimate of 3MH production was 30% higher (Rathmacher et al., 1994), and the estimates of production were highly correlated ($r=0.93$, $p < 0.0001$).

Model Advantages

The structural configurations of these models are not unique, and alternative arrangements may also be compatible with the data. The present models represent a framework and methodological approach describing steady-state 3MH kinetics in the whole animal and constitutes a working theory for testing by further experimentation with designs which alter muscle protein breakdown. The rate of 3MH production is an important tool in understanding the regulation of muscle protein degradation. The advantages of these models are that: 1) it does not necessitate quantitative urine collection (plasma model); 2) it reduces error due to the frequency of plasma sampling versus the infrequency of urine collection in the other models; 3) it is more quantitative and it measures the total production rate independent of the determination of free or conjugated forms; 4) it gives information about pool size and transfer rates; 5) it establishes a relationship to muscle mass; 6) it provides a method for direct measurement of muscle proteolysis in swine and sheep; and 7) it does not require restraint of the animals for long periods.

PREDICTION OF MUSCLE MASS USING 3MH KINETIC PARAMETERS

The meaningful explanation of protein kinetics measurement is dependent on accurate measurement of body muscle. The body is usually divided into two large fractions, fat and fat-free mass (FFM). Fat-free mass is further divided into skeletal muscle (the largest (50%) and most variable fraction) and other body soft tissues (skin and other non-muscle tissues). Techniques to measure body composition are numerous and include: total body water, total body potassium, urinary creatinine excretion, underwater weighing, neutron activation analysis conductivity, and bioelectrical impedance (Wolfe, 1992). The current methods available to measure body composition only provide estimates of fat and fat-FFM. None estimate muscle mass directly. The major problem with all these methods is that skeletal muscle can be a highly variable component of FFM. Although there may be a relationship between 3MH (Lukaski and Mendez, 1980; Lukaski et al., 1981; Mendez et al., 1984; Wang et al., 1996) and other muscle metabolites (creatinine), it is clear whether changes in muscle metabolism, specifically muscle proteolysis, will bias these estimates of muscle mass and yield unreliable results. However, by the use of an isotope of 3MH, a three-compartment model has been developed which yields results that are directly correlated with muscle mass in animals (Table 4). We propose to use this model along with other conventional methods to predict body composition in order to estimate muscle proteolysis and total muscle mass and will attempt to express muscle proteolysis on a more exacting basis than ever before.

Rathmacher et al. (1996) reported on a human study, where normal subjects were given a bolus of 3-[methyl-$^2H_3$]-methylhistidine (0.13 μmol/kg) into the forearm vein of the dominant arm. Blood samples were then taken from 1 to 4320 minutes. Four 24-h urine collections were made starting the 24 h before the kinetic study started and throughout blood sampling. Body composition was determined by underwater weighing. The data from a single bolus dose of tracer can be described by a kinetic model of 3MH metabolism in humans. Plasma enrichment of the tracer was described by a linear time-invariant three-compartment model. The model defines masses and fluxes of 3MH within the subjects, and in particular, the intracellular *de novo* production of 3MH. The relationship between these steady-state values and leand body mass were evaluated.

From the three-compartment model of 3MH metabolism, the model parameters ($L_{i,j}$), steady-state pool size of these compartments (M_i), steady-state mass transfer rates ($R_{i,j}$), and an estimate of proteolysis (U_i) were calculated. The first pool is the compartment where the tracer is introduced and has volume and mass comparable to plasma plus extracellular fluid. Compartments 2 and 3 are the intracellular pools of 3MH in muscle tissues. When the myofibrilar tissue in muscle is degraded, the 3MH released from this tissue enters these tissue compartments. Prior to degradation, 3MH is a part of the proteins actin and myosin and is not described by the model compartments. We would expect the amount of free 3MH located in these compartments to be proportional to the muscle mass of humans and animals, because more than 90% of the 3MH bound to protein is found in this tissue. As the mass of skeletal muscle becomes larger, the pool of 3MH should also become larger – if we can assume that the *de novo* production of 3MH, its fractional and mass transfer rates, and its concentration within the pools all remain relatively constant. Unfortunately, these assumptions cannot always be made, and thus these known variables would have to be included in any predictive model.

Table 4. Relationships between model parameters and muscle mass.

Correlation	Correlation coefficient, r	P-value
M_3[a] vs. kg muscle, swine	0.59	0.006
$R_{3,2}$[b] vs. kg muscle, swine	-0.64	0.002
Plasma 3MH vs. kg muscle, swine	-0.52	0.02
Multiple regression model vs. kg muscle, swine	0.95	0.0001
M_2 vs. muscle, humans	0.91	0.09
M_3 vs. muscle, humans	0.56	0.44
U_3 vs. muscle, humans	0.74	0.026
$L_{2,1}$[c] vs. muscle, humans	--0.92	0.08
Multiple regression model vs. muscle, humans	0.98	0.1

[a]M_{ii} = compartment mass i.
[b]R_{ij} = mass transfer rate from compartment j to compartment i.
[c]Fractional transfer rate from compartment j to i.

This approach has been taken in subsequent analysis of the data in swine (Figure 14) and in humans. In Table 4, we have correlated the compartment mass and other parameters of a particular species with muscle mass. In addition, other model parameters have been found to be indicative of muscle mass. In the first example, the mass of compartment 3 was positively correlated with muscle mass in swine, while the mass transfer rate from compartment 2 to 3 was negatively correlated. Plasma concentration of 3MH in swine was also negatively correlated to the mass of muscle.

A study was conducted in twenty crossbred barrows (67.3 ± 1 kg) where tissue composition was measured (Rathmacher et al., 1996). The 3MH kinetic data were fitted to a three-compartment model of 3MH metabolism. The swine were then killed and the carcasses dissected into fat, lean, and bone. The lean tissues were analyzed for protein, fat, and moisture. It was assumed that 95% of the skeletal muscle was represented by this muscle mass. Stepwise multiple regression was conducted on model parameters and live bodyweight to predict muscle mass; variables were added such that R^2 was maximized. The R^2 was maximized at .74 for a one-variable model and at .99 for a 12-variable model. A representative model is presented in Figure 14 to demonstrate the predictive power of the multiple regression equations. The best four-variable model is as follows: Muscle mass (kg) = –10.9 + 0.00000354 (mass of compartment 3, nmol) – 0.000704 (mass transfer rate from compartment 2 to 3, nmol/min) + .476 (weight of the pig) + 0.0067 (3MH production, nmol/min) (R^2 = .90, p < .0001, sd = 1.34 kg). A similar approach was taken with the human data set to predict FFM. The model included the variables M_2 (Compartment masses of pools 2) and U_3 (estimate of proteolysis).

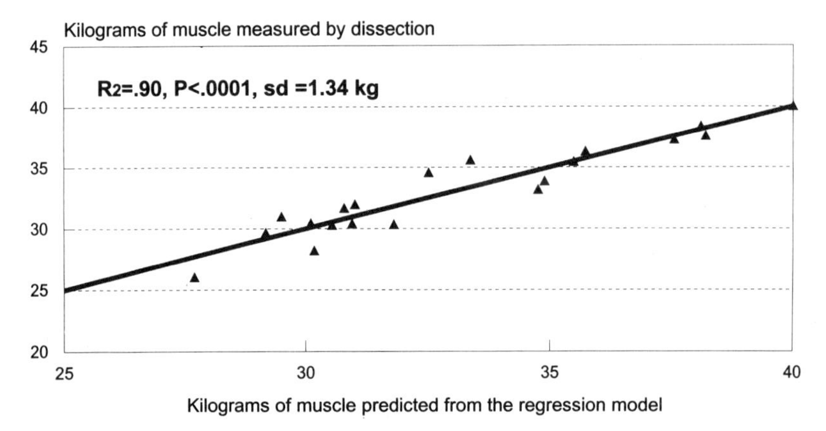

Figure 14. Stepwise multiple regression was conducted on 3-methylhistidine compartmental model parameters and body weight to predict muscle mass. Variables were added such that R^2 was maximized. The R^2 was maximized at .74 for a one-variable model and .99 for a 12-variable model. The best five-variable model is as follows: Muscle mass (kg) = – 10.9 + 0.00000354 (mass of compartment 3, nmol) – 0.000704 (mass transfer rate from compartment 2 to 3, nmol/min) + 0.476 (pig weight, kg) + 0.0067 (3MH production, nmol/min). R^2=.90, p <.0001.

KINETIC MODEL APPLICATIONS

The role of muscle protein breakdown in the regulation of net protein loss (or gain) of muscle is poorly understood. Muscle is degraded in response to many metabolic situations including: starvation, infection, surgery, diabetes, nutrition level, hormonal, exercise and stress conditions. Although general effects on muscle mass have been quantitated, advances defining mechanisms of muscle loss are limited due to limitations in methodology available to quantitate myofibrillar protein turnover versus protein turnover of muscle as a whole. The protein reserve of skeletal muscle is composed of two distinct fractions; myofibrillar protein, the structural component, and non-myofibrillar, non-structural component. The myofibrillar protein makes up 60% of the skeletal muscle protein with turnover slower than non-myofibrillar protein (Bates et al., 1983). There is also evidence to show that myofibrillar and non-myofibrillar are not under the same metabolic control nor degraded by the same mechanism (Goodman, 1987). Therefore, experimental designs where total proteolysis and synthesis measurements are made may reflect primarily non-myofibrillar breakdown and synthesis.

Skeletal muscle protein is a major tissue in whole-body protein turnover. However, little information is available for the maintenance of the protein reserve of skeletal muscle and the contribution of this tissue to overall protein turnover under different nutritional, hormonal and disease states. The 3MH kinetic model would be useful tool in evaluating muscle proteolysis directly in the situations illustrated in Table 5. The model has been used

Table 5. Applications of the 3-methylhistidine compartmental model in situations of altered muscle protein turnover.

Application	Response	Reference
Comparison between 3MH production and proteinase activities in muscle of protein-deficient barrows.	Myofibrillar proteolysis was increased by 27% in protein deficient barrows; no direct relationship between myofibrillar proteolysis and in vitro proteinase activity	Van den Hemel-Grooten et al., 1995.
Realimentation from a protein-deficient diet on muscle proteolysis in growing swine.	Myofibrillar proteolysis was not different than controls swine during the protein refeeding period.	Van den Hemel-Grooten et al., 1996.
The effect of the β-agonist, ractopamine(RAC), and dietary protein on muscle protein turnover in finishing swine	An increase in dietary protein (DP) from 10% to 18% decreased muscle proteolysis; RAC increased muscle proteolysis by 10%when 10% DP was fed and decreased muscle proteolysis 20% when 18% DP was fed.	Rathmacher, 1993.
The effect of increased dietary energy during gestation on muscle protein turnover during lactation in sows	Sows on a higher energy diet during gestation have an increased muscle proteolysis rate during the 1st week of lactation.	Trottier et al., 1995.
The effect of esrogenic and androgenic implants on 3MH production in growing beef cattle	There was a 20% decrease in 3MH production in the trenebolone acetate implanted cattle, but when combined with an estrogen implant the decrease was prevented	Rathmacher et al., 1993.
The effect of limiting one amino acid on muscle proteolysis in growing swine	There was a 6% improvement in the FBR of swine fed an adequate dietary lysine	Rathmacher et al., 1995.
Muscle protein turnover in large and small frame cattle	The large frame cattle gained weight faster and had a higher FBR when adjusted for the difference in the rate of gain	Rathmacher et al, 1993.
The effect of nutrition following surgical stress on muscle proteolysis	In terms of post-surgical nutrition, meeting the protein requirement is critical in minimizing muscle protein catabolism and hyper supplementation of both energy and protein has little affect.	Rathmacher, 1993.

to examine muscle proteolysis in the following examples: nutrition following surgical stress, varying protein and energy nutrition, lactation, hormonal intervention, limiting one amino acid in the diet, dietary realimentation, and differences due to genetics.

We have developed compartmental models for describing 3MH metabolism and these models are unique for differences seen in cattle, dogs, and humans as compared to sheep and swine. The models were validated in humans and cattle by comparing the model 3MH production to that of urinary production. This observation indirectly validated the models for sheep and swine. A minimal model approach appears to be related to the complete three-compartment model and may ease the measurement of muscle proteolysis. The model parameters may have the potential of predicting muscle mass and the 3MH kinetic model can be used in experimental situations where muscle proteolysis is altered.

In conclusion, the structural configurations of these models are not unique, and alternative arrangements may also be compatible with the data. The presented models represent a methodological approach to describe the steady-state 3MH kinetics in the whole animal, and provide a working model for testing designs which alter muscle protein breakdown. The determination of the rate of 3MH production is an important tool in understanding the regulation of muscle protein degradation and mass. This model offers advantages over the traditional model of urinary 3MH production.

ACKNOWLEDGMENTS

The authors extend thanks to Greg Link, Allen Trenkle, Richard Seagrave, Rick Sharp, Paul Flakoll, David Anderson, and Ray Paxton for their efforts in the development of the kinetic models and to Deb Webb, Connie Coates, Becky Zenkovich and Donna Rice for aid in data collection for the model development. We gratefully acknowledge the research contributions of Naji Abumrad, Patricia Molina, Miguel Molina, Javier Bonilla, Robert Easter, Natalie Trottier, Jan Gärssen, Hënriette van den Hemel, M.W.A. Verstegen, J.T. Yen, and M. Koohmaraie.

CORRESPONDING AUTHOR

Please address all correspondence to:
John A. Rathmacher, PhD
Research Scientist
Metabolic Technologies, Inc.
Iowa State University Research Park
2625 North Loop Drive, Suite 2150
Ames IA 50010
ph: 515-296-9916
fax: 515-296-0908
rathmacher@mti-hmb.com

REFERENCES

Bates PC; Grimble GK; Sparrow MP; Millward DJ. Myofibrillar protein turnover: synthesis of protein-bound 3-methylhistidine, actin, myosin heavy chain and aldolase in rat skeletal muscle in the fed and starved states. *Biochem J,*1983, 214:593-605.

Berman M; Weiss MF. *SAAM Manual.* US Department of Health, Education andWelfare Publication No.(NIH) 78-180. Washington, DC: US Government Printing Office. 1978.

Boston RC; Grief PC; Berman M. Conversational SAAM: An inter-reactive program for kinetic analysis of biological systems. *Comp Prog Biomed,* 1981, 13:111-119.
Fryburg DA; Barrett EJ; Louard RJ; Gelfand RA. Effect of starvation on human muscle protein metabolism and its response to insulin. *Am J Physiol (Endocrinol Metab),* 1990, 259:E477-E482.
Garlick PJ. Assessment of protein metabolism in the intact animal, in: *Protein Deposition in Animals.* Buttery PJ; Lindsay DB; Eds. Butterworths: London. 1980.
Garlick PJ; Clugston GA. Measurement of whole body protein turnover by constant infusion of carboxyl-labeled leucine, in: *Nitrogen Metabolism in Man.* Waterlow JC; Stephen JML; Eds. Applied Science: London, New Jersey. 1981, pp. 303-322.
Garlick PJ; Wernerman J; McNurlan MA; Essèn P; Lobley GE; Milne E; Calder GA; Vinnars E. Measurement of the rate of protein synthesis in muscle of postabsorptive young men by injection of 'flooding dose' of [1-13C]Leucine. *Clin Sci,* 1989, 77:329-336.
Garlick PJ; McNurlan MA; Essèn P; Wernerman J. Measurement of tissue protein synthesis rates in vivo: A critical analysis of contrasting methods. *Am J Physiol,* 1994, 29:E287-E297.
Goodman MN. Differential effects of acute changes in cell Ca^{2+} concentration on myofibrillar and non-myofibrillar protein breakdown in the rat extensor digitorum longus muscle in vitro: Assessment by production of tyrosine and N-tau-methylhistidine. *Biochem J,* 1987, 241:121-127.
Harris CI. Reappraisal of the quantitative importance of non-sleletal-muscle source of N-tau-methylhistidine in urine. *Biochem J,* 1981, 194:1011-1014.
Harris CI; Milne G. The urinary excretion of Nt-methyl histidine in sheep: An invalid index of muscle protein breakdown. *Br J Nutr,* 1980, 44:129-140.
Harris CI; Milne G. The urinary excretion of N-tau-methyl histidine by cattle: Validation as an index of muscle protein breakdown. *Br J Nutr,* 1981a, 45:411-422.
Harris C I; Milne G. The inadequacy of urinary (N-tau)-methyl histidine excretion in the pig as a measure of muscle protein breakdown. *Br J Nutr,* 1981b, 45:423-429.
Harris CI; Milne G. The identification of the N-methyl histidine-containing dipeptide, balenine, in muscle extracts from various mammals and the chicken. *Comp Biochem Physiol,* 1987, 86B(2):273-279.
Harris CI; Milne G; Lobley GE; Nicholas GA. 3-Methylhistidine as a measure of skeletal-muscle protein catabolism in the adult New Zealand white rabbit. *Biochem Soc Trans,* 1977, 5:706-708.
Haverberg LN; Omstedt PT; Munro HN; Young VR. Nt-Methylhistidine content of mixed proteins in various rat tissues. *Biochem Biophys Acta,* 1975, 405:67-71.
Johnson P; Harris CI; Perry SV. 3-Methylhistidine in actin and other muscle proteins. *Biochem J,* 1967, 105:361-370.
Long CL; Dillard DR; Bodzin JH; Geiger JW; Blakemore WS. Validity of 3-methylhistidine excretion as an indicator of skeletal muscle protein breakdown in humans. *Metabolism,* 1988, 37:844-849.
Long CL; Haverberg LN; Young VR; Kinney JM; Munro HN; Geiger JW. Metabolism of 3-methylhistidine in man. *Metabolism,* 1975, 24:929-935.
Lukaski H; Mendez J. Relationship between fat-free weight and urinary 3-methylhistidine excretion in man. *Metabolism,* 1980, 29:758-761.
Lukaski HC; Mendez J; Buskirk ER; Cohn SH. Relationship between endogenous 3-methylhistidine excretion and body composition. *Am J Physiol (Endocrinol. Metab),* 1981, 240(3):E302-E307.
Mendez J; Lukaski HC; Buskirk ER. Fat-free mass as a function of maximal oxygen consumption and 24-hour urinary creatinine, and 3-methylhistidine excretion. *Am J Clin Nutr,* 1984, 39:710-714.
Millward DJ; Garlick PJ; Stewart RJC; Nnanyelugo DO; Waterlow JC. Skeletal-muscle growth and protein turnover. *Biochem J,* 1975, 150:235-243.
Nishzawa M; Shimbo M; Hareyama S. Fractional catabolic rates of myosin and actin estimated by urinary excretion of N-methyl histidine: The effect of dietary protein level on catabolic rates under conditions of restricted food intake. *Br J Nutr,* 1977, 37:345-353.
Rathmacher JA; Link GA; Nissen SL. Measurement of 3-methylhistidine production in lambs by using compartmental-kinetic analysis. *Br J Nutr,* 1993, 69:743-755
Rathmacher J; Nissen S. Rate of 3-methylhistidine (3MH) exchange between tissues does not explain non-quantitative urinary excretion of 3MH in swine and sheep. *FASEB J,* 1992, 6(5):A196.(Abstract)
Rathmacher J; Trenkle A; Nissen S. The use of compartmental models of 3-methylhistidine flux to evaluate skeletal muscle protein turnover in implanted steers. *J Anim Sci,* 1993, 71:135.(Abstract)
Rathmacher JA; Link GA; Flakoll PJ; Nissen SL. Gas chromatographic-mass spectrometric analysis of stable isotopes of 3-methylhistidine in biological fluids: application to plasma kinetics in vivo. *Biol Mass Spectrom,* 1992, 21:560-566.
Rathmacher JA; Link GA; Nissen SL. Technical Note: The use of a compartmental model to estimate the de novo production rate of Nt-methylhistidine in cattle. *J Anim Sci,* 1992, 70:2104-2108.
Rathmacher JA; Flakoll PJ; Nissen SL. A compartmental model of 3-methylhistidine metabolism in humans. *Am J Physiol,* 1995, 269(Endocrinol. Metab. 32):E193-E198.

Rathmacher JA; Roy N; Yen JT; Bernier JF; Lapierre H; Nissen SL. The effect of dietary lysine supply on whole-body myofibrillar protein turnover in pigs. *J Anim Sci,* 1995, 73(Suppl. 1):139.

Rathmacher JA; Nissen SL; Paxton RE; Anderson DB. Estimation of 3-methylhistidine production and muscle mass in swine by compartmental analysis. *J Anim Sci,* 1996, 74:46-56.

Rathmacher JA. Comparative evaluation of muscle proteolysis by a compartmental model of 3-methylhistidine. Dissertation: Iowa State University, Ames, Iowa, 1993.

Reeds PJ. Regulation of protein turnover, in: *Animal Growth Regulation.* Campion DR; Hausmann GJ; Martin RJ; Eds. Plenum Press: New York. 1989. p. 183.

Schoenheimer R; Rittenberg D. The study of intermediary metabolism of animals with the aid of isotopes. *Physiol Rev,* 1940, 20:218.

Simon O. Metabolsim of proteins and amino acids in: *Protein Metabolism in Farm Animals.* Bock H-D; Eggum BO; Low AG; Simon O; Zebrowska T; Eds . VEB Deutscher Landwirtschaftsveriag: Berlin, Germany. 1989. p. 273.

Sjölin J; Stjernström H; Arturson G; Andersson E; Friman G; Larsson J. Exchange of 3-methylhistidine in the splanchnic region in human infection. *Am J Clin Nutr,* 1989, 50:1407-1414.

Sjölin J; Stjernström H; Friman G; Larsson J; Wahren J. Total and net muscle protein breakdown in infection determined by amino acid effluxes. *Am J Physiol (Endocrinol Metab),* 1990, 258:E856-E863.

Trottier NL; Easter RA; Rathmacher JA; Nissen SL. Effect of energy intake during gestation on feed intake, body weight change, and protein metabolism during lactation in primiparous sows. *J Anim Sci,* 1995,73(Suppl. 1):185.

Van den Hemel-Grooten HNC; Koohmaraie M; Yen JT; Arbona JR; Rathmacher JA; Nissen SL; Fiorotto ML; Garssen GJ; Verstegen MWA. Comparison between 3-methylhistidine production and proteinase activity in skeletal muscle during protein deficiency in growing barrows. *J Anim Sci,* 1995, 73:2272-2281.

Van den Hemel-Grooten HNA; Rathmacher JA; Garssen GJ; Schreurs VVAM; Verstegen MWA. Contribution of gastrointestinal tract to whole-body 3-methylhistidine production in growing pigs. *J Anim Physiol Anim Nutr,* 1997, 77:84-90.

Van den Hemel-Grooten HNA. 3-Methylhistidine production and muscle proteinase activity in growing pigs: protein metabolism as a tool for growth modulation. Thesis Landbouwuniversiteit Wageningen, 1996.

Wang ZM; Matthews DE; Heymsfield SB. Total body skeletal muscle mass: Evaluation of 24 hour urinary 3-methylhistidine excretion method by computerized tomography. *FASEB J,* 1996, 10(3):A734.

Wolfe R. *Radioactive and stable isotopes tracers in biomedicine: Principles and practice of kinetic analysis.* Wiley-Liss: New York. 1992. pp. 145-165.

Young VR. The role of skeletal and cardiac muscle in the regulation of protein metabolism, in: *Mammalian Protein Metabolism.* Munro HN; Ed. Academic: New York. 1970. pp. 585-674.

Young VR; Alex SD; Baliga BS; Munro HN; Muecke W. Metabolism of administered 3-methylhistidine: Lack of muscle transfer ribonucleic acid charging and quantitative excretion as 3-methylhistidine and its N-acetyl derivative. *J Biol Chem,* 1972, 217:3592-3600.

Young VR; Haverberg LN; Bilmazes C; Munro HN. Potential use of 3-methylhistidine excretion as an index of progressive reduction in muscle protein catabolism during starvation. *Metabolism,* 1973, 22:1429-1436.

Young VR; Munro HN. Nt-Methylhistidine (3-methylhistidine) and muscle protein turnover: An overview. *Fed Proc,* 1978, 37:2291-2300.

Zhang XJ; Chinkes DL; Sakurai Y; Wolfe RR. An isotopic method for measurement of muscle protein fractional breakdown rate in vivo. *Am J Physiol,* 1996, 270:E759-E767.

MODELING RUMINANT DIGESTION AND METABOLISM

R.L. Baldwin and K.C. Donovan

Department of Animal Science
University of California, Davis
Davis, CA 95616-8521

INTRODUCTION

Our general objective in this chapter is to provide an overview of our research program which for the past 25 years has involved the use and coordination of both experimental and modeling research approaches. Our overall goal has been to advance our understanding of animal digestion (particularly in ruminants) and metabolism, and devise objective and quantitative means to bridge the gap between our knowledge of basic animal functions and animal performance. Early on we came to the view that systems analysis may be the only approach available which enables rigorous quantitative evaluations of our knowledge of animal functions and full utilization of this knowledge in the solution of problems in animal production. This approach is compatible with the view of science depicted in schematic form (Fig. 1) by Thornley and France (1984) and discussed in greater detail by Forbes and France (1993) and Baldwin (1995).

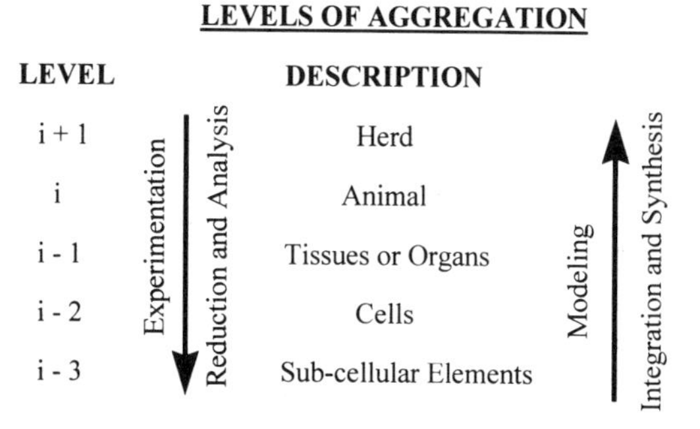

Figure 1. Levels of aggregation (adapted from Thornley and France, 1984).

Mathematical Modeling in Experimental Nutrition
Edited by Clifford and Müller, Plenum Press, New York, 1998

Basically, this representation indicates that our understanding of animal functions advances through looking downward throughout the hierarchy of animal organization from the organ to the cellular to sub-cellular and molecular levels as required to explain and understand basic functions. This process requires experimentation and is referred to as reduction and analysis. Many scientists have highly successful and productive careers in this portion of the scientific process. However, in context with our above-stated objective of effectively utilizing knowledge of lower level functions in improving animal production, it is essential that the second portion of the scientific process, namely integration and synthesis through modeling, be undertaken.

Models can be classified in a number of different ways. The classification we prefer in accord with our objectives follows (adapted from Thornley and France, 1984):

Dynamic	vs	Static
Deterministic	vs	Stochastic
Mechanistic	vs	Empirical

In this context, *dynamic* models are made up of differential equations of the form $dx/dt = F(1) + F(2) ..... F(i)$. Models made up of equations of this form allow users to trace the behavior of a system through time such that simulations may accommodate both the quantitative and dynamic domains. This is important in modeling animal systems because animals change over time and, many times, past or current management decisions influence subsequent function. *Static* models are usually algebraic in form, contain no time-dependent variables, and are solved once for a given set of conditions specified as input. Current feeding systems used by human and animal dietitians are static. Current body weight, activity, and rate of production (growth, milk, pregnancy, etc.) are input. Then the energy requirements of the individual are calculated for that animal on that day. Such models have proven to be quite useful over the past 100 years but have a number of severe limitations which have been discussed elsewhere (Thornley and France, 1984; Robson and Poppi, 1990; Forbes and France, 1993; Baldwin, 1995; and others).

Deterministic models yield a single answer for each simulation run because all parameters are entered as exact values. The simulation outputs are nominally considered to represent the average animal of the population. The term *stochastic* is usually taken to reflect either uncertainty regarding cause or effect relationships within the system or true sources of variance in a population of animals. In both cases, a model becomes stochastic when parameters are specified as a mean value ± a standard deviation, and the value used in the model is allowed to vary randomly in sequential runs where ranges specified by the standard deviation and a random number generator define the actual parameter values used. Stochastic models must be run a number of times to obtain an estimate of the population mean. Such models also yield estimates of variance within the simulated population.

The implication that the introduction of stochastic elements be used to reflect uncertainty or lack of understanding of cause and effect relationships within the system is totally unacceptable given our research objectives. On the other hand, given a basically sound deterministic model and specific knowledge that one or more specific (genetic) traits are major causes of variation in responses of individual animals, it can be quite instructive to introduce these as stochastic elements into the model and ask the question: How much of the total variance in a population can be explained on the bases of these one or two variable traits?

The term *empirical* usually applies to regression equations which are used to describe a relationship between two or more variables observed at a level i (Fig.1). The only constraint imposed is that the equation represent a statistical best fit of a data set. Coefficients and the like in the equations need not and most often do not imply anything about underlying relationships. Such equations must be applied carefully to assure that they are not applied to situations not defined by the data set upon which the equation is based. Most current models used in formulating nutritional recommendations for animals and humans are based upon empirical equations. *Mechanistic* equations, sometimes also called theoretical equations, are derived not from a particular set of data but rather from concepts regarding the fundamental nature of the system. The concepts upon which mechanistic equations and, ideally, the parameter values used to implement the equations, arise from studies of lower level functions (i-1, i-2, etc.). Well-conceived models based upon mechanistic equations should have explanatory power and apply generally (see France and Thornley (1984) and Baldwin (1995) for more detailed discussions).

Our current model of digestion and metabolism in lactating cows will be used in subsequent sections to illustrate the approaches we have developed and utilized in the modeling component of our research program. This and related models are dynamic, deterministic, and mechanistic in keeping with the objectives of our total research program. We will not discuss the digestive element of the model in any detail as this would not be expected to be of interest to participants in the symposium. However, we will discuss the use of an algebraic or static model to deduce parameter values for carbohydrate fermentation which cannot be measured directly. Then we will discuss both the animal element of the model in general terms, and some of the types of data we have had to collect to advance the program in terms of selection of appropriate equation forms and the generation of required numerical inputs. Finally, we will discuss model behavior, the introduction of a stochastic element, and the potential applied use of the current model.

RUMEN FERMENTATION MODEL

We have known for 50 to 60 years that the several rumen microbes ferment specific chemical entities—cellulose, hemicellulose, pectin, starch, organic acids, sugars, etc. However, we have been unable to define, quantitatively, a relationship between the chemical composition of diets and the fermentation products formed. There have been several reasons for this. One has been that concentrations of mono- and disaccharides in the rumen are very low and turnover rates very high, so that results from experiments using radioisotope tracers to investigate metabolism have been difficult to interpret—the concentrations of labeled substrates have been too high or the periods of measurement have been too long relative to turnover rate. Another problem has been that purification of insoluble plant components (e.g., cellulose, hemicellulose, etc.) changes the physical characteristics of those components in terms of crystallinity and association with other cell-wall components which, in turn, changes the mix of microbes fermenting these substrates. Koong et al. (1975) proposed a modeling approach to the problem of relating diet composition to fermentation products. The basic approach involved expression of stoichiometric relationships such as

$$\text{Hexose} \rightarrow 2\ \text{Pyruvate} + 2\ \text{ATP} + 2H_2 \tag{1}$$

$$2\ \text{Pyruvate} \rightarrow 2\ \text{Acetate} + 2\ \text{ATP} + 2H_2 + 2CO_2 \tag{2}$$

in the form of algebraic equations (in Fortran)

```
ACETF(I) = 2.0 * A(I) * ALPHA(I) * CH2O(I)                    (3)
```

where ACETF(I) is acetate formed as a result of fermentation of a specific carbohydrate I (CH2O(I)), A(I) is the proportion of carbohydrate I converted to acetate and ALPHA(I) is the portion of used carbohydrate I which was fermented as compared to that used to make microbial cell components. Similar equations were formulated for propionate, butyrate, valerate, hydrogen, CO_2, and methane formation during fermentation, thus assuring that the law of conservation of matter was satisfied in all solutions of the model. Murphy et al. (1982) gathered two very large data sets representing a wide range of differences in diet composition, one for forage-fed animals and the other for animals fed diets containing significant amounts of concentrate feeds. Then a computer algorithm was used to adjust estimates of A(I), B(I), C(I), etc., until the weighted sum of squares of residuals (WSQR) between the observed and the predicted estimates of products produced were minimized. A partial listing of the results obtained is presented in Table 1.

Table 1. Estimated rumen fermentation parameters.[a]

		Proportion of carbohydrate converted to:			
Substrate	Diet	Acetate A(I)	Propionate B(I)	Butyrate C(I)	Valerate D(I)
Soluble Carbohydrate[b]	R[c]	0.69 ± 0.06	0.20 ± 0.01	0.10 ± 0.05	0.0 ± 0.00
	C[d]	0.45 ± 0.03	0.21 ± 0.04	0.30 ± 0.04	0.04 ± 0.00
Starch	R	0.59 ± 0.04	0.14 ± 0.02	0.20 ± 0.05	0.06 ± 0.00
	C	0.40 ± 0.01	0.30 ± 0.01	0.20 ± 0.00	0.10 ± 0.01
Hemicellulose	R	0.57 ± 0.06	0.18 ± 0.03	0.21 ± 0.05	0.05 ± 0.00
	C	0.56 ± 0.01	0.26 ± 0.01	0.11 ± 0.00	0.07 ± 0.00
Cellulose	R	0.66 ± 0.10	0.09 ± 0.00	0.23 ± 0.09	0.03 ± 0.00
	C	0.79 ± 0.00	0.06 ± 0.00	0.06 ± 0.00	0.09 ± 0.00

[a]From Murphy et al. (1982).
[b]Soluble carbohydrate fraction includes organic acids and pectin in this analysis.
[c]R codes roughage rations.
[d]C codes rations containing more than 50% of a cereal-based concentrate ration.

Confidence limits for the deduced parameter values were estimated using the jackknife procedure (Miller, 1974). In this procedure, ten percent of the original data set was removed on a random basis to form ten and eight data subsets for the roughage and concentrate groups, respectively. These data subsets were then used to estimate the parameter values. Variance among these was used to calculate the standard errors of estimate presented in Table 1. Standard errors were usually below 10% or in range with biological variance. There were some systematic errors of prediction and these were or have subsequently been addressed.

When non-linear least squares methods are used for parameter estimation, there is a possibility that the solution obtained is nonunique, i.e., that more than one combination of parameter values can yield a similar WSQR. The test for uniqueness adopted was to set the proportion of propionate formed from a substrate (B(I)) at values different from the deduced value and allow the program to vary the other 15 parameter values during iterative runs. This procedure was applied to all five substrates. Two extreme results are shown in Figure 2. The sharply convex relationship found for soluble carbohydrates—and the other carbohydrates—indicated uniqueness. The absence of such a relationship for protein indicates that the deduced parameter values were nonunique. The latter observation reflects the fact that protein content did not vary significantly among the diets for which data were available. When variance in a diet component is small, unique solutions cannot be obtained. This is why parameter values for protein are not presented in Table 1.

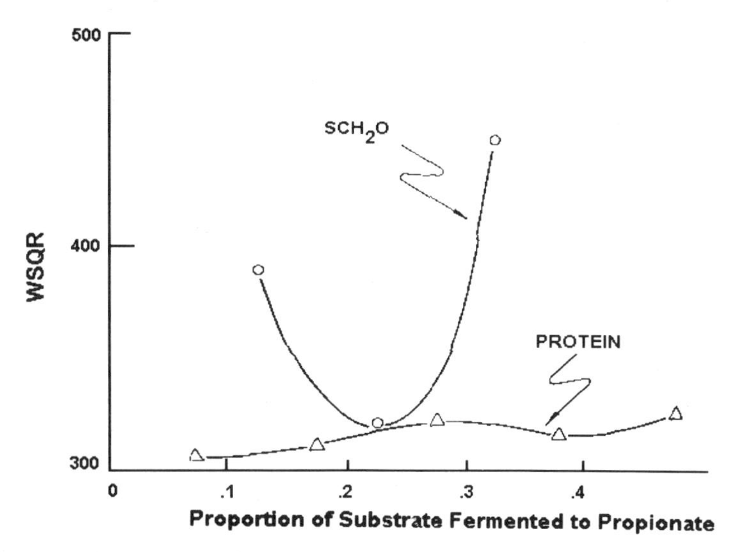

Figure 2. Sum of weighted squares of residual (observed-predicted; WSQR) as influenced by changes in the proportion of soluble carbohydrates (SCH20) or protein fermented away from the deduced parameter value. Deduced parameter values for soluble carbohydrates and protein were 0.205 and 0.302, respectively (adapted from Murphy et al., 1982).

Molar proportions of volatile fatty acids (VFA) were the primary data used to deduce the parameter values presented in Table 1. Data on VFA production rates from 48 experimental animals were not used for parameter estimation and thus could be used for parameter evaluation. This was achieved by incorporating the deduced stoichiometric coefficients into our mechanistic, dynamic model of ruminant digestion (Baldwin, 1995) and simulating the VFA production data. Example results of this challenge are presented in Figure 3. Very good accuracy of prediction of VFA production data was observed. Respective coefficients of determination for acetate, propionate, and butyrate were 0.97, 0.95, and 0.95. When only sheep data (rates of production less than 6 mol/d) were used, the R^2 value was 0.84.

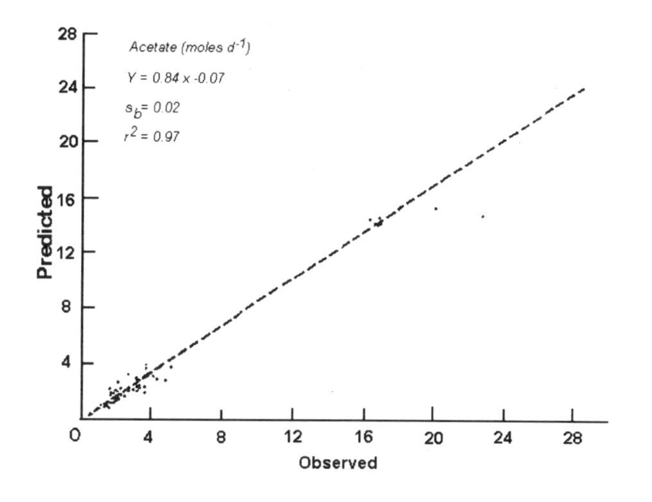

Figure 3. Predicted vs observed production of acetate.

Also, using these stoichiometric coefficients, very acceptable estimates of the metabolizable energy values of a wide range of feeds were obtained (Fig 4). The R^2 for the relationship depicted in Figure 4 was 0.85 and is in range with experimental variance associated with repeated measures of ME values on the same feeds.

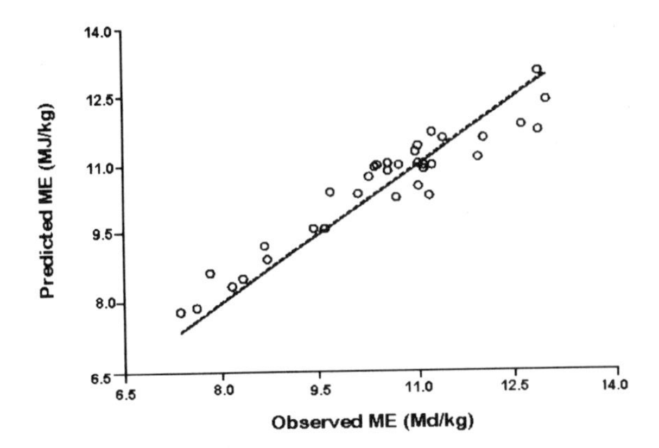

Figure 4. Comparison of predicted vs observed estimates of metabolizable energy. Regression equation for best fit line (—) was y = a + bx with an R^2 value of 0.85. Regression equation for line of equality (---) was y = bx (adapted from Baldwin, 1995).

ANIMAL ELEMENT OF THE LACTATING COW MODEL

A block diagram of our current model of the metabolism of a lactating cow is presented in Figure 5. The model looks quite complex at first glance because of interactions among nutrients and pools, but in truth is quite simple as only 11 state variables are represented: blood, including acetate, amino acids, lipids (as fatty acids), urea, and glucose; body and visceral protein; storage triacylglyceride; the lactation hormone complex; udder enzymatic capacity; and milk in the udder.

Initial formulation of the model required an extensive survey of the literature to ascertain relative organ weights and energy expenditures, blood and tissue metabolite levels under varying conditions, rates of nutrient turnover and oxidation in intact animals and individual tissues, arteriovenous differences for specific organs, and many more data (Baldwin, 1995). In all modeling studies, problems with the availability of data required for unique definition of equation forms and parameter values are encountered. Some of these limitations compromise the concept stated above that all concepts and numerical inputs to mechanistic models must be firmly based upon experimental data. We are constantly confronted with this problem even though our experimental program for the past 25 years has been dedicated to the collection of the required data. We have discussed numerous examples of cases where our mechanistic models were found to be inadequate at either the conceptual or numerical input level—or both—as well as the experiments we have undertaken to alleviate these problems (1995). Only one example of such an approach will be presented here. Note in the block diagram (Fig. 5) that four equations determine milk synthesis (one arrow equals one equation):

I. An equation for milk protein (Pm) synthesis

$$U_{Aa,Pm} = V_{Aa,Pm} * UENZ/(1.0 + k_{Aa,Pm}/cAa) \quad (4)$$

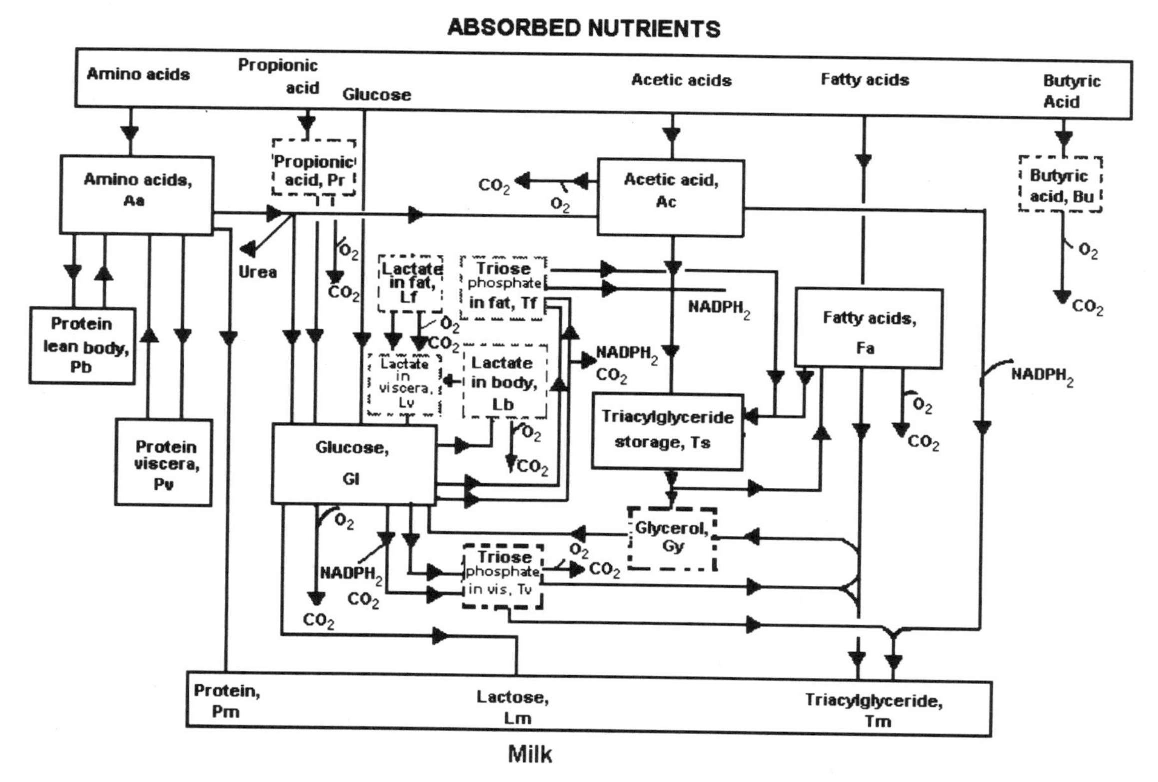

Figure 5. Diagrammatic representation of parts of the metabolic element of the lactating dairy cow model. Entities enclosed in boxes represent state variables and arrows indicate fluxes (adapted from Baldwin, 1995).

where $U_{Aa,Pm}$ is the rate of amino acid (Aa) incorporation into milk protein, $V_{Aa,Pm}$ is the scalar used to convert UENZ to the current mammary gland enzymatic capacity for protein synthesis to the maximal capacity (Vmax) for the reaction, $k_{Aa,Pm}$ is the apparent affinity of the udder for amino acid incorporation into protein and cAa is the current concentration of amino acid in blood.

II. An equation for incorporation of blood lipids (FA) and acetate (Ac) into milk fat (Tm)

$$U_{FA,Tm} = V_{FA,Tm} * UENZ/(1.0 + k_{FA,Tm}/cFA + k1_{FA,Tm}/cGl) \quad (5)$$

$$U_{Ac,Tm} = V_{Ac,Tm} * UENZ/(1.0 + k_{Ac,Tm}/cAc + k1_{Ac,Tm}/cGl) \quad (6)$$

where $V_{FA,Tm}$ and $V_{Ac,Tm}$ are scalars for UENZ, $k_{FA,Tm}$ and $k_{Ac,Tm}$ are apparent udder affinities for blood lipids and acetate, $k1_{FA,Tm}$ and $k1_{Ac,Tm}$ are apparent affinities of the udder for glucose use for esterification of FA and the use of glucose as a source of $NADPH_2$ for fatty acid formation from acetate and the esterification of resulting fatty acids, respectively.

III. An equation for lactose (Lm) synthesis

$$U_{Gl,Lm} = V_{Gl,Lm} * UENZ/(1.0 + k_{Gl,Lm}/cGl + k1_{Gl,Lm}/cAa) \quad (7)$$

where $U_{Gl,Lm}$ is the rate of utilization of glucose (Gl) for lactose (Lm) synthesis, $V_{Gl,Lm}$ is a scalar, $k_{Gl,Lm}$ is the affinity of the udder for glucose use in lactose synthesis and $k1_{Gl,Lm}$ is the affinity of the udder for amino acid use for α-lactalbumin synthesis which is, in turn, essential to lactose synthesis.

These equations are clearly highly aggregated as is appropriate for animal level models, but parameterization of these equations is not at all straightforward nor possible on the bases of whole-animal input:output data. In order to help resolve this dilemma in parameterization of the whole-animal model, two approaches have been utilized. The first has been to conduct extensive studies of mammary tissue metabolism using mainly *in vitro* radiotracer techniques and *in vivo* arteriovenous difference studies across the udder in order to establish appropriate input:output relationships and to resolve such issues as pathways of nutrient utilization under various conditions and interactions among nutrients.

The second approach has been to utilize these and other (enzyme, metabolite, etc.) data to formulate detailed models of tissue metabolism which can then be used to identify the dominant features of tissue metabolism which must be captured in equations used at the animal level and to help parameterize these highly aggregated representations. A block diagram of our current model of cow mammary gland metabolism is presented in Figure 6 by way of illustration.

The model has been published, discussed in detail, and fully documented (Hannigan and Baldwin, 1994; Baldwin, 1995). We are not going to elaborate here the coding conventions used, nor the reasons why some intermediary metabolites are included (for example, glucose-6-phosphate (G6P)), while others (such as phosphoenolpyruvate (PEP)) are not. Our objectives in presenting the model diagram are two-fold. First, to illustrate that tissue models capture a great deal more detail than animal models do as is appropriate to the dictates of system theory regarding scale and detail. Second, the potential for identifying and evaluating in a quantitative fashion primary or dominant effectors of overall tissue function are present in such tissue models.

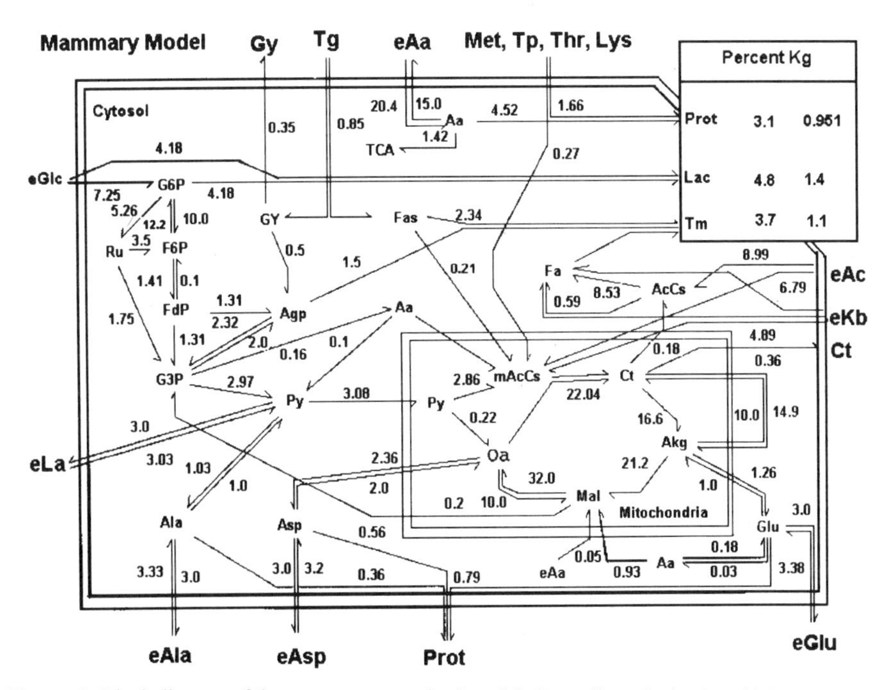

Figure 6. Block diagram of the cow mammary gland model. Cytosolic and mitochondrial spaces are bounded by solid double lines. Pools are identified by two- three- or four-letter codes. Arrows between pools indicate fluxes and numbers associated with the arrows are the fluxes at steady-state concentrations (M/day). Extracellular space is that area outside of the cytosolic space. Pool sizes of extracellular nutrients do not change through a particular simulation, while pool sizes of intracellular nutrients and products do change dependent upon the relevant inputs and outputs (fluxes) to the pool. Abbreviations are milk fat (Tm), cytosolic acetyl-CoA (AcCs), mitochondrial acetyl-CoA (mAcCs) (adapted from Baldwin, 1995).

The *in vitro* data summarized in Figures 7 to 10 and Table 2 are presented to illustrate some utility of such data in modeling tissue metabolism. Figures 7 and 8 illustrate a common result in that most tissues exhibit saturation kinetics *in vitro*.

Such observations clearly suggest that Michaelis-Menten (saturation kinetic) type equations are appropriate to represent most tissue level transactions even though concentrations required to saturate the system may be considerably above concentrations normally encountered under normal physiological conditions. The data presented in Figures 7 and 8 indicate a feature of experimental design which is essential to gaining insights to critical tissue metabolic properties. This is the use of radioisotope tracers labeled in specific positions. The difference in the patterns of lactate-1-^{14}C and lactate-2-^{14}C oxidation reflects the fact that most of the conversion of lactate-1-^{14}C to CO_2 is due to the action of pyruvate dehydrogenase while most of the oxidation of lactate-2-^{14}C reflects the conversion of lactate to acetyl-CoA, which is subsequently oxidized via the tricarboxylic acid cycle. The difference between the oxidation of the two radiolabled carbons of lactate also indicates how much acetyl-CoA formed from lactate via the pyruvate dehydrogenase reaction is metabolized via pathways other than the tricarboxylic acid pathway.

Table 2. Effect of acetate on apparent kinetic parameters for glucose-1-^{14}C use.

Product	Acetate	K_{glc} (mM)[a]	Vmax[b]
Lactose	2	3.4	2.3
	8	1.9	1.9
Glycerol	2	1.6	1.2
	8	0.8	0.6
Citrate	2	8.1	1.3
	8	0.9	0.4

[a]Apparent affinity for conversion of glucose to product.
[b]Apparrent maximum capacities for conversion of glucose-1-^{14}C to product expressed as μg atoms tracer converted /(g * hr).

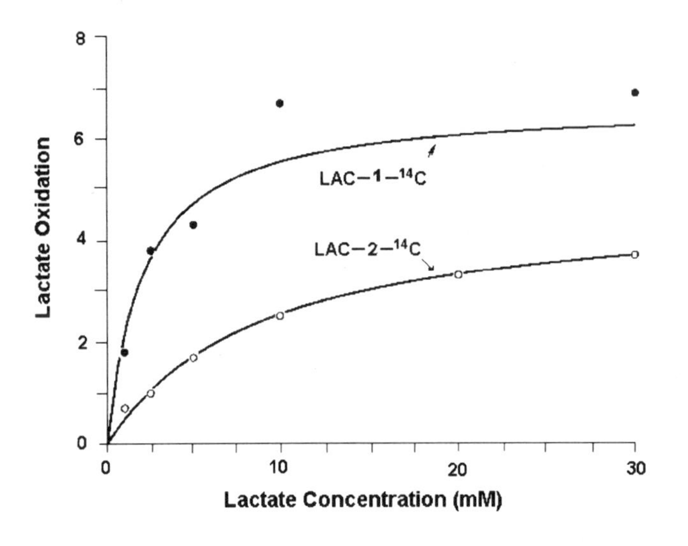

Figure 7. Kinetic data on cow mammary metabolism *in vitro* (adapted from Forsberg et al., 1985b).

Similarly, the differences between glucose-1-^{14}C and glucose-2-^{14}C oxidation depicted in Figure 8 provide insight regarding the relative use of glucose-6-P via the pentose phosphate and Embden-Meyerhof pathways and, in this case, the effect of acetate availability upon the contributions of these two alternate pathways. These types of data can be evaluated quanti-tatively and objectively using tissue level models and are very helpful or even essential in formulating equations for animal level models.

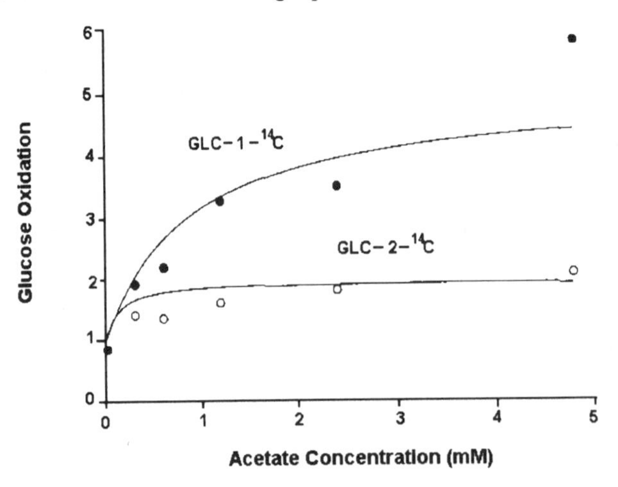

Figure 8. Effects of glucose and acetate concentrations upon rates of glucose oxidation (adapted from Forsberg et al., 1985a).

Figures 9 and 10 illustrate the importance of considering nutrient interactions in tissue level models.

The data in Figure 9 clearly indicate that rates of lactate conversion to fatty acids in cow mammary tissue are linear functions of lactate concentrations up to supraphysiological concentrations. However, at physiological concentrations of acetate lipogenesis from lactate is strongly inhibited due to the negative feedback of acetyl-CoA on the pyruvate dehydrogenase reaction. Several additional interactions are summarized in Table 2. These interactions must be accounted for in tissue-level models and when found to be quantitatively important under physiological conditions, must be captured in equations forms adopted for use in whole-animal models.

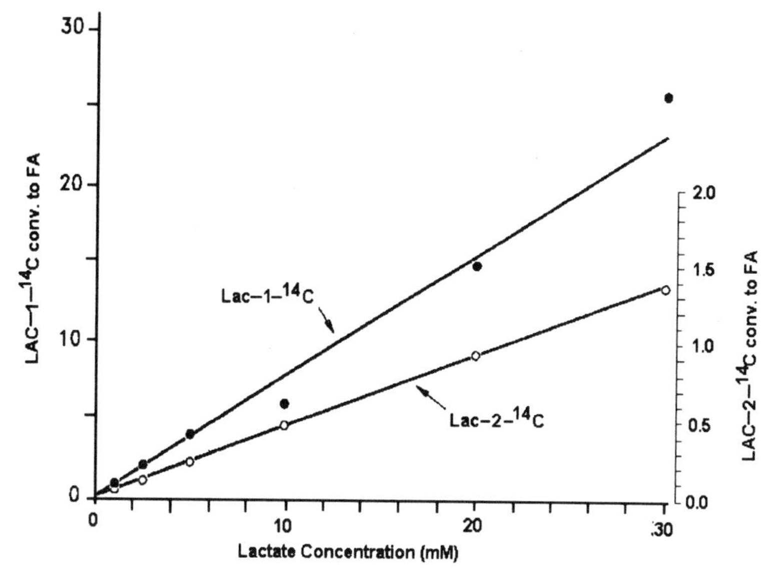

Figure 9. Conversion of lactate carbons 1 and 2 (Lac-1-^{14}C and Lac-2-^{14}C) to fatty acids (FA) in the presence of 2 mM glucose and 2 mM acetate (adapted from Forsberg et al., 1985b).

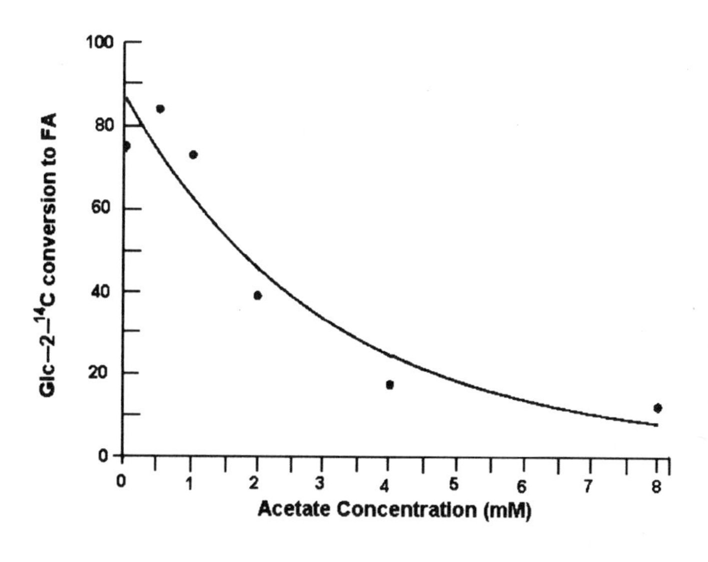

Figure 10. Effect of acetate on conversion of [2-^{14}C] glucose (Glc-2-^{14}C) to fatty acids. Glucose concentration was 16 mM. Means are of three observations ± SE (adapted from Forsberg et al., 1985a).

The data summarized in Figs. 11 to 16 and Tables 3 and 4 arose from an arteriovenous difference study of the uptakes of nutrients by cow mammary glands to support the development of both our mammary-tissue model and our lactating-cow model. These data were essential to the formulation of equation forms and the parameterization of both models (Baldwin, 1995). The data summarized in Figures 11 to 14 indicate clearly that uptakes of most nutrients are a linear function of their concentrations in arterial blood.

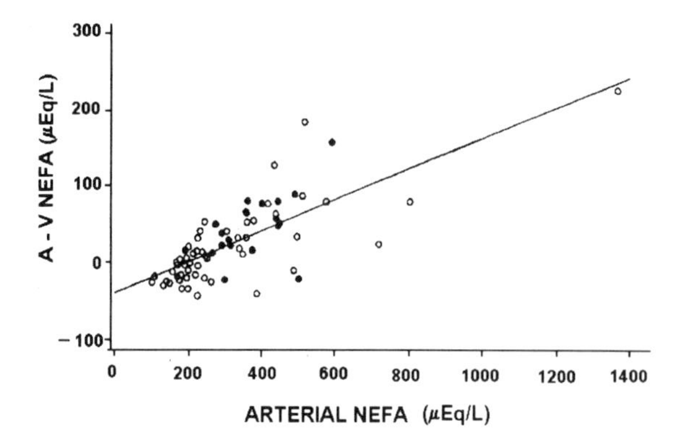

Figure 11. Arterio-venous differences for NEFA taken up by the mammary glands of lactating cows (adapted from Miller et al., 1991).

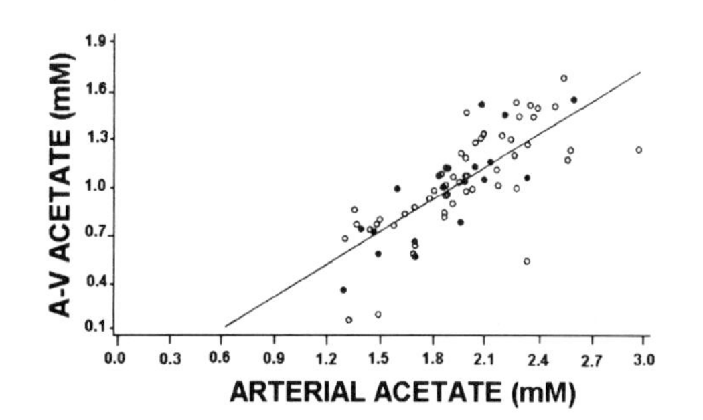

Figure 12. Arterio-venous differences for acetate taken up by the mammary glands of lactating cows (adapted from Miller et al., 1991).

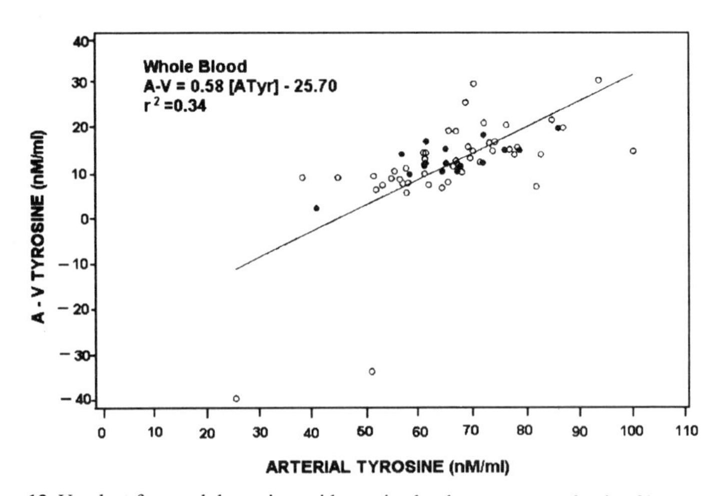

Figure 13. Uptake of several the amino acid tyrosine by the mammary glands of lactating dairy cows (adapted from Hanigan et al., 1992).

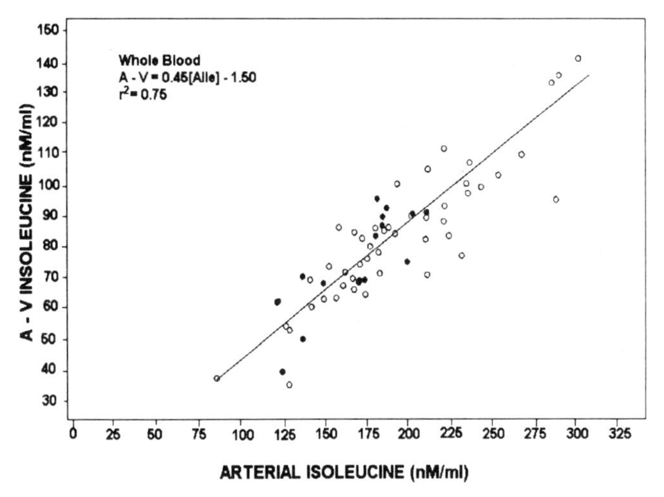

Figure 14. Uptake of the amino acid isoleucine by the mammary glands of lactating dairy cows (adapted from Hanigan et al., 1992).

This should not be taken to imply that the kinetics of use of these nutrients are also linear functions of their arterial concentrations. The *in vitro* data discussed above and many other observations discussed elsewhere (Baldwin, 1995) indicate that this is not true. Uptakes of nutrients are not all linear functions of their arterial concentrations, as is illustrated for glucose in Figure 15.

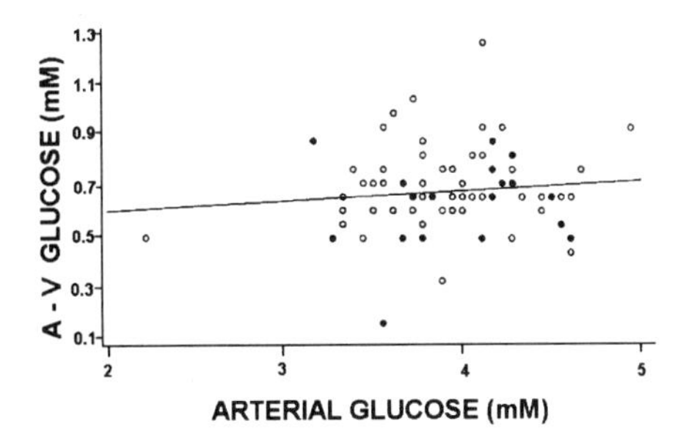

Figure 15. Arterio-venous differences for glucose blood metabolite taken up by the mammary gland of lactating cows (adapted from Miller et al., 1991).

This observation suggests that factors other than arterial concentration govern the uptake of glucose by the udder. Acetate availability is one of the factors (Figure 16) as is also implied in Figure 8.

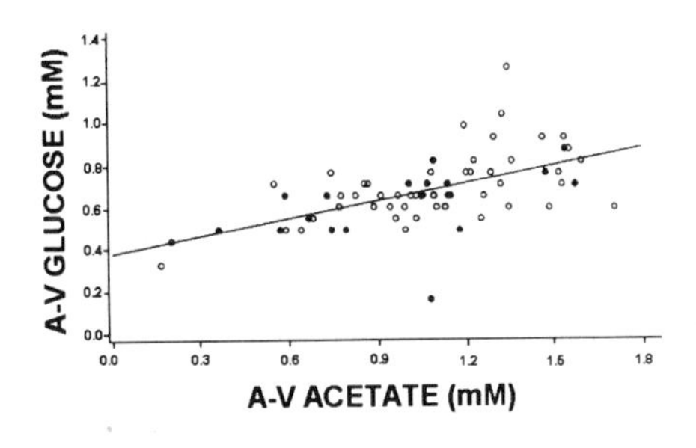

Figure 16. Arterio-venous differences for glucose vs acetate for the mammary glands of lactating cows (adapted from Miller et al., 1991).

Because blood nutrient concentrations vary throughout the lactation cycle, one would expect that patterns of nutrient availabilities to the mammary glands for milk synthesis would also change. This is reflected in the crude summary of nutrient concentrations in blood and uptakes by the udders of early and mid-lactation cows presented in Table 3.

Table 3. Effect of stage of lactation on concentrations and arterio-venous differences across the mammary glands for key metabolites.[a]

	Early-lactation cows		Mid-lactation cows	
Metabolite	Concentration (mM)	Uptake (mM/d)	Concentration (mM)	Uptake (mM/d)
Glucose	3.83	0.71	4.09	0.70
Acetate	1.95	1.04	2.11	1.15
D-beta-hydroxybutyrate	1.56	0.46	1.04	0.28
Non-esterified fatty acids	347	38	224	-0.5
Triacylglycerol	6.97[b]	2.72[c]	9.147[b]	3.65[c]

[a]From Baldwin, 1995.
[b]Concentration expressed as μEq/dL.
[c]Uptake expressed as mg/dL.

We have found in the course of our development of mechanistic models of animal metabolism that, regardless of one's overall objective, the first basic step in formulation of a model should be the definition of a standard or reference condition for input:output relationships. In this case, the reference conditions set were for a perceived average tissue or cow. This situation is very useful in the initial parameterization of the model, and when used to solve for a steady-state condition (constant inputs, concentrations, etc.) it can help to identify errors in coding and in calculations of numerical values within the model. The reference condition set for inputs and outputs for our mammary gland model and mammary elements in the lactating cow model are presented in Table 4.

Table 4. Uptake:output balance for mammary glands of a model cow producing 30 kg of milk per day.

	Uptake/d		Milk	Output/d	
Nutrient	Moles	MJ	Component	Moles	MJ
Glucose	11.4	32.1	Lactose	4.10	23.2
Amino Acids	8.3	21.6	Protein	7.86	20.4
Acetate	13.0	11.4	TAG	4.80	45.6
βHBA	0.94	1.9	Citrate	0.34	0.7
Stearate	3.09	35.0			
Glycerol	-0.83	-1.4			

Mammary efficiency = MJ output/MJ uptake = 0.89 (from Baldwin, 1995).

This reference condition is fictitious but is a useful one as noted above and discussed in detail by Baldwin (1995). If the model is well-conceived, and the approaches used are capable of simulating reality, responses in blood nutrient concentrations as influenced by alternative diets, feed intakes, stages of lactation etc. should occur in the animal and cause the expected (observed) differences in nutrient availabilities to the udder and the composition and yield of milk. This is further illustrated in Figure 17 which is a plot of radiolabled fatty acid oxidation rates as a function of fatty acid concentrations in blood.

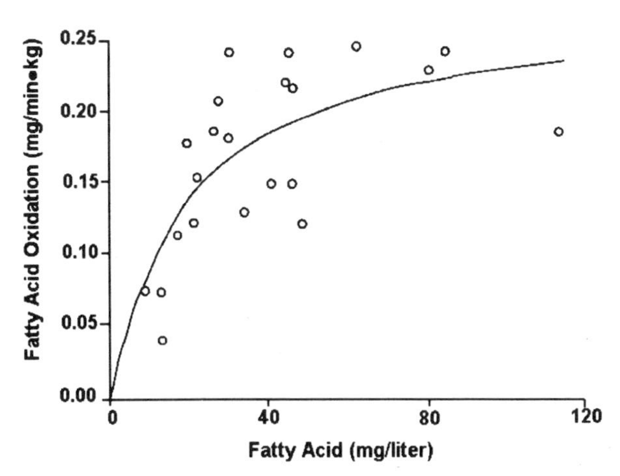

Figure 17. Representation of the behavior of an equation commonly used in modeling animal systems. The figure represents the large variance around the relationship that can arise when such equation forms are applied to whole-animal data (from Baldwin, 1995).

The point in presenting the Figure 17 is not to illustrate that rates of blood nutrient oxidation *in vivo* exhibit saturation kinetics but rather to depict the tremendous variance about the statistical best-fit relationship. If the empirical equation derived from the statistical exercise were to be used in a model, the model would not be able to simulate or account for variance relative to the best-fit equation when modeling animal systems. On the other hand, a good mechanistic model would be expected to reproduce and explain this variance. In this case, the primary causes of the observed variance were found to be differences in the physiological and nutritional states of the individual animals studied and the effects of these differences upon concentrations of alternate oxidizible substrates in blood such as acetate, ketone bodies, amino acids, and glucose.

BEHAVIOR OF THE LACTATING COW MODEL

In our introduction, we emphasized that we prefer dynamic models over the static models in current use for estimating nutrient requirements of animals because "animals change over time and, many times, past or current management decisions influence subsequent function." Broster and Broster (1984) summarized the results of a large number of full-lactation studies with lactating dairy cows subjected to alternative feeding strategies. A summary figure from their analysis is presented in Figure 18 to illustrate the effects of feeding strategy on current productivity and subsequent performance. A clear-cut carry-over or residual effect of high rates of feeding early in lactation upon subsequent performance is evident .

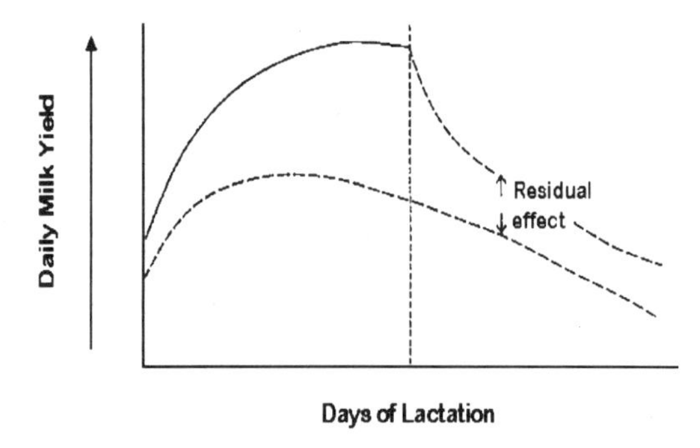

Figure 18. Long-term effects of high (—) and low (---) planes of nutrition on milk production of dairy cows (adapted from Broster and Broster, 1984).

It should be clear that to be acceptable, a dynamic cow model should exhibit these responses. A modeling analysis of effects of feeding cows at low (L_) and high (H_) energy intakes and medium (_M) and high (_H) protein rations upon lactation performance is presented in Figure 19.

The model does exhibit the required characteristics. In fact, simulated responses were very close to observed values (Baldwin et al, 1987). Another attribute a mechanistic model should have is the capacity to evaluate a hypothesis for probable adequacy as explanations of observed responses. The simulated and observed results of such an evaluation are presented in Figure 20.

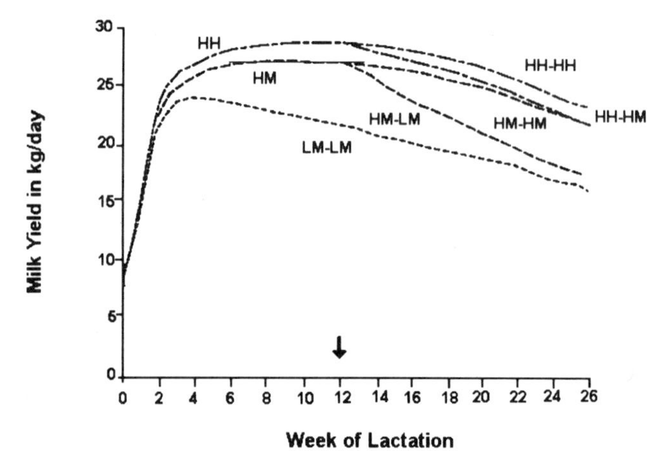

Figure 19. Effect of different feeding strategies upon lactational performance. L_ indicates a feeding rate of 5 kg per day plus 1 kg feed per 3 kg milk averaged over the previous 3 weeks. H_ indicates a feeding rate of 8 kg per day plus 1 kg feed per 3 kg average daily milk yield. _M indicates the standard forage:concentrate (50:50) ration of 15% crude protein. _H indicates standard ration was adjusted to 18% crude protein with fishmeal. Simulated changeovers of diet and feeding strategy occurred at week 12 of lactation (from Baldwin, 1995).

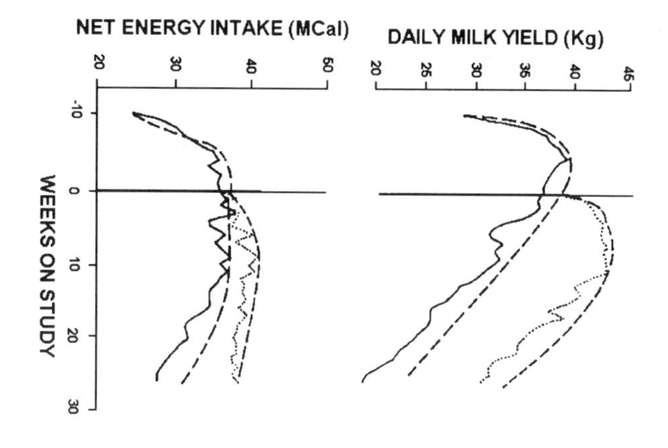

Figure 20. Comparison of effects of growth hormone administration *in vivo* and lactational hormone(s) administration in model.

Several hypotheses considered as possible mechanisms for explaining responses of lactating cows to administration of recombinant bovine somatotropin (rBST) were evaluated. The only hypothesis adequate to explain the response was that rBST acted directly or indirectly to enhance the metabolic capacity of the udder. The simulated response are presented in Figure 20.

The model has been subjected to a large number of data set challenges and has generally survived well. Both experimental and modeling efforts continue to depict improvements of several elements which behave poorly under certain circumstances. This is to be expected because research also continues to improve upon the static, largely, empirical models in current use for estimating human and animal nutrient requirements. On the other hand, we continue to gain confidence in the model and have started to evaluate it in terms of its potential use in animal agriculture to support management decision making. Results of a study of model-predicted milk production and the profitability of several alternative feeding strategies in 100 cow herds are presented in Table 5.

In these simulations, a stochastic variable defining the genetic capacities for milk production of cows in the herd was introduced as a mean value ± a standard deviation. This was done to evaluate effects of diet and feeding strategy for each cow in the herd rather than the average cow. In this fashion, the effects of overfeeding poor cows with high and medium quality diets (diet 1 & 2), underfeeding superior cows with medium and poor quality diets (diet 1 & 3), and the alternative feeding strategies (all diets and feeding strategies simulated are used in practice) upon overall herd profitability could be assessed.

Two trends are evident in the simulation outputs. Starting lactations with diet 2 instead of diet 1 resulted in higher averages for milk production and profit (Table 5). The increased protein input during early lactation to cows of above average genetic potential paid dividends. The second trend that was evident was that the longer a more nutrient-dense diet was fed, the greater the resulting increases in performance and profit. With slightly different feed costs, the ranking of the feeding strategies could be affected.

This limited evaluation demonstrates the value of the use of dynamic models to evaluate alternative feeding strategies and to enable researchers or managers to test them in the real world. As there are an infinite number of feeding strategies, this approach could be used to identify those that are most optimal in terms of profit or other criteria.

Table 5. Evaluation of alternative herd feeding strategies.[a]

Strategy	Diet and Strategy	Milk (kg)	FCM (lb)	DMI (kg)	EBW (kg)	Profit ($)
1	1	9044[b]	19648[a]	6295[c]	690[bc]	1888[c]
2	2	1027[g]	20007[a]	6619[e]	701[f]	2179[i]
3	3	8324[e]	18155[c]	6069[d]	585[g]	1708[h]
4	1 → 3 T = 98	8502[d]	18776[d]	6123[d]	598[d]	1766[g]
5	1 → 3 M = 20	9015[b]	19580[a]	6285[c]	689[b]	1882[c]
6	2 → 1 T = 98	9899[ac]	19761[a]	6447[a]	694[c]	2061[e]
7	2 → 1 → 3 T = 98; T = 210	9829[a]	19654[a]	6455[a]	653[a]	2022[a]
8	2 → 1 M = 29.5	10079[c]	19709[a]	6483[a]	690[b]	2035[ad]
9	2 → 1 → 3 M = 29.5 M = 20	10095[c]	19747[a]	6483[a]	689[b]	2031[ae]
Pooled standard error		63	155	21	1	14

[a]The first three strategies involved feeding the simulated herd of 100 cows diets 1, 2, or 3 for the whole lactation. Diets 1 and 2 were 60% concentrate diets with crude protein contents of 15 and 18%, respectively. Diet 3 was 30% concentrate with a crude protein content of 15%. Respective prices of the three diets were 0.123, 0.138, and 0.11 $/kg. Strategy 4 involved feeding diet 1 for 98 days and diet 3 for the remainder of lactation. Strategy 5 involved starting lactation with diet 1 and switching to diet 3 after milk production decreased to 20 kg/d. Strategy 6 simulated feeding individual cows diet 2 until day 98 of lactation and diet 1 thereafter. Strategy 7 involved changing from diet 2 to diet 1 at day 98 of lactation and diet 3 at day 210 of lactation. Strategy 8 involved changing from diet 2 to diet 1 when milk production dropped below 29.5 kg/day. Strategy 9 involved a shift from diet 2 to 1 when milk production dropped below 29.5 kg/day and to diet 3 after milk production decreased to 20 kg/d.

Values with differing superscripts within a column were significantly different from on another according to the Student-Newman-Kuhl test ($P<0.5$). Strategy 7 was arbitrarily assigned the superscript 'a' for the purposes of a common reference point.

Dynamic, mechanistic lactating dairy cow models allow the generation of data for evaluations of animal performance and managerial decisions. A large number of variables can be examined to give direction to decision-making for researchers and managers. As teachers of animal science and lactation courses, we wish to note the benefits that can accrue to students who can undertake performance evaluations such as those presented in Table 5. They will gain an understanding that extends significantly beyond what one can compile from available literature, and will thus be better equipped to respond to the complex matrix of decisions a dairy manager or nutritionist must address. Current practitioners may address these issues intuitively or through the use of static equation systems, but both approaches are, at most, semiquantitative when used in evaluations of lactation and economic performance. In our view, dynamic models are absolutely essential to economic evaluations of the risk associated with current decisions because these decisions clearly influence subsequent performance. When a proven, dynamic, mechanistic model becomes available to students and practitioners to generate data such as those presented in Table 5 and, further, to enable cause-and-effect analyses of the underlying reasons for observed responses, we will have created a superior instrument for both the teaching and application of our science. This is the challenge we pose, to ourselves and others who share this vision.

CORRESPONDING AUTHOR

Please address all correspondence to:
R.L. Baldwin
Department of Animal Sciences
University of California, Davis
One Shields Avenue
Davis, CA 95616-8521

REFERENCES

Baldwin RL. *Modeling Digestion and Metabolism*. Chapman & Hall: UK. 1995.

Baldwin RL; France J; Beever DE; Gill M; Thornley JHM. Metabolism of the lactating cow, III. Properties of mechanistic models suitable for evaluation of energetic relationships and factors involved in the partition of nutrients. *J Dairy Res,* 1987, 54:133-145.

Broster WH; Broster VJ. Reviews of the progress of dairy science: Long-term effects of plane of nutrition on performance of the dairy cow. *J Dairy Res,* 1984, 51:149-163.

Forbes JM; France J. *Quantitative Aspects of Ruminant Digestion and Metabolism*. CAB International: UK. 1993.

Forsberg NE; Baldwin RL; Smith NE. Roles of glucose and its interactions with acetate in maintenance and biosynthesis in bovine mammary tissue. *J Dairy Sci,* 1985a, 68:2544-2549.

Forsberg NE; Baldwin RL; Smith NE. Roles of lactate and its interactions with acetate in maintenance and biosynthesis in bovine mammary tissue. *J Dairy Sci,* 1985b, 68:2550-2556.

France J; Thornley JHM. *Mathematical Models in Agriculture*. Butterworth: London. 1984.

Hanigan MD; Baldwin RL. A mechanistic model of mammary gland metabolism in the lactating cow. *Ag Sys,* 1994, 45:369-419.

Hanigan MD; Calvert CC; DePeters EJ; Reis BL; Baldwin RL. Kinetics of amino acid extraction by lactating mammary glands in control of sometribove-treated Holstein cows. *J Dairy Sci,* 1992, 75:161-173.

Koong LJ; Baldwin RL; Ulyatt MJ; Charlesworth TJ. Iterative computation of metabolic flux and stoichiometric parameters for alternate pathways of rumen fermentation. *Comp Prog Biomed,* 1975, 4:209-213.

Miller RG. The jackknife – A review. *Biometrika,* 1974, 61:1-20.

Miller PS; Reis BL; Calvert CC; DePeters EJ; Baldwin RL. Patterns of nutrient uptake by the mammary glands of lactating dairy cows. *J Dairy Sci,* 1991, 74:3791-3799.

Murphy MR; Baldwin RL; Koong LJ. Estimation of stoichiometric parameters for rumen fermentation of roughage and concentrate diets. *J Anim Sci,* 1982, 55:411-421.

Robson AB; Poppi DP. *Third International Workshop on Modelling Ruminant Digestion and Metabolism in Farm Animals*. Lincoln University, Canterbury, New Zealand. 1990.

Thornley JHM; France J. Role of modeling in animal production research and extension work, in: *Modeling Ruminant Digestion and Metabolism*. Baldwin RL; Bywater AC; Eds. Department of Animal Science, University of California, Davis, CA. 1984.

DESIGNING A RADIOISOTOPE EXPERIMENT USING A DYMAMIC, MECHANISTIC MODEL OF PROTEIN TURNOVER

H.A. Johnson, C.C. Calvert, and R.L. Baldwin

Animal Sciences Department
University of California
Davis, CA 95616

INTRODUCTION

Many experiments require preliminary results in order to determine the best experimental design to fulfill the research objectives. Ranges of data values, sampling intervals and numbers of test subjects need to be determined before the experiment is conducted so that differences between treatments can be detected at the appropriate level of significance. Especially in radioisotope experiments where exposure and contamination need to be minimized, previous knowledge of experimental conditions can be invaluable in planning experiments. A model which represents current knowledge of the animal or system to be studied can be an important tool with which to identify appropriate experimental designs.

Beginning with the simplest possible description of the system to be studied, a model is created and refined based on previous experimental data. The model is used to generate data sets for different levels of the unknown parameter which the experiment will be designed to measure. Random variations due to experimental error are added to the data. Then the data are fit to the model to see if the unknown parameter can be predicted from the data and if the prediction of the unknown parameter by the model is unique. If the predictions are successful, subsets of the data can be fit to determine the minimum measurements necessary to identify the unknown parameter. In this chapter, a four-pool model of protein turnover is used to define an experimental design for estimating protein synthesis rate and the source of amino acids used for protein synthesis.

BACKGROUND

Amino acids for protein synthesis via the aminoacyl tRNA pool can arise from an extracellular pool (channeling), an intracellular pool, or from protein degradation (recycling). Figure 1A shows the possible sources of amino acids due to channeling and

Figure 1B shows the sources due to recycling. A high rate of channeling indicates that the amino acids in the aminoacyl tRNA pool come from extracellular sources and a low rate of channeling indicates amino acids for protein synthesis come from the intracellular pool. A high rate of recycling indicates amino acids from protein degradation go directly to protein synthesis through the aminoacyl tRNA pool without mixing with the intracellular pool. A low rate of recycling indicates amino acids from protein degradation go to the intracellular pool where they can be oxidized or charge tRNA for protein synthesis.

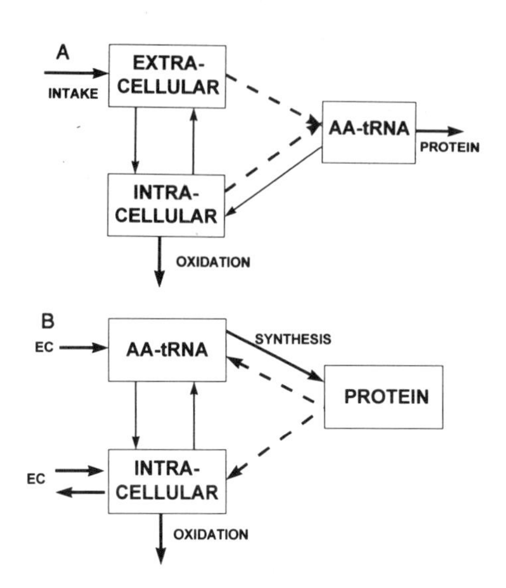

Figure 1. Sources of amino acids for protein synthesis. Possible sources of amino acids for protein synthesis are defined in (1A) by the dashed lines. Channeling is the flow of amino acids from the extracellular pool to the aminoacyl tRNA (AA-tRNA) pool. Possible routes of amino acids from protein degradation are defined in (1B) by the dashed lines. Recycling is the flow of amino acids from protein (degradation) to the AA-tRNA pool without passing through the intracellular pool.

High rates of channeling suggest that tRNA is associated with a structure or a localized concentration within the cell which transfers amino acids from the extracellular compartment to the protein synthetic machinery without mixing with other pools of amino acids. Evidence for channeling is based on the equilibration of radiolabeled amino acid between amino acid pools and seems to differ for different tissues. Barnes et al. (1994) and Van Venrooij et al. (1974) found in certain tissues in chickens and in HeLa cells, respectively, the specific radioactivity of the tRNA pool was closer to the specific radioactivity of the extracellular pool than the intracellular pool. In addition, Negrutskii and Deutscher (1992) found that only aminoacyl-tRNAs synthesized within the cell were resistant to RNAse degradation whereas exogenously (from the media) supplied aminoacylated tRNAs were distributed equally between the intracellular and extracellular compartments. Therefore aminoacyl-tRNA is probably associated with a structure in the cell for protein synthesis which protects the aminoacyl-tRNA from degradation. Evidence for the intracellular pool as the protein precursor pool, however, is supported by the work of Barnes et al. (1994), Wettenhall and London (1975), and Kipnis et al. (1961), who found in other tissues of the chicken, porcine lymphocytes, and rat diaphragm, respectively, that the specific radioactivity of the tRNA pool was closer to the specific radioactivity of the intracellular pool than the extracellular pool.

Because the protein pool contains large amounts of unlabeled amino acid, it would be expected that the greater the rate of recycling, the greater the similarity of the specific radioactivity of the precursor pool to the protein pool. Therefore, most of the evidence for recycling comes from the comparison of the specific radioactivities of the tRNA, extracellular, intracellular, and protein pools. Barnes et al. (1992) incubated chicken macrophages with ^{3}H leucine and found that tRNA specific radioactivity was much lower than the intracellular and extracellular (medium) specific radioactivities. After the medium was switched to contain unlabeled amino acid, the tRNA specific radioactivity was higher than the extracellular or intracellular specific radioactivities and was very close to the protein specific radioactivity. The most likely source of dilution of the tRNA pool for the pre-incubation period and the most likely source of radiolabel from the incubation period were the protein pools. Smith et al. (1988) developed an autoradiographic method in conjunction with a mathematical method to estimate protein recycling in the rat brain. Using this method, it was estimated that 40-50% of the leucine from protein was recycled in the rat brain (Sun et al., 1992) and approximately 50% of the leucine was recycled in the rat liver (Smith and Sun, 1995). Because the calculation of the fractional synthesis rate (FSR), the percentage of protein synthesized per day, is based on an accurate determination of the specific radioactivity of the precursor and product pools, it is important to determine the influence of recycling and channeling on specific radioactivity changes in the tRNA, extracellular, intracellular, and protein pools.

DESCRIPTION OF THE MODEL

The model consists of three free amino acid pools and a protein pool and represents leucine kinetics in the whole animal (Figure 2). The whole-body free leucine pool is divided among an extracellular leucine pool (EC), an intracellular leucine pool (IC), and a leucyl tRNA pool (tRNA). The extracellular pool is assumed to be homogenous and representative of the free amino acid in the extracellular space and plasma. The influx of leucine into the extracellular pool (OE) is absorption and the efflux from the intracellular pool (IO) is oxidation. The aminoacylated tRNA pool represents leucine covalently bound to tRNA and can be derived from the extracellular pool (CH-channeling), the intracellular pool (IT), or protein degradation (RE-recycling). Although aminoacyl tRNA may be involved in processes other than protein synthesis, it is assumed that leucyl tRNA cannot be deacylated. The flux from the extracellular to the intracellular pool (EI) represents the transport of leucine into the intracellular compartment. Intracellular leucine can be oxidized, transported back to the extracellular pool (IE) or undergo acylation. Leucine from degradation can go to the intracellular (PI) or leucyl tRNA pools (RE). The flux of leucine from leucyl tRNA to leucine in protein pool (P) is protein synthesis (SYN). Protein synthesis can only occur via aminoacyl tRNA. The intracellular pool size is leucine content in the intracellular space for all cells. The extracellular pool includes all leucine in the extracellular space and plasma, while the aminoacyl tRNA pool is total leucyl tRNA in the mouse.

The leucine in the protein pool size was based on the protein composition of a 30g reference mouse (Johnson, 1997). The average protein content of organs was assumed to contain 6% leucine (Waterlow et al., 1978) on a dry basis. Therefore when corrected for water content, the leucine in protein content was 2%. Protein synthesis flux (SYN) was calculated by averaging the protein fractional synthesis rate (FSR) estimates from the literature (Johnson, 1997) and converting them to μmols leucine/min. In a non-growing rodent, protein synthesis and degradation are equal, therefore SYN = PI + RE. The percent leucine recycling was estimated by assuming that a percentage of the leucine from protein degradation would either go back to leucyl tRNA or to the intracellular leucine pool. Similarly, the percent channeling was estimated by assuming that a percentage of the leucine for protein synthesis would be supplied by either the extracellular pool or intracellular pool.

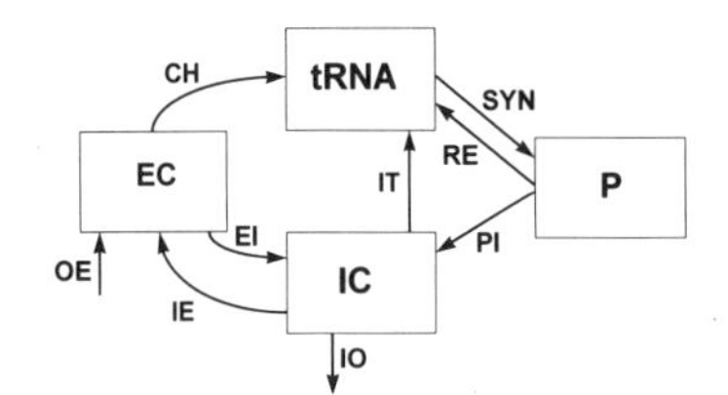

Figure 2. The whole-mouse model of protein turnover. See text for explanation of labeling.

Calculations to estimate initial leucine content in the extracellular, intracellular, and tRNA pools were based on body weight. To calculate the extracellular leucine content, it was assumed that the concentration of leucine in the plasma was the same as the extracellular space and that there was rapid equilibration between the two spaces. Therefore the amount of leucine in the extracellular pool was the amount in the extracellular space plus the amount in the plasma (Hider et al., 1969; Hider et al., 1971a; Waterlow et al., 1978). The leucine content of the intracellular pool was estimated from Obled et al. (1989). To calculate the leucyl tRNA pool size, total tRNA from liver and muscle was used to estimate the amount of tRNA in the whole mouse (Vinayak, 1987). The number of leucine codons (6) was divided by the number of total codons (60) to estimate that 10% of the total tRNA was leucyl tRNA. Assuming 88% of tRNA is charged (Palmiter, 1975), the total charged leucyl tRNA was 0.88 times 0.1.

Some literature values were available to estimate leucine fluxes between the free amino acid pools. If literature values were not available, assumptions about the relationships between fluxes were used (Johnson, 1997). Estimates were based on the assumption that the rodent is essentially not growing over the 15–30 minute experimental time period. Therefore oxidation was set equal to intake and the leucyl tRNA and leucine in protein pools were expected to remain in steady state. Figure 3 shows pool sizes and fluxes which were used in the model.

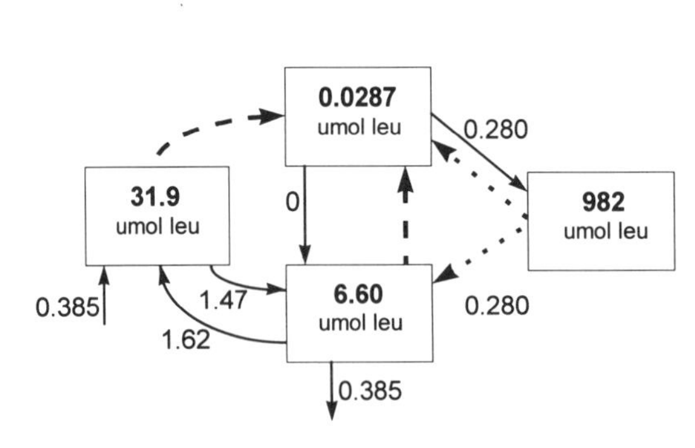

Figure 3. Pool sizes and fluxes of a whole-mouse model of protein turnover. Pool sizes are in μmols of leucine and fluxes are in μmols leucine/min. Dashed lines represent the possible sources of amino acids for protein synthesis and dotted lines represent the possible routes for amino acids from protein degradation. The percent recycling was estimated by assuming that a percentage of the total leucine from protein degradation would go back to leucyl tRNA. Similarly, the percent channeling was estimated by assuming that a percentage of the total leucine for protein synthesis would be supplied by the extracellular pool.

Model Evaluation with Independent Data

Two data sets with whole-body protein and amino acid specific radioactivities were simulated to find the fluxes EI, IE, IT, PI, CH, and RE. The first data set is a flooding dose experiment in 30 g mice given 10 μCi ^{14}C leucine/100 μmol leucine/100 g bodyweight by Bernier and Calvert (1987). Whole-body specific radioactivity measurements were taken at

2, 5, 10, 15, 20, and 30 minutes in the free amino acid pool (acid supernatant) and protein pool (acid precipitate). The second data set is also a flooding dose experiment, in which 70 g rats given 23.7 μCi ^{14}C leucine/140 μmol leucine/70 g bodyweight by Obled (1996). Whole-body specific radioactivity measurements were taken at 5, 7, 9, 11, 13, and 15 minutes in the plasma, free amino acid pool (acid supernatant), and protein pool (acid precipitate). Fluxes were fit using ACSL (MGA, 1995) to have the model produce specific radioactivities as close to the data as possible.

In general, the fluxes determined from the Bernier and Obled data sets were fairly close to each other (EI, IE, IT, PI, CH, and RE; Figure 4). The percent error of prediction of specific radioactivities for the Obled data set were less than 7% and less than 4% for the Bernier data. The fits from both data sets indicated that a high percent of channeling was occurring. If CH was forced to equal zero, the protein specific radioactivity could not get high enough to match the protein specific radioactivity of the data. Figure 4 shows graphs of the observed protein specific radioactivities vs. fits of the data if the IC (I) pool supplies all of the leucine or if the EC (E) pool supplies all of the leucine for protein synthesis.

(4A) Bernier & Calvert Protein Specific Radioactivity

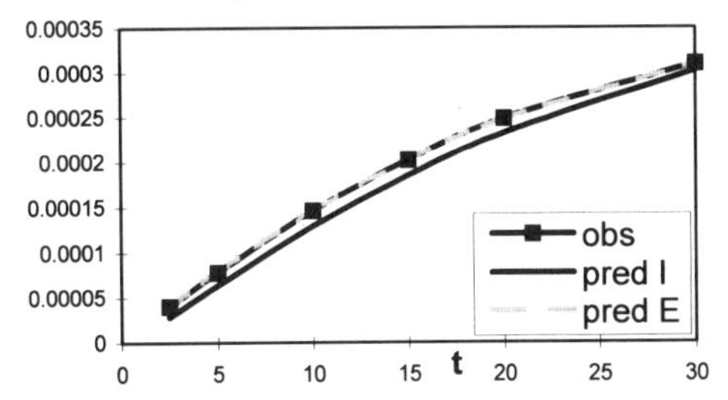

(4B) Obled Protein Specific Radioactivity

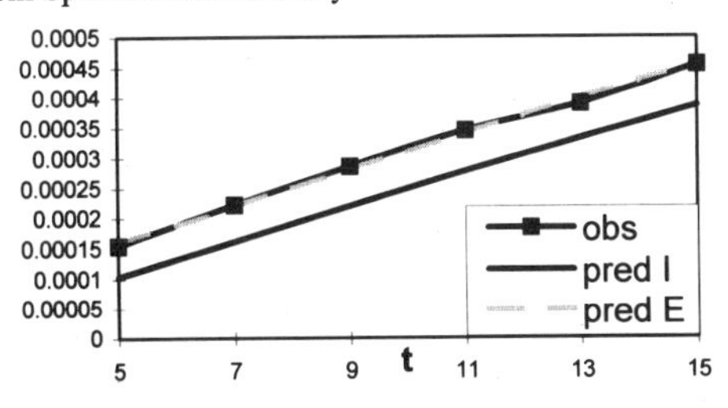

Figure 4. Protein specific radioactivity from the (4A) Bernier and Calvert data (1987) and (4B) Obled data (1996). Observed data vs. the predicted extracellular (E) or intracellular (I) pools, if each were the sole source of leucine for protein synthesis.

Recycling did not improve the fit of the data from Obled and Bernier (not shown) in that the percent error for the specific radioactivity predictions for free leucine and protein stayed approximately the same. Therefore the data could not establish if recycling was actually occurring. However, recycling did increase the percent protein synthesized per minute and the total μmoles protein synthesized per minute for both data sets. But, the overall fractional synthesis rates were unchanged. Therefore recycling enabled more protein to be synthesized without increasing the FSR calculated experimentally.

Sensitivity Analysis

A sensitivity analysis was conducted to evaluate the model's response to a percent change in the model parameters. In order to determine which method would be most sensitive for measuring the percent channeling and recycling, four experimental methods for determining FSR were examined. Changes in specific radioactivity were examined at four levels of the tRNA, extracellular, and protein pools: 0% recycling and 100% channeling, 100% recycling and 100% channeling, 100% recycling and 0% channeling, and 0% recycling and 0% channeling. Because the intracellular leucine pool is difficult to measure and therefore is usually estimated by difference from the total free leucine, extracellular leucine, and leucyl tRNA, specific radioactivity changes in the intracellular pool were ignored. The injection specific radioactivity (3.0 μCi ^{14}C leucine/30 μmol leucine for 30 minutes) and the experimental protocol for the flooding dose method (according to Bernier and Calvert (1987) for a 30 g mouse) was used . The injection specific radioactivity and experimental protocols according to Pomposelli et al. (1985) and Peters and Peters (1972) were used for the continuous infusion and pulse dose methods, respectively. A pulse dose of 2 μCi ^{3}H leucine followed by a chase of 100 μmol of unlabeled leucine at 180 minutes was used for the prelabeled protein method. Values for the pulse dose (3.0 μCi ^{14}C leucine/0.0091 μmol leucine for 60 minutes) and continuous infusion (1.0 μCi ^{14}C leucine/0.02 μmol leucine for 180 minutes) specific radioactivities had to be adjusted to a 30 g mouse. The fluxes used for the model are those from Figure 3.

The pulse dose method showed the greatest differences in specific radioactivities for high and low rates of channeling (Figure 5). There were large differences in the specific radioactivities of the leucyl tRNA, extracellular, and protein pools between the 0% channeling and 100% channeling curves. However, most of the differences are gone 10-15 minutes after the dose is given. Therefore, measurements to predict the percent channeling should be taken in these initial 15 minutes. If the specific radioactivities at 0 and 100% channeling are compared at 30 minutes, leucyl tRNA is decreased by 80%, extracellular is increased by 50%, and protein is increased by 28%. Because experimental error is between10-15%, the measurement of the specific radioactivities of all three pools should contribute significantly to the estimation of the rate of channeling.

The prelabeled protein method resulted in the greatest differences in specific radioactivities due to high and low rates of recycling. Figure 6 shows changes in specific radioactivities due to 0 and 100% rates of channeling and recycling at the time that the chase is given. Most of the changes take place during the chase, therefore measurements of specific radioactivity just before, during, and after the chase would result in better predictions of the percent recycling. If measurements of specific radioactivity are taken at 210 minutes at 0 and 100% recycling, the specific radioactivity of the leucyl tRNA pool is increased 300%, specific radioactivity of the extracellular pool is increased 20%, and the specific radioactivity of the protein is decreased 10%. Therefore, measuring the specific radioactivity of the leucyl tRNA pool is the most critical for determining the percent recycling.

The pulse and prelabeled protein method could be combined in a pulse-chase experiment. Measurements before 15 minutes would be used to estimate the percent channeling and measurements from 180-210 minutes could be used to estimate the percent recycling. The number of time points, number of animals per time point and ability of the model to predict channeling, recycling, and protein synthesis rate still need to be determined.

IDENTIFYING THE EXPERIMENTAL DESIGN

The pulse-chase protocol was a pulse dose of 1.5 μCi of ^{14}C leucine per 100 g body weight and 6.7 μCi ^{3}H leucine per 100 g body weight given at 0 minutes and 100 μmol of unlabeled leucine given at 180 minutes. Four specific radioactivity data sets were generated from the first model: 10% recycling of the protein pool and 100% channeling,

90% recycling and 100% channeling, 10% channeling and 0% recycling, and 90% channeling and 0% recycling. Zero to fifteen percent errors were randomly added or subtracted from the generated data points to simulate experimental variation of measurements. Four measurements were considered possible: the total specific radioactivity of the free amino acid pools (extracellular, intracellular, and tRNA), the specific radioactivity of plasma which was used to estimate the specific radioactivity of the extracellular pool, the specific radioactivity of the tRNA, and the specific radioactivity of the protein pool. Overall, FSR was 32%/day. The data sets were fitted to the model to determine which and how many measurements were needed to predict FSR, the rate of recycling and channeling. In Table 1, the model settings for FSR, recycling and channeling are compared to those predicted from fitting the data generated by the model with 0-15% variation on the generated data. Pool sizes from the model were also compared to those from the predicted rates.

(5A) Pulse Dose Leucyl tRNA Specific Radioactivity

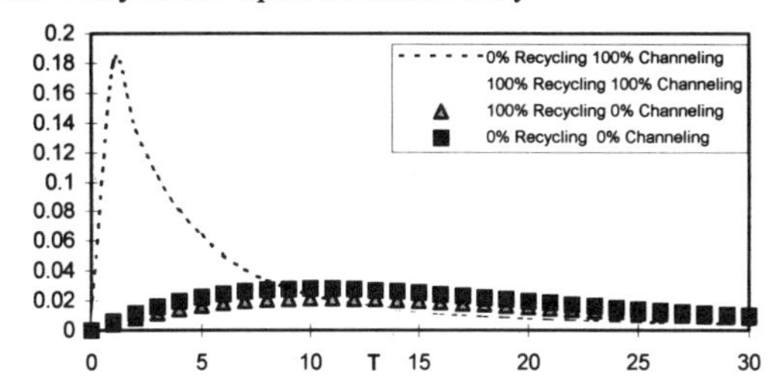

(5B) Pulse Dose Extracellular Specific Radioactivity

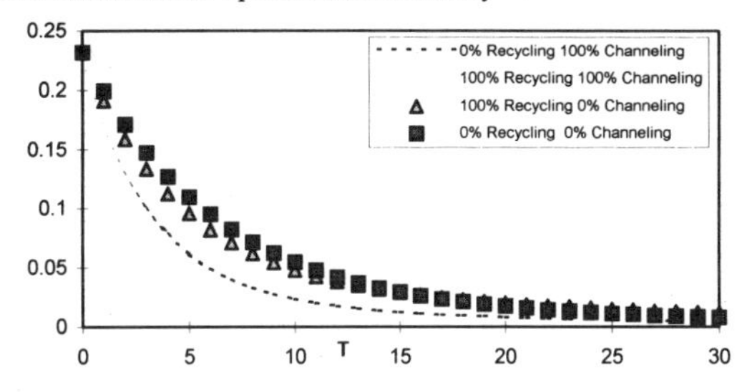

(5C) Pulse Dose Protein Specific Radioactivity

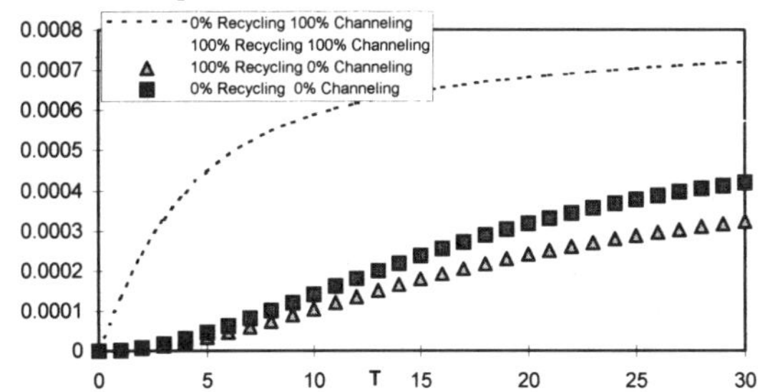

Figure 5. Changes in specific radioactivities (μCi/μmol) in the (5A) leucyl tRNA pool, (5B) extracellular pool, and (5C) protein pool for the pulse dose at high and low rates of channeling and recycling. Time (T) is in minutes.

(6A) Prelabeled Protein leucyl tRNA Specific Radioactivity

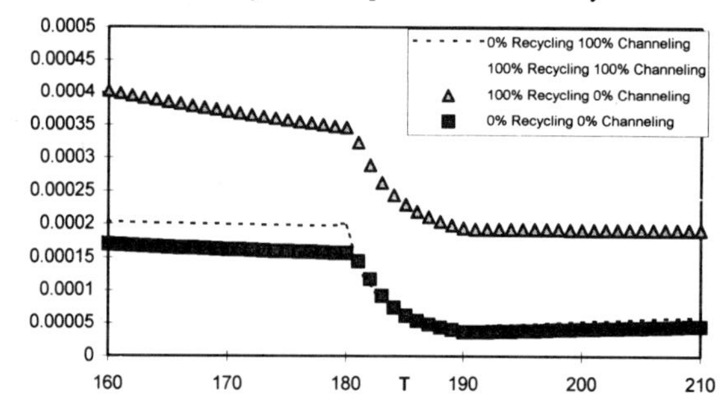

(6B) Prelabeled Protein Extracellular Specific Radioactivity

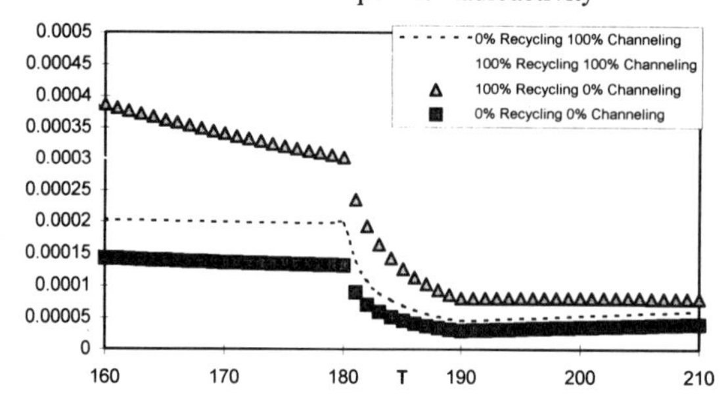

(6C) Prelabeled Protein Specific Radioactivity

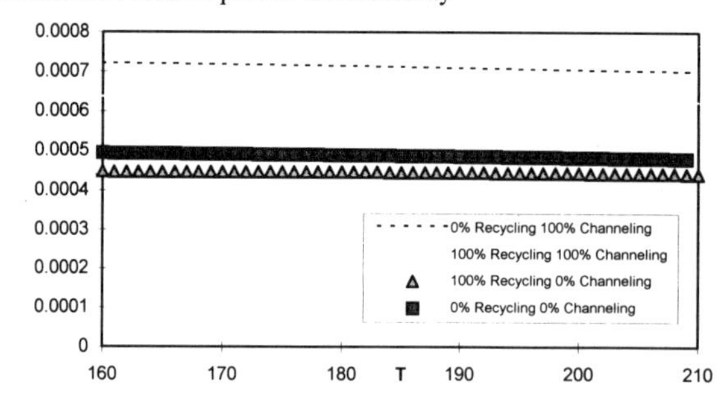

Figure 6. Changes in specific radioactivities (μCi/μmol) in the (6A)leucyl tRNA pool, (6B) extracellular pool, and (6C) protein pool for the prelabeled protein method at high and low rates of channeling and recycling. Time (T) is in minutes.

Predictions of recycling rate, channeling rate, and FSR were very close to those used to generate the data. The prediction of percent channeling when recycling was 0% had the greatest standard deviation, and only the intracellular free leucine pool changed in size. In Table 2, the actual specific radioactivities are compared to the specific radioactivities predicted from fitting the data from the ^{14}C leucine pulse dose and the ^{3}H leucine pulse dose. Because percent errors are recorded for each specific radioactivity measurement, only the worst prediction, protein specific radioactivity, is shown. The observed percent error was the error added to or subtracted from the generated data to add variation to the data. The predicted percent error was the percent error of the predicted specific radioactivity relative to the observed specific radioactivity.

Table 1. Results of fitting model data at low and high rates of channeling (0% recycling).

	10% Channeling			90% Channeling		
	Observed	Predicted	SD	Observed	Predicted	SD
FSR (%)	32.0	32.0	0.654	32.0	32.2	0.532
Recycling (%)	0.0	-0.01	1.43	0.0	0.988	0.897
Channeling (%)	10.0	11.9	0.472	90.0	93.0	3.56
IC (μmol leu)	3.14	3.14		3.14	3.13	
tRNA (μmol leu)	0.0287	0.0287		0.0287	0.0287	
EC (μmol leu)	1.36	1.36		1.36	1.36	
Protein (μmol leu)	982	982		982	982	

Table 2. Variation in model data (observed) and variation predicted by fitting for the specific radioactivity of the protein pool at low and high rates of channeling (0% recycling). T is the time of measurement (minutes).

	10% Channeling		90% Channeling	
T (min)	Observed % error	Predicted % error	Observed % error	Predicted % error
2	7.09	-6.94	4.95	2.22
4	-9.88	-23.1	13.3	9.71
10	-7.31	-13.7	-7.94	-10.3
180	-12.8	-15.8	-8.55	-9.96
190	7.84	6.31	-4.83	-5.66
210	6.48	5.11	-14.4	-17.5

The percent errors on the predicted specific radioactivities for 10% and 90% channeling from the data were within 6% of the percent error randomly assigned to the generated data, except for the 4 minute measurement at 10% channeling, 0% recycling. Most predicted percent errors were within 1-2% of the error of the generated data. However, the model over-predicted the specific radioactivity of the protein pool from 2-10 minutes at low rates of channeling (with low recycling) with both the ^{14}C leucine data and ^{3}H leucine data. An additional data point early on in the pulse dose experiment did not improve the predicted percent error or the predictions of recycling and channeling rates. Table 3 lists the results from a comparison of rates, pools sizes, and specific radioactivities generated by the model to those predicted by fitting the pulse-chase data at different rates of recycling.

Predictions of synthesis, recycling, and channeling rates were very close to those observed (set) in the model. The predicted 100% channeling rate had the greatest standard deviations and only the intracellular leucine pool size changed slightly.

In Table 4, an example of the actual specific radioactivities were compared to the specific radioactivities predicted from fitting the data for protein to predict the rate of recycling. Most of the percent error of the specific radioactivities predicted by the model were within 3% of the randomly assigned error of the generated data. Increasing the number of mice for each data point to three, decreased the standard variation of the predictions of recycling, channeling, and fractional synthesis rate by half but did not improve the percent errors on the specific radioactivities predictions. Therefore because the predictions of channeling and recycling rates were very close to the values set in the simulation for all four data sets generated from the model, this experimental protocol was the one most likely to accurately estimate the rates of channeling and recycling.

Table 3. Results of fitting model data at low and high rates of recycling (100% channeling.)

	10% Recycling			90% Recycling		
	Observed	Predicted	SD	Observed	Predicted	SD
FSR (%)	32.0	32.2	0.727	32.0	31.4	1.05
Recycling (%)	10.0	10.2	1.51	90.0	89.8	0.313
Channeling (%)	100	100	4.51	100	100	4.63
IC (μmol leu)	2.76	2.77		0.259	0.258	
tRNA (μmol leu)	0.0287	0.0287		0.0287	0.0287	
EC (μmol leu)	1.40	1.40		1.77	1.77	
Protein (μmol leu)	982	982		982	982	

Table 4. Variation in model data (observed) and variation predicted by fitting for the specific radioactivity of the protein pool at low and high rates of recycling (100% channeling). T is the time of measurement (minutes).

	10% Recycling		90% Recycling	
T (min)	Observed % error	Predicted % error	Observed % error	Predicted % error
2	-4.76	-5.47	3.55	3.82
4	3.11	2.61	-5.32	-5.27
10	-15.0	-18.1	-11.2	-12.2
180	-12.9	-15.0	1.68	1.91
190	0.110	-0.0663	6.52	6.37
210	7.98	7.23	-11.4	-12.6

CONCLUSIONS

According to the model, protein synthesis rate, recycling, and channeling can be determined simultaneously using a pulse dose of ^{14}C leucine with a bolus of unlabeled leucine three hours later. Channeling is predicted from specific radioactivity measurements taken at early time points (2, 4, and 10 minutes) with 2–6 measures at each time point. Recycling is predicted from specific radioactivity measurements taken during the chase (180, 190, and 210 minutes) with 2 to 6 measures (1–3 animals) per time point. However, the successful estimation of synthesis rate, channeling, and recycling is really dependent on how well the model represents protein turnover in a whole animal. Sensitivity analysis will indicate the most important model elements which need to be measured accurately and the least important (insensitive) elements which do not contribute to model function. If elements do not add to model function, they should be eliminated. Insensitive elements will only contribute to a false sense of security in the solutions and increase the likelihood of multiple solutions given the same model structure. The data generated by the model and fitting the data to the model are dependent on the model structure, then model validation with independent data is extremely important.

Pool and Flux Estimates

The lack of data available to define the free amino acid pool sizes and kinetics made the comparison of the fitted fluxes to the values derived from the literature difficult. Because the amount of extracellular space seems to be relatively constant (Roberts and Morelos, 1965; Hider et al., 1969; Hider et al., 1971a; Hider et al., 1971b; Khairallah et al., 1977) and the tRNA pool size should not change over a short period of time, the weakest estimation of free leucine is the value for intracellular pool size.

In the flooding dose method, the intracellular pool is relatively small compared to the influx of unlabeled leucine but because the initial intracellular pool is larger than the extracellular pool (without flood dose) and tRNA pool. An inaccurate estimate of the respective sizes of the intracellular and extracellular pools could alter the specific radioactivity predicted by the model. In addition, the intracellular pool is likely to be affected by the previous metabolic state of the animal and is the most difficult pool size to determine.

In cell culture, the intracellular pool is distinctly different and fairly simple to separate from the extracellular pool. However, on a whole tissue or whole body basis, the intracellular pool is usually assumed to be the tissue homogenate which also includes the extracellular pool and so must be corrected for extracellular space.

For the purposes of this model, it has been assumed that the metabolism of the rat and mouse are essentially the same when comparisons are based on body size. If the fluxes are changed to a μmole per gram basis, the fluxes (EI, CH, IT, IE) and the μmoles of protein synthesized per minute are very close for the Bernier and Obled data fits (based on rat and mouse models, respectively). Also, the effect of a large dose of amino acid on the system is unknown. The extracellular pool may expand, intracellular pool may increase, or oxidation may increase. It unlikely that protein synthesis is increased by excess leucine (Tovar et al., 1992). However, excess leucine may increase the competition for transport of other amino acids which use the L system and the oxidation enzymes may be able to adjust to the large influx of amino acid in a short period of time (Calvert et al., 1982).

Channeling

Only two sets of flooding dose data were available to test the model. The fluxes and specific radioactivity changes predicted by the model from the Bernier and Obled data sets indicate that 100% of the leucine for tRNA charging (protein synthesis) is from the extracellular pool. The added dilution of the unlabeled leucine from the flooding dose in the intracellular pool appears to be great enough to prevent the specific radioactivity of the protein pool to increase fast enough to match the observed values (Figures 4A and 4B). However, some leucine could be transported by the red blood cell to the intracellular or extracellular pool (Hanigan et al., 1991). The available leucine to the cell from the blood could be much higher and not accounted for in the estimation of the extracellular specific radioactivity using only plasma. In addition, because the flooding dose was used, leucine was not in short supply and the use of the intracellular pool (if it is thought of as a 'buffer') was probably not necessary. The high rate of channeling observed with the flooding dose method may not be consistently true when other methods are used. The estimation of channeling is dependent on a high specific radioactivity of the protein pools and a higher specific radioactivity of the source of amino acid for protein synthesis. The estimation of the specific radioactivity of the intracellular and extracellular pools by the model is dependent on the kinetics from the flooding dose data. Because the flooding dose method uses concentrations of leucine which are higher than physiological levels, the resulting predictions of fluxes may not be the same across methods. Therefore, additional data which included specific radioactivity changes over time for at least protein and plasma using continuous infusion and pulse dose methods would be useful to further validate the model.

Two conditions help to define channeling: (1) the presence of microcompartmentation through a multienzyme complex (physical structure) or a localized concentration of substrates; and (2) an association with a cellular structure (Ovadi, 1991). Many of the aminoacyl-tRNA synthetase enzymes have been shown to form complexes and associate with the cytoskeletal framework of the cell and endoplasmic reticulum for protein synthesis (Hershey, 1991). However an association of the cell membrane or amino acid transport system with the aminoacyl-tRNA synthetase enzyme complexes is possible but has not been shown. Negrutskii and Deutscher (1991, 1994), using Chinese hamster ovary cells, showed that aminoacyl-tRNA is probably associated with a structure to supply amino acids for protein synthesis. In addition, the high specific radioactivity of the tRNA pool compared to the intracellular pool implies at least one compartment with a high localized concentration of amino acid for protein synthesis (Barnes et al., 1994; Khairallah and Mortimore, 1976). Therefore, there is substantial evidence to support the concept that amino acids go to protein synthesis from an organized source (extracellular or intracellular) based on the dynamics and structures within a cell. Definitive proof, however, of the extracellular pool being the exclusive source is not available.

Recycling

Data from the Obled and Bernier experiments were not conclusive for determining if recycling was occurring. If recycling was forced into the solution, only 11% was predicted to be occurring, a level which would be difficult to separate from experimental error. The fact that the high level of channeling was still maintained in the solution could have caused the recycling prediction to be artificially low because the flooding dose method cannot predict the rate of channeling if recycling is high. Because the turnover rates of the protein pools were based on estimates of FSR using the flooding dose for the individual protein pools, they would underestimate the actual synthesis rate if significant recycling was occurring.

Recycling may have several metabolic benefits, especially in fast protein turnover tissues. Recycling could decrease the amino acid requirement, save energy associated with transportation and charging of aminoacyl-tRNA and account for the linkage between rates of protein synthesis and protein degradation. The amino acid requirement for protein synthesis can be decreased in two ways. The more recycling, the lower the intake of amino acid needed to synthesize protein because the same amino acid can be 'reused' for synthesis. Thus a substrate cycle is possible in protein turnover. The energy requirement associated with protein metabolism could be lowered due to decreased charging or transport requirements and greater control over protein turnover is implied. An animal's metabolism could be more resilient because there would be a lag effect on protein deficiencies or amino acid imbalances.

The amino acid requirement would also be lowered by a decreased amino acid oxidation because they are less available to the degradative enzyme systems. Energy would be saved because oxidation would be lowered, thus decreasing the need to transport more amino acids into the cell. In addition, tRNA may play a role in degradation, reducing the energy associated with forming an initiation complex. The benefits due to recycling suggest an organized structure or system linking protein degradation to tRNA or the protein synthesis machinery.

Protein degradation is carried out by several different systems which could be coupled to protein synthesis. The ubiquitin-proteosome system targets unassembled, damaged, or misfolded proteins for degradation in the cytosol. Ubiquitin has been identified as a heat shock protein and in some cases requires tRNA to bind to mark protein substrates (Dice, 1987). Therefore tRNA may be a cofactor of protein degradation or may aid in linking protein degradation products from the ubiquitin pathway to protein synthesis.

Within the endoplasmic reticulum (ER), proteins which are misformed or misfolded, signal peptides from protein synthesis, and some enzymes involved in the control of metabolic pathways in the ER are rapidly degraded (Klausner and Sitia, 1990). Amino acids from degradation in the ER could easily be reused for synthesis due to the closer proximity of protein synthesis while translocation into the ER is taking place. It remains to be proven if there is a link between elongation factors at the ER membrane and amino acids from degradation inside the ER.

Lysosomes in the cytoplasm contain several different acid hydrolases for processing proteins, recycling receptors, and degrading proteins. Transport proteins have been found in the membranes of lysosomes which may transport the products of degradation to the intracellular compartment or directly to the protein synthetic machinery (Dice, 1987). The lysosomal degradation pathway is probably important during times of nutrient deprivation (Rivett, 1990) and during chronic sepsis (Voisin et al., 1996). Possible links for protein degradation to be directly connected to protein synthesis or tRNA without mixing with the intracellular milieu exist, however a definite structure or compartmentalization has not been established.

There is some interaction between recycling and channeling (Figures 5 and 6). Low rates of recycling do not cause changes in specific radioactivity equivalent to those seen with high rates of channeling because at these high rates, the tRNA and fast turnover protein pool specific radioactivities are the highest. Similarly, at low rates of channeling and high rates of recycling, the specific radioactivity of the protein and tRNA is the lowest. However, the specific radioactivity of the tRNA pool is the same as the intracellular pool at 0% recycling and 0% channeling. But, at these low rates of channeling and low rates of recycling, the intracellular pool specific radioactivity is at its lowest. At high rates of recycling and high rates of channeling, the changes in specific radioactivities of the pools are similar to low recycling, high channeling but are lower due to the dilution effect of recycling. Therefore, in order to identify recycling rate and channeling rate simultaneously, the method and measurement time points should be chosen so that the greatest differences between pool-specific radioactivities exists.

Accurate estimates of FSR are dependent on the specific radioactivity of the pool which is the source of amino acid for tRNA charging and on the amount of amino acids which are recycled to protein synthesis without mixing with the amino acid in the intracellular pool. Because rates of recycling and channeling can vary between tissues and with the amino acid used as a tracer, it is imperative that limitations associated with each of the methods is known for individual tissues and whole body estimations.

CORRESPONDING AUTHOR

Please address all correspondence to:
Heidi Johnson
Animal Science Department
University of California, Davis
One Shields Avenue
Davis, CA 95616

REFERENCES

Barnes DM; Calvert CC; Klasing KC. Source of amino acids for tRNA acylation. *Biochem J*, 1992, 283:583-589.

Barnes DM; Calvert CC; Klasing KC. Source of amino acids for tRNA acylation in growing chicks. *Amino Acids,* 1994, 7:67-278.

Bernier JF; Calvert CC. Effect of a major gene for growth on protein synthesis in mice. *J Ani Sci,* 1987, 65:982-995.

Calvert CC; Klasing KC; Austic RE. Involvement of food intake and amino acid catabolism in the branched-chain amino acid antagonism in chicks. *J Nutr*, 1982, 112:627-635.

Dice JF. Molecular determinants of protein half-lives in eucaryotic cells. *FASEB J*, 1987, 1:349-357.

Hanigan MD; Calvert CC; DePeters EJ; Reis BL; Baldwin RL. Whole blood and plasma amino acid uptakes by lactating bovine mammary glands. *J Dairy Sci,* 1991, 74:2484-2490.

Hershey JWB. Translational control in mammalian cells. *Ann Rev Biochem,* 1991, 60:717-755.

Hider RC; Fern EB; London DR. Relationship between intracellular amino acids and protein synthesis in the extensor *digitorum longus* muscle of rats. *Biochem J,* 1969, 114:171-178.

Hider RC; Fern EB, London DR. Identification in skeletal muscle of a distinct extracellular pool of amino acids, and its role in protein synthesis. *Biochem J,* 1971a, 121:817-827.

Hider RC; Fern EB; London DR. The effect of insulin on free amino acid pools and protein synthesis in rat skeletal muscle *in vitro*. *Biochem J,* 1971b, 125:751-756.

Johnson HA. A modeling investigation of whole body protein turnover based on leucine kinetics in rodents. Dissertation. 1997.

Khairallah EA; Airhart J; Bruno MK; Puchalsky D; Khairallah L. Implications of amino acid compartmentation for the determination of rates of protein catabolism in livers in meal fed rats. *Acta Biol Med Germ,* 1977, 36:1735-1745.

Khairallah EA; Mortimore GE. Assessment of protein turnover in perfused rat liver. *J Biol Chem,* 1976, 251:1375-1384.

Kipnis DM; Reiss E; Helmreich E. Functional heterogeneity of the intracellular amino acid pool in mammalian cells. *Biochimica et Biophysica Acta,* 1961, 51:519-524.

Klausner RD; Sitia R. Protein degradation in the endoplasmic reticulum. *Cell*, 1990, 62:611-614.

Mitchell and Gauthier Assoc. Inc. *ACSL: Advanced Continuous Simulation Language.* MGA Inc.: Concord, Massachusetts. 1995.

Negrutskii BS; Deutscher MP. Channeling of aminoacyl-tRNA for protein synthesis *in vivo. Proc Natl Acad Sci USA,* 1991, 88:4991-4995.

Negrutskii BS; Deutscher MP. A sequestered pool of aminoacyl-tRNA in mammalian cells. *Proc Natl Acad Sci USA,* 1992, 89:3601-3604.

Negrutskii BS; Stapulionis R; Deutscher MP. Supramolecular organization of the mammalian translation system. *Proc Natl Acad Sci USA,* 1994, 91:964-968.

Obled C. Personal communication. 1996.

Obled C; Barre F; Millward DJ; Arnal M. Whole body protein synthesis: Studies with different amino acids in the rat. *Am J Physiol,* 1989, 257:E639-E646.

Ovadi J. Physiological significance of metabolic channeling. *J Theor Biol,* 1991, 152:1-22.

Palmiter RD. Quantitation of parameters that determine the rate of ovalbumin synthesis. *Cell,* 1975, 4:189-197.

Peters T; Peters JC. The biosynthesis of rat serum albumin. *J Biol Chem,* 1972, 247:3858-3863.

Pomposelli JJ; Palombo JD; Hamawy KJ; Bistrian BR; Blackburn GL; Moldawer LL. Comparison of different techniques for estimating rates of protein synthesis *in vivo* in healthy and bacteraemic rats. *Biochem J,* 1985, 226:37-42.

Rivett JA. Intracellular protein degradation. *Essays Biochem,* 1990, 25:39-73.

Roberts S; Morelos BS. Regulation of cerebral metabolism of amino acids-IV. Influence of amino acid levels on leucine uptake, utilization and incorporation into protein *in vivo*. *J Neurochem,* 1965, 12:373-387.

Smith CB; Deibler GE; Eng N; Schmidt K; Sokoloff L. Measurement of local cerebral protein synthesis *in vivo*: Influence of recycling of amino acids derived from protein degradation. *Proc Natl Acad Sci USA,* 1988, 85:9341-9345.

Smith CB; Sun Y. Influence of valine flooding on channeling of valine into tissue pools and on protein synthesis. *Am J Physiol,* 1995, 268:E735-E744.

Sun Y; Deibler GE; Sokoloff L; Smith CB. Determination of regional rates of cerebral protein synthesis adjusted for regional differences in recycling of leucine derived from protein degradation into the precursor pool in conscious, adult rats. *J Neurochem,* 1992, 59:863-873.

Tovar AR; Tews JK; Torres N; Madsen DC; Harper AE. Competition for transport of amino acids into rat heart: Effect of competitors on protein synthesis and degradation. *Metab Clin Exper,* 1992, 41:925-933.

VanVenrooij WJ; Moonen H; VanLoon-Klaassen L. Source of amino acids used for protein synthesis in HeLa cells. *Eur J Biochem,* 1974, 50:297-304.

Vinayak M. A comparison of tRNA populations of rat liver and skeletal muscle during aging. *Biochem Internat,* 1987, 15:279-285.

Voisin L; Breuille D; Combaret L; Pouyet C; Taillandier D; Aurousseau E; Obled C; Attaix D. Muscle wasting in a rat model of long-lasting sepsis results from the activation of lysosomal, Ca-activated, and ubiquitin-proteosome proteolytic pathways. *J Clin Invest,* 1996, 97:1610-1617.

Waterlow JC; Garlick PJ; Millward DJ. *Protein Turnover in Mammalian Tissues and in the Whole Body.* Amsterdam: North-Holland. 1978. p. 119.

Wettenhall REH; London DR. Incorporation of amino acids into protein from an intracellular pool of lymphocytes. *Biochimica et Biophysica Acta,* 1975, 390:363-373.

Part IV

CHEMICAL / PHYSICAL

ANALYTICAL METHODS

PROTOCOL DEVELOPMENT FOR BIOLOGICAL TRACER STUDIES

Stephen R. Dueker,[1] A. Daniel Jones,[2] Andrew J. Clifford[1]

[1]Department of Nutrition
[2]Facility for Advanced Instrumentation
University of California, Davis
Davis, CA 95616

ABSTRACT

Improved instrumentation and the increased availability of labeled compounds have democratized the application of isotope-dilution (tracer) methodology in nutrient metabolism. Still, the most challenging aspects of tracer experimentation reside in the steps that precede the measurement of an isotopically labeled tracer, i.e. the design of a suitably labeled tracer and its isolation and purification from complex biological matrices. Construction of useful mathematical models of nutrient dynamics require methodologies that guarantee that the integrity of the tracer is maintained across the entire sampling and analyte isolation protocol. The ability to provide accurate and reliable data highlights a need for analytical chemists to play a central role in these studies. In this regard, examples and discussion of issues relevant to stable-isotope experimentation are provided.

INTRODUCTION

Tracer kinetic studies provide the most quantitative means for assessing human *in vivo* nutrient metabolism and nutritional requirements because these nutrients are already present in the fluids and tissues of study participants. Tracer studies are based upon the administration of a nutrient enriched with a heavy isotope at one or more of its nuclides to produce a heavy analog (tracer) of the dietary nutrient (tracee). An optimally designed tracer will be metabolized identically to the compound it is meant to follow, but will be easily detectable by mass spectrometric or radiometric methods. Time-dependent tracer/tracee ratios from the sampling of extracellular compartments (blood and urine) can serve as the basis for compartmental modeling and can afford prediction of metabolic events within inaccessible tissues and organs (Bier, 1997; Cobelli et al., 1987). Such insight cannot be obtained from measurement of static circulating metabolite levels.

The term tracer has been synonymous with the high incorporation of radioactive isotopes, for many investigations of organic nutrient dynamics this has meant carbon-14

Mathematical Modeling in Experimental Nutrition
Edited by Clifford and Müller, Plenum Press, New York, 1998

and tritium. Despite a growing interest in stable isotope technology in the late 1930s and early 1940s, the widespread use of stable isotopes was sidelined by the greatly increased availability of carbon-14 and tritium in the late 1940s and the availability of scintillation (decay) counting (Bier, 1987). Interest in stable isotopes resurfaced in the 1970s, driven by concern over radiation exposure and facilitated by the increased availability of affordable stable isotopes and improvements in mass spectrometers.

Modern mass spectrometers are significantly improved in both their capabilities and ease-of-use over their older counterparts, and are fast becoming essential bioanalytical tools for many disciplines. With fewer technological obstacles, isotope-dilution experimentation is finding an increasing number of research and clinical applications. Still, the most challenging aspects of tracer experimentation reside in those steps that precede measurement of an isotopically labeled tracer, i.e. the design of a suitably labeled tracer and its isolation and purification from complex biological matrices. The methodology must guarantee that the integrity of the label is maintained across the sampling and analyte isolation protocol to ensure the interpretability of the data with a high degree of confidence. The need for accurate and reliable data highlights a central role for analytical chemists in studies of nutrient dynamics.

NUCLEAR CHEMISTRY

The basic building blocks of atomic nuclei are the proton and the neutron. In a neutral atom, the number of protons is equal to the number of electrons. Early mass spectrometry experiments discovered that certain chemically pure gases contained components of more than one molecular mass. Nuclei of a given element having different mass numbers differ in their total number of neutrons and such mass variants of chemically similar atoms are referred to as isotopes (Sears et al., 1982).

Isotopes fall into two further categories—radioactive or stable—depending on the stability of their nuclear configuration. Many of the naturally occurring and man-made isotopes undergo radio-active decay, a process that results in emission of a charged particle from the nucleus and the transmutation of the original nuclide to a different, more stable element in the periodic table. Some common radioactive elements used in biology are ^{14}C, ^{32}P, ^{35}S, ^{3}H, and ^{131}I. Stable isotopes have nuclear configurations that do not transmute and are not radioactive. A compound can be chemically, though rarely, isotopically pure due to the natural occurrence of stable nuclides. Stable isotopes have greater natural abundances than their radioactive analogs, and their predictable occurrence facilitates deduction of elemental composition by mass spectrometry (Table 1).

Table 1. Natural isotopic abundances of common elements. Most abundant = 100.

^{1}H	100	^{2}H	0.016		
^{12}C	100	^{13}C	1.1	^{14}C	10^{-14}
^{14}N	100	^{15}N	0.37		
^{16}O	100	^{17}O	0.04	^{18}O	0.20
^{28}Si	100	^{29}Si	5.1	^{30}Si	3.4
^{32}S	100	^{33}S	0.79	^{34}S	4.4
^{35}Cl	100	^{37}Cl	32.6		

Advantages of Stable vs. Radioactive Isotopes in Tracer Experimentation

At first glance, the indisputable benefits afforded by the use of stable isotopes are the elimination of exposure to ionizing radiation, and of the associated handling and disposal costs of radioactive materials. Beyond this, several advantages arise from the specificity of mass spectrometric relative to radiometric detection, including:

1. Simultaneous determination of the tracer/tracee molar ratios (% enrichment).
2. Specific detection of the analyte according to its mass and mass fragmentation profile and known chromatographic behavior (hyphenated GC and LC/MS systems).
3. Potential for administration of several different tracers both simultaneously and repeatedly. As a result, more information can be obtained from a single experiment, which maximizes the benefit-to-cost ratio of human studies.

Radioactive decay counting will determine the tracer concentration without providing information on the endogenous substrate (tracee). Because the tracer/tracee ratios are essential to realizing the full potential of compartmental modeling, the endogenous content will need to be assessed by a separate assay. The accuracy of the final ratio is thus a product of the imprecision of the numerator (tracer) and denominator (tracee) terms. As such, the propagation of measurement error will compromise the accuracy of the ratio determination and the resulting kinetic curves will appear rough and lacking in a reproducible fine structure. Stable isotope dilution experimentation avoids similar problems by directly determining the tracer enrichment in a single measurement.

Confidence in an assay stems from a series of measures taken that select for the analyte of interest away from potential interferences. With decay counting, a disintegration event is recorded regardless of the chemical form of the analyte. Thus, there is no direct means of confirming that the counts recorded come only from the substrate in question (Bier, 1987). This is true even following specific isolation procedures. With mass spectrometric analysis, the specific detection of an analyte according to its mass and mass fragmentation pattern will add a level of compound specificity to the analysis that is not obtainable with radiometric detection. Mass spectrometric specificity is further enhanced via interfacing to efficient separation instruments such as gas and liquid chromatographs and augmented by the discriminating characteristics of the MS ionization mode. The additive potential of several layers of selectivity virtually eliminates analyte ambiguity in the analysis.

The high specificity of mass spectrometry is a liability when unknown metabolites are present. Without prior knowledge of the chemical structure of putative metabolites, biologically important compounds could be overlooked by the analysis. Decay counting of a radiolabeled sample prior to sample fractionation would provide an inventory of total activity. If the total activity were not accounted for following fractionation, other metabolites would be indicated. Modifications could then be pursued until there was full accounting of the missing activity.

Mass Spectrometry

Mass spectrometers will detect stable and radioactive isotopes equally well, but radiation concerns limit the application of radioactive compounds to healthy human experimentation. The counting of extremely low concentrations of rare radioactive isotopes is the domain of Accelerator Mass Spectrometry (AMS) and is discussed by Vogel and Turteltaub elsewhere in this volume. Unless otherwise specified, the term mass spectrometry in this chapter refers to 'conventional' (quadrupole and magnetic sector) mass spectrometry instrumentation and does not include accelerator equipped instruments.

There are several analytical parameters to be considered when evaluating a detection method. Most pertinent are measurement precision, quantitation limit, and instrument dynamic range. *Measurement precision* is the ability to reproduce a value for a set of observations and is often expressed as the Relative Standard Deviation (RSD) or the Coefficient of Variation (CV). Measurement precision is important in kinetic modeling where data points are assigned weights that are proportional to the inverse of the imprecision (standard deviation) in the measurement. The *quantitation limit* is the lowest concentration or quantity of an analyte that the analytical process can measure with statistically defined precision and accuracy. The quantitation limit is often calculated as the analyte concentration that yields a signal-to-noise ratio of 10. This differs from limit of detection which is the smallest amount of compound that can be detected and is generally taken to equal the blank signal plus three standard deviations of the blank (Green, 1996). The third term, *dynamic range*, is defined as the analyte concentration range over which the calibration or instrument response curve is linear. This term becomes significant in tracer studies where the tracer enrichments can be exceedingly small. An additional parameter that is not always appreciated is the robustness of the method, which can be gauged by sample throughput rates and the resistance of the analysis to inaccuracy arising from minor perturbations in the protocol.

While mass spectrometers are increasingly interfaced to liquid chromatographs, most applications employ gas-phase interfaces using capillary column gas chromatographs. Gas chromatography (GC) not only provides the highest resolving power of common chromatographic options, but also is compatible with the standard ionization modes used in mass spectrometry. The general operation of a gas chromatography/mass spectrometer is summarized as follows:

1. chromatographic resolution of analyte(s) of interest from potential interferences;
2. creation of gas-phase ions;
3. separation of ions in space or time based on their mass-to-charge ratio; and
4. measurement of the quantity of each mass to charge (m/z) species.

A mass spectrometer consists of an ion source, a mass-selective analyzer, and an ion detector. Common mass analyzers include quadrupole, ion trap, magnetic sector, and time-of-flight instruments. Quadrupole and ion trap instruments separate intact molecules using electrodynamic fields and routinely achieve femtomolar levels of sensitivity. Higher precision measurements of stable isotopes are typically achieved using magnetic sector analyzers and multiple ion collectors. Such instruments are fundamentally different in approach because the analytes are combusted to simple gases for isotope enrichment determination. They are less sensitive but achieve greater precision than quadrupole and ion trap instruments (Bier, 1997).

Quadrupole. The quadrupole mass filter consists of four parallel hyperbolic rods that radiate an electrodynamic field (see Figure 1). Through manipulation of alternating direct current (dc) and radio frequency (rf) voltages across the rods, a stable ion trajectory is created for ions within a chosen m/z range, which allows these ions to pass on for detection (Watson, 1997). The voltages across the rods can be changed rapidly which makes these instrument well suited to GC separations, which require rapid scanning abilities for measurement of multiple ions during the elution of analytes from the GC column. These rapid scan capabilities are needed for selected ion monitoring (SIM), where multiple specific ions are chosen for analysis. High sensitivity is derived from the use of high gain electron multipliers, which comes at the expense of instrument precision. RSDs of 2–5% for repeat measurements are routinely observed in the literature.

Specificity is further enhanced though a choice of ionization modes. Electron impact (EI) involves the stripping of an electron from the gas-phase molecule by bombardment with an electron beam. This mode produces significant decomposition of the molecule, which can serve as an aid towards deducing molecular structure. The site(s) of label incorporation can often be determined from fragment analysis. A different approach that minimizes fragmentation, chemical ionization (CI), uses an externally added reagent gas to the ionization chamber. These reagent ions produce a host of charged species that react via charge or ion transfer reactions with the analyte. A feature of CI is the strong molecular ion (actually, protonated molecule $[M+H]^+$) observed in the spectra which aids in compound identification.

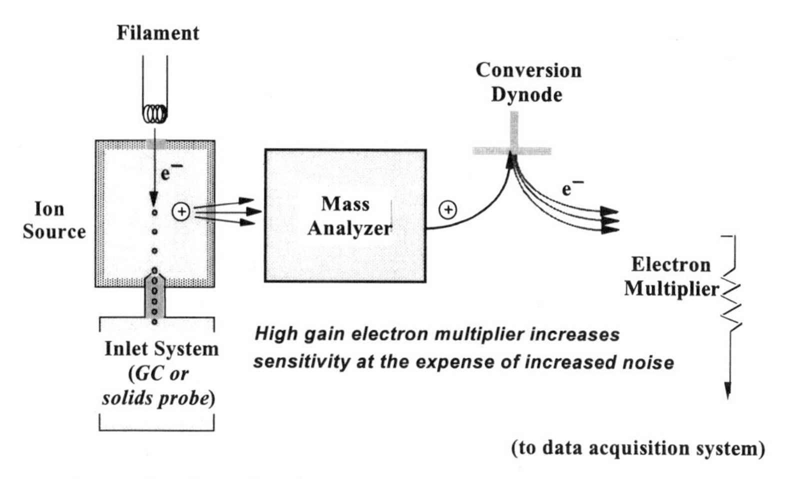

Figure 1. Components of a quadrupole mass spectrometer.

Ion Trap Detection. An ion trap mass spectrometer is a cylindrical quadrupole that uses three electrodes to trap ions in a small volume. A mass spectrum is obtained by increasing the electrode voltages to eject ions of increasing mass sequentially from the trap. The advantages of the ion trap MS include its compact size, and its ability to trap and accumulate ions to increase the S/N ratio of a measurement. Ionization can be effected using both EI and CI modes.

Originally ion trap instruments were not well suited for tracer studies because they displayed a limited dynamic range due to the requirement that the total number of ions in the trap not exceed $\sim10^5$ (March and Todd, 1996). Two features commercialized in the last decade, the selected ion storage (SIS) scan function and the automatic gain control (AGC), have significantly improved the versatility of these instruments for quantitative analysis. The SIS scan function enables the selective storage and/or ejection of single ion species using multifrequency waveforms (Buttrill et al., 1992). This feature can increase instrument sensitivity through the removal of unwanted ions from the sample matrix and column bleed. Automatic gain control prevents the buildup of excessive space charge by maintaining the correct number of ions in the trap through the adjustment of ionization time, leading to a greatly improved dynamic range. The unique ion trapping capabilities of these instruments also permits experiments akin to tandem MS/MS.

GC-C-IRMS. Gas Chromatography-Combustion-Isotope Ratio Mass Spectrometry (GC-C-IRMS) is a technique in which a mixture of compounds are separated on a gas chromatograph and combusted to simple gases such as $C0_2$ and quantified by an on-line multicollector isotope ratio mass spectrometer (see Figure 2) (Tobias and Brenna, 1996; Metges and Petzke, 1997). GC-C-IRMS can be used to measure natural and extremely small variations in isotopic enrichment due to their high measurement precision capabilities. For this reason, they have been applied to study the natural isotopic fractionation of isotopes through the biosphere as a measure of ecological and environmental processes. GC-C-IRMS measures the lowest tracer enrichments (~0.001 atom % enrichment) (Koletzko et al., 1997) with the highest precision of conventional mass spectrometers, although at the expense of sensitivity and the loss of intramolecular position information of the label (Bier, 1997). Specificity is derived from the known chromatographic behavior of the analyte. This technology is most developed for $^{13}C/^{12}C$ determinations (as CO_2) and biomedical studies using ^{13}C-labeled tracers.

The high precision of GC-C-IRMS is related to two features: (1) The simultaneous measurement of two or more ion beams which corrects for time-dependent variations in ionization efficiency. (2) The use of Faraday cup detectors dedicated to each m/z channel for the measurement of ion-beam current; a signal from the Faraday cup does not undergo the same amplification process used in single collector mass spectrometers, and thus electronic noise is not introduced as in the use of electron multiplier/photomultiplier tubes with conventional MS instruments.

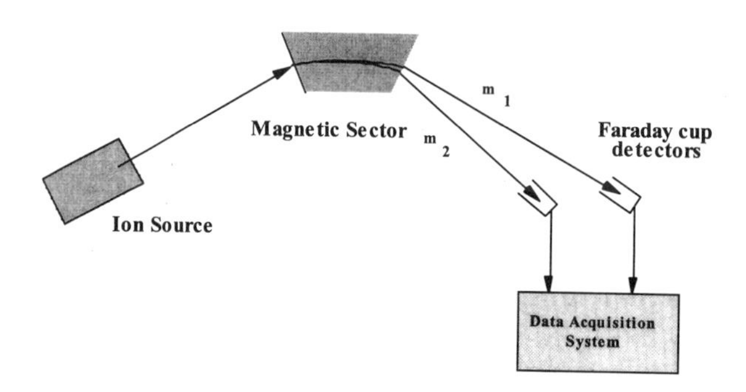

Figure 2. Multiple collector isotope ratio mass spectrometer.

TRACER SELECTION

Heavy stable isotopes are available for many of the common organic nuclides though ^{13}C and/or deuterium are preferred with most applications. In the planning of a tracer, the following considerations are applicable.

Number of Heavy Isotopes per Compound.

The degree of isotope incorporation into a tracer will have significant bearing on the quality of the final measurement. Because stable isotopes are naturally abundant, molecular enrichments of the biological metabolites must be sufficiently above natural levels to achieve reliable quantitative outcomes. This is illustrated in the mass profile of a volatile silyl derivative of retinol (vitamin A) in Figure 3a. Contributions from naturally occurring heavy isotopes ^{13}C, ^{29}Si, and ^{30}Si lead to a series of heavy isotopomers above the monoisotopic mass at m/z 400. For traditional mass spectrometric analyses, an optimally designed tracer at this mass should incorporate a minimum of four isotope substitutions so as to minimize background signal from the parent compound (0.21% at m/z 404). Of course, this is a function of the level of enrichment expected in the biological samples. Large changes at greater than 5% enrichment could accurately be determined using a tri-substituted molecule. As a rule, a minimum of a 3 Dalton increase of the tracer relative to the tracee is required for most biological studies where 1—2% plasma enrichments are targeted (Wolf, 1992).

The contribution of natural abundances can be fully appreciated from the isotopic profile of a rat liver carboxylesterase in Figure 3c. In this example, the monoisotopic peak is insignificant and a tracer would require substantial enrichment (> 25 amu) to be quantitatively useful. Conversely, if the protein (or the retinol derivative) were combusted to CO_2, the isotopic ratio ($^{13}C/^{12}C$) would mimic other naturally occurring biomolecules, regardless of mass (Figure 3b). Herein lies the advantage of GC-C-IRMS analyses: a simple and predictable isotopic background where low enrichments can be determined with high precision.

For GC-C-IRMS applications, multilabeled or perlabeled compounds can considerably reduce the amount of tracer administered in the GC-C-IRMS experiments by eliminating intramolecular isotope dilution that occurs during the combustion process (Pont et al., 1997; Guo et al., 1997). Minor natural variations in ^{13}C background can introduce significant error into low enrichment determinations and one approach to minimize error is to use non-traced sample analytes to predict the changes in carbon-13 background enrichment (Guo et al., 1997).

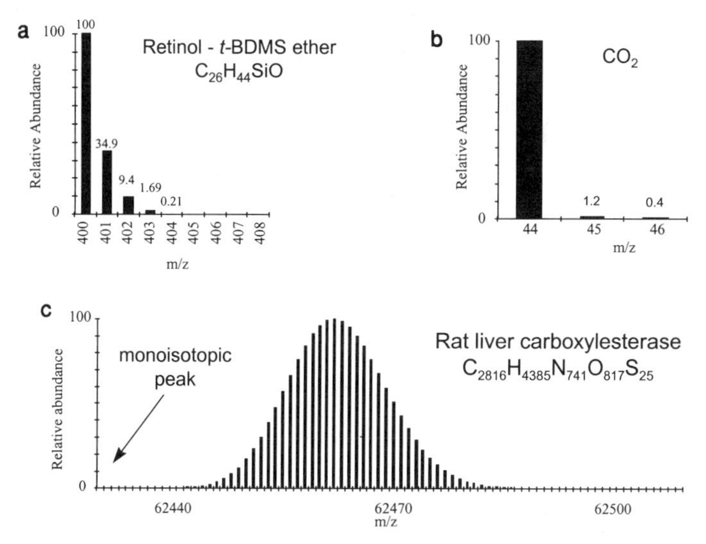

Figure 3. The abundance of heavy isotopomers increases with molecular mass. Contributions from naturally abundant heavy nuclides produce a family of heavy isotopomers for a compound.

Isotope Placement/Position Stability

The isotope should be placed at chemically inert sites so as to minimize both the loss of label to the surrounding media and any exchanges with reactive biomolecules during metabolism and sample workup. This is a major concern at deuterium and oxygen-18 substitutions because labeled compounds are often prepared using reversible reactions. Carbon and nitrogen isotopes are less likely to exchange unless extensive metabolic transformation of the tracer occurs.

The rapid exchangeability of hydrogen isotopes attached to heteroatoms (O, N, S) is well known, examples including carboxylic acids, amines, and sulfhydryl groups. Such active positions would not be suitable sites for deuterium substitution. Hydrogen isotope transfer reactions from a carbon atom are in most cases much slower, but proceed at a significant rate when the resulting carbanion formed from proton removal is stabilized by electronic resonance.

Significant potential for deuterium exchange exists at positions adjacent to carbonyl functions and, more generally, whenever a heteroatom is attached to a double bond, due to rapid equilibration in solution of tautomeric forms. Tautomerization is a type of isomerization that involves a rapid proton shift among atoms of a molecule with a corresponding change in placement of the double bond. This reaction is subject to both acid and base catalysis, as might occur in the stomach or during sample processing for analysis.

Tautomeric equilibration predicts a loss of deuterium label at the 3' and 5' positions from the urinary metabolite folic acid, p-aminobenzoyl glutamic acid (Figure 4). The susceptibility of these position to exchange was verified in our synthesis of [2',3',5',6'-$^{2}H_4$]folic acid (Dueker et al., 1995) from perdeuterated toluene and observed by Santhosh-Kumar at al. (1995) during acid hydrolysis of a [3',5-'$^{2}H_4$]folate. The reaction scheme also illustrates exchangeable ^{18}O positions at the acyl carbon. This reaction is also subject to acid and base catalysis.

Several purification protocols report the use of strong anion exchange resins eluted with 0.1 N HCl to isolate these compounds from urine (Gregory et al., 1990; Santhosh-Kumar at al., 1995). Volatile derivatives for GC/MS analysis where also prepared using both acidic conditions and high temperature conditions. The use of such conditions appears to present conditions conducive to label exchange in light of the observations described above.

Many of the reaction mechanisms that lead to label loss are the same reactions used to prepare labeled molecules (Evans, 1979). Exchange of active protons in deuterium oxide is often the easiest and thus most economical means for synthesis of deuterated tracers. The variable and often incomplete labeling with deuterium is a result of the method of label incorporation, base- and/or acid-catalyzed exchange with D_2O. This is an equilibrium process that is never theoretically complete due to the dilution of the D_2O with the exchanged protons, although this problem can be minimized by using large excess of D_2O. It is of considerable value to review the reactions for a synthetic scheme, understanding that most reactions are reversible to some degree and the ease of synthesis can also forecast the ease of label loss.

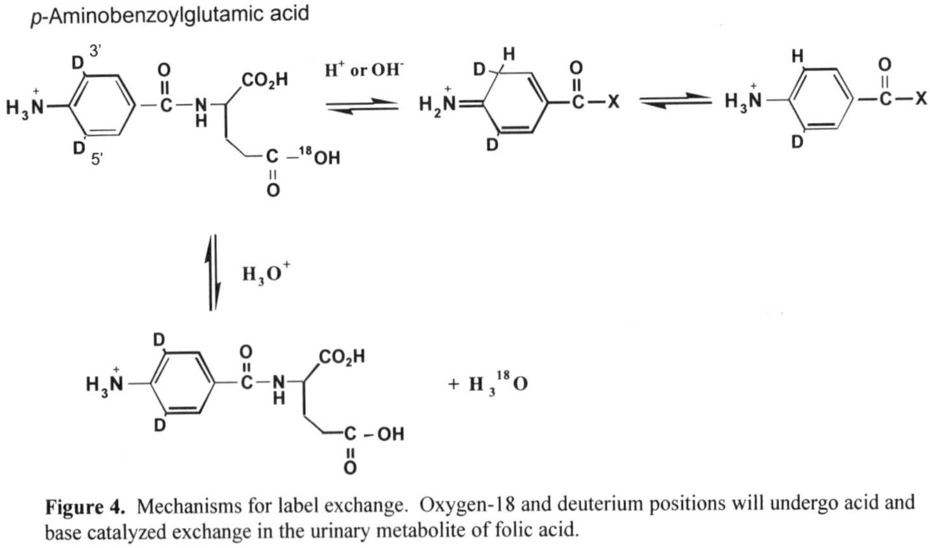

p-Aminobenzoylglutamic acid

Figure 4. Mechanisms for label exchange. Oxygen-18 and deuterium positions will undergo acid and base catalyzed exchange in the urinary metabolite of folic acid.

Isotopic Effects on Reaction Rates/Adverse Health Effects

Heavy isotopes can have considerable affect upon the vibrational modes involved in the making or breaking of bonds and will exert some effects upon the reaction rates in biological systems (Katz and Crespi, 1970). The most pronounced effects are seen with the hydrogen isotopes, protium and deuterium, where the mass ratio of $^{2}H/^{1}H$ is the largest possible for any pair of stable isotopes of the same element.

The kinetic isotope effect, a change of rate that occurs upon isotopic substitution, is a widely used tool for elucidating reaction mechanism in chemistry and biological systems (Korzekwa et al., 1995). With isotopes, the electronic structure does not change and the strength of a bond is solely attributable to the change in mass, which manifests itself primarily in the frequencies of vibrational modes: the greater the mass, the stronger the bond (Figure 5) (Lowry and Richardson, 1987). In living organisms, replacement of deuterium for protium will markedly suppress reaction rates when the point of substitution is at a reaction center. The extent to which ^{2}H can replace ^{1}H in living organisms has been reviewed (Katz and Crespi, 1970), and while certain plant microorganisms and bacteria can tolerate complete replacement with deuterium, animals will tolerate a maximum of ~ 35% with signs of toxicity appearing at less than half that level (Greg et al., 1973). Higher plants tolerate somewhat higher levels. Organisms grown in the presence of other isotopes (^{15}N, $^{18}0$, ^{13}C) are completely viable, as expected with much smaller kinetic isotope effects. Mice raised and maintained on a diet that averaged 80 atom % ^{13}C displayed no ill effects over 234 days (Greg et al., 1973). Highly enriched ^{13}C algae have been grown (Uphaus et al., 1970) and the kinetics of biosynthetically-labeled algal [U-^{13}C]β-carotene have been investigated in human subjects without evidence of biological discrimination (Parker et al., 1993)

It is highly improbable that a toxic threshold would be approached in kinetic tracer studies of micronutrients. The important point is to have awareness that isotopic substitution can affect the metabolic course of a labeled substrate, producing some fractionation among tissue compartments of the tracer. Unless the site of catalysis is the information of interest, it is important to substitute a heavy isotope away from a reactive center in the molecule.

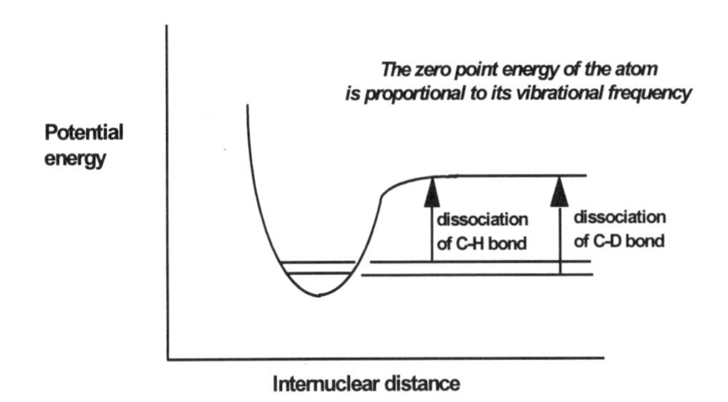

Figure 5. Kinetic isotope effects of deuterium. The activation energy for breaking a C-D bond is greater than that for a C-H bond due to differences in the zero point energies of the bonds.

Isotope Costs

Under many conditions, compound cost and availability may be the overriding criteria in the selection of the exact isotope. The degree to which precursor costs can affect isotope selection is illustrated in the synthesis of stable-isotopically labeled folic acid. To achieve a ring-labeled folate with deuterium, a quantity of 30g of [$^{2}H_8$]toluene was used at a cost of $150.00 (Dueker et al., 1995). The expense for a comparable ^{13}C-labeled compound starting with [$^{13}C_6$]toluene is about $124,800. As a rule, ^{13}C will present a much greater expense than deuterated compounds. It can be expected that as the market demand for such compounds expands, the price of this isotope will be driven down. Presently ^{13}C compounds are used in clinical settings to investigate gastrointestinal malfunctions, determine bacterial loads, and assess liver function (Rating and Langhans 1997).

CHEMICAL PURITY/ISOTOPIC PURITY OF TRACERS

No tracer study should be initiated without thorough characterization of the chemical and isotopic purity of the compound. *Chemical purity* describes the degree that the tracee dose is a single compound, whereas *isotopic purity* represents a single isotopomer. In many cases it will be necessary to commission a custom synthesis from an outside laboratory. A custom synthesis may represent a first attempt by the manufacturer, and with this attempt come the usual pitfalls that are only discovered empirically. Our independent analyses of commercially available labeled compounds have yielded results quite removed from those of the manufacturer. This discrepancy is explained in certain cases where proton or ^{13}C nuclear magnetic resonance (NMR) spectra are provided as a measure of isotopic and chemical purity. Integration of the area under an NMR peak provides an estimate of the relative number of atoms but is not quantitative, thus relative integrals can be in error by as much as 10%. The real strength of NMR is in locating the label position within the molecule. NMR cannot serve as an accurate measure of small differences in isotopic substitution or of low concentrations of chemical impurities. Mass spectrometric analysis is the most accurate means for the assessment of isotopic purity.

Chemical impurities often arise from side reactions that occur in the synthesis process and can range from stereoisomers of the compound to entirely unrelated chemicals. In many ways the stereo- and structural isomers can pose a greater risk to the data set than unrelated compounds, due to competition for the same biological receptors and metabolic pathways. The chemical purity of all compounds should be verified, and this is best accomplished using high performance liquid chromatography (HPLC) coupled to a 'Universal' detector, such as Refractive Index and Evaporative Light-Scattering detection devices, that are independent of a sample's optical properties. If these were not available, it may suffice to monitor the HPLC eluent by UV/Vis at several wavelengths, including the low UV region between 215 and 220 nm that will detect the presence most organic compounds.

ISOLATION AND PURIFICATION ISSUES

Tracer studies are difficult due to the low concentrations at which the tracer exists relative to the tracee. The susceptibility of the measurement to low-level interferences necessitates more stringent preparation practices and additional measures than analyses not involving tracers. For example, some common interferences are derivatives of fatty acids and industrial pthalates that seem to be present ubiquitously on laboratory glass and plasticware. As a rule, we pre-wash glassware and derivatization vials with hexane to remove such interferences and avoid fingerprint contamination by wearing latex gloves.

The parallel processing of blank samples—samples that do not contain tracer—should be incorporated as a standard practice to verify that the method delivers samples free from such interferences. Standards do not serve as surrogates because they are usually not taken along the same path as 'true' biologically derived blanks.

Maintaining the integrity of the label requires an understanding of the reactivity of the label position as addressed above. It is possible, if not likely, that methods developed for extraction of the endogenous compound would not be suitable when the tracer label occurs in even moderately exchangeable positions. In that regard, steps may be necessary to limit sample exposure to acid and alkaline conditions and high temperatures, conditions that encourage reactivity.

Finally, thought should be given to sample throughput capabilities for the method, as well as the stability of the analyte in the final, analysis-ready form. If rates of sample analysis are limited by access to the mass spectrometer, can samples be stored for extended periods without compromising the accuracy of the analysis?

STABLE ISOTOPES IN NUTRIENT METABOLISM STUDIES

The following two examples represent our recent work with β-carotene and vitamin A tracer experimentation in humans. The first example illustrates how new technologies leveraged with a fundamental understanding of chemistry can afford advancement in the power of isotope-dilution experimentation. The second example shows how primary isotope effects on the partitioning of deuterated carotenoids can be exploited using liquid chromatography to determine mol% enrichments from human plasma. This represents a novel method for the analysis of isotopomers and may be expandable to more situations as research in this area progresses.

Example 1: Retinol Analysis Using Linear and Quadrupole Detection:

In the following example taken from vitamin A (retinol) isotope dilution experiments conducted in humans, we aimed to determine the lowest enrichment that the tracer, retinol-d_4, could be reliably measured. Our experience using linear-beam quadrupole instruments found that high and variable artifactual signal was present at the mass of the deuterated isotopomer (m/z 259) when measuring unlabeled retinol (m/z 255) and this signal prevented reliable tracer/tracee ratio measurements below 1-2 mol % (Dueker et al., 1993,1994; Handelman et al., 1993). A typical response for tBDMS-retinol (tBDMS refers to a silyl derivative of retinol) using a linear-beam quadrupole instrument is shown Figure 6. Retinoids were known to be easily protonated and preliminary studies using hydrogen as a carrier gas showed extensive hydrogenation of retinol during MS analysis. We suspected ion-molecule reactions of a similar nature were occurring in the ionization source during quadrupole analysis even when helium was the carrier gas. A strategy to minimize these reactions was needed and the solution was found in the application of ion trap instrumentation operated in the selected ion storage (SIS) mode (Dueker et al., 1997).

The SIS scan function of the ion trap enabled the selective storage and/or ejection of single ion species using multifrequency waveforms. With this feature, the target retinol isotopomers were retained while other ions were expelled from the trap. By removing unwanted reactive fragment ions, ion-molecule reactions were minimized and artifactual chemical noise eliminated. This reduction in noise expanded the analytic range to isotopic enrichment levels lower than those previously obtainable on linear-beam instruments. Triplicate analyses of a biological sample showed good precision with a RSD $< 4\%$ ($n = 3$) at an enrichment of 0.64%. Analysis of standards showed linearity down to 0.25% enrichments, a level approaching the contribution of naturally occurring isotopes in the molecule. This example illustrates the following issues:

1. The magnitude of the background signal in the tracer channel limits the dynamic range of the method.
2. Variability in this signal is especially detrimental to the sensitivity of the analysis.
3. A fundamental understanding of chemistry combined with new technologies can offer solutions to experimental problems.

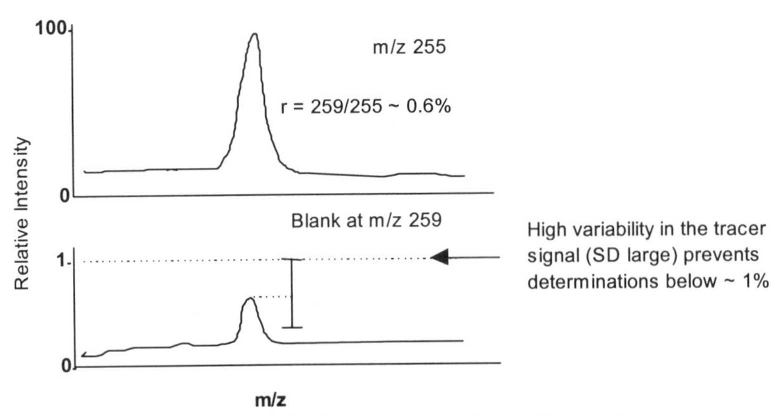

Figure 6. Ion-molecule interactions in the source can produce artifactual signal. The accurate determination of retinol-d_4 in the presence of excess retinol (m/z 255) is limited by the magnitude and variability of the signal at the mass of the deuterated isotopomer (m/z 259) during the analysis of blank samples containing no retinol-d_4.

Example 2. Analysis of Deuterated Isotopomers of β-Carotene Using HPLC.

β-Carotene is a phytochemical polyene and possibly essential micronutrient that is cleaved enyzmatically to vitamin A (retinol) and other metabolites including retinoic acid. To establish the role of dietary β-carotene and vitamin A in human health, stable isotope-labeled β-carotene was administered orally and the resulting time-dependent concentrations of plasma metabolites such as retinol were determined using ion trap GC/MS. Such investigations are facilitated by use of β-carotene-d_8, which is metabolized to retinol-d_4 (Dueker et al., 1994, 1993). This data has been used to construct compartmental models of β-carotene metabolism that can predict the fate and distribution of β-carotene and its metabolite vitamin A in inaccessible tissues and organs (Novotny et al., 1995, 1996). The use of multiple heavy isotopes minimizes interferences from isotopomers containing naturally occurring ^{13}C, ^{2}H, ^{29}Si, and ^{30}Si (in silyl derivatives), and improves measurements at low enrichments.

Having methods previously developed for the analysis of retinol isotopomers by GC/MS, we turned our attention to β-carotene isotopomer analysis. The susceptibility of the β-carotene isotopomers to decompostion precluded their analysis by GC/MS. This left the option of using a direct exposure probe for introduction into the ion source. It was established that rapid evaporation of the carotenoids was effected by passing a current of 1 A through a probe filament loaded with the carotenes. MS analysis was performed on a ZAB-2F-HS mass spectrometer (VG Analytical) with 100 eV ionization. The direct probe method does not provide chromatographic separation as do GC and LC interfaces, and exhaustive measures using multiple chromatographic operations were performed to achieve a level of chemical purity compatible with direct probe analysis. Still, single-stage MS analysis using probe introduction often showed the presence of contaminants including cholesterol ester (characterized by a peak at m/z 367) and a fatty acyl ester of glycerol (numerous peaks between m/z 520 and 560, some of which interfered with the β-carotene molecular ions). To overcome these interferences, a tandem MS/MS using linked metastable scans at constant B/E was tested (Budzikiewicz, 1982). The improved selectivity of a second mass analyzer facilitated enrichment determinations to < 1%.

However, we were not fully satisfied with analysis by MS/MS due to the rigor of the isolation protocol. Preliminary HPLC analysis of the β-carotene isotopomers showed a difference in retention times for β-carotene-d_8 and β-carotene which were attributed to a small isotope effect upon the partitioning of the two isotopomers between the stationary and the mobile phase. The isotopomers were fully resolved by isocratic reverse-phase HPLC as shown in the Figure 7. Deuteration decreases the retention times in RP-HPLC, presumably by effecting a small change in hydrophobic surface area available to bind the stationary phase, as predicted in the stoichiometric displacment model of reverse-phase chromatography (Geng and Regnier, 1985). According to this rationale, substitution at olefinic positions would be expected to have a greater effect on retention times than at sp^3 hybridized carbons (methyl groups of β-carotene).

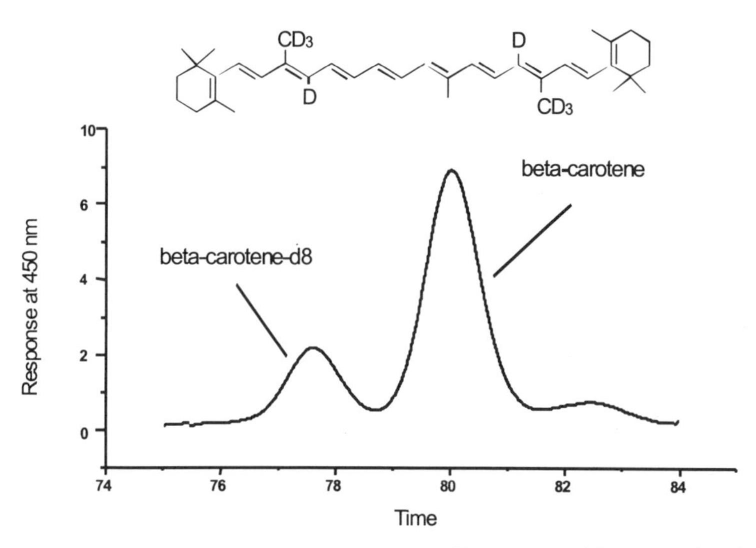

Figure 7. HPLC chromatogram showing the resolution of β-carotene and β-carotene-d_8. Absorbance was monitored at 450 nm.

The retention model is supported by evidence from recent human studies using β-carotene-d_6 in place of the octa-deuterated compound. In the β-cartotene-d_6 molecule, the label resides entirely at the methyl positions. Similar chromatographic systems to the one described for the β-carotene-d_8 separations were unable to baseline resolve the isotopomers, even under conditions of increased retention (weaker mobile phase composition). Traces of standard mixtures of the isotopomers are presented in Figure 8 as the shaded inserts. Further studies are needed with compounds labeled at the olefinic positions to better understanding the effect of isotopic substitution in relation to orbital hybridization and molecular position. The relative ease of HPLC compared to MS methods makes this mode of analysis an extremely attractive alternative at moderate enrichments (≥4%).

Although the superior analytical results are obtained when isotopomers are baseline resolved, there are ways to work with multiple overlapping peaks. In the case of the d_6/d_0 analyses, we applied a Gaussian curve-fitting routine (Origin Graphic Package, by Microcal) to the overlapping peaks (Figure 7) with excellent results. With the software, the user defines the peak apex and an approximate value for the peak half-width; the peaks are then fitted, and an integrated area value is obtained for the mathematically resolved peaks. The shaded insets represent the chromatographic profile of the unresolved isotopomers at 450 nm. The large traces are the mathematically resolved peaks following a Gaussian fit of the data. This approach produced reproducible calibration curves from 5 to 100% enrichments. At enrichments below ~5% it is difficult to determine the retention position of the deuterated isotopomer and the accuracy of the determination is decreased and tandem MS/MS is recommended as described previously.

The separation of deuterium-enriched isotopes by HPLC represents an alternative to MS determinations when larger, pharmacological dosing is either desirable or not fatal to the goals of the study. The magnitude of the phenomenon is quite small but still exploitable by the high-resolving power of RP-HPLC and the availability of 3 μm bonded silicas which minimize longitudinal diffusion (source of band broadening) and improve the resolvability of closely related isotopomers.

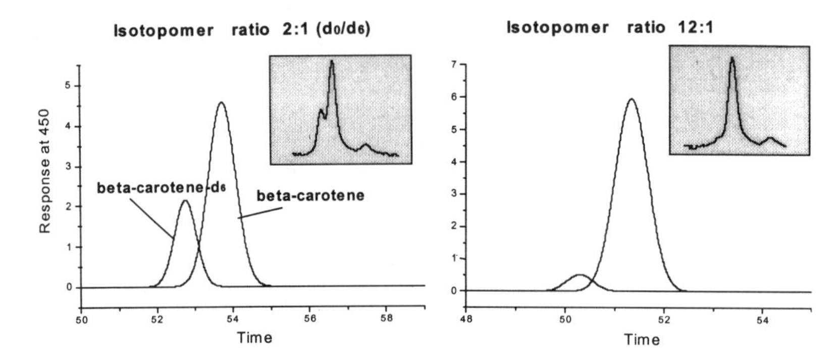

Figure 8. Mathematical resolution of β-carotene-d_6 and β-carotene standard mixtures using a Gaussian curve-fitting routine. Inserts represent chromatographic traces monitored at 450 nm.

CONCLUSIONS

Stable isotope tracers are increasing well-accepted as tools for elucidating the dynamics of nutrient metabolism in humans. Human studies are expensive and require careful consideration of analytical details before dosing begins. In this report we have discussed our observations from the viewpoint of the analytical chemist, whose central role is to deliver data that is accurate, precise, and reproducible.

ACKNOWLEDGEMENTS

We wish to thank Drs. Yumei Lin and Roger Mercer for their input into the development of this manuscript. Much of this work was supported by NIH grants RO1 DK 48307, RO1 DK 45939, R13 DK-53081. The Varian Saturn Mass Spectrometer was purchased in part with funds from NIH grants ES04699 and ES05707.

CORRESPONDING AUTHOR

Please address all correspondence to:
Dr. Stephen Dueker
Department of Nutrition
University of California, Davis
One Shields Avenue
Davis, CA 95616-8669

REFERENCES

Bier DM. The use of stable isotopes in metabolic investigation. *Bailliere's Clin Endocrinol Metab,* 1987, 4 :817-836.

Budzikiewicz H. Mass spectra of carotenoids — Labelling studies, in: *Carotenoid Chemistry and Biochemistry.* Britton G; Goodwin TW; Eds. Pergamon Press: Oxford, U.K. 1982. pp. 155-65.

Buttrill SE Jr; Shaffer B; Karnicky J; Arnold JT. Selected ion trapping to increase sensitivity for low level samples in complex matrices. *Proc 40th ASMS Conf Mass Spectrom Allied Topics.* Washington DC. 1992. pp. 1015-1016.

Cobelli C; Toffolo G; Bier DM; Nosadini R. Models to interpret kinetic data in stable isotope tracer studies. *Am J Phys,* 1987, 253:E551-564.

Dueker SR; Jones AD; Smith GM; Clifford AJ. Stable isotope methods for the study of beta-carotene-d8 metabolism in humans utilizing tandem mass spectrometry and high-performance liquid chromatography. *Anal Chem,* 1994, 66(23):4177-85.

Dueker SR; Jones AD; Smith GM; Clifford AJ. Preparation of [2',3',5',6'-$^{2}H_4$]Pteroylglutamic acid. *J Labeled Comp Radiopharm,* 1995, 36, 10:981-991.

Dueker SR; Lunetta JM; Jones AD; Clifford AJ. Solid-phase extraction protocol for isolating retinol-d4 and retinol from plasma for parallel processing for epidemiological studies. *Clin Chem,* 1993, 39:2318-22.

Dueker, SR; Mercer, RS; Jones, AD; Clifford, AJ. An ion trap method for determination of retinol-d_4 at low enrichments. *Anal Chem,* (in press).

Evans, EA. Isotopic labeling with carbon-14 and tritium, in: *Principles of Radiopharmacology.* Colombetti LG; Ed. CRC Press: Boca Raton. 1979. pp. 11-25.

Geng X; Regnier FE. Stoichiometric displacement of solvent by non-polar solutes in reversed-phase liquid chromatography. *J Chromatog,* 1985, 332:147-168.

Green, JM. A practical guide to analytical method validation. *Anal Chem,* 1996, 68:A305-A309.

Gregg CT; Hutson JY; Prine JR; Ott DG; Furchner JE. Substantial replacement of mammalian body carbon with carbon-13. *Life Sci,* 1973, 13(7):775-82.

Gregory JF 3d; Bailey LB; Toth JP; Cerda JJ. Stable-isotope methods for assessment of folate bioavailability. *Am J Clin Nut,* 1990, 51(2):212-5.

Guo Z K; Nielsen S; Burguera B; Jensen MD. Free fatty acid turnover measured using ultralow doses of [U-^{13}C]palmitate. *J Lipid Res,* 1997, 38, 1888-1895.

Handelman GJ; Haskell MJ; Jones AD; Clifford AJ. An improved protocol for determining ratios of retinol-d_4 to retinol isolated from human plasma. *Anal Chem,* 1993, 65(15):2024-2028.

Katz JJ; Crespi HL. Isotope effects in Biological Systems, in: *Isotope Effects in Chemical Reactions.* Series title: ACS monograph, 167. Collins CJ; Bowman NS; Eds. Van Nostrand Reinhold: New York. 1970. pp. 286-363.

Koletzko B; Sauerwald T; Demmelmair H. Safety of stable isotope use. *Eur J Pediatr,* 1997, 156:S12-7.

Korzekwa KR; Gillette JR; Trager WF. Isotope effect studies on the cytochrome P450 enzymes. *Drug Metab Rev,* 1995, 27(1-2):45-59.

Lowry TH; Richardson KS. *Mechanism and Theory in Organic Chemistry. 3rd Ed.* Harper & Row: New York. 1987.

March RE; Todd JFJ. *Practical Aspects of Ion Trap Mass Spectrometry. Vol. I-III.* CRC Press: New York. 1996.

Metges CC; Petzke KJ. Measurement of $^{15}N/^{14}N$ isotopic composition in individual plasma free amino acids of human adults at natural abundance by gas chromatography-combustion isotope ratio mass spectrometry. *Anal Biochem,* 1997, 247(1):158-64.

Novotny JA; Dueker SR; Zech LA; Clifford AJ. Compartmental analysis of the dynamics of beta-carotene metabolism in an adult volunteer. *J Lipid Res,* 1995, 36(8):1825-38.

Novotny JA; Zech LA; Furr HC; Dueker SR; Clifford AJ. Mathematical modeling in nutrition: Constructing a physiologic compartmental model of the dynamics of beta-carotene metabolism. *Adv Food Nut Res,* 1996, 40:25-54.

Parker RS; Swanson JE; Marmor B; Goodman KJ; Spielman AB; Brenna JT; Viereck; SM; Canfield WK. Study of beta-carotene metabolism in humans using ^{13}C-beta-carotene and high precision isotope ratio mass spectrometry. *Ann NY Acad Sci,* 1993, 691:86-95.

Pont F; Duvillard L; Maugeais C; Athias A; Persegol L; Gambert P; Verges B. Isotope ratio mass spectrometry, compared with conventional mass spectrometry in kinetic studies at low and high enrichment levels: Application to lipoprotein kinetics. *Anal Biochem,* 1997, 248, 277-87.

Rating D; Langhans CD. Breath tests: Concepts, applications and limitations. *Eur J Pediatr,* 1997, 156:S18-23.

Santhosh-Kumar CR; Deutsch JC; Hassell KL; Kolhouse NM; Kolhouse JF. Quantitation of red blood cell folates by stable isotope dilution gas chromatography-mass spectrometry utilizing a folate internal standard. *Anal Biochem,* 1995, 10, 225(1):1-9.

Sears, FW; Zemansky, MW; Young, HD. *University Physics. 6th Ed.* Addison-Wesley Pub. Co: Reading, Mass. 1982.

Tobias HJ; Brenna JT. High-precision D/H measurement from organic mixtures by gas chromatography continuous-flow isotope ratio mass spectrometry using a palladium filter. *Anal Chem,* 1996, 68(17):3002-3007.

Uphaus RA; Flaumenhaft E; Katz JJ. Isotope biology of ^{13}C. Extensive incorporation of highly enriched ^{13}C in the alga Chlorella vulgaris. *Biochem Biophys Acta,* 1970, 215:421.

Watson JT *Introduction to Mass Spectrometry, 3rd Ed.* Lippincott-Raven: Philadelphia. 1997.

Wolfe RR. *Radioactive and Stable Isotope Tracers in Bio-Medicine–Principles and Practice of Kinetic Analysis.* John Wiley and Sons: New York. 1992.

PLASMA SOURCE MASS SPECTROMETRY IN EXPERIMENTAL NUTRITION

Ramon M. Barnes

Department of Chemistry
Lederle Graduate Research Center
University of Massachusetts
Amherst, Massachusetts

ABSTRACT

The development and commercial availability of plasma ion source, specifically inductively coupled plasma, mass spectrometers (ICP-MS) have significantly extended the potential application of stable isotopes for nutritional modeling. The status of research and commercial ICP-MS instruments, and their applications and limitations for stable isotopic studies are reviewed. The consequences of mass spectroscopic resolution and measurement sensitivity obtainable with quadrupole, sector, time-of-flight, and trap instruments on stable isotope analysis are examined. Requirements for reliable isotope measurements with practical biological samples including tissues and fluids are considered. The possibility for stable isotope analysis in chemically separated compounds (speciation) also is explored. On-line compound separations by chromatography or electrophoresis, for example, have been combined instrumentally with ICP-MS. Some possibilities and requirements are described for stable isotope speciation analysis.

INTRODUCTION

Predicting the fate of chemicals in biological, environmental, and nutritional systems is a powerful motivation in many fields of science today (Vighi and Calamari, 1993). The form, concentration, and distribution of organic and inorganic chemicals in living botanical and biological species, the environment, the atmosphere, and geosphere need to be quantified reliably. The ultimate goal of many scientists is to make evaluated predictions based upon mathematical models and experimental data describing these macro and micro systems. One of the weakest aspects in predicting the fate of chemicals is the availability of reliable data and suitable models based upon them. Often the chemical analyst is challenged by either the low concentration, evasive nature or inaccessibility, or reactivity of the analyte. To meet these challenges for elements, in particular for metals and metalloids,

Mathematical Modeling in Experimental Nutrition
Edited by Clifford and Müller, Plenum Press, New York, 1998

and their isotopes, inductively coupled plasma mass spectrometry (ICP-MS) has become a valuable tool for modern biological, medical, nutritional, and environmental chemical analyses.

Applications in nutritional research that would be otherwise difficult or impossible are feasible with the unique features of ICP-MS. High sensitivity, characterized spectral interferences, rapid mass scanning, and individual isotope measurements are now combined with sophisticated sample preparation, separations, or stable isotope additions to achieve rapid semi-quantitative analysis, element speciation, and high accuracy. The semi- and quantitative analyses of materials, the separation and detection of macromolecules in blood and other tissues, and the tracking of stable isotopes added either purposely or inadvertently are important applications of ICP-MS than can affect nutritional modeling.

Advances in ICP-MS for human nutrition and toxicology were described earlier (Barnes, 1993, 1996). These articles include references to useful texts and reviews. Monographs (Evans et al., 1995; Prichard et al. 1996) and reviews (Taylor et al., 1997; Delves, 1995; Mellon et al., 1993; Nuttall et al., 1996) have been published recently.

In the elemental analysis of biological, biomedical, and nutritional materials, three primary objectives are considered: (1) identification and quantification of elemental concentrations, (2) identification and quantification of compounds containing metals, and (3) evaluation of their bioavailability, mobility, and/or toxicity. Typically, when a sample in a nutritional study is analyzed, the elemental composition of the material is first sought, and then the concentration of an individual element or group of elements is judged to be normal or abnormal. Information about the spatial and temporal distributions of an element within a sample magnifies the analysis requirements. Because nutritional models are keyed to the kinetics among compartments, measuring temporal changes in total and isotopic compositions are essential (Siva Subramanian and Wastney, 1995; Buckley, 1996). The response of the element concentration and distribution to system perturbations (e.g., disease, diet, environment, poisons) (Taylor, 1996) extends simple analysis to involve critical and sometimes unknown *in vivo* processes. Furthermore, the identification of particular metal-containing compounds besides, or instead of, total metal concentration is desirable but often is restricted by the sensitivity of element-specific detection. These metal-containing compounds are usually separated from the matrix and other similar complex materials, typically with chromatographic or electrophoretic techniques (Caroli, 1996; Cornelis, 1996; Feldmann, 1996; Pergantis et al., 1997). The metal-containing molecules are then identified, generally with the aid of on-line spectroscopic detectors. Identifying the chemical form of the metal in the compound also is critical, because changes in chemical form can often alter the role played by the species in biological and environmental systems. Although molecular and mass spectroscopic techniques provide invaluable information about metal species, applications are limited by the sensitivity of the techniques (Mellon et al., 1993).

Mineral and trace metal bioavailability and nutrition are established with animal and human subjects through element balance, isotope tracer, or radioisotope studies (Buckley, 1966). The chemical form of the metal in these investigations is of growing concern, especially when the mechanisms of metabolism or toxicity are to be characterized. In each of these aspects of biological, clinical, nutritional, and environmental material analysis, ICP-MS can provide a valuable and sometimes unique contribution.

Stable isotope analysis methods for biological and nutritional materials are well established with conventional mass spectrometric and nuclear approaches (Crews et al., 1994; Mellon et al., 1993; Prichard et al., 1996; Alfassi, 1994; Caroli, 1996; Cornelius, 1996). Typically, stable isotopes are employed to characterize absorption, bioaccumulation, excretion, and kinetics of trace minerals. An extensive literature of isotope methodology already exists for conventional ion source mass spectrometry. Accepted techniques such as stable isotope dilution analysis (IDA) or administration of multiple isotopes of the same element, and valuable applications with these conventional isotope techniques are being extended by ICP-MS (Klinkenberg et al., 1996).

The incentive to apply stable isotope techniques to nutritional research is growing as their advantages and novel instrumentation capabilities are recognized by researchers. Inductively coupled plasma MS and new isotope measurement technologies enhance the other approaches ordinarily used, because the sample preparation is easy, the analysis rapid, and the precision sufficient for most biomedical, environmental, and nutritional applications. Furthermore, nutritional, pharmacological, and toxicological models of animals and humans and ecotoxicological risk models can be expanded with sufficient data. The sample throughput of ICP-MS significantly increases the range and usefulness of stable isotope measurements for these kinetic models.

In most studies, the interactions and co-operation among professionals such as biologists, environmentalists, nutritionists, toxicologists, clinicians, and analytical chemists, are critical for maximizing the utility of stable isotope methods with biomedical and nutritional materials.

INDUCTIVELY COUPLED PLASMA – MASS SPECTROMETRY

The general applications, instrumentation, principles, and features of ICP-MS are examined in a recent book (Tanner and Holland, 1997) and reviews (Hieftje, 1996; Newman, 1996; Nuttall et al., 1996; Mahoney et al., 1997). After almost 15 years of commercial ICP-MS instrumentation and with a growing number of ICP-MS systems in practical laboratories, its applications, maturity, limitations, and future have been examined (Gray, 1994). Compared to ICP atomic emission spectrometry (ICP-AES), ICP-MS provides much lower limits of detection, simpler spectral interpretation with fewer spectral interferences, and reliable isotope ratio and isotope dilution analysis results. On the other hand, ICP-AES compared to ICP-MS uses simpler and lower cost instrumentation, operates with lower daily running costs, displays better tolerance to total sample salt content, exhibits lower matrix effects, and draws on a large, mature applications database. For example, limits of detection reported by ICP-MS manufactures with current instrumentation are orders of magnitude better than either ICP-AES or earlier ICP-MS systems. More than 50 elements can be detected below the 10 pg/mL level, which translates into reliable isotope ratio determination of many mineral and trace elements of concern in human nutrition and environmental studies. Improved precision permits smaller, more physiologically relevant doses of tracer elements to be applied.

Thus, higher sensitivity with current ICP-MS systems improves the measurement precision. With the ICP quadrupole MS, the precision range of stable isotope measurements is typically better (0.1–2% relative standard deviation, (RSD) (Begley and Sharp, 1997)) than either neutron activation analysis (NAA) or gas chromatography mass spectrometry (GC-MS) (Moens and Dams, 1995) as illustrated in Figure 1. However, the precision is poorer than thermal ionization mass spectrometry (TI-MS) with a magnetic sector MS (Newman, 1996b). Quadrupole TI-MS and fast atom bombardment MS (FAB-MS) appear to be comparable with ICP-MS (Briche et al., 1997; Crews et al., 1994; Mellon et al., 1993). Moreover, ICP-MS with a high-resolution, double-focusing, magnetic-sector mass analyzer equipped with multiple detectors matches the accuracy and precision (typically 0.01–0.05%) of TI-MS stable isotope measurement (Halliday et al., 1995; Hirata, 1996; Lee and Halliday, 1995). A double-focusing, magnetic-sector ICP-MS in a scanning mode also provides significant improvement in isotope ratio precision compared to quadrupole-based ICP-MS (e.g., ≤0.05% for $^{25}Mg^+/^{26}Mg^+$ and $^{206}Pb^+/^{207}Pb^+$ at signal intensity values of ≥500000 counts/s) (Vanhaecke et al., 1996, 1997). At low analyte concentrations and with sufficient resolution to separate spectral interferences (R = 3000), the isotope ratio precision of sector-field ICP-MS is somewhat degraded (e.g., $^{63}Cu/^{65}Cu$ RSD of ≤0.6% in human serum reference material) (Vanhaecke et al., 1997).

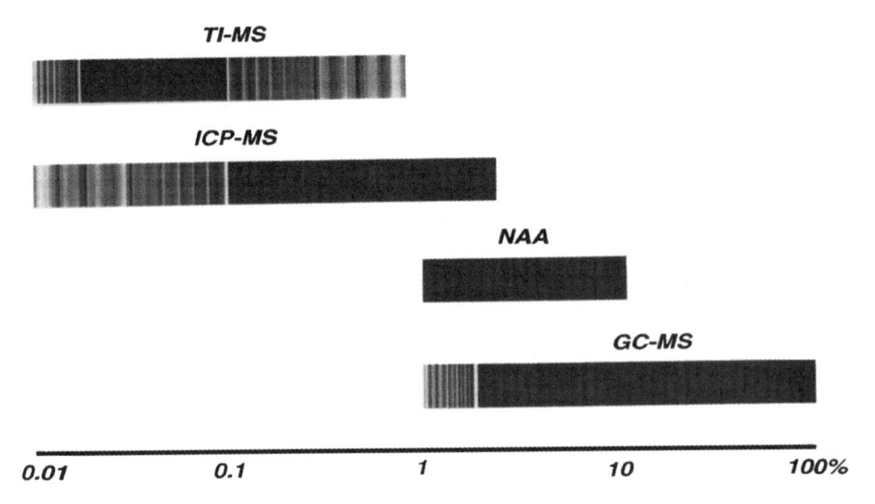

Figure 1. Stable isotope analysis precision. The range of isotope analysis precision is represented for thermal ionization (TI), inductively coupled plasma (ICP), gas chromatography (GC) mass spectrometry (MS) and neutron activation analysis (NAA). The two ranges given for TI-MS and ICP-MS correspond to magnetic-sector (lower range) and quadrupole (upper range) MS systems.

Some of the serious limitations of earlier commercial quadrupole ICP-MS systems (Gray, 1994) such as matrix suppression or enhancement, noise and precision, stability, tolerance to total dissolved solids, reliability, and mass resolution and spectral interferences, have been resolved in third-generation instruments. For example, a signal stability of 4% RSD over 4 hours was reported recently for 24 elements during a 12 h test. Matrix effects resulting from space charge in the ion optic system have also been reduced by automatically tuning the lens focusing voltage with mass (Wolf and Denoyer, 1995; Denoyer et al., 1995). A criterion for comparing the features of ICP-MS instrumentation has been reported (Analytical Methods Committee, 1997).

In the applications of ICP-MS to biological and botanical, and clinical–medical materials (Taylor et al., 1997; Subramanian, 1996), both elemental and isotope ratio analyses furnish survey and quantitative results. Semiquantitative analysis provides survey data of the total element content of biological and environmental materials (Amarasiriwardena et al., 1997). Quantitative analysis of major, minor, and trace impurities depends upon external calibration, standard additions, or isotope dilution techniques. Obtaining an elemental mass spectrum of a biological or nutritional sample is essentially simple; however, the spectrum requires interpretation especially with low-resolution quadrupole MS instruments. Isobaric species are not resolved, and polyatomic ion species from the matrix, sample preparation steps, and the argon plasma cause interferences when coincident with analyte ions at common resolutions of R = 300. Polyatomic ions arising from biological materials often include combinations of calcium, cadmium, chlorine, sulfur, oxygen, phosphorus, potassium, and sodium. Spectral overlaps by polyatomic species from sample matrix and argon plasma background limit the method for some critical major isotopes of nutritional interest, such as chromium, iron, manganese, vanadium, cadmium, and selenium. Considerable effort has targeted approaches to reduce or eliminate these interferences. These methods range from distinct sample preparation methods to alternative gas and mixed gas plasmas, desolvation, multivariate correction methods, selective removal of plasma matrix ions (Eiden et al., 1996), or high resolution mass spectrometers (Vanhaecke et al., 1996, 1997). Operating the ICP at lower than normal power (600 W *vs.* 1200 W) can reduce or eliminate some polyatomic spectral background from argon ion species by shifting the dominant background ion to NO^+ (Tanner, 1995). With these operating conditions, detection limits reported for ^{7}Li, ^{23}Na, ^{39}K, ^{40}Ca, and ^{56}Fe are improved considerably.

The rapid, multi-element capability of ICP-MS is an ideal match for the requirements of many biological and nutritional analyses, and semi-quantitative analysis software is included with modern systems (Amarasiriwardena et al., 1997). Polyatomic ion interferences limit the number of elements that can be identified with a quadrupole MS, however. Subtraction of interfering signals in the blank is sufficient to correct some spectral interferences, but the approach can yield low values for other elements (such as boron, magnesium, and aluminum) that result from signal suppression (i.e., non-spectral interferences (Vanhaecke, Riondato et al., 1996)). Reliable concentration estimates to ±20–30% are achieved for numerous elements with matrix matched standards, internal reference elements, or by adding a multi-element standard to a sample aliquot.

The literature contains many examples of quantitative multi-element analysis of biologicals by means of high-resolution ICP-MS. Vanhoe et al. (1994) determined 11 ultratrace elements (lithium, boron, molybdenum, cadmium, tin, cesium, barium, mercury, lead, and bismuth) and Moens et al. (1994) determined vanadium, iron, copper, zinc, and silver; while Yoshinaga (1995) evaluated the determination of aluminum, vanadium, chromium, and manganese in human serum using high-resolution ICP-MS. Nixon and Moyer (1966) developed routine clinical determinations of arsenic, cadmium, lead, and thallium in urine and whole blood. Schemes for the determination of aluminum, silicon, titanium, chromium, cerium, lanthanum, neodynium, and zirconium from food and feces are given in Figure 2 (Lásztity et al., 1996). When arsenic is determined simultaneously, the protocol is modified, so that arsenic is not lost during ashing (Lásztity et al., 1995).

For certification of biological and environmental reference materials, ICP-MS compares well with radiochemical neutron activation analysis and other techniques (Moens and Dams, 1995) and is accepted by international standards producing organizations (Örnemark et al., 1997). Among the quantitative single-element ICP-MS studies are the determination of ultratrace levels of platinum and gold in plasma and animal cells for the pharmacokinetic evaluation of cancer drugs (Balcerzak, 1997; Begerow et al., 1996; Begerow and Dunemann, 1996; Christodoulou et al., 1966); the determination of selenium in serum (Delves and Sieniawska, 1997; Rayman et al., 1996) and whole blood (Mestek et al., 1997); of copper and zinc in plasma and urine (Szpunar et al., 1997), lead in blood (Paschal et al., 1995), nickel in blood, feces, and urine (Patriarca et al., 1996), molybdenum in plasma (Luong et al., 1997), and thorium and uranium in urine (Ting et al., 1996).

To achieve accurate ICP-MS element concentrations and isotope ratio analyses in complex clinical and biological samples, preconcentration and separation techniques are applied either to remove the matrix or to concentrate the analyte (Caroli, 1996; Prichard et al., 1996; Feldmann, 1966; Luong et al., 1997; Veillon et al., 1966). Often the analyte is separated chemically from the sample before determination. Various sample preparation and separation schemes have been developed for stable isotope analyses. Organic samples generally require complete digestion before separation and preconcentration steps to free organically bound elements. Typically, this preparation is less complicated and shorter than required for TI-MS.

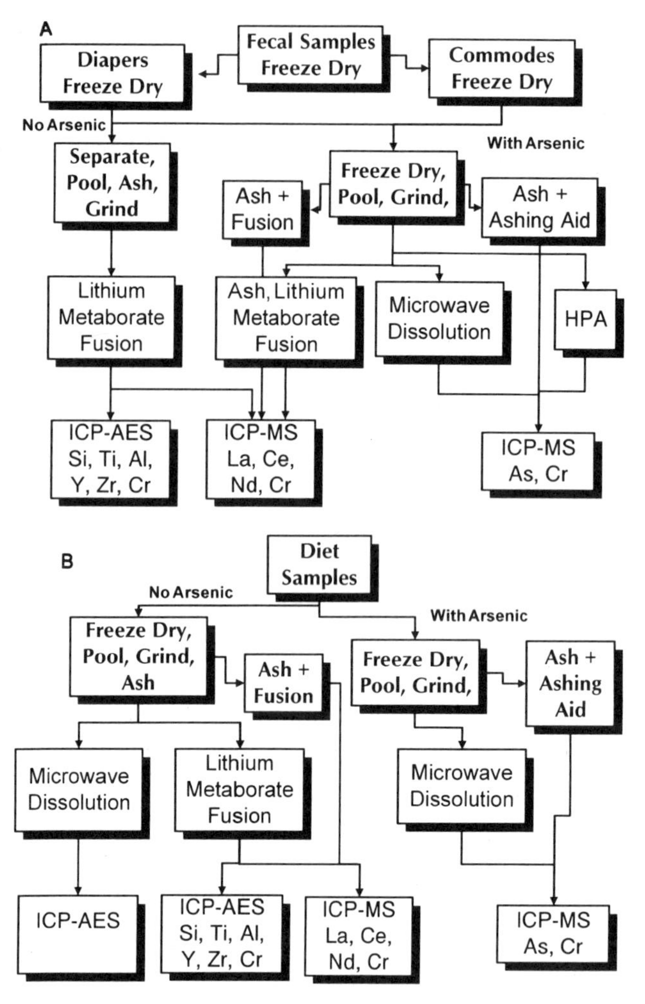

Figures 2A and 2B. Outline of sample preparation and analysis procedures for (A) infant feces and (B) food with (right side) and without (left side) arsenic determinations (Lásztity et al., 1995, 1996).

Slight improvement in the mass spectrometric resolution is sufficient to minimize spectral interferences. Most commercial quadrupole ICP-MS systems allow some adjustment to the resolution (Δamu = 1–0.6, resolving power R = 300) with a commensurate sacrifice in sensitivity. Research quadrupoles have been operated at high resolving power (R = 9000), but they are not available commercially. Practical sector-based ICP-MS instruments provide both reduced background (< 1 cps) and high resolving power (R ≥ 7500). For these sector-based ICP-MS systems, ion transmission is reduced as resolution is increased, and the instrument cost is considerably higher than quadrupole instruments.

Contamination also limits the accuracy of ICP-MS determinations (Prichard et al., 1996). Appropriate sample preparation and field sampling blanks are essential, and clean-room facilities and pure reagents are needed to obtain accurate results. Another practical problem arises from sample carry over in the ICP-MS sample introduction system. For a few elements such as gold, boron, iodine, mercury, and osmium, extended cleanout periods are required to wash the nebulizer and spray chamber.

The quantitative analysis of trace impurities by ICP-MS employs either external calibration, standard additions, or isotope dilution analysis. External calibration results in the highest sample throughput, but the accuracy and precision obtained are best at low concomitant concentrations. Performance is improved with matrix-matched standards or internal reference element additions. The method of additions reduces the sample throughput, but good accuracy results without spectral interferences. Isotope dilution analysis (IDA) also reduces sample throughput, but very good accuracy and precision are obtainable (Beary et al., 1997; Klinkenberg et al., 1996; Luong et al., 1997; Örnemark et al., 1997; Paschal et al., 1995; Patriarca et al., 1996).

In vivo uptake and retention studies in living organisms by ICP-MS has been considered as one of the most exciting and potentially useful areas of isotope dilution analysis (Crews et al., 1994; Mellon et al., 1993). The *in vitro* stable isotope dilution method involves the addition of a known amount of highly enriched isotope to the biological sample, which differs from the *in vivo* tracer. When these two techniques are combined, at least three stable isotopes are required and at least two isotope ratios are measured.

In summary, the advantageous characteristics of ICP-MS for stable isotope ratio analysis include the rapid sample analysis (5 min/sample) and high throughput, minimum sample preparation, broad multi-element capability, and the determination of elements with high ionization potential (Lyon et al., 1966). The most serious problem with quadrupole MS is the relatively limited precision (0.6–0.1% RSD) in the best measurement case and the spectral overlap interference in the worst case. Ultimate isotope ratio precision can be achieved with high-resolution sector-based ICP-MS.

On the horizon, researchers are adapting other MS instrumentation to the ICP ion source. For example, time of flight (TOF) (Mahoney et al., 1997), ion trap (IT) (Barinaga et al., 1996; Eiden et al., 1996a, 1996b), and Fourier transform ion cyclotron resonance (FTICR) instruments, and double-sector MS with array detectors (Burgoyne et al., 1997; Cromwell and Arrowsmith, 1996) have been tested. A unique feature of ICP-TOF-MS is the high ion-utilization efficiency and rapid, complete mass spectrum produced. This essentially simultaneous detection and high repetition sampling rate makes ICP-TOF-MS a preferred approach for the analysis of transient signals from laser ablation, electrothermal vaporization, flow injection, chromatographic columns, and other time-dependent sample introduction devices. Furthermore, higher resolution than quadrupole MS can be obtained with an ion reflectron arrangement, and isotope ratio precision is similar to values expected from counting statistics (Mahoney et al., 1997). With the IT, resolving powers higher than the quadrupole can be obtained, and spectroscopic interferences (e.g., Ar^+, ArH^+, ArO^+, $ArCl^+$, Ar_2^+) also can be reduced by collisional dissociation inside the trap. Commercial versions of one or more of these could reach the marketplace in 1998. The FT-ICR, still in early stages of development with an ICP ion source, exhibits extremely high resolving powers.

STABLE ISOTOPES FOR TRACER STUDIES

A variety of methods exist for the measurement of stable isotopes: NAA, TI-MS, GC-MS, fast atom bombardment mass spectrometry (FAB-MS), ICP-MS, electron impact and chemical ionization mass spectrometry (EI and CI-MS), and field desorption mass spectrometry (FD-MS) (Crews et al., 1994; Mellon et al., 1993). Isotope ratio measurement with elemental-mode electrospray mass spectrometry (EME-MS) (Ketterer and Guzowski, 1966), by contrast, is in its infancy. Selection of the instrumental approach applied depends upon the instrument availability, the isotope's analytes, and the precision and sensitivity required. Thermal ionization MS is the definitive and most precise method for most analyses (Abrams, 1994), but ICP-MS provides shorter sample preparation and analysis times at lower cost with a precision that is generally adequate for most tracer studies (Diemer and Heumann, 1997; Lyon et al., 1996).

Stable isotope tracers in nutritional studies permit following the dynamic/equilibrium transport of chemical elements or their compounds. This can aid the understanding of the metabolism of minerals in foods or those administered as therapeutic agents under a broad range of conditions of health or disease, or the mobilization of metals in the environment. Thus, stable isotope tracers are useful for metabolic modeling, especially with infants and women to whom administration of a radioisotope is undesirable. Numerous multiple-isotope labeling options exist, and no limit is fixed for the period of the experiment, as might occur with a short-lived radioisotope (Crews et al., 1994; Patriarca, 1996).

The application of ICP-MS for stable isotope tracer nutritional studies is developing (Crews et al., 1994; Fischer et al., 1997; Patriarca, 1996). Target human population groups include both normal and pre-term infants, healthy adults, patients, and women of child-bearing age. There are several examples of studies on the metabolism of several minerals and trace elements (including cadmium, magnesium, iron, copper, zinc, molybdenum, lead, and selenium): such as the dietary availability of iron to infants and animals (Kim et al., 1996); the exchangeability of dietary pools of zinc; selenium- and magnesium-turnover studies; and Cu transport in animals and humans (Lyon et al., 1996; Buckley, 1966). Other elements for which ICP-MS stable isotope ratio analyses have been applied with non-clinical samples include: lithium, boron, magnesium, chorine, potassium, chromium, iron, copper, zinc, selenium, bromine, rubidium, strontium, rhenium, osmium, lead, uranium, and thorium.

Stable isotope applications can have several limitations. For example, there may be only one stable isotope for some elements. It may be difficult to obtain the level of enrichment of the tracer in the endogenous pools needed for detection. Stable isotopes are accessible from a limited number of suppliers, sometimes in restricted quantities. Furthermore, the costs of the stable isotope, sample preparation, and type of mass spectrometer can also constrain isotope selection (Abrams, 1994). Isotope ratio analysis is more difficult than a simple quantitative measurement, and the measurement precision ultimately limits the analysis utility. Unlike radioisotopes, *in vivo* localization of the stable isotope is not readily detected without taking a sample for external analysis.

Two important requirements need be considered before applying stable isotopes to tracer studies. First, the measurement precision and accuracy should be evaluated for the isotope enrichment achievable after *in vivo* administration of the stable isotope dose. Enrichment should be sufficient to provide signal count differences adequate to be measured statistically. Second, the feasibility of achieving acceptable isotopic enrichment in the target fluid, tissue, or subcompartment should be evaluated to quantify the tracer with the requisite accuracy and precision. The degree of isotopic enrichment in biological systems depends upon the amount of isotope administered, the efficacy of absorption and retention, the specific compartment distribution, the amount of element in the target compartment, distribution of the chemical forms in the compartment, and the natural stable isotopic abundances and composition. The cost of the stable isotope to achieve 10% enrichment and the desire to give physiological doses comparable to the daily dietary intake are considerations.

A number of recent applications with animals and humans demonstrate the power of ICP-MS (Crews et al., 1994; Fischer et al., 1997; Patriarca, 1996). Typically, iron, selenium, copper, magnesium, and zinc have been examined for human metabolism studies by ICP-MS. The investigation of zinc metabolism in pre-term infants by ICP-MS is especially appropriate, because some of the sample amounts are limited, the use of stable isotope addition is desirable, and potential interferences may exist in the measurements (Veillon et al., 1996; Wastney et al., 1996). Work to determine the composition of the optimal diet for premature infants has focused on achieving normal postnatal development and nutrient accretion, but the definition of this diet requires the measurement of infant dietary mineral bioavailability and retention. Such sample analyses once performed with NAA have been replaced by ICP-MS analysis. Elements such as zinc, copper, slenium, and iron have been investigated, and extrinsic stable isotope tag and standard nutrient balance methods were found to correlate well.

To obtain quantitative descriptions of zinc metabolism, a stable zinc isotope tracer (^{70}Zn) is administered orally or intravenously to infants, and plasma, red blood cells, urine, feces, and nutrients are sampled. For example, total zinc by ICP-AES and zinc isotope ratios were measured by ICP-MS (Fig. 3; Amarasiriwardena et al., 1992), and the data obtained were evaluated with a model of zinc metabolism (Wastney et al., 1996).

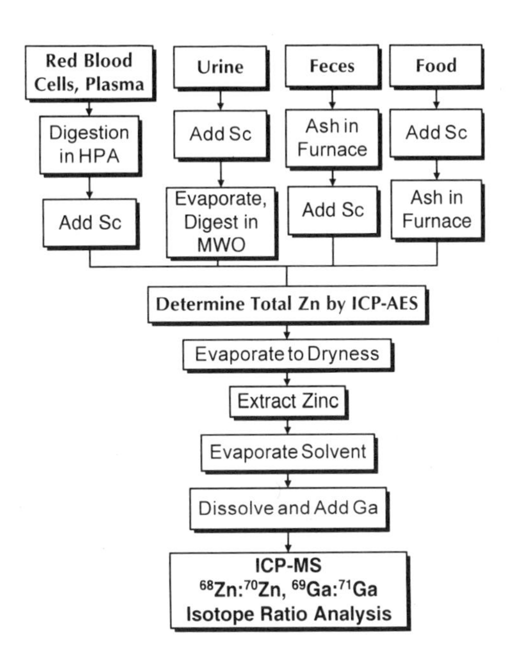

Figure 3. An outline of sample preparation and analysis procedures for total and isotopic zinc in blood, feces, food, and urine by ICP spectrometry (Amarasiriwardena et al., 1992). Scandium is added as an internal reference element for ICP-AES determinations. Samples were digested by high pressure ashing (HPA) and closed microwave heating (MWO).

Toxicology, Ecotoxicology, and Toxicokinetics

Toxicology, ecotoxicology, and toxicokinetics represent fields of significant potential development for stable isotope analysis by ICP-MS. Yet, relatively little experience has developed, probably because of the ethical question of conducting human experiments with heavy metal poisons. Lead is an exception, owing to the interest in lead poisoning, and models of childhood lead exposure are now available (Goodrum et al., 1996). A valuable toxicological application of ICP-MS isotope analysis has been source identification involved in childhood lead poisoning cases. The determination of lead isotope ratios can be used for source and pathway characterization of lead in the environment. Lead isotope ratios in gasoline and environmental samples can be determined effectively by ICP-MS (Quétel et al., 1997). Whenever unique environmental lead sources such as paint, dust, or soil are ingested by children, the lead isotope composition in the child's blood and feces changes to reflect the isotope abundance of the source. Particular sources can be pin-pointed by their congruency with the blood lead isotopic composition. With source identification, lead abatement can be directed to the appropriate specific lead supply.

Assigning source-specific contributions to blood lead on a global rather than individual basis can be equivocal, however (Gulson et al., 1996). Measurement of lead flux from bone to blood has been investigated with stable lead isotopes measured by TI-MS (Inskip et al., 1996) and ICP-MS (Laughlin et al., 1997; Seaton et al., 1997).

An extension of lead identification possible only with ICP-MS is the measurement of lead composition and isotope abundances in chromatographically separated macromolecules (Bergdahl et al., 1996). On-line size-exclusion chromatographic (SEC) separation of proteins with ICP-MS detection should permit the determination of the metal content and isotopic ratios in each separated fraction. This approach has been applied to the identification of lead-bound proteins in human and animal sera, for example, and demonstrates the value of chemical speciation in biomedical studies. The technique has been employed with other metal-binding biomolecules using a sector-field ICP-MS instrument wherein elements especially subject to spectral interference can be measured unambiguously.

Kinetic Modeling with Stable Isotope Analysis

In pharmaco- and toxicodynamic studies, the primary objects are to identify the system sinks and sources, establish the dynamics and kinetics between them, define their physiological relationships, and characterize the biochemical mechanisms (Buckley, 1966). Kinetic modeling has become an important component in nutrition, pharmacology, and ecotoxicology (Siva Subramanian and Wastney, 1965). Mineral studies are especially appropriate to compartmental modeling. Models permit investigators to view system components in perspective, provide insight into the system performance, and represent a means to assemble and correlate information systematically (Buckley, 1966). Aspects of the biological system considered that might be otherwise overlooked can become apparent, and gaps in knowledge and data are revealed. Simulation of a biological system before experimentation also improves experimental design and planning, tracer techniques, and sampling protocols.

Biokinetic parameters can be determined with stable and radioactive isotopes by measuring the change of tracer concentration in body fluids or tissues after *in vivo* tracer administration. Mathematical functions are fitted to concentration-time data to characterize kinetic uptake and loss parameters. For example, Buckley (1996) demonstrated changes in tissue pool sizes, balance determinations, adaptation in absorption endogenous excretion, and non-steady state modeling for copper metabolism in adults. Kinetic models have been studied widely using isotope tracers such as ^{62}Ni, ^{65}Zn, ^{70}Zn, and more of these studies are using ICP-MS for isotopic analysis. For example, ICP-MS zinc analyses was used to support kinetic modeling of zinc metabolism in neonates (Wastney et al., 1996). Total zinc was determined by ICP-AES and $^{70}Zn/^{68}Zn$ was measured by ICP-MS in infant serum, urine, feces, and food (cf. Fig. 3). Appropriate sample preparation steps and contamination control in a cleanroom environment were essential. The determination of ^{70}Zn in serum with ICP quadrupole MS also requires sample preconcentration and separation (Veillon et al., 1996). The future challenge in these studies is to separate the Zn-binding biomolecules on-line and to quantify their Zn isotope ratios with ICP-MS.

Equally challenging is to extend ICP-MS isotope ratio analysis to nuclides with isobaric or polyatomic interferences (e.g., lithium, silicon, sulfur, calcium, titanium, vanadium, chromium, and iron) in biofluids and tissues using high-resolution ICP-MS (Yoshinaga, 1995). Accurate and precise determination of $^{42}Ca/^{43}Ca/^{44}Ca/^{48}Ca$ and $^{54}Fe/^{56}Fe/^{57}Fe$, for example, by ICP-MS would simplify calcium (Smith et al., 1996) and iron kinetic studies.

Chemical Speciation and Stable Isotope Analysis

The metabolic behavior or toxicity of an element, especially at trace concentration levels, depends on its chemical forms in biological systems (Lobinski, 1997; Caroli, 1996; Cornelis, 1996). Similarly, the environmental partitioning, toxicity, bioaccumulation, persistence, and mobility of an environmental hazard is determined by the chemical form (Vighi and Calamari, 1993). Speciation of a trace element in a biological matrix represents the identification of the biologically active compounds to which the element is bound and the quantification of the element in the particular compound (Cornelis, 1996; Gardiner and Delves, 1994). Toxic effects generally involve speciation of small molecules (e.g., methyl mercury, triethyltin, tetraethyllead) or of free ionic forms, while biological functions concern evaluation of large molecules (e.g., metallo-enzymes, metallo-proteins) as illustrated in Figure 4. Either direct species determination or analysis after separation are employed. Speciation studies of elements in biological and environmental samples, however, are prone to errors as a result of changes of analyte *in vitro*, during and after sampling, and at the time of the measurement (Cornelis, 1996; Quevauviller, 1996).

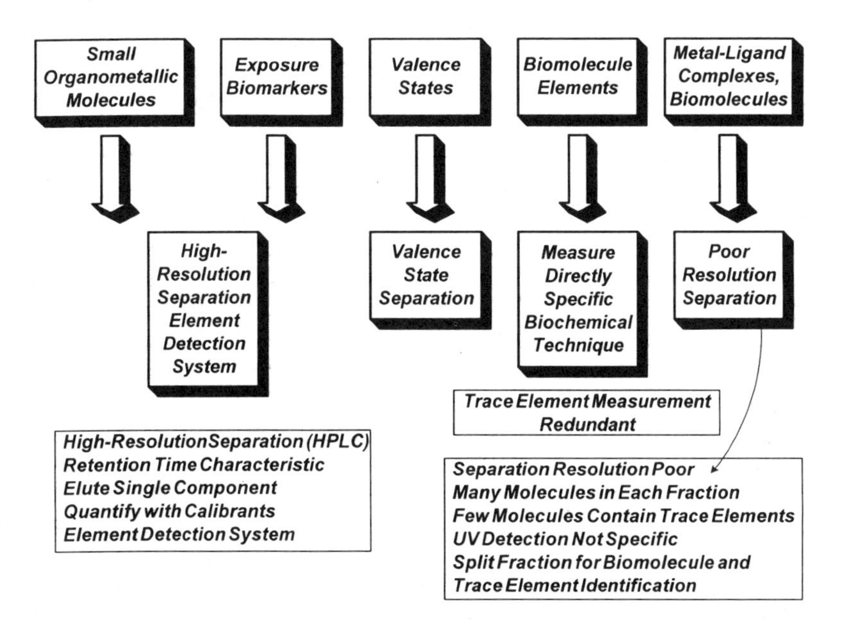

Figure 4. Classification of chemical species into five groups including small organometallic molecules, exposure biomarkers that are changed from inorganic to small organometallic compounds, elements with different valence states, trace elements in biomolecules, and macromolecules including unidentified proteins and trace element complexes with low molecular weight compounds (Cornelis, 1996). Features of each group are summarized.

Elemental speciation with plasma source MS has become popular during the past decade, and numerous combinations of separation and detection approaches are available for chemical speciation analysis (Caroli, 1996; Donard and Lobinski, 1996; Gardiner and Delves, 1994; Lobinski, 1997; Pergantis et al., 1997). Among plasma sources (e.g., ICP-AES, microwave-induced plasma), ICP-MS has become a valuable partner in chemical speciation of metals especially, because of its very low detection limits and high selectivity.

Most fractionating techniques are unable to distinguish among the large number of molecular components in biological fluids (Cornelis, 1996). However, when they are combined with an element-specific detector like ICP-MS, the added detection selectively sometimes relaxes the requirement for high resolution separation. Separation of bound-metal species before ICP-MS detection is advantageous because the matrix electrolytes (e.g., sodium) are separated and do not generate an MS signal suppression during the analysis (Feldmann, 1996). Active development is underway with HPLC-ICP-MS systems for the separation and identification of metal-containing molecules (Taylor et al., 1997). For example, organic forms of arsenic (Magnuson et al., 1996), cadmium, chromium, iron, lead (Bergdahl et al., 1996), mercury (Wan et al., 1997), platinum (Balcerzak, 1997), rare earth elements, selenium (Bird et al., 1997; Muñoz Olivas et al., 1996), tellurium, tin, vanadium, and zinc have been separated and/or quantified in environmental and biological materials.

Stable isotope ratio analysis of chromatographically separated, metal-bound molecules is not well developed, and only a limited number of studies have been described (Crews et al., 1994; Owen et al., 1992; Taylor et al., 1997), such as the isotopic determination of proteins by ICP-MS after a size exclusion or liquid chromatographic separation. One potential limitation of this approach is the analytical precision obtained for isotope ratio analysis resulting from the transient chromatographic signal, especially for low metal concentrations or low ion abundances. Isotope ratio analysis for samples introduced by transient techniques exhibit a precision of 1 to 3%. This is poorer than achieved by continuous nebulization, because the sample measurement time is short or signal averaging is not effective. A primary factor in determining the precision and accuracy of isotope ratio measurements with transient sampling techniques is the amount of material injected. These measurements could certainly benefit from the high sensitivity of third-generation ICP-MS instrumentation (Newman, 1996). Furthermore, software for obtaining data and calculating isotope ratios from on-line chromatographic signals requires continued development.

Biological applications of combined separation and ICP-AES techniques, like element-specific detection of metallodrugs and their metabolites, generally can be extended with ICP-MS. For example, pharmacokinetic studies of platinum-based drugs by HPLC-ICP-AES and their independent determination by ICP-MS established the basis for HPLC-ICP-MS of platinum-based drugs and their metabolites (Balcerzak, 1997). Similar analyses have been demonstrated for gold-drug therapy. Stimulated by the hazard of lead poisoning of children through environmental exposure, the application of stable isotope analysis by ICP-MS together with high resolution protein separation might be combined effectively with physiologically based toxicokinetic models and applied to human children (Bergdahl et al., 1996). The ICP-MS measurement of lead isotope ratios can provide the essential data, but quantitative determination of lead and/or lead isotope ratios by HPLC-ICP-MS has yet to be reported.

Electrophoretic methods are among the most widely used high-resolution techniques for analytical and preparative separations (Baker, 1995). Electrokinetic capillary techniques include numerous separation principles that can be applied to a variety of low- and high-molecular mass compounds. Detection systems from on-column direct and indirect spectrophotometry to off-column MS are available commercially (Cai and Henion, 1995). Capillary electrophoresis (CE), a technique for high-resolution, low flow-rate separations of (macro)molecules, is applied typically in biomedical, clinical, forensic, nutrition, and pharmaceutical fields (Figure 5) (Landers, 1995; Lehman et al., 1996; Lindeberg, 1996; St. Claire, 1966). However, common flow rates are too low for stable operation of most MS ion sources, including continuous fast atom bombardment, electrospray, and ICP (Cai and Henion, 1995). In these cases, an appropriate interface design using a make-up solvent is required (Lu et al., 1995). The small sample volumes and flows used with CE also challenge the design for an interface for conventional ICP-MS systems. However, capillary electrophoresis has been interfaced with microwave plasma, flame photometric, and ICP-MS detectors (Barnes, 1997; Lu and Barnes, 1996, 1997). Capillary electrophoresis – ICP-MS, for example, has been applied to selenium species in milk fractions and platinum species from soil (Michalke and Schramel, 1997; Michalke et al., 1997) and copper, cadmium, and zinc in metallothioneins (Lu et al., 1995, Lu and Barnes, 1996). Magnuson et al. (1997) modified an ion chromatography – hydride generation – ICP-MS interface (Magnuson et al., 1996) for CE and determined four arsenic compounds in water.

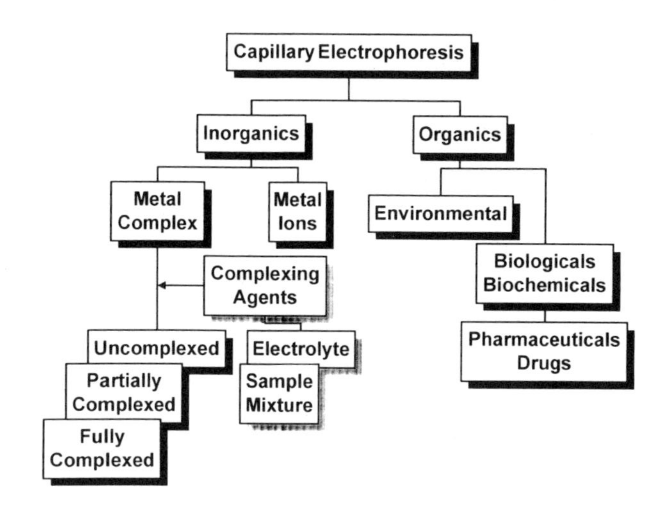

Figure 5. Application areas of capillary electrophoresis include inorganic and organic molecules.

At this early stage of CE-ICP-MS development, the prospect is good for applying effective CE-ICP-MS systems for trace metal speciation to biological, medical, and nutritional problems (Figure 5). The manufacture of a commercial CE-ICP interface for MS is the first major step in the growth of the technique. With a simple interface available, ICP-MS users are expected to extend HPLC-ICP-MS and generate novel applications. However, as with size exclusion chromatography (SEC) of metal containing compounds (Gardiner and Delves, 1994), instrumental, column, or packing material artifacts in CE for complex polymer systems, colloids, or biomolecules could limit trace and ultra-trace metal quantitative analysis. Eventually techniques to minimize or eliminate metal binding artifacts for biomolecules in SEC (Cornelis, 1996) and CE must be developed.

CONCLUSION

Inductively coupled plasma MS can have a significant impact on nutritional studies. Kinetic modeling with stable isotope tracers is an established and productive approach for a limited number of elements. With ICP-MS data can be collected for new nutritional experiments with elements including the nontraditional ones that might be encountered in pharmacokinetics (e.g., gold, gadolinium, platinum, technetium). For example, liquid chromatographic separation of ^{99m}Tc-labelled human serum albumin (Verbeke and Verbruggen, 1996; Harms et al., 1996) could well be monitored on-line by ICP-MS. Furthermore, with ICP-MS instruments capable of high resolving power especially with multi-collector detection, some nutritional tracer studies performed previously only by TI-MS isotope analysis (e.g., ^{42}Ca/^{43}Ca/^{44}Ca/^{48}Ca and ^{54}Fe/^{56}Fe/^{57}Fe) probably can be evaluated by ICP-MS (Crews et al., 1994). Moreover, quantification of metal binding molecules in nutrition is under-utilized. A major undeveloped topic in chemical speciation of mineral-binding forms is nutritional modeling.

The combination of stable isotope tracers with chemical speciation also is an unexplored field for nutritional modeling. Preliminary experiments with metal-binding proteins detected by ICP-MS indicate a reasonable likelihood that isotope ratios can be determined in separated species but with a sacrifice of precision. Sensitive, ICP – sector-field MS instruments give higher counts and better precision than quadrupole systems. It is probable that a satisfactory combination of accuracy, precision, and sensitivity for CE stable isotope tracer chemical speciation will require the high resolving power ICP-MS. Column artifacts, however, have yet to be overcome, and the resulting sensitivity may be insufficient for some metals. By including isotope tracers of specific chemical species in the modeling process, ICP-MS detection can be expected to extend the potential of biokinetic models.

ACKNOWLEDGMENT

Preparation of this paper was supported by the *ICP Information Newsletter*.

REFERENCES

Abrams SA. Clinical studies of mineral metabolism in children using stable isotopes. *J Ped Gast Nutr,* 1994, 19:151-163.

Alfassi ZB. Ed. *Determination of Trace Elements.* VCH: Weinheim, Germany. 1994.

Amarasiriwardena C; Krushevska A; Foner H; Argentine MD; Barnes RM. Sample preparation for inductively coupled plasma mass spectrometric determination of the zinc-70 to zinc-80 isotope ratio in biological samples. *J Anal At Spectrom,* 1992, 7:915-921.

Amarasiriwardena D; Durrant SF; Lásztity A; Krushevska A; Argentine MD; Barnes RM. Semiquantitative analysis of biological materials by inductively coupled plasma – mass spectrometry. *Microchem J,* 1997, 56:352-372.

Analytical Methods Committee. Report by the analytical methods committee evaluation of analytical instrumentation. Part X inductively coupled plasma mass spectrometers. *Analyst,* 1997, 122:393-408.

Bacon JR; Crain JS; McMahon AW; Williams JG. Atomic spectrometry updates – Atomic mass spectrometry. *J Anal At Spectrom,* 1996, 11:355R-393R.

Baker DR. *Capillary Electrophoresis.* John Wiley: New York. 1995.

Balcerzak M. Analytical methods for the determination of platinum in biological and environmental samples. *Analyst,* 1997, 122:67R-74R.

Barinaga CJ; Eiden GC; Alexander ML; Koppenaal DW. Analytical atomic spectroscopy using ion trap devices. *Fres J Anal Chem,* 1996, 355:487-493.

Barnes RM. Advances in inductively coupled plasma mass spectrometry: human nutrition and toxicology. *Anal Chimica Acta,* 1993, 283:115 - 130.

Barnes RM. Analytical plasma source mass spectrometry in biomedical research. *Fres J Anal Chem,* 1996, 355:433-441.

Barnes RM. Capillary electrophoresis and ICP-spectrometry: Status report. *Fres J Anal Chem,* 1997, in press.

Beary ES; Paulsen PJ; Jassie LB; Fassett JD. Determination of environmental lead using continuous-flow microwave digestion isotope dilution inductively coupled plasma mass spectrometry. *Anal Chem,* 1997, 69:758-766.

Begerow J; Dunemann L. Mass spectral interferences in the determination of trace levels of precious metals in human blood using quadrupole and magnetic sector field inductively coupled plasma mass spectrometry. *J Anal At Spectrom,* 1966, 11:303-306.

Begerow J; Turfeld M; Dunemann L. Determination of physiological platinum levels in human urine using magnetic sector field inductively coupled plasma mass spectrometry in combination with ultraviolet photolysis. *J Anal At Spectrom,* 1966, 11:913-916.

Begley IS; Sharp BL. Characterisation and correction of instrumental bias in inductively coupled plasma quadrupole mass spectrometry for accurate measurement of lead isotope ratios. *J Anal At Spectrom,* 1997, 12:395-402.

Bergdahl IA; Schütz A; Grubb A. Application of liquid chromatography–inductively coupled plasma mass spectrometry to the study of protein-bound lead in human erythrocytes. *J Anal At Spectrom,* 1996, 11:735-738.

Bird SM; Uden PC; Tyson JF; Block E; Denoyer E. Speciation of Selenoamino acids and organoselenium compounds in selenium-enriched yeast using high-performance liquid chromatography–inductively coupled plasma mass spectrometry. *J Anal At Spectrom,* 1997, 12:785-788.

Briche CSJ; Taylor PDP; De Bièvre P. Measurement of platinum isotope amount ratios by negative ionization mass spectrometry using a thermionic quadrupole mass spectrometer. *Anal Chem,* 1997, 69:791-793.

Buckley WT. Application of compartmental modeling to determination of trace element requirements in humans. *J Nutr,* 1996, 126:2312S-2319S.

Burgoyne TW, Hieftje GM; Hites RA. Design and performance of a plasma-source mass spectrograph. *J Am Soc Mass Spec,* 1996, 8:307-318.

Cai J; Henion J. Capillary electrophoresis–mass spectrometry, *J Chromatog,* A, 1995, 703:667-692.

Caroli S. Ed. Element speciation in bioinorganic chemistry, in: Chemical Analysis, Vol. 135. Winefordner JD; Ed. John Wiley: New York, NY. 1996.

Christodoulou J; Kashani M; Keohane BM; Sadler PJ. Determination of gold and platinum in the presence of blood plasma proteins using inductively coupled plasma mass spectrometry with direct injection nebulization. *J Anal At Spectrom,* 1966, 11:1031-1035.

Cornelis R. Involvement of analytical chemistry in chemical speciation of metals in clinical samples. *Ann Clin Lab Sci,* 1996, 26:252-263.

Crews HM; Ducros V; Eagles J; Mellon FA; Katenmayer P; Luten JP; McGaw BA. Mass spectrometric methods for studying nutrient mineral and trace element absorption and metabolism in humans using stable isotopes. A review. *Analyst,* 1994, 119:2491-2514.

Cromwell EF; Arrowsmith P. Novel multichannel plasma-source mass spectrometer. *J Am Soc Mass Spectrom,* 1996, 7:458-466.

Delves HT. Biological monitoring of trace elements to indicate intake and uptake from foods and beverages, in: *Biomarkers in Food Chemical Risk Assessment.* Crews HM; Hanley AB; Eds. The Royal Society of Chemistry: Cambridge, UK. 1995. pp. 27-38.

Delves HT; Sieniawska CE. Simple method for the accurate determination of selenium in serum by using inductively coupled plasma mass spectrometry. *J Anal At Spectrom,* 1997, 12:387-389.

Denoyer DR; Jacques D; Debrah E; Tanner SD. Determination of trace elements in uranium: Practical benefits of a new ICP-MS lens system. *At Spectrom,* 1995, 16:1-6.

Diemer J; Heumann KG. Bromide/bromate speciation by NTI-IDMS and ICP-MS coupled with ion exchange chromatography. *Fres J Anal Chem,* 1997, 357:74-79.

Donard OFX; Lobinski R. Plasma spectrometry and molecular information. *J Anal At Spectrom,* 1996, 11:871-876.

Eiden GC; Barinaga CJ; Koppenaal, DW. Plasma source ion trap mass spectrometry: enhanced abundance sensitivity by resonant ejection of atomic ions. *J Am Soc Mass Spectrom,* 1996, 7:1161-1171.

Eiden GC; Barinaga CJ; Koppenaal DW. Selective removal of plasma matrix ions in plasma source mass spectrometry, *J Anal At Spectrom,* 1996, 11:317-322.

Evans EH; Giglio JJ; Castillano TM; Caruso JA. *Inductively Coupled and Microwave Induced Plasma Sources for Mass Spectrometry.* The Royal Society of Chemistry: Cambridge, UK. 1995.

Feldmann J. Ion chromatography coupled with inductively-coupled argon plasma mass spectrometry: Multielement speciation as well as on-line matrix separation technique. *Anal Comm,* 1996, 33:11-13.

Fischer PWF; L'Abbé MR; Coclell KA; Gibson RS; Eds. *Trace Elements in Man and Animals - 9. Proceedings of the Ninth International Symposium on Trace Elements in Man and Animals.* NRC Research Press: Ottawa, Canada. 1997. 667 p.

Gardiner PHE; Delves HT. The chemical speciation of trace elements in biomedical specimens: Analytical techniques. Techniques in instrumental analytical chemistry, in: *Trace Element Analysis in Biological Specimens, Vol. 15.* Chapter 9, pp. 185-212. 1994.

Goodrum, PE; Diamond GL; Hassett JM; Johnson DL. Monte Carlo modeling of childhood lead exposure: Development of a probabilistic methodology for use with the USEPA IEUBK model for lead in children. *Human Ecol Risk Assess,* 1996, 2(4):681-708.

Gray AL. Inductively coupled plasma mass spectrometry in maturity – What problems remain? *Anal Proc Anal Comm,* 1994, 31:371 - 375.

Gulson, BL; Pisaniello D; McMichael AJ; Mizon KJ; Korsch MJ; Luke C; Ashbolt R; Pederson DG; Vimpani G; Mahaffey KR. Stable lead isotope profiles in smelter and general urban communities: A comparison of environmental and blood measures. *Environ Geochem Health,* 1966, 18:147-163.

Halliday AN; Lee DC; Christensen JN; Walder AJ; Freedman PA; Jones CE; Hall CM; Yi W; Teagle D. Recent developments in inductively coupled plasma magnetic sector multiple collector mass spectrometry. *Inter J Mass Spectrom Ion Proc,* 1995, 146/147: 21-33.

Harms AV; van Elteren JT; Claessens HA. Technetium speciation: Non-size effects in size-exclusion chromatography. *J Chromatog A,* 1996, 755:219-225.

Hieftje, G. The future of plasma spectrochemical instrumentation. *J Anal At Spectrom,* 1996, 11:613-621.

Hirata T. Lead isotope analyses of NIST standard reference materials using multiple collector inductively coupled plasma mass spectrometry coupled with a modified external correction means for mass discrimination effect. *Analyst,* 1996, 121:1407-1411.

Inskip MJ; Franklin CA; Baccanale CL; Manton WI; O'Flaherty EJ; Edwards CMH; Blenkinsop JB; Edwards EB. Measurement of the flux of lead from bone to blood in a nonhuman primate (*Macaca fascicularis*) by sequential administration of stable lead isotopes. *Fund Appl Tox,* 1966, 33:235-245.

Ketterer ME; Guzowski Jr JP. Isotope ratio measurements with elemental-mode electrospray mass spectrometry. *Anal Chem,* 68:883-887.

Kim N; Atallah MT; Amarasiriwardena C; Barnes RM. Pectin with low molecular weight and high degree of esterification increases absorption of ^{58}Fe in growing rats. *J Nutr,* 1996, 126:1883-1890.

Klinkenberg H; Van Borm W; Souren F. A theoretical adaptation of the classical isotope dilution technique for practical routine analytical determinations by means of inductively coupled plasma mass spectrometry. *Spectrochim Acta,* 1996, B51:139-153.

Landers JP. Clinical capillary electrophoresis. *Clin Chem,* 1995, 41(4):495-509.

Lásztity A; Kotrebai M; Barnes RM. Inductively coupled plasma spectrometry determination of marker elements for childhood soil ingestion. *Microchem J,* 1996, 54:452-464.

Lásztity A; Krushevska A; Kotrebai M; Barnes RM; Amarasiriwardena D. Arsenic determination in environmental, biological and food samples by inductively coupled plasma mass spectrometry. *J Anal At Spectrom,* 1995, 10:505-510.

Laughlin NK; Smith DR; Fowler BA; Flegal AR; Luck ML; The rhesus monkey as an animal model to evaluate the efficacy of succimer (Chemet®) chelation therapy. *Proceedings of the 26th Annual Meeting of the Society of Toxicology.* Cincinnati, OH. 1997. Paper 586.

Lee DC; Halliday AN. Precise determinations of the isotopic compositions and atomic weights of molybdenum, tellurium, tin and tungsten using ICP magnetic sector multiple collector mass spectrometry. *Inter J Mass Spectrom Ion Proc,* 1995, 146/147: 35-46.

Lehmann R; Liebich HM; Voelter W. Applications of capillary electrophoresis in clinical chemistry – Developments from preliminary trials to routine analysis. *J Cap Electroph,* 1996, 3:89-110.

Lindeber J. Capillary electrophoresis in food analysis. *Food Chem,* 1996, 55:73-94,95-101.

Lobinski R. Elemental speciation and coupled techniques. *Appl Spectrosc,* 1997, 51:260A-278A.

Lobinski R; Marczenko Z. Spectrochemical trace analysis for metals and metalloids, in: *Wilson & Wilson's Comprehensive Analytical Chemistry, Vol. 30.* Weber SG; Ed. Elsevier Science: Amsterdam, The Netherlands. 1996.

Lu Q; Barnes RM. Current status of capillary electrophoresis and inductively coupled plasma spectrometry interface and applications to metal-binding proteins. *Spectrochim Acta B,* 1997, submitted.

Lu Q; Bird SM; Barnes RM. Interface for capillary electrophoresis and inductively coupled plasma mass spectrometry. *Anal Chem,* 1995, 67:2949-2956.

Luong ET; Houk SR; Serfass RE. Chromatographic isolation of molybdenum from human blood plasma and determination by inductively coupled plasma mass spectrometry with isotope dilution. *J Anal At Spectrom,* 1997, 12:703-708.

Lyon TDB; Fletcher S; Fell GS; Patriarca M. Measurement and application of stable copper isotopes to investigations of human metabolism. *Microchem J,* 1996, 54:236-245.

Magnuson ML; Creed JT; Brockhoff CA. Speciation of arsenic compounds by ion chromatography with inductively coupled plasma mass spectrometry detection utilizing hydride generation with a membrane separator. *J Anal At Spectrom,* 1996, 11:893-898.

Magnuson ML; Creed JT; Brockhoff CA. Speciation of arsenic compounds in drinking water by capillary electrophoresis with hydrodynamically modified electroosmotic flow detected through hydride generation inductively coupled plasma mass spectrometry with a membrane gas-liquid separator. *J Anal At Spectrom,* 1997, 12:689-695.

Mahoney PP; Ray SJ; Hieftje GM. Time-of-flight mass spectrometry for elemental analysis. *Appl Spectrosc,* 1997, 51:16A-28A.

Mellon FA; Eagles J; Fox TE; Fairweather-Tait SJ. *Anal Chim Acta,* 1993, 283: 190-198.

Mestek O; Suchánek M; Vodicková Z; Zemanová B; Zíma T. Comparison of the suitability of various atomic spectroscopic techniques for the determination of selenium in human whole blood. *J Anal At Spectrom,* 1997, 12:85-89.

Michalke B; Lustig S; Schramel P. Analysis of the stability of platinum-containing species in soil samples using capillary electrophoresis interfaced on-line with inductively coupled plasma mass spectrometry. *Electrophoresis,* 1977, 18:196-201.

Michalke B; Schramel P. Coupling of capillary electrophoresis with ICP-MS for speciation investigations. *Fres J Anal Chem,* 1997, 357:594-599.

Moens L; Dams R. NAA and ICP-MS: A comparison between two methods for trace and ultra-trace element analysis. *J Radioanal Nucl Chem, Articles,* 1995, 192:29-38.

Moens L; Verrept P; Dams R; Greb U; Jung G; Laser B. New high resolution inductively coupled plasma mass spectrometry technology applied for the determination of V, Fe, Cu, Zn, and Ag in human serum *J Anal At Spectrom,* 1994, 9:1075-1078.

Muñoz Olivas R; Donard OFX; Gilon N; Potin-Gautier M. Speciation of organic selenium compounds by high-performance liquid chromatography – Inductively coupled plasma mass spectrometry in natural samples. *J Anal At Spectrom,* 1996, 11:1171-1176.

Newman A. Elements of ICPMS. *Anal Chem News Features,* 1996, 69:46A-51A.

Newman A. The precise world of isotope ratio mass spectrometry. *Anal Chem News Features,* 1996, 69:373A-377A.

Nixon DE; Moyer TP. Routine clinical determination of lead, arsenic, cadmium, and thallium in urine and whole blood by inductively coupled plasma mass spectrometry. *Spectrochim Acta,* 1966, B51:13-25.

Nuttall, KL; Gordon, WH; Ash KO. Inductively coupled plasma mass spectrometry for trace element analysis in the clinical laboratory. *Ann Clin Lab Sci,* 1995, 25(3):264-271.

Örnemark U; Taylor PDP; De Bièvre P. Certification of the rubidium concentration in water materials for the international measurement evaluation programme (IMEP) using isotope dilution inductively coupled plasma mass spectrometry. *J Anal At Spectrom,* 1997, 12:567-572.

Owen LMW; Crews HM; Hutton RC; Walsh A. Preliminary study of metals in proteins by high-performance liquid chromatography – Inductively coupled plasma mass spectrometry using multi-element time-resolved analysis. *Analyst,* 1992, 117:649-655.

Paschal DC; Caldwell KL: Ting BG. Determination of lead in whole blood using inductively coupled argon plasma mass spectrometry with isotope dilution. *J Anal At Spectrom,* 1995, 10:367-370.

Patriarca M; Lyon TD; McGaw B; Fell GS. Determination of selected nickel isotopes in biological samples by inductively coupled plasma spectrometry with isotope dilution. *J Anal At Spectrom,* 1996, 11:297-302.

Patriarca M. The contribution of inductively coupled plasma mass spectrometry to biomedical research. *Microchem J,* 1996, 54:263-271.

Pergantis SA; Winnik W; Benowski D. Determination of ten organoarsenic compounds using microbore high-performance liquid chromatography coupled with electrospray mass spectrometry - mass spectrometry. *J Anal At Spectrom,* 1997, 12:531-536.

Prichard E; MacKay GM; Points J; Eds. *Trace Analysis: A Structured Approach to Obtaining Reliable Results.* The Royal Society of Chemistry: Cambridge, UK. 1996.

Quétel CR; Thomas B; Donard OFX; Grousset FE. Factorial optimization of data acquisition factors for lead isotope ratio determination by inductively coupled plasma mass spectrometry. *Spectrochim Acta,* 1997, B52:177-187.

Quevauviller P. Atomic spectrometry hyphenated to chromatography for elemental speciation: Performance assessment within the standards, measurements and testing programme (Community Bureau of References) of the European Union. *J Anal At Spectrom,* 1996, 11:1225-1231.

Rayman, MP; Abou-Shakra, FR; Ward, NI. Determination of selenium in blood serum by hydride generation inductively coupled plasma mass spectrometry. *J Anal At Spectrom,* 1996, 11:61-68.

Seaton CL; Osterloh J; Smith DR. The skeleton as an endogenous source of Pb exposure and the effects of bone Pb and therapeutic treatments on bone loss due to osteopenia. *Proceedings of the 26th Annual Meeting of the Society of Toxicology.* Cincinnati, OH. 1997. Paper 1713.

Siva Subramanian KNS; Wastney ME; Eds. *Kinetic Models of Trace Element Metabolism During Development.* CRC Press: Boca Raton, Florida. 1965.

Smith SM; Wastney ME; Nyquist LE; Shih C-Y; Wiesmann H; Nillen JE; Lane HW. Calcium kinetic with microgram stable isotope does and saliva sampling. *J Mass Spec,* 1996, 31:1265-1270.

St. Claire, III RL. Capillary electrophoresis. *Anal Chem,* 1966, 68:569R-586R.

Subramanian KS. Determination of metals in biofluids and tissues: sample preparation methods for atomic spectroscopic techniques. *Spectrochim Acta,* 1996, B51:291-319.

Szpunar J; Bettner J; Robert M; Chassaigne H; Cammann K; Lobinski R; Donard OFX. Validation of the determination of copper and zinc in blood plasma and urine by ICP MS with cross-flow and direct injection nebulization. *Talanta,* 1997, 44:1389-1396.

Tanner SD. Characterization of ionization and matrix suppression in inductively coupled 'cold' plasma mass spectrometry. *J Anal At Spectrom,* 1995, 10:905-921.

Tanner SD; Holland, JG; Eds. *Plasma Source Mass Spectrometry Developments and Applications.* The Royal Society of Chemistry: Cambridge, UK. 1997.

Taylor A. Detection and monitoring of disorders of essential trace elements. *Ann Clin Biochem,* 1996, 33:486-510.
Taylor A; Branch S; Crews HM; Halls DJ; Owen LMW; White M. Atomic spectrometry update – Clinical and biological materials, foods and beverages. *J Anal At Spectrom,* 1997, 12:119R-221R.
Ting BG; Paschal DC; Caldwell KL. Determination of thorium and uranium in urine with inductively coupled plasma mass spectrometry. *J Anal At Spectrom,* 1996, 11:339-342.
Vanhaecke F; Moens L; Dams R; Papadakis I; Taylor P. Applicability of high-resolution ICP-mass spectrometry for isotope ratio measurements. *Anal Chem,* 1997, 69:268-273.
Vanhaecke F; Moens L; Dams R; Taylor P. Precise measurement of isotope ratios with a double-focusing magnetic sector ICP mass spectrometer. *Anal Chem,* 1996, 68:567-569.
Vanhaecke F; Riondato J; Moen L; Dams R. Non-spectral interferences encountered with a commercially available high resolution ICP-mass spectrometer. *Fres J Anal Chem,* 1996, 355:397-400.
Vanhoe H; Dams R; Versieck J. Use of inductively coupled plasma mass spectrometry for the determination of ultra-trace elements in human serum. *J Anal At Spectrom,* 1994, 9:23-31.
Veillon C; Patterson KY; Moser-Veillon PB. Digestion and extraction of biological materials for zinc stable isotope determinations by inductively coupled plasma mass spectrometry. *J Anal At Spectrom,* 1996, 11:727-730.
Verbeke K; Verbruggen A. Usefulness of fast protein liquid chromatography as an alternative to high performance liquid chromatography of ^{99m}Tc-labelled human serum albumin preparations. *J Pharm Biomed Anal,* 1996, 14:1209-1213.
Vighi M; Calamari D. Prediction of the environmental fate of chemicals. *Annli dell'Istituto Superiore di Sanità,* 1993, 29:209-223.
Wan CC; Chen CS; Jian SJ. Determination of mercury compounds in water samples by liquid chromatography–inductively coupled plasma mass spectrometry with an in situ nebulizer/vapor generator. *J Anal At Spectrom,* 1997, 12:683-687.
Wastney ME; Angelus P; Barnes RM; Siva Subramanian KNS. Zinc kinetics in preterm infants: A compartmental model based on stable isotope data. *Am J Phys,* 1996, 40:R1452-R1459.
Wolf RE; Denoyer ER. Design and performance criteria for a new ICP-MS for environmental analysis. *At Spectrom,* 1995, 16:22 - 27.
Yoshinaga J. Analysis of trace elements in clinical samples by plasma ion source MS. *Bunseki Kagaku,* 1995, 44:895-903.

ACCELERATOR MASS SPECTROMETRY AS A BIOANALYTICAL TOOL FOR NUTRITIONAL RESEARCH

John S. Vogel[1] and Kenneth W. Turteltaub[2]

[1]Center for Accelerator Mass Spectrometry
[2]Biology and Biotechnology Research Program
Lawrence Livermore National Laboratory
7000 East Avenue, L-397
Livermore, CA 94551

ABSTRACT

Accelerator Mass Spectrometry is a mass spectrometric method of detecting long-lived radioisotopes without regard to their decay products or half-life. The technique is normally applied to geochronology, but is also available for bioanalytical tracing. AMS detects isotope concentrations to parts per quadrillion, quantifying labeled biochemicals to attomole levels in milligram-sized samples. Its advantages over non-isotopic and stable isotope labeling methods are reviewed and examples of analytical integrity, sensitivity, specificity, and applicability are provided.

INTRODUCTION

Validation of nutritional modeling is derived from the location and quantification of nutrients or their metabolites in isolatable body pools. The analytical method used in this process must be specific and sensitive. Specificity provides a confident signal above the complex biological matrix, while sensitivity allows tracing nutrients at relevant levels that do not saturate or perturb metabolic pathways. The method should also depend on an intrinsic property of the nutrient that is not affected by metabolism or the biological context.

Isotopic labeling emerged 60 years ago with the first mass spectrometers and the new understanding of isotopic masses (Mellon and Sandtröm, 1996). The discovery of radiocarbon fifty years ago and the intense development of decay counting techniques introduced radioisotope labeling as the one method which combines the maximal properties of specificity and sensitivity in an intrinsic label. The more recent development of biochemical separation techniques that depend on molecular structure, such as luminescence or fluorescence, permit a less specific quantitation . High sensitivity is possible because specificity comes from the known response of the metabolites in the separation process. These less general methods became even more prevalent as the effort to remove radioactive materials from common laboratory use gained popularity (Garman, 1997). However, no non-isotopic labeling method combines the specificity, sensitivity, and integrity across all metabolites of a wide variety of nutrients as well as radioisotope labeling. In particular, isotopic labeling can be more independent of structure than the newer methods, an important property if the identity and structure of all possible metabolites is still unknown for the compound under study.

Radioisotopes are distinctive and specific because they are very rare in natural materials. Any radioisotope-labeled compound has a very high abundance-to-background ratio in biological systems (but the poor signal to noise of the isotope detector may mask this fact). For example, ^{14}C has a natural level of abundance due to cosmic radiation at 1.2 parts per trillion. The "rare" stable isotope of carbon, ^{13}C, has a natural abundance of 1.1%. A part per million concentration of a ^{13}C-labeled compound (assume 200 g/mol) will change the concentration of ^{13}C by only 0.3 per mil, measurable under good conditions using an excellent mass spectrometer. The same material labeled with ^{14}C changes the concentration of that isotope in the biological sample by a factor of 3 million. Efficient detection of radioisotopes is a key to using this specificity.

However, radioisotopes that have a half-life short enough to be efficiently detected by their decay produce high radiation hazards in the laboratory. Compounds labeled with such isotopes (^{32}P is one example) must be locally synthesized to take advantage of the high count rates before the activity decays. Isotope handling and radiation exposure are minimized by simple synthesis processes, usually direct exchange. Compounds labeled through exchange can lose the label through exchange, resulting in loss of the desired integrity of the isotope signal. It is this series of detriments, all arising from the need for short and easily detectable half-lives, that has fueled the move away from radioisotope labeling.

Radioisotopes that have longer half-lives, such as ^{14}C(5730 years), are inefficiently detected by decays. Measuring even 0.1% of the ^{14}C in a sample requires uninterrupted counting for 8.3 years [0.1% x 5730 years / ln(2)]. The sensitivity and specificity of the radioisotope label are thus wasted in detecting decays. In the late 1970s and throughout the '80s a mass spectrometric method for directly detecting ^{14}C and other long-lived isotopes was developed in low-energy nuclear physics laboratories. The technique, accelerator mass spectrometry (AMS), is primarily used in radiocarbon dating for geochronology studies, but has recently been applied by several AMS facilities to the quantification of long-lived isotopes for biochemical labeling and tracing. We recently reviewed the applications of the technique across several disciplines (Vogel et al., 1995), and discussed its use in biological tracing (Vogel and Turteltaub, 1992).

METHODS AND EXAMPLES OF ^{14}C ANALYSIS

Briefly, AMS is a type of tandem isotope ratio mass spectrometry in which a low energy (tens of keV) beam of negative atomic and small molecular ions are mass analyzed to 1 AMU resolution (mass 14, for example). These ions are then attracted to a gas or solid foil collision cell that is held at VERY high positive potential (2–10 megaVolts). In passing through the foil or gas, two or more electrons are knocked from the atomic or molecular ion, making it positive in charge. These positive ions then accelerate away from the positive potential to a second mass analyzer where an abundant charge state (4+ in the case of a 7MV dissociation) is selected. The loss of 4 or more electrons (to a charge state of 3+ or more positive) in the collision cell destroys all molecules, leaving only nuclear ions at relatively high energies (20–100 MeV) that can be individually and uniquely identified by several properties of their interaction with detectors. These are the core "tricks" of AMS: molecular dissociation to remove molecular isobars, and ion identification to distinguish nuclear isobars. Beyond these two fundamental properties of AMS, special tricks unique to each element must also be developed: ^{14}N is separated from ^{14}C in the ion source because nitrogen does not make a negative ion, ^{63}Cu is strongly suppressed in ^{63}Ni samples through a volatile carbonyl separation since copper does not form such a gas, ^{10}B is absorbed more rapidly than ^{10}Be in a foil before the final detector because the absorbing interaction is dependent on the square of the atomic number, etc.

Because AMS was initially developed for the difficult task of geochronology, in which the highest level of the isotope is a function of its natural production, the sensitivity of AMS stretches from parts per billion to parts per quadrillion. At these low concentrations, even the spectrometry sectors specified above are insufficient. A diagram of the Lawrence Livermore National Laboratory (LLNL) AMS (Figure 1) shows that several magnetic and electric sectors are needed to reduce ion counts to low enough rates that the ion identification techniques can operate.

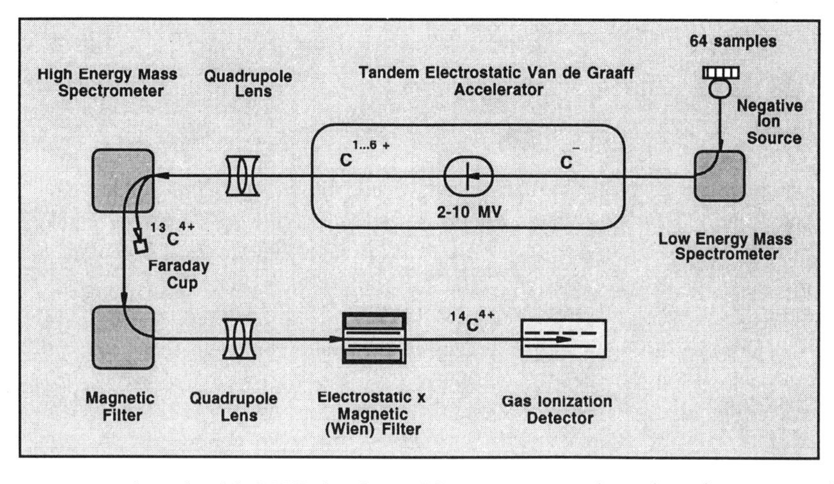

Figure 1. A schematic of the LLNL Accelerator Mass spectrometer shows the main components from the negative ion source, through the low energy spectrometer, the collision cell at megaVolt potential, multi-sectior high energy spectrometer, to one of the several final ion identification detectors.

Most AMS systems are hardly "benchtop" or even "laboratory-sized." The LLNL AMS facility, for example, fills a 2000 sq. ft. area. This is a general purpose research AMS system, however, that is being used for analyses from mass 3 to mass 240. There are now much smaller spectrometers being designed specifically for biological tracing applications with tritium and radiocarbon. These have a smaller, but still finite, footprint approaching 100 sq. ft. (Hughey et al., 1997; Mous et al., 1997; Suter et al., 1997). The present cost of an AMS spectrometer does not yet scale with its footprint, with the custom-built LLNL system costing about $7 million to assemble, and the new small spectrometers expected to cost ≈$1 million for the near term.

AMS quantification is not tied to a radioactive decay rate, but the desired counting precision still dictates the measurement time, as with decay counting. The measurement time for a given precision is inversely proportional to the intensity of the ion beam derived from the sample material. A cesium fast-atom-bombardment ion source produces maximum (100–300 μAmp) negative carbon atomic ions from a thermally and electrically conducting material such as graphite. Using graphite, 10,000 counts of ^{14}C are obtained from 0.1–1 mg of graphite in only a few minutes. Multiple measurements interspersed with standards allow us to obtain the ^{14}C-concentrations of 15 to 25 samples per hour.

All biological samples are combusted to CO_2 in individual sealed tubes, and the CO_2 is reduced to graphite on an iron or cobalt catalyst in a second sealed tube, which also contains zinc metal and titanium hydride (Vogel, 1992). This process is most reliable if the sample contains approximately 1 mg of carbon as CO_2. Smaller biological samples, such as purified DNA aliquots, or individual HPLC peaks are mixed with a "carrier" containing 1 mg of carbon that is low in ^{14}C. AMS is essentially an isotope ratio mass spectrometer. Careful carbon inventory of the sample prior to combustion is required to unfold the resultant ratio and deduce the total amount of label found in the original sample.

Further details of experiment design and methods for ^{14}C-AMS are found in a number of papers by the Molecular Toxicology research group and the Center for AMS at LLNL (Buchholz et al., 1997; Creek et al., 1994; Frantz et al., 1995; Turteltaub et al., 1992; Kautiainen et al., 1997). The first uses of ^{14}C-AMS in human toxicology and metabolism were also performed by the Center for AMS with the LLNL Health and Ecological Assessment Division (Bogen et al., 1995; Bogen et al., 1996; Keating et al., 1995; Williams et al., 1992). An important demonstration of AMS in human breath analysis comes from Lund University (Stenstrom et al., 1997)

AMS PROPERTIES

Chemicals can be synthesized with long-lived isotope labels in a stable and significant molecular location. Most of the compounds used at LLNL have been ring-labeled or have had similarly stable incorporation. The loss of label due to exchange can be all but eliminated. Chemical specialists can make the compound at a laboratory properly equipped for complex synthesis and safety, because the slow decay of the label means that little is lost during transportation and storage. Thus, long-lived isotopes provide higher integrity to the isotope signal, but this integrity is useless without the sensitivity of AMS.

AMS sensitivity for radiocarbon is emphasized by Figure 2 which shows the isotope ratio of $^{14}C/C$ over the range available through AMS. Units of ^{14}C concentration that may be more familiar to chemists (mol), health scientists (Curies), and laboratory biologists (dpm) are also shown. Because AMS came out of the ^{14}C-dating community, the unit "Modern" is also introduced. This is the concentration of ^{14}C that would be present in the quiescent atmosphere due only to cosmic radiation. Two anthropogenic effects have had profound effects on the atmospheric concentration of ^{14}C in the past century. The burning of fossil fuels to power the industrial revolution increased the amount of ^{14}C-free CO_2 in the atmosphere from the mid-1800s. The atmospheric testing of nuclear weapons then greatly increased the amount of ^{14}C in the atmosphere, doubling its concentration by the year 1963. This huge excess of ^{14}C has been drawn out of the atmosphere and into the oceans with an uptake half-life of 15 years since the atmospheric test ban treaty was signed in 1964. For this reason, the current atmosphere has radiocarbon equivalent to 1.1 Modern. A Modern is defined as 13.56 dpm per gram carbon and is equivalent to 97.8 attomoles (amoles) of ^{14}C/mg carbon or 6.11 femtoCurie/mg (fCi/mg) carbon.

Radiocarbon dating with AMS extends from ^{14}C concentrations equal to Modern back to approximately 50,000 "radiocarbon" years, which corresponds to 0.1% Modern. The less sophisticated processing of samples to graphite for biochemical measurements gives a more conservative lower limit at 1% Modern, or 1 amol ^{14}C/mg carbon. The upper limit of AMS ^{14}C counting is related to specifics of the spectrometer, especially the ion source and the ion identification counter. Count rates in the latter begin to blur the distinction between ion signals at 1000 Modern, and the direst consequence of pushing the AMS envelope is a badly contaminated ion source that must be dismantled and laboriously cleaned. This occurs at thousands Modern and is to be strictly avoided through careful design and testing of experimental protocols. In practice, AMS experiments are designed to provide samples between 1% and 100 Modern, or 1 amol to 10 fmol of isotope label per sample. Measurements are done to 3-5% precision as measured by the standard deviation of 3 or more measurements of the ^{14}C concentration. AMS is one of the few methods for quantitating molecules precisely over this range. Radiocarbon dating is a much more stringent application of AMS, and an International Intercomparison has shown that AMS is more precise than liquid scintillation, and as accurate as CO_2 proportional counting. (Scott 1990)

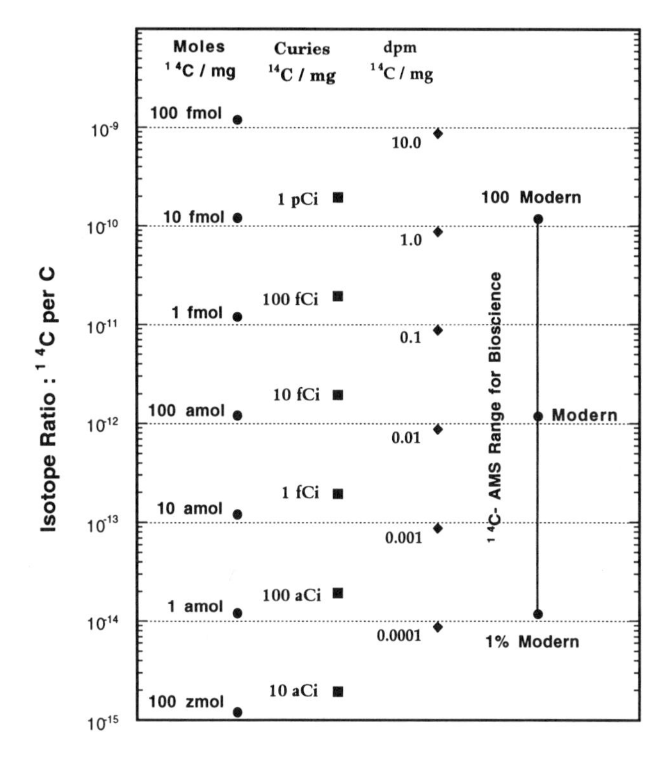

Figure 2. The isotope ratio of radiocarbon to stable carbon is presented in comparison to other measures of radiocarbon concentration, including the unit "Modern" which approximates the natural level of radiocarbon in living materials. Contamination of the equipment limits AMS samples to 1000 Modern or less.

The integrity of the label, the high sensitivity of AMS, and precision in measurement lead fundamentally to the high specificity of the method. The dosing compound is needed in such low quantities that the entire dose is easily made radio-pure through analytical HPLC. Low doses of stably-labeled compounds do not saturate metabolism and do not randomly exchange the isotope so that the presence of the isotope is a specific indicator of the chemical or its derivatives.

Figure 3 shows the chromatogram of rat urine integrated over a 24-hour collection after dosing at 100 pg/kg of methyl imidazo quinoxaline (2-amino-3,8-dimethyl-imidazo[4,5-*f*]quinoxaline or MeIQx) labeled with ^{14}C at 7.24 mCi/mmol. In this and the following discussion, MeIQx is used as an example of tracing metabolism and its effects.

While MeIQx is a toxin rather than a nutrient, it is used here as an example of a specific compound whose uptake is at least dietary in origin.

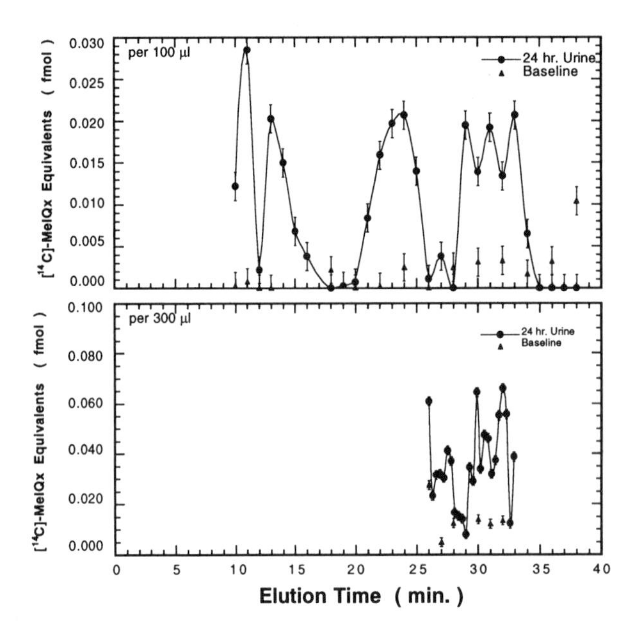

Figure 3. An HPLC analysis of urine from a rat obtained over 24 hours after an exposure to 100 pg/kg of [^{14}C]-MeIQx (methyl imidazo quinoxaline), a potent mutagen found in cooked meats. This dose corresponds to the amount ingested by a normal person in a single bite of a hamburger. The unidentified and unexpected metabolites after 25 minutes are seen to be real by the repetition of the HPLC in the lower frame at a higher time resolution. Sensitivity to 10 amol, with 2 amol precision, is demonstrated.

This data arose from a study of dietary mutagens, which form a focus of the molecular toxicology studies performed at LLNL (Turteltaub et al., 1990; Frantz et al., 1995; Turteltaub et al., 1992). The total dose in this metabolic study corresponds to 12 fmol (1pCi) of ^{14}C, 2.6 fmol of which appears in the metabolites visible in the upper trace of Figure 3. The first 3 peaks are known metabolites from high dose studies, but the O-sulfamate that appeared at 19 minutes is not present at this dose. Four (or maybe six) unexpected and undetermined new metabolites are seen at >26 minutes elution. The detectability of such unexpected metabolites is an inherent strength of isotopic labeling in metabolism research. The initial HPLC elution at 1 ml/min with 1 minute collection times was repeated at 20 second collection times at >26 minutes as shown in the lower part of Figure 3. This was done to confirm the suspicious up-down behavior seen in the late peaks. Clearly, the new metabolite profiles are real, perhaps even more complex, and stand out well over baseline ^{14}C levels that are also shown in both traces of the Figure. Baseline sensitivity from our HPLC is approximately 3 ± 2.5 amol per ml.

HPLC fractions such as these have too little carbon for direct processing to graphite, so 1 mg or more carbon is added to the dried eluents in the form of 50 μl of 33.3 mg/ml tributyrin in methanol. The methanol is then removed in a vacuum concentrator. These procedures have been found to be primary sources of random contamination of individual samples, especially in the vacuum concentrator. The baseline sensitivity for HPLC separation will fall down to the 100s of zeptomoles when an interface to take HPLC eluent directly to the ion source, without dilution and reduction, is completed. Specificity of quantitation at very low doses is assured with AMS measurement.

The comfortable range for AMS quantification spans four orders of magnitude, but is easily expanded upward in experimental design by dilution of the labeled compound with unlabeled equivalent. This dilutes the specific activity of the dosing solution, but the sensitivity is so great that even low specific activity can be traced in biological systems. Figure 4 demonstrates the range of AMS quantitation with the compound at a fixed specific activity. This data represents the tissue concentrations found in rat organs after 42 days of sub-chronic dosing with labeled MeIQx impregnated into the rat chow. Three animals are represented at each dose, with error bars equal to the standard deviation of the separate measurements. This standard deviation is smaller than the data point in some instances. One surprise from the development of laboratory methods for AMS quantitation is that the procedures must be done so carefully to avoid accidental contamination at the fCi level, that many expected "natural" variations, such as those between animals, are greatly reduced. The high sensitivity of AMS demands procedures be scrutinized beyond just "Good Laboratory Practice" to "Paranoid Laboratory Practice" (PLP).

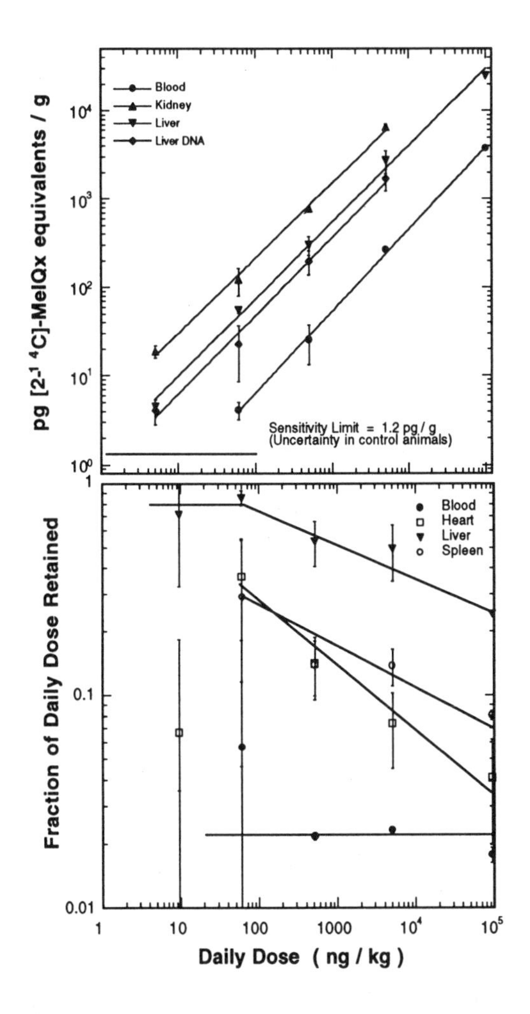

Figure 4. The quantity of [^{14}C]-MeIQx and its metabolites in rat tissues after 42 days of sub-chronic exposure in daily chow is shown as a function of dose. The upper frame indicates a linear response as a 45° line on a log-log plot. The data are shown as the average response of 3 rats at each dose. The sensitivity limit was derived from ^{14}C measured in control animals. Dividing the tissue data by the applied dose shows a non-linear response in the lower frame at a greater sensitivity than can be seen in log plot.

The value of PLP is seen in the physiological range that is available under AMS measurement. The high dose point was immeasurable for the kidneys because the tissues were well above 1000 Modern. However, the less affected organs from the same rats were quantified at two and sometimes three orders of magnitude lower in ^{14}C concentrations. It is impossible to overemphasize how difficult such clean dissections and fraction definitions are to perform while maintaining the distinct physiological signal.

The upper plot of Figure 4 shows the usual log-log plot of tissue dose response. Linearity of dose response is difficult to recognize with such scales. In order to test for linearity, the data in the upper plot has been replotted as fraction of the daily applied dose in the lower plot. The uncertainties in the data become larger at lower dose due to the division, but remain significant enough to recognize a distinct non-linear response in the retention of MeIQx. A truly linear response would produce horizontal lines in the lower plot. However, we see that the tissues, except for possibly the blood, retain higher fractions of the consumed dose at low doses. This is perhaps the opposite of what might be expected for unsaturated metabolic responses, and certainly the opposite of the desired response to toxins!

APPLICABILITY TO HUMAN STUDY

AMS can reinvigorate the use of isotope labeled chemicals, even in this day of intense development of non-radioactive tracing using stable isotopes (Mellon and Sandtröm, 1996) or fluorescent tagging (Garman, 1997). Despite the very common and real concern over radioactive waste and handling, a major impediment to use of radioactive isotopes, even at the low levels attained with AMS, is the exposure of human subjects to radiation. The risks due to radiation within experimental subjects can only be judged in relation to other, often unavoidable radiation, exposures. The least avoidable exposure is that due to the cosmic rays and their products within the atmosphere. Only life several meters underground escapes this form of radiation, which deposits a dose of approximately 300 μSievert per year per person at the latitudes of the continental United States. A Sievert is a deposited energy equivalent to a joule per kilogram. Each ^{14}C decay releases an average of 8 picojoules per decay. Figure 5 shows the radiation exposure received by a 70 kg research subject as a function of the elimination mean life of the ^{14}C-labeled compound. An administered dose was calculated that produced a peak concentration of ^{14}C label in the blood that was equal to the natural ^{14}C, providing an easily measured blood level of 2 Modern, generally about 1 hour after dosing (assuming a 0.2 hr uptake time constant). This level of ^{14}C also provided an easily distinguished signal 16% above natural ^{14}C at two mean elimination lives from dosing. Thus, the calculated initial ^{14}C dose shown in the Figure is

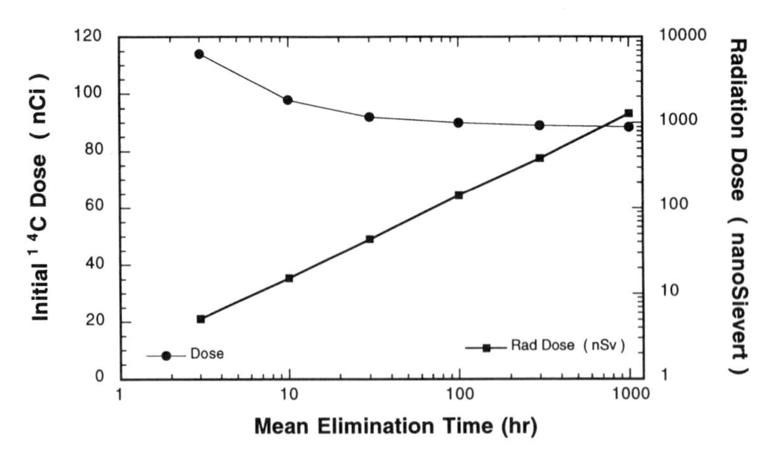

Figure 5. The radiation exposure to a human subject from a ^{14}C-labeled compound is a function of the total initial radioactive dose and the biological elimination time of the compound. The figure shows the initial ^{14}C dose required to double the amount of ^{14}C naturally present in the blood, and the total integrated radiation dose above natural levels that the host receives as that dose is eliminated, both as functions of the biological elimination mean life.

sufficient to determine the kinetic parameters of the labeled compound. It should be noted that the largest integrated radiation dose received by the subject for the 1000 hour case is equal to the dose received from cosmic rays in a single day, approximately a μSievert.

The issues of radioactive laboratory or waste handling are also significant restrictions on the use of radioactive labels and elements in research. However, the Consolidated Federal Register defines radioactive waste as materials containing 50 nCi per gram (CFR 20.2005). Because Figure 5 shows that 100 nCi total ^{14}C is sufficient to perform detailed kinetic experiments in 70 kg humans under the assumption of total body dispersal ($V_{distribution}$ = 42 liters), it is unlikely that any component of the dose, the biological sample, or the excreta, could attain the level requiring handling or disposal as radioactive.

AMS COSTS

The low level of radioactive label needed in AMS-supported nutrition studies has the effect of driving down the costs of certain experiments in perhaps unexpected ways. Because the AMS is sensitive to vastly lowered isotope concentrations, the specific activity (amount or isotope label incorporated into the chemical to be traced) can be much lower than is commonly used. If every molecule of a compound contains one ^{14}C, the specific activity is 62.4 mCi/mmol (or nCi/nmol). Commercially available labeled compounds are more often available with activities of 10-20 mCi/mmol, meaning that only one in 3 to 5 molecules will have a ^{14}C substituted within them.

The cost of synthesizing a labeled compound is often directly related to the specific activity requested. The relation seems exponential rather than linear, as well. With AMS sensitivity, specific activities may be reduced to 1, 0.1, 0.01 mCi/mmol, and lower. Less efficient, and more cost-effective chemical syntheses may be able to produce these lower activities. One research group at UC Davis will attempt to obtain inexpensive labeled compounds, especially natural phytochemicals, by growing plants in an enclosed atmosphere of enhanced $^{14}CO_2$. A related effort at producing inexpensive biochemicals with ^{14}C labels will use the resultant plant material and other inexpensive labeled stocks to produce the needed compound through bio-expression.

The long-term study of chronic exposure to low-levels of food-derived mutagens mentioned above emphasizes the hidden cost-effectiveness of AMS quantification. Figure 6 shows a preliminary study done at only 2 doses, roughly equivalent to the MeIQx found in 5 and 10 fried hamburgers. It is clear that many tissues come to a static concentration of MeIQx by week 3, but the study was concerned with the macromolecular damage to be found in the DNA. As the second plot of Figure 6 shows, the DNA extracted from liver tissue displays a much slower uptake. The mean uptake times derived from least-square fits to exponential asymptotes were 5.7 days and 15.4 days for the higher dosed liver and the liver DNA respectively. With these uptake times, 95% of the saturated asymptote would be reached in 17 and 46 days for the tissue and the associated DNA, necessitating a chronic feeding regimen of at least 7 weeks.

Rats were fed chow impregnated with ^{14}C-labeled MeIQx at 5 doses for up to 7 weeks with 2 sets of rats at the 5 doses continuing on undosed chow. Sets of rats were harvested and analyzed at 1 day, 1 week, 3 weeks, 7 weeks, 8 weeks, and 9 weeks. Five dose points at 6 time points with 3 rats at each point, plus controls and repeats required a protocol of over 100 rats and several kg of chow. If these studies had been performed at ^{14}C levels measurable with scintillation counting, the ^{14}C-labeled compound would have cost over $2 billion, making the absurd assumption of no price break for quantity purchase. Even with a more reasonable assumption about quantity pricing, it is clear that long-term chronic dosing strategies can be accomplished in numerous animals with only tens or hundreds of micrograms of labeled compound, a low enough quantity that quality assurance procedures such as HPLC purification of the administered product are possible, even using analytical-sized instruments. Such quality assurance capability is especially important in human research and also results in high cost effectiveness derived from confident interpretation of the data.

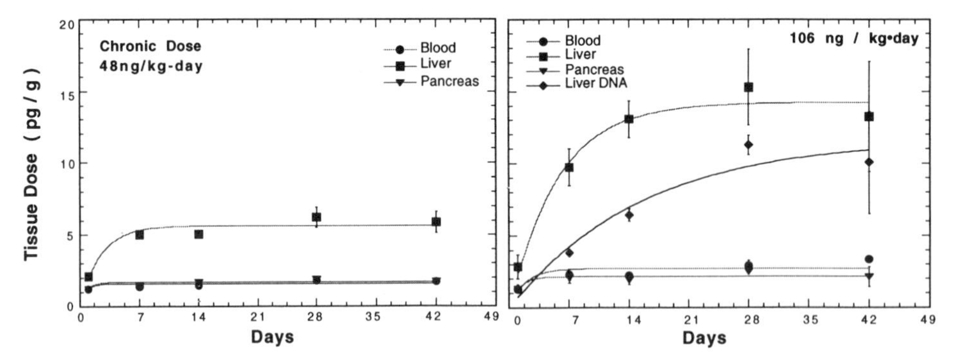

Figure 6. The exponential rise to asymptotic saturation of rat tissues during sub-chronic exposure to [^{14}C]-MeIQx is shown as a function of exposure time. The doses correspond to the amount of toxin in 5 (left) or 10 (right) hamburgers a day. Equilibrium is faster at low dose, and saturation is not reached within the separated DNA of liver until 7 weeks. Such long studies would use prohibitive amounts of labeled compound without AMS detection.

ELEMENTAL TRACING

Although the tracing capabilities for ^{14}C-labeled compounds is the most developed of the nutritional tools, particularly at LLNL, there are isotopes of other elements that are available for high sensitivity, low radioactivity applications using AMS. The periodic table of the elements in Figure 7 shows only elements that have isotopes with half-lives between 10 years and 100 Myr. These are the isotopes for which AMS offers a large gain in sensitivity. There are a number or rare earth elements that fit in this category, but those elements are of minor interest in nutrition studies. Major, minor, and putative nutrient elements are highlighted by various shadings in the Figure, and the 8 elements for which CAMS/LLNL has developed routine preparation chemistry and AMS measurement are circled. Other AMS laboratories may have specific capabilities beyond ours, but the same 8 elements are generally the most utilized in the AMS community.

Tritium (Roberts et al., 1994) and radiocarbon are obvious labels in organic nutrition. The only other bulk nutrient elements having long-lived isotopes are chlorine and calcium. While measurement of ^{36}Cl is routine in many AMS facilities, it is primarily used in earth sciences. However, with its 350,000 year half-life, it is ideally suited for application to chlorine studies in animals, humans or plants. ^{41}Ca is one of the most exciting isotopes to be made available to the nutrition and health communities. Although calcium isotopes are commonly used already, these are the short-lived isotopes ^{45}Ca and ^{47}Ca, with 165 and 4.5 day half-lives. Even with these high specific radioactivities, these isotopes have long been used in metabolism research and clinical testing. ^{41}Ca is essentially stable within human life spans ($t_{1/2}$ = 116,000 years) and is being developed as an aid in osteoporosis research and treatment. (Elmore et al. 1990; Fink et al., 1990; Freeman et al., 1997).

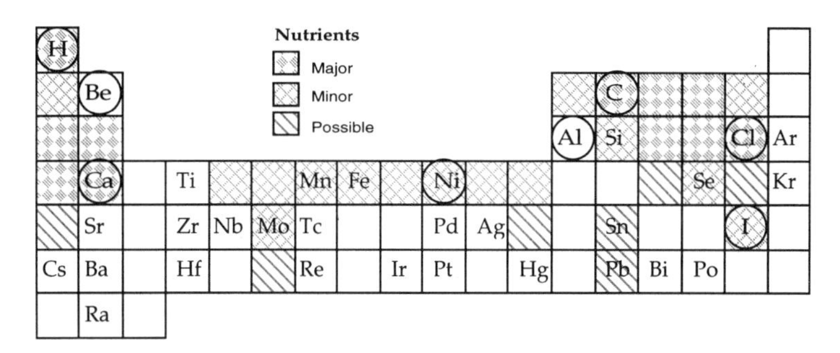

Figure 7. The periodic table of the elements is shown with only the elements that have a long-lived isotope that could be quantified efficiently using AMS. The elements for which CAMS/LLNL has routine chemical preparation and measurement capabilities are circled. The major, minor, and putative nutrients are identified.

Among the trace nutrients, nickel is the most developed, with relatively simple sample definition and preparation for ^{59}Ni (Marchetti et al., 1997; McAninch et al., 1997). While AMS detection of ^{53}Mn and ^{60}Fe has been accomplished in meteoritic samples (Gartenmann et al., 1997; Korschinek et al., 1990; Zoppi et al., 1994; Nishiizumi et al., 1981), these trace isotopes – that would be so valuable for trace nutrient studies – are not available from any commercial source and may be best obtained from extraction out of meteorites. Thus, of the mid-group trace elements, ^{59}Ni is the one readily available isotope for nutritional and detailed biochemical studies.

Selenium is a trace element nutrient that has received intense development at CAMS/LLNL both because it is toxic at levels not much higher than nutrient levels, and because it is found in nearly toxic levels in portions of California's Central Valley. The element has been studied using the short-lived isotope ^{75}Se, but human research would benefit from the use of ^{79}Se, whose half-life has recently been determined to be 2 Myr (Jiang et al., 1997). Unfortunately, the nuclear isobar ^{79}Br contaminates our general purpose ion source to a very high level due to measurements of ^{36}Cl. Until we obtain an ion source that is dedicated to this and other compatible elements, our development is halted. ^{79}Se will be available for tracing purposes, because it is a fission-fragment isotope that is present from nuclear reactors as a by product of the fission process.

The final trace metal nutrient, ^{93}Mo, is also available from fission processes and the AMS detection of this isotope will be possible in the future. CAMS is also developing ^{129}I for protein labeling to replace the short-lived iodine isotopes in some applications, but also for completely new studies of peptide distribution within humans (Vogel et al., 1997). This is the last of the trace nutrients traceable with AMS. These isotopes have half-lives greater than tens of thousands years, but are detectable at atto- and femtomole levels, making human use safe beyond the concern of even the most sceptical informed critic.

AMS also measures several non-nutrient elements. Aluminum is the most important of these, because life evolved on a planet where aluminum is copious but only existed as a chemically and biologically inert oxide. In the past century, however, humans learned to refine the metal from the oxide and we now use it in everything from structural components, to cooking pots, to underarm deodorants, to emulsifiers in infant formula. The biological effects of this active chemical element in humans is currently being studied using ^{26}Al, whose 750,000-year half-life is more amenable to biological experimentation than the next longest-lived isotope, ^{29}Al (6.5 minutes). (Barker et al., 1992; Day et al., 1994; Harris et al., 1996; Jouhanneau et al., 1993; Priest et al., 1995; Priest et al., 1996; King et al., 1997; Kislinger et al., 1997; Steinhausen et al., 1996).

Table 1. Half-lives of selected isotopes with natural abundances at less than parts-per-trillion.

Element	Mass	$t_{1/2}$ (years)
Hydrogen	3	12.3
Beryllium	10	1.6×10^6
Carbon	14	5730
Aluminum	26	720,000
Silicon	32	~ 130
Chlorine	36	301,000
Calcium	41	116,000
Manganese	53	3.7×10^6
Iron	60	100,000
Nickel	59	110,000
Selenium	79	2×10^6
Molebdynum	93	3500
Tin	126	100,000
Iodine	129	16×10^6
Lead	202	53,000

CONCLUSION

Accelerator mass spectrometry is a new method for the study of both organic and elemental nutrients with sufficient sensitivity, specificity, range, and integrity to provide input data to nutritional models at chemical and radiation doses that are safe and within the normal daily doses of humans.

ACKNOWLEDGMENTS

Some data examples were taken from works published or in preparation by colleagues from the Biology and Biotechnology Research Program at LLNL, including Christopher Frantz, Robert Mauthe, and Esther Fultz. Work was performed under the auspices of the U.S. Department of Energy at Lawrence Livermore National Laboratory under contract W-7405-ENG-48.

CORRESPONDING AUTHOR

Please address all correspondence to:
John S. Vogel
Center for Accelerator Mass Spectrometry
Lawrence Livermore National Laboratory
7000 East Avenue, L-397
Livermore, CA 94551

REFERENCES

Barker JJ; Day JP; Priest ND; Newton D; Drumm PV; Lilley JS; Newton GWA. Development of Al-26 accelerator mass spectrometry for aluminum absorption experiments in humans. *Nuc Inst & Meth,* 1992, B68:319-322.

Bogen KT; Keating GA; Vogel JS. In vitro kinetics for non-steady-state uptake of chlorinated solvents from dilute aqueous solutions into human skin. *The Toxicologist,* 1995, 15:318.

Bogen KT; Keating GA; Vogel JS. Chloroform and trichloroethylene uptake from water into human skin *in vitro:* Kinetics and risk implications. *Pred Percut Penet,* 1996, 4B:195-198.

Buchholz BA; Pawley NH; Vogel JS; Mauthe RJ. Pyrethroid decrease in CNS from nerve-agent pretreatment. *J App Tox,* 1997, 17:231-4.

Creek MR; Frantz CE; Fultz E; Haack K; Redwine K; Shen N; Turteltaub KW; Vogel JS. 14C AMS quantification of biomolecular interactions using microbore and plate separations. *Nuc Inst & Meth,* 1994, B92:454-458.

Day JP; Barker J; King SJ; Miller RV; Templar J; Lilley JS; Drumm PV; Newton GWA; Fifield K; Stone JOH; Allan GL; Edwardson JA; Moore PB; Ferrier IN; Priest ND; Newton D; Talbot RJ; Brock JH; Sanchez L; Dobson CB; Itzhaki RF; Radunovic A; Bradbury MWB. Biological chemistry of aluminum studied using Al-26 and accelerator mass spectrometry. *Nuc Inst & Meth,* 1994, B92:463-468.

Elmore D; Bhattacharyya MH; Sacco-Gibson N; Peterson DP. Ca-41 as a long-term biological tracer for bone resorption. *Nuc Inst & Meth,* 1990, B52:531-5.

Fink D; Middleton R; Klein J; Sharma P. /sup 41/Ca: Measurement by accelerator mass spectrometry and applications. *Nuc Inst & Meth,* 1990, B47:79-96.

Frantz CE; Bangerter C; Fultz E; Mayer KM; Vogel JS; Turteltaub KW. Dose-Response studies of MeIQx in rat liver and liver DNA at low doses. *Carcinogenesis,* 1995, 16:367-373.

Freeman SPHT; King JC; Vieira NE; Woodhouse LR; Yergey AL. Human calcium metabolism including bone resorption measured with /sup 41/Ca tracer. *Nuc Inst & Meth,* 1997, B123:266-70.

Garman, A. *Non-radioactive Labeling: A Practical Introduction. Vol 1: Biological Techniques.* Sattelle DB; Ed. Academic Press Ltd: London. 1997.

Gartenmann P; Schnabel C; Suter M; Synal HA. /sup 60/Fe measurements with an EN tandem accelerator. *Nuc Inst & Meth,* 1997, B123:132-6.

Harris WR; Berthon G; Day JP; Exley C; Flaten TP; Forbes WF; Kiss T; Orvig C; Zatta PF. Speciation of aluminum in biological systems. *J Tox Envir Health,* 1996, 48:543-68.

Hughey BJH; Klinkowstein RE; Shefer RE; Skipper PL; Tannenbaum SR; Wishnok JS. Design of a compact 1MV AMS system for biomedical research. *Nuc Inst & Meth,* 1997, B123:153-158.

Jiang, Songsheng, Jingru Guo, Shan Jiang, Chunsheng Li, Anzhi Cui, Ming He, Shaoyong Wu, and Shilin Li. Determination of the half-life of /sup 79/Se with the accelerator mass spectrometry technique. *Nuc Inst & Meth,* 1997, B123:405-9.

Jouhanneau P; Lacour B; Raisbeck G; Yiou F; Banide H; Brown E; Drueke T. Gastrointestinal absorption of aluminum in rats using 26Al and accelerator mass spectrometry. *Clin Nephr,* 1993, 40:244-8.

Kautiainen A; Vogel JS; Turteltaub KW. Dose-dependent trichloroethylene binding to hepatic DNA and protein at low dose. *Chemico-Bio Int,* 1997, 106:109-121.

Keating GA; Naik A; McKone TE; Guy RH; Vogel JS. Assessment of dermal exposure to drinking water contaminants – New measurements and models, in: *Proc Int Symp: Assessing and Managing Health Risks from Drinking Water Contamination: Approaches and Applications.* Reichard EG; Zapponi GA; Eds. IAHS Press. 1995.

King SJ; Day JP; Oldham C; Popplewell JF; Ackrill P; Moore PB; Taylor GA; Edwardson JA; Fifield LK; Liu K; Cresswell RG. The influence of dissolved silicate on the physiological chemistry of aluminium, studied in humans using tracer /sup 26/Al and accelerator mass spectrometry. *Nuc Inst & Meth,* 1997, B123:254-8.

Kislinger G; Steinhausen C; Alvarez-Bruckmann M; Winklhofer C; Ittel TH; Nolte E. Investigations of the human aluminum biokinetics with /sup 26/Al and AMS. *Nuc Inst & Meth,* 1997, B123:259-65.

Korschinek G; Muller D; Faestermann T; Gillitzer A; Nolte E; Paul M. Trace analysis of Fe-55 in biosphere and technology by means of AMS. *Nuc Inst & Meth,* 1990, B52:498-501.

Marchetti AA; Hainsworth LJ; McAninch JE; Leivers MR; Jones PR; Proctor ID; Straume T. Ultra-separation of nickel from copper metal for the measurement of /sup 63/Ni by AMS. *Nuc Inst & Meth,* 1997, B123:230-4.

McAninch JE; Hainsworth LJ; Marchetti AA; Leivers MR; Jones PR; Dunlop AE; Mauthe R; Vogel JS; Proctor ID; Straume T. Measurement of 63Ni and 59Ni by AMS using characteristic projectile X-rays. *Nuc Inst & Meth,* 1997, B123:137-143.

Mellon FA; Sandtröm B. *Stable Isotopes in Human Nutrition.* Academic Press Ltd.: London. 1996.

Mous DJW; Purser KH; Fokker W; Van den Broek R; Koopmans RB. A compact ^{14}C isotope ratio mass spectrometer for biomedical applications. *Nuc Inst & Meth,* 1997, B123:159-162.

Nishiizumi K; Murrell MT; Arnold JR; Elmore D; Ferraro RD; Gove HE; Finkel RC. Cosmic-ray-produced /sup 36/Cl and /sup 53/Mn in Allan Hills-77 meteorites. *Earth and Planetary Science Letters,* 1981, 52:31-8.

Priest ND; Newton D; Day JP; Talbot RJ; Warner AJ. Human metabolism of aluminium-26 and gallium-67 injected as citrates. *Hum Ex Tox,* 1995, 14:287-93.

Priest ND; Taltot RJ; Austin JG; Day JP; King SJ; Fifield K; Cresswell RG. The bioavailability of 26Al-labelled aluminum citrate and aluminum hydroxide in volunteers. *Biometals,* 1996, 9:221-8.

Robert ML; Velsko C; Turteltaub KW. Tritium AMS for biomedical applications. *Nuc Inst & Meth,* 1994, B92:459-462.

Scott EM. Proccedings: International workshop on intercomparison of radiocarbon laboratories, at Glasgow, Scotland. *Radiocarbon*, 1990, 32 253-397.

Steinhausen C; Gerisch P; Heisinger B; Hohl C; Kislinger G; Korschinek G; Niedermayer M; Nolte E; Dumitru M; Alvarez-Brueckmann M; Schneider M; Ittel TH. Medical application of /sup 26/Al. *Nuc Inst & Meth,* 1996, B113:479-83.

Stenstrom K; Leide-Svegborn S; Erlandsson B; Hellborg R; Skog G; Mattsson S; Nilsson LE; Nosslin B. A program for long-term retention studies of /sup 14/C-labelled compounds in man using the Lund AMS facility. *Nuc Inst & Meth,* 1997, B123:245-8.

Suter M; Jacob S; Synal HA. AMS of /sup 14/C at low energies. *Nuc Inst & Meth,* 1997, B123:148-52.

Turteltaub KW; Vogel JS; Balhorn R; Gledhill BL; Southon JR; Caffee MW; Finkel RC; Nelson DE; Proctor ID; Davis JC. Accelerator mass spectrometry in the biomedical sciences: applications in low-exposure biomedical and environmental dosimetry. *Nuc Inst & Meth,* 1990, B52:517-23.

Turteltaub KW; Vogel JS; Frantz CE; Shen N. Fate and distribution of 2-amino-1-methyl-6-phenyl-imidazo[4,5-b]pyridine (PhIP) in mice at a human dietary equivalent dose. *Cancer Res,* 1992, 52:4682-4687.

Vogel JS. Rapid production of graphite without contamination for biomedical AMS. *Radiocarbon,* 1992, 34:344-350.

Vogel JS; Freeman SPHT; McAninch JE. Elements in biological AMS. *Nuc Inst & Meth,* 1997, B123:241-244.

Vogel JS; Turteltaub KW. Biomolecular tracing through accelerator mass spectrometry. *Trends in Analytical Chemistry,* 1992, 11:142-149.

Vogel JS; Turteltaub KW; Finkel R; Nelson DE. Accelerator mass spectrometry – Isotope quantification at attomole sensitivity. *Analyt Chem,* 1995, 67:A353-A359.

Williams ML; Vogel JS; Ghadially R; Brown BE; Elias PM. Exogenous origin of n-alkanes in pathologic scale. *Arch Derm,* 1992, 128:1065-71.

Zoppi U; Suter M; Synal HA. Isobar separation with gas ionization counters in accelerator mass spectrometry. *Nuc Inst & Meth,* 1994, B89:262-265.

INDEX